SELECTED CHAPTERS FROM MONTGOMERY: APPLIED STATISTICS AND PROBABILITY FOR ENGINEERS, 4TH EDITION

A Wiley Canada Custom Publication
for the

University of Alberta

STAT 235

Copyright © 2008 by John Wiley & Sons Canada, Ltd

Copyright © 2007 by John Wiley & Sons, Inc. All rights reserved. No part of this work covered by the copyright herein may be reproduced, transmitted, or used in any form or by any means—graphic, electronic or mechanical—without the prior written permission of the publisher. Any request for photocopying, recording, taping, or inclusion in information storage and retrieval systems of any part of this book shall be directed to The Canadian Copyright Licensing Agency (Access Copyright). For an Access Copyright licence, visit www.accesscopyright.ca or call toll-free, 1-800-893-5777. Care has been taken to trace ownership of copyright material contained in this text. The Publishers will gladly receive any information that will enable them to rectify any erroneous reference or credit line in subsequent editions.

Cover Photo Credit: Norm Christiansen

Marketing Manager: Anne-Marie
Custom Coordinator: Rachel Coffey

Printed and bound in the United States of America
10 9 8 7 6 5 4 3 2

John Wiley & Sons Canada, Ltd
6045 Freemont Blvd.
Mississauga, Ontario
L5R 4J3
Visit our website at: www.wiley.ca

Applied Statistics and Probability for Engineers

Fourth Edition

Douglas C. Montgomery
Arizona State University

George C. Runger
Arizona State University

John Wiley & Sons, Inc.

To:

Meredith, Neil, Colin, and Cheryl

Rebecca, Elisa, George, and Taylor

EXECUTIVE PUBLISHER *Bruce Spatz*
ASSOCIATE PUBLISHER *Daniel Sayre*
ACQUISITIONS EDITOR *Jennifer Welter*
PRODUCTION EDITOR *Nicole Repasky*
MARKETING MANAGER *Phyllis Cerys*
SENIOR DESIGNER *Kevin Murphy*
MEDIA EDITOR *Stefanie Liebman*
EDITORIAL ASSISTANT *Mary Moran-McGee*
PRODUCTION SERVICES MANAGEMENT *mb editorial services*
COVER IMAGE *Norm Christiansen*

This book was set in 10/12 pt. Times Roman by TechBooks and printed and bound by R.R. Donnelley/Willard Division. The cover was printed by Phoenix Color.

This book is printed on acid-free paper. ∞

Copyright © 2007 John Wiley & Sons, Inc. All rights reserved.

No part of this publication may be reproduced, stored in a retrieval system or transmitted in any form or by any means, electronic, mechanical, photocopying, recording, scanning or otherwise, except as permitted under Sections 107 or 108 of the 1976 United States Copyright Act, without either the prior written permission of the Publisher, or authorization through payment of the appropriate per-copy fee to the Copyright Clearance Center, Inc. 222 Rosewood Drive, Danvers, MA 01923, website www.copyright.com. Requests to the Publisher for permission should be addressed to the Permissions Department, John Wiley & Sons, Inc., 111 River Street, Hoboken, NJ 07030-5774, (201) 748-6011, fax (201) 748-6008, website http://www.wiley.com/go/permissions.
To order books or for customer service please, call 1-800-CALL WILEY (225-5945).

Library of Congress Cataloging-in-Publication Data:

Montgomery, Douglas C.
    Applied statistics and probability for engineers/Douglas C. Montgomery, George C. Runger.—4th ed.
      p. cm.
    Includes index.
    ISBN-13: 978-0-471-74589-1
    ISBN-10: 0-471-74589-8
      1. Statistics. 2. Probabilities. I. Runger, George C. II. Title.

QA276.12.M645 2006
519.5—dc22

2005056261

Printed in the United States of America

10 9 8 7 6 5 4 3 2 1

The Role of Statistics in Engineering

CHAPTER OUTLINE

1-1 THE ENGINEERING METHOD AND STATISTICAL THINKING

1-2 COLLECTING ENGINEERING DATA

    1-2.1 Basic Principles

    1-2.2 Retrospective Study

    1-2.3 Observational Study

    1-2.4 Designed Experiments

    1-2.5 Observing Processes Over Time

1-3 MECHANISTIC AND EMPIRICAL MODELS

1-4 PROBABILITY AND PROBABILITY MODELS

LEARNING OBJECTIVES

After careful study of this chapter you should be able to do the following:
1. Identify the role that statistics can play in the engineering problem-solving process
2. Discuss how variability affects the data collected and used for making engineering decisions
3. Explain the difference between enumerative and analytical studies
4. Discuss the different methods that engineers use to collect data
5. Identify the advantages that designed experiments have in comparison to other methods of collecting engineering data
6. Explain the differences between mechanistic models and empirical models
7. Discuss how probability and probability models are used in engineering and science

1-1 THE ENGINEERING METHOD AND STATISTICAL THINKING

An engineer is someone who solves problems of interest to society by the efficient application of scientific principles. Engineers accomplish this by either refining an existing product or process or by designing a new product or process that meets customers' needs. The **engineering,** or **scientific, method** is the approach to formulating and solving these problems. The steps in the engineering method are as follows:

1. Develop a clear and concise description of the problem.
2. Identify, at least tentatively, the important factors that affect this problem or that may play a role in its solution.

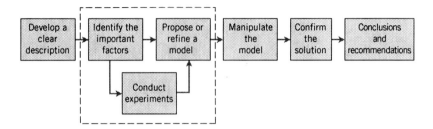

Figure 1-1 The engineering method.

3. Propose a model for the problem, using scientific or engineering knowledge of the phenomenon being studied. State any limitations or assumptions of the model.
4. Conduct appropriate experiments and collect data to test or validate the tentative model or conclusions made in steps 2 and 3.
5. Refine the model on the basis of the observed data.
6. Manipulate the model to assist in developing a solution to the problem.
7. Conduct an appropriate experiment to confirm that the proposed solution to the problem is both effective and efficient.
8. Draw conclusions or make recommendations based on the problem solution.

The steps in the engineering method are shown in Fig. 1-1. Notice that the engineering method features a strong interplay between the problem, the factors that may influence its solution, a model of the phenomenon, and experimentation to verify the adequacy of the model and the proposed solution to the problem. Steps 2–4 in Fig. 1-1 are enclosed in a box, indicating that several cycles or iterations of these steps may be required to obtain the final solution. Consequently, engineers must know how to efficiently plan experiments, collect data, analyze and interpret the data, and understand how the observed data are related to the model they have proposed for the problem under study.

The field of **statistics** deals with the collection, presentation, analysis, and use of data to make decisions, solve problems, and design products and processes. Because many aspects of engineering practice involve working with data, obviously some knowledge of statistics is important to any engineer. Specifically, statistical techniques can be a powerful aid in designing new products and systems, improving existing designs, and designing, developing, and improving production processes.

Statistical methods are used to help us describe and understand **variability**. By variability, we mean that successive observations of a system or phenomenon do not produce exactly the same result. We all encounter variability in our everyday lives, and **statistical thinking** can give us a useful way to incorporate this variability into our decision-making processes. For example, consider the gasoline mileage performance of your car. Do you always get exactly the same mileage performance on every tank of fuel? Of course not—in fact, sometimes the mileage performance varies considerably. This observed variability in gasoline mileage depends on many factors, such as the type of driving that has occurred most recently (city versus highway), the changes in condition of the vehicle over time (which could include factors such as tire inflation, engine compression, or valve wear), the brand and/or octane number of the gasoline used, or possibly even the weather conditions that have been recently experienced. These factors represent potential **sources of variability** in the system. Statistics gives us a framework for describing this variability and for learning about which potential sources of variability are the most important or which have the greatest impact on the gasoline mileage performance.

We also encounter variability in dealing with engineering problems. For example, suppose that an engineer is designing a nylon connector to be used in an automotive engine application. The engineer is considering establishing the design specification on wall thickness at 3/32 inch but is somewhat uncertain about the effect of this decision on the connector pull-off force. If the pull-off force is too low, the connector may fail when it is installed in an engine. Eight prototype units are produced and their pull-off forces measured, resulting in the following data (in pounds): 12.6, 12.9, 13.4, 12.3, 13.6, 13.5, 12.6, 13.1. As we anticipated, not all of the prototypes have the same pull-off force. We say that there is variability in the pull-off force measurements. Because the pull-off force measurements exhibit variability, we consider the pull-off force to be a random variable. A convenient way to think of a random variable, say X, that represents a measurement, is by using the model

$$X = \mu + \epsilon \qquad (1\text{-}1)$$

where μ is a constant and ϵ is a random disturbance. The constant remains the same with every measurement, but small changes in the environment, test equipment, differences in the individual parts themselves, and so forth change the value of ϵ. If there were no disturbances, ϵ would always equal zero and X would always be equal to the constant μ. However, this never happens in the real world, so the actual measurements X exhibit variability. We often need to describe, quantify and ultimately reduce variability.

Figure 1-2 presents a **dot diagram** of these data. The dot diagram is a very useful plot for displaying a small body of data—say, up to about 20 observations. This plot allows us to see easily two features of the data; the **location,** or the middle, and the **scatter or variability.** When the number of observations is small, it is usually difficult to identify any specific patterns in the variability, although the dot diagram is a convenient way to see any unusual data features.

The need for statistical thinking arises often in the solution of engineering problems. Consider the engineer designing the connector. From testing the prototypes, he knows that the average pull-off force is 13.0 pounds. However, he thinks that this may be too low for the intended application, so he decides to consider an alternative design with a greater wall thickness, 1/8 inch. Eight prototypes of this design are built, and the observed pull-off force measurements are 12.9, 13.7, 12.8, 13.9, 14.2, 13.2, 13.5, and 13.1. The average is 13.4. Results for both samples are plotted as dot diagrams in Fig. 1-3. This display gives the impression that increasing the wall thickness has led to an increase in pull-off force. However, there are some obvious questions to ask. For instance, how do we know that another sample of prototypes will not give different results? Is a sample of eight prototypes adequate to give reliable results? If we use the test results obtained so far to conclude that increasing the wall thickness increases the strength, what risks are associated with this decision? For example, is it possible that the apparent increase in pull-off force observed in the thicker prototypes is only due to the inherent variability in the system and that increasing the thickness of the part (and its cost) really has no effect on the pull-off force?

Often, physical laws (such as Ohm's law and the ideal gas law) are applied to help design products and processes. We are familiar with this reasoning from general laws to specific cases. But it is also important to reason from a specific set of measurements to more general cases to answer the previous questions. This reasoning is from a sample (such as the eight connectors) to a population (such as the connectors that will be sold to customers). The

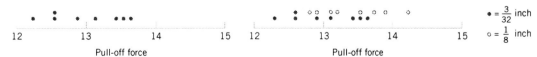

Figure 1-2 Dot diagram of the pull-off force data when wall thickness is 3/32 inch.

Figure 1-3 Dot diagram of pull-off force for two wall thicknesses.

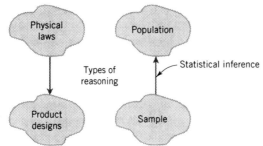

Figure 1-4 Statistical inference is one type of reasoning.

reasoning is referred to as **statistical inference.** See Fig. 1-4. Historically, measurements were obtained from a sample of people and generalized to a population, and the terminology has remained. Clearly, reasoning based on measurements from some objects to measurements on all objects can result in errors (called sampling errors). However, if the sample is selected properly, these risks can be quantified and an appropriate sample size can be determined.

1-2 COLLECTING ENGINEERING DATA

1-2.1 Basic Principles

In the previous section, we illustrated some simple methods for summarizing data. In the engineering environment, the data is almost always a sample that has been selected from some population. Three basic methods of collecting data are

- A **retrospective** study using historical data
- An **observational** study
- A **designed experiment**

An effective data collection procedure can greatly simplify the analysis and lead to improved understanding of the population or process that is being studied. We now consider some examples of these data collection methods.

1-2.2 Retrospective Study

Montgomery, Peck, and Vining (2001) describe an acetone-butyl alcohol distillation column for which concentration of acetone in the distillate or output product stream is an important variable. Factors that may affect the distillate are the reboil temperature, the condensate temperature, and the reflux rate. Production personnel obtain and archive the following records:

- The concentration of acetone in an hourly test sample of output product
- The reboil temperature log, which is a plot of the reboil temperature over time
- The condenser temperature controller log
- The nominal reflux rate each hour

The reflux rate should be held constant for this process. Consequently, production personnel change this very infrequently.

A retrospective study would use either all or a sample of the historical process data archived over some period of time. The study objective might be to discover the relationships among the two temperatures and the reflux rate on the acetone concentration in the output product stream. However, this type of study presents some problems:

1. We may not be able to see the relationship between the reflux rate and acetone concentration, because the reflux rate didn't change much over the historical period.
2. The archived data on the two temperatures (which are recorded almost continuously) do not correspond perfectly to the acetone concentration measurements (which are made hourly). It may not be obvious how to construct an approximate correspondence.
3. Production maintains the two temperatures as closely as possible to desired targets or set points. Because the temperatures change so little, it may be difficult to assess their real impact on acetone concentration.
4. Within the narrow ranges that they do vary, the condensate temperature tends to increase with the reboil temperature. Consequently, the effects of these two process variables on acetone concentration may be difficult to separate.

As you can see, a retrospective study may involve a lot of **data,** but that data may contain relatively little useful **information** about the problem. Furthermore, some of the relevant data may be missing, there may be transcription or recording errors resulting in outliers (or unusual values), or data on other important factors may not have been collected and archived. In the distillation column, for example, the specific concentrations of butyl alcohol and acetone in the input feed stream are a very important factor, but they are not archived because the concentrations are too hard to obtain on a routine basis. As a result of these types of issues, statistical analysis of historical data sometimes identifies interesting phenomena, but solid and reliable explanations of these phenomena are often difficult to obtain.

1-2.3 Observational Study

In an observational study, the engineer observes the process or population, disturbing it as little as possible, and records the quantities of interest. Because these studies are usually conducted for a relatively short time period, sometimes variables that are not routinely measured can be included. In the distillation column, the engineer would design a form to record the two temperatures and the reflux rate when acetone concentration measurements are made. It may even be possible to measure the input feed stream concentrations so that the impact of this factor could be studied. Generally, an observational study tends to solve problems 1 and 2 above and goes a long way toward obtaining accurate and reliable data. However, observational studies may not help resolve problems 3 and 4.

1-2.4 Designed Experiments

In a designed experiment the engineer makes *deliberate* or *purposeful changes* in the controllable variables of the system or process, observes the resulting system output data, and then makes an inference or decision about which variables are responsible for the observed changes in output performance. The nylon connector example in Section 1-1 illustrates a designed experiment; that is, a deliberate change was made in the wall thickness of the connector with the objective of discovering whether or not a greater pull-off force could be obtained. Experiments designed with basic principles such as **randomization** are needed to establish **cause and effect** relationships.

Much of what we know in the engineering and physical-chemical sciences is developed through testing or experimentation. Often engineers work in problem areas in which no scientific or engineering theory is directly or completely applicable, so experimentation and observation of the resulting data constitute the only way that the problem can be solved. Even when there is a good underlying scientific theory that we may rely on to explain the phenomena of interest, it is almost always necessary to conduct tests or experiments to confirm that the theory is indeed operative in the situation or environment in which it is being applied. Statistical thinking and statistical methods play an important role in planning, conducting, and analyzing the data from engineering experiments. Designed experiments play a very important role in engineering design and development and in the improvement of manufacturing processes.

For example, consider the problem involving the choice of wall thickness for the nylon connector. This is a simple illustration of a designed experiment. The engineer chose two wall thicknesses for the connector and performed a series of tests to obtain pull-off force measurements at each wall thickness. In this simple **comparative experiment,** the engineer is interested in determining if there is any difference between the 3/32- and 1/8-inch designs. An approach that could be used in analyzing the data from this experiment is to compare the mean pull-off force for the 3/32-inch design to the mean pull-off force for the 1/8-inch design using statistical hypothesis testing, which is discussed in detail in Chapters 9 and 10. Generally, a hypothesis is a statement about some aspect of the system in which we are interested. For example, the engineer might want to know if the mean pull-off force of a 3/32-inch design exceeds the typical maximum load expected to be encountered in this application, say 12.75 pounds. Thus, we would be interested in testing the hypothesis that the mean strength exceeds 12.75 pounds. This is called a **single-sample hypothesis testing problem.** Chapter 9 presents techniques for this type of problem. Alternatively, the engineer might be interested in testing the hypothesis that increasing the wall thickness from 3/32- to 1/8-inch results in an increase in mean pull-off force. It is an example of a **two-sample hypothesis testing problem.** Two-sample hypothesis testing problems are discussed in Chapter 10.

Designed experiments are a very powerful approach to studying complex systems, such as the distillation column. This process has three factors, the two temperatures and the reflux rate, and we want to investigate the effect of these three factors on output acetone concentration. A good experimental design for this problem must ensure that we can separate the effects of all three factors on the acetone concentration. The specified values of the three factors used in the experiment are called **factor levels.** Typically, we use a small number of levels for each factor, such as two or three. For the distillation column problem, suppose we use a "high," and "low," level (denoted $+1$ and -1, respectively) for each of the factors. We thus would use two levels for each of the three factors. A very reasonable experiment design strategy uses every possible combination of the factor levels to form a basic experiment with eight different settings for the process. This type of experiment is called a **factorial experiment.** Table 1-1 presents this experimental design.

Table 1-1 The Designed Experiment (Factorial Design) for the Distillation Column

Reboil Temp.	Condensate Temp.	Reflux Rate
-1	-1	-1
$+1$	-1	-1
-1	$+1$	-1
$+1$	$+1$	-1
-1	-1	$+1$
$+1$	-1	$+1$
-1	$+1$	$+1$
$+1$	$+1$	$+1$

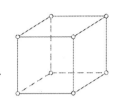 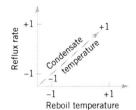

Figure 1-5 The factorial design for the distillation column.

Figure 1-5 illustrates that this design forms a cube in terms of these high and low levels. With each setting of the process conditions, we allow the column to reach equilibrium, take a sample of the product stream, and determine the acetone concentration. We then can draw specific inferences about the effect of these factors. Such an approach allows us to proactively study a population or process.

An important advantage of factorial experiments is that they allow one to detect an **interaction** between factors. Consider only the two temperature factors in the distillation experiment. Suppose that the response concentration is poor when the reboil temperature is *low*, regardless of the condensate temperature. That is, the condensate temperature has no effect when the reboil temperature is *low*. However, when the reboil temperature is *high*, a *high* condensate temperature generates a good response, while a *low* condensate temperature generates a poor response. That is, the condensate temperature changes the response when the reboil temperature is *high*. The effect of condensate temperature depends on the setting of the reboil temperature and these two factors would interact in this case. If the four combinations of *high* and *low* reboil and condensate temperatures were not tested, such an interaction would not be detected.

We can easily extend the factorial strategy to more factors. Suppose that the engineer wants to consider a fourth factor, type of distillation column. There are two types: the standard one and a newer design. Figure 1-6 illustrates how all four factors, reboil temperature, condensate temperature, reflux rate, and column design, could be investigated in a factorial design. Since all four factors are still at two levels, the experimental design can still be represented geometrically as a cube (actually, it's a *hypercube*). Notice that as in any factorial design, all possible combinations of the four factors are tested. The experiment requires 16 trials.

Generally, if there are k factors and they each have two levels, a factorial experimental design will require 2^k runs. For example, with $k = 4$, the 2^4 design in Fig. 1-6 requires 16 tests. Clearly, as the number of factors increases, the number of trials required in a factorial experiment increases rapidly; for instance, eight factors each at two levels would require 256 trials. This quickly becomes unfeasible from the viewpoint of time and other resources. Fortunately, when there are four to five or more factors, it is usually unnecessary to test all possible combinations of factor levels. A fractional factorial experiment is a variation of the basic

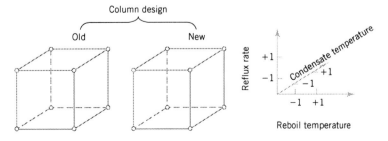

Figure 1-6 A four-factorial experiment for the distillation column.

8   CHAPTER 1 THE ROLE OF STATISTICS IN ENGINEERING

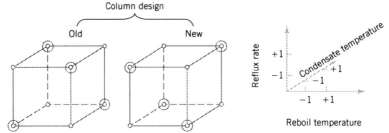

Figure 1-7 A fractional factorial experiment for the connector wall thickness problem.

factorial arrangement in which only a subset of the factor combinations are actually tested. Figure 1-7 shows a fractional factorial experimental design for the four-factor version of the distillation experiment. The circled test combinations in this figure are the only test combinations that need to be run. This experimental design requires only 8 runs instead of the original 16; consequently it would be called a **one-half fraction.** This is an excellent experimental design in which to study all four factors. It will provide good information about the individual effects of the four factors and some information about how these factors interact.

Factorial and fractional factorial experiments are used extensively by engineers and scientists in industrial research and development, where new technology, products, and processes are designed and developed and where existing products and processes are improved. Since so much engineering work involves testing and experimentation, it is essential that all engineers understand the basic principles of planning efficient and effective experiments. We discuss these principles in Chapter 13. Chapter 14 concentrates on the factorial and fractional factorials that we have introduced here.

1-2.5 Observing Processes Over Time

Often data are collected over time. In this case, it is usually very helpful to plot the data versus time in a **time series plot.** Phenomena that might affect the system or process often become more visible in a time-oriented plot and the concept of stability can be better judged.

Figure 1-8 is a dot diagram of acetone concentration readings taken hourly from the distillation column described in Section 1-2.2. The large variation displayed on the dot diagram indicates a lot of variability in the concentration, but the chart does not help explain the reason for the variation. The time series plot is shown in Figure 1-9. A shift in the process mean level is visible in the plot and an estimate of the time of the shift can be obtained.

W. Edwards Deming, a very influential industrial statistician, stressed that it is important to understand the nature of variability in processes and systems over time. He conducted an experiment in which he attempted to drop marbles as close as possible to a target on a table. He used a funnel mounted on a ring stand and the marbles were dropped into the funnel. See Fig. 1-10. The funnel was aligned as closely as possible with the center of the target. He then used two different strategies to operate the process. (1) He never moved the funnel. He just

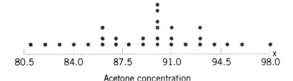

Figure 1-8 The dot diagram illustrates variation but does not identify the problem.

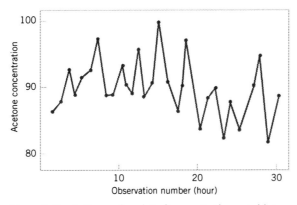

Figure 1-9 A time series plot of concentration provides more information than the dot diagram.

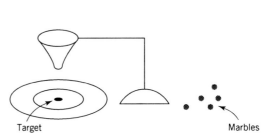

Figure 1-10 Deming's funnel experiment.

dropped one marble after another and recorded the distance from the target. (2) He dropped the first marble and recorded its location relative to the target. He then moved the funnel an equal and opposite distance in an attempt to compensate for the error. He continued to make this type of adjustment after each marble was dropped.

After both strategies were completed, he noticed that the variability of the distance from the target for strategy 2 was approximately 2 times larger than for strategy 1. The adjustments to the funnel increased the deviations from the target. The explanation is that the error (the deviation of the marble's position from the target) for one marble provides no information about the error that will occur for the next marble. Consequently, adjustments to the funnel do not decrease future errors. Instead, they tend to move the funnel farther from the target.

This interesting experiment points out that adjustments to a process based on random disturbances can actually *increase* the variation of the process. This is referred to as **overcontrol** or **tampering**. Adjustments should be applied only to compensate for a nonrandom shift in the process—then they can help. A computer simulation can be used to demonstrate the lessons of the funnel experiment. Figure 1-11 displays a time plot of 100 measurements (denoted as y) from a process in which only random disturbances are present. The target value for the process

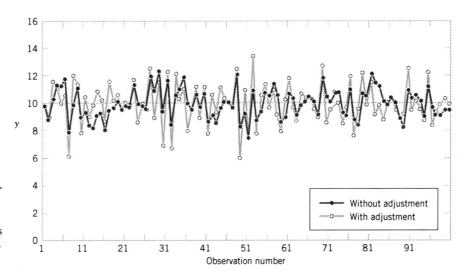

Figure 1-11 Adjustments applied to random disturbances overcontrol the process and increase the deviations from the target.

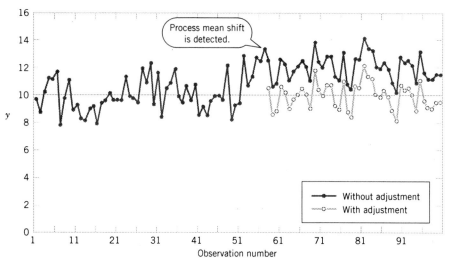

Figure 1-12 Process mean shift is detected at observation number 57, and one adjustment (a decrease of two units) reduces the deviations from target.

is 10 units. The figure displays the data with and without adjustments that are applied to the process mean in an attempt to produce data closer to target. Each adjustment is equal and opposite to the deviation of the previous measurement from target. For example, when the measurement is 11 (one unit above target), the mean is reduced by one unit before the next measurement is generated. The overcontrol has increased the deviations from the target.

Figure 1-12 displays the data without adjustment from Fig. 1-11, except that the measurements after observation number 50 are increased by two units to simulate the effect of a shift in the mean of the process. When there is a true shift in the mean of a process, an adjustment can be useful. Figure 1-12 also displays the data obtained when one adjustment (a decrease of two units) is applied to the mean after the shift is detected (at observation number 57). Note that this adjustment decreases the deviations from target.

The question of when to apply adjustments (and by what amounts) begins with an understanding of the types of variation that affect a process. A **control chart** is an invaluable way to examine the variability in time-oriented data. Figure 1-13 presents a control chart for the concentration data from Fig. 1-9. The **center line** on the control chart is just the average of the concentration measurements for the first 20 samples ($\bar{x} = 91.5$ g/l) when the process is stable. The **upper control limit** and the **lower control limit** are a pair of statistically derived

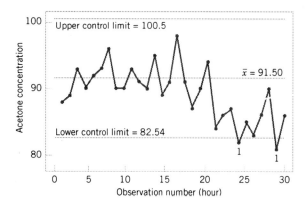

Figure 1-13 A control chart for the chemical process concentration data.

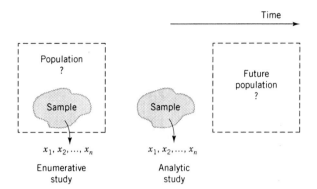

Figure 1-14 Enumerative versus analytic study.

limits that reflect the inherent or natural variability in the process. These limits are located three standard deviations of the concentration values above and below the center line. If the process is operating as it should, without any external sources of variability present in the system, the concentration measurements should fluctuate randomly around the center line, and almost all of them should fall between the control limits.

In the control chart of Fig. 1-13, the visual frame of reference provided by the center line and the control limits indicates that some upset or disturbance has affected the process around sample 20 because all of the following observations are below the center line and two of them actually fall below the lower control limit. This is a very strong signal that corrective action is required in this process. If we can find and eliminate the underlying cause of this upset, we can improve process performance considerably.

Furthermore, W. Edwards Deming pointed out that data from a process are used for different types of conclusions. Sometimes we collect data from a process to evaluate current production. For example, we might sample and measure resistivity on three semiconductor wafers selected from a lot and use this to evaluate the lot. This is called an **enumerative** study. However, in many cases we use data from current production to evaluate future production. We apply conclusions to a conceptual, future population. Deming called this an **analytic** study. Clearly this requires an assumption of a **stable** process and Deming emphasized that control charts were needed to justify this assumption. See Fig. 1-14 as an illustration.

Control charts are a very important application of statistics for monitoring, controlling, and improving a process. The branch of statistics that makes use of control charts is called **statistical process control**, or **SPC**. We will discuss SPC and control charts in Chapter 16.

1-3 MECHANISTIC AND EMPIRICAL MODELS

Models play an important role in the analysis of nearly all engineering problems. Much of the formal education of engineers involves learning about the models relevant to specific fields and the techniques for applying these models in problem formulation and solution. As a simple example, suppose we are measuring the flow of current in a thin copper wire. Our model for this phenomenon might be Ohm's law:

$$\text{Current} = \text{voltage}/\text{resistance}$$

or

$$I = E/R \tag{1-2}$$

We call this type of model a **mechanistic model** because it is built from our underlying knowledge of the basic physical mechanism that relates these variables. However, if we performed this measurement process more than once, perhaps at different times, or even on different days, the observed current could differ slightly because of small changes or variations in factors that are not completely controlled, such as changes in ambient temperature, fluctuations in performance of the gauge, small impurities present at different locations in the wire, and drifts in the voltage source. Consequently, a more realistic model of the observed current might be

$$I = E/R + \epsilon \qquad (1\text{-}3)$$

where ϵ is a term added to the model to account for the fact that the observed values of current flow do not perfectly conform to the mechanistic model. We can think of ϵ as a term that includes the effects of all of the unmodeled sources of variability that affect this system.

Sometimes engineers work with problems for which there is no simple or well-understood mechanistic model that explains the phenomenon. For instance, suppose we are interested in the number average molecular weight (M_n) of a polymer. Now we know that M_n is related to the viscosity of the material (V), and it also depends on the amount of catalyst (C) and the temperature (T) in the polymerization reactor when the material is manufactured. The relationship between M_n and these variables is

$$M_n = f(V, C, T) \qquad (1\text{-}4)$$

say, where the *form* of the function f is unknown. Perhaps a working model could be developed from a first-order Taylor series expansion, which would produce a model of the form

$$M_n = \beta_0 + \beta_1 V + \beta_2 C + \beta_3 T \qquad (1\text{-}5)$$

where the β's are unknown parameters. Now just as in Ohm's law, this model will not exactly describe the phenomenon, so we should account for the other sources of variability that may affect the molecular weight by adding another term to the model; therefore

$$M_n = \beta_0 + \beta_1 V + \beta_2 C + \beta_3 T + \epsilon \qquad (1\text{-}6)$$

is the model that we will use to relate molecular weight to the other three variables. This type of model is called an **empirical model**; that is, it uses our engineering and scientific knowledge of the phenomenon, but it is not directly developed from our theoretical or first-principles understanding of the underlying mechanism.

To illustrate these ideas with a specific example, consider the data in Table 1-2. This table contains data on three variables that were collected in an observational study in a semiconductor manufacturing plant. In this plant, the finished semiconductor is wire bonded to a frame. The variables reported are pull strength (a measure of the amount of force required to break the bond), the wire length, and the height of the die. We would like to find a model relating pull strength to wire length and die height. Unfortunately, there is no physical mechanism that we can easily apply here, so it doesn't seem likely that a mechanistic modeling approach will be successful.

Figure 1-15 presents a three-dimensional plot of all 25 observations on pull strength, wire length, and die height. From examination of this plot, we see that pull strength increases as

1-3 MECHANISTIC AND EMPIRICAL MODELS

Table 1-2 Wire Bond Pull Strength Data

Observation Number	Pull Strength y	Wire Length x_1	Die Height x_2
1	9.95	2	50
2	24.45	8	110
3	31.75	11	120
4	35.00	10	550
5	25.02	8	295
6	16.86	4	200
7	14.38	2	375
8	9.60	2	52
9	24.35	9	100
10	27.50	8	300
11	17.08	4	412
12	37.00	11	400
13	41.95	12	500
14	11.66	2	360
15	21.65	4	205
16	17.89	4	400
17	69.00	20	600
18	10.30	1	585
19	34.93	10	540
20	46.59	15	250
21	44.88	15	290
22	54.12	16	510
23	56.63	17	590
24	22.13	6	100
25	21.15	5	400

both wire length and die height increase. Furthermore, it seems reasonable to think that a model such as

$$\text{Pull strength} = \beta_0 + \beta_1(\text{wire length}) + \beta_2(\text{die height}) + \epsilon$$

would be appropriate as an empirical model for this relationship. In general, this type of empirical model is called a **regression model**. In Chapters 11 and 12 we show how to build these models and test their adequacy as approximating functions. We will use a method for

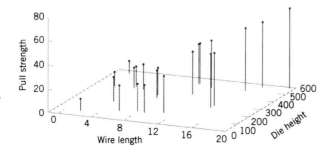

Figure 1-15 Three-dimensional plot of the wire bond pull strength data.

14 CHAPTER 1 THE ROLE OF STATISTICS IN ENGINEERING

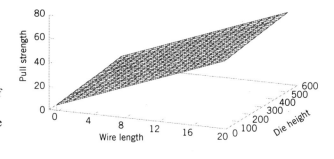

Figure 1-16 Plot of predicted values of pull strength from the empirical model.

estimating the parameters in regression models, called the method of least squares, that traces its origins to work by Karl Gauss. Essentially, this method chooses the parameters in the empirical model (the β's) to minimize the sum of the squared distances between each data point and the plane represented by the model equation. Applying this technique to the data in Table 1-2 results in

$$\widehat{\text{Pull strength}} = 2.26 + 2.74(\text{wire length}) + 0.0125(\text{die height}) \tag{1-7}$$

where the "hat," or circumflex, over pull strength indicates that this is an estimated or predicted quantity.

Figure 1-16 is a plot of the predicted values of pull strength versus wire length and die height obtained from Equation 1-7. Notice that the predicted values lie on a plane above the wire length–die height space. From the plot of the data in Fig. 1-15, this model does not appear unreasonable. The empirical model in Equation 1-7 could be used to predict values of pull strength for various combinations of wire length and die height that are of interest. Essentially, the empirical model could be used by an engineer in exactly the same way that a mechanistic model can be used.

1-4 PROBABILITY AND PROBABILITY MODELS

In Section 1-1, it was mentioned that decisions often need to be based on measurements from only a subset of objects selected in a sample. This process of reasoning from a sample of objects to conclusions for a population of objects was referred to as statistical inference. A sample of three wafers selected from a larger production lot of wafers in semiconductor manufacturing was an example mentioned. To make good decisions, an analysis of how well a sample represents a population is clearly necessary. If the lot contains defective wafers, how well will the sample detect this? How can we quantify the criterion to "detect well"? Basically, how can we quantify the risks of decisions based on samples? Furthermore, how should samples be selected to provide good decisions—ones with acceptable risks? Probability models help quantify the risks involved in statistical inference, that is, the risks involved in decisions made every day.

More details are useful to describe the role of probability models. Suppose a production lot contains 25 wafers. If all the wafers are defective or all are good, clearly any sample will generate all defective or all good wafers, respectively. However, suppose only one wafer in the lot is defective. Then a sample might or might not detect (include) the wafer. A probability model, along with a method to select the sample, can be used to quantify the risks that the defective wafer is or is not detected. Based on this analysis, the size of the sample might be increased (or decreased). The risk here can be interpreted as follows. Suppose a series of lots, each with exactly one defective wafer, are sampled. The details of the method used to select the sample are postponed until randomness is discussed in the next chapter. Nevertheless, assume that the same size sample (such as three wafers) is selected in the same manner from each lot. The proportion of the lots in which the defective wafer is included in the sample or, more specifically, the limit of this proportion as the number of lots in the series tends to infinity, is interpreted as the probability that the defective wafer is detected.

A probability model is used to calculate this proportion under reasonable assumptions for the manner in which the sample is selected. This is fortunate because we do not want to attempt to sample from an infinite series of lots. Problems of this type are worked in Chapters 2 and 3. More importantly, this probability provides valuable, quantitative information regarding any decision about lot quality based on the sample.

Recall from Section 1-1 that a population might be conceptual, as in an analytic study that applies statistical inference to future production based on the data from current production. When populations are extended in this manner, the role of statistical inference and the associated probability models becomes even more important.

In the previous example, each wafer in the sample was only classified as defective or not. Instead, a continuous measurement might be obtained from each wafer. In Section 1-2.5, concentration measurements were taken at periodic intervals from a production process. Figure 1-8 shows that variability is present in the measurements, and there might be concern that the process has moved from the target setting for concentration. Similar to the defective wafer, one might want to quantify our ability to detect a process change based on the sample data. Control limits were mentioned in Section 1-2.5 as decision rules for whether or not to adjust a process. The probability that a particular process change is detected can be calculated with a probability model for concentration measurements. Models for continuous measurements are developed based on plausible assumptions for the data and a result known as the central limit theorem, and the associated normal distribution is a particularly valuable probability model for statistical inference. Of course, a check of assumptions is important. These types of probability models are discussed in Chapter 4. The objective is still to quantify the risks inherent in the inference made from the sample data.

Throughout Chapters 6 through 15, decisions are based on statistical inference from sample data. Continuous probability models, specifically the normal distribution, are used extensively to quantify the risks in these decisions and to evaluate ways to collect the data and how large a sample should be selected.

IMPORTANT TERMS AND CONCEPTS

Analytic study	Fractional factorial	Population	Statistical inference
Cause and effect	experiment	Probability model	Statistical process
Designed experiment	Hypothesis testing	Problem-solving	control
Empirical model	Interaction	method	Statistical thinking
Engineering method	Mechanistic model	Randomization	Tampering
Enumerative study	Observational study	Retrospective study	Time series
Factorial experiment	Overcontrol	Sample	Variability

Probability

CHAPTER OUTLINE

2-1 SAMPLE SPACES AND EVENTS
    2-1.1 Random Experiments
    2-1.2 Sample Spaces
    2-1.3 Events
    2-1.4 Counting Techniques

2-2 INTERPRETATIONS OF PROBABILITY
    2-2.1 Introduction
    2-2.2 Axioms of Probability

2-3 ADDITION RULES

2-4 CONDITIONAL PROBABILITY

2-5 MULTIPLICATION AND TOTAL PROBABILITY RULES
    2-5.1 Multiplication Rule
    2-5.2 Total Probability Rule

2-6 INDEPENDENCE

2-7 BAYES' THEOREM

2-8 RANDOM VARIABLES

LEARNING OBJECTIVES

After careful study of this chapter you should be able to do the following:

1. Understand and describe sample spaces and events for random experiments with graphs, tables, lists, or tree diagrams
2. Interpret probabilities and use probabilities of outcomes to calculate probabilities of events in discrete sample spaces
3. Use permutation and combinations to count the number of outcomes in both an event and the sample space
4. Calculate the probabilities of joint events such as unions and intersections from the probabilities of individual events
5. Interpret and calculate conditional probabilities of events
6. Determine the independence of events and use independence to calculate probabilities
7. Use Bayes' theorem to calculate conditional probabilities
8. Understand random variables

2-1 SAMPLE SPACES AND EVENTS

2-1.1 Random Experiments

If we measure the current in a thin copper wire, we are conducting an experiment. However, in day-to-day repetitions of the measurement the results can differ slightly because of small variations in variables that are not controlled in our experiment, including changes in ambient temperatures, slight variations in gauge and small impurities in the chemical composition of the wire if different locations are selected, and current source drifts. Consequently, this experiment (as well as many we conduct) is said to have a random component. In some cases, the random variations, are small enough, relative to our experimental goals, that they can be ignored. However, no matter how carefully our experiment is designed and conducted, the variation is almost always present, and its magnitude can be large enough that the important conclusions from our experiment are not obvious. In these cases, the methods presented in this book for modeling and analyzing experimental results are quite valuable.

Our goal is to understand, quantify, and model the type of variations that we often encounter. When we incorporate the variation into our thinking and analyses, we can make informed judgments from our results that are not invalidated by the variation.

Models and analyses that include variation are not different from models used in other areas of engineering and science. Figure 2-1 displays the important components. A mathematical model (or abstraction) of the physical system is developed. It need not be a perfect abstraction. For example, Newton's laws are not perfect descriptions of our physical universe. Still, they are useful models that can be studied and analyzed to approximately quantify the performance of a wide range of engineered products. Given a mathematical abstraction that is validated with measurements from our system, we can use the model to understand, describe, and quantify important aspects of the physical system and predict the response of the system to inputs.

Throughout this text, we discuss models that allow for variations in the outputs of a system, even though the variables that we control are not purposely changed during our study. Figure 2-2 graphically displays a model that incorporates uncontrollable inputs (noise) that combine with the controllable inputs to produce the output of our system. Because of the uncontrollable inputs, the same settings for the controllable inputs do not result in identical outputs every time the system is measured.

Random Experiment

> An experiment that can result in different outcomes, even though it is repeated in the same manner every time, is called a random experiment.

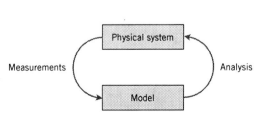

Figure 2-1 Continuous iteration between model and physical system.

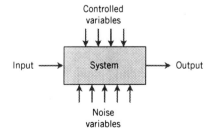

Figure 2-2 Noise variables affect the transformation of inputs to outputs.

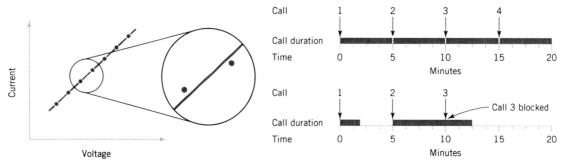

Figure 2-3 A closer examination of the system identifies deviations from the model.

Figure 2-4 Variation causes disruptions in the system.

For the example of measuring current in a copper wire, our model for the system might simply be Ohm's law. Because of uncontrollable inputs, variations in measurements of current are expected. Ohm's law might be a suitable approximation. However, if the variations are large relative to the intended use of the device under study, we might need to extend our model to include the variation. See Fig. 2-3.

As another example, in the design of a communication system, such as a computer or voice communication network, the information capacity available to service individuals using the network is an important design consideration. For voice communication, sufficient external lines need to be purchased from the phone company to meet the requirements of a business. Assuming each line can carry only a single conversation, how many lines should be purchased? If too few lines are purchased, calls can be delayed or lost. The purchase of too many lines increases costs. Increasingly, design and product development is required to meet customer requirements *at a competitive cost*.

In the design of the voice communication system, a model is needed for the number of calls and the duration of calls. Even knowing that on average, calls occur every five minutes and that they last five minutes is not sufficient. If calls arrived precisely at five-minute intervals and lasted for precisely five minutes, one phone line would be sufficient. However, the slightest variation in call number or duration would result in some calls being blocked by others. See Fig. 2-4. A system designed without considering variation will be woefully inadequate for practical use. Our model for the number and duration of calls needs to include variation as an integral component. An analysis of models including variation is important for the design of the phone system.

2-1.2 Sample Spaces

To model and analyze a random experiment, we must understand the set of possible outcomes from the experiment. In this introduction to probability, we make use of the basic concepts of sets and operations on sets. It is assumed that the reader is familiar with these topics.

Sample Space

> The set of all possible outcomes of a random experiment is called the **sample space** of the experiment. The sample space is denoted as S.

A sample space is often defined based on the objectives of the analysis.

EXAMPLE 2-1
Molded Plastic Part

Consider an experiment in which you select a molded plastic part, such as a connector, and measure its thickness. The possible values for thickness depend on the resolution of the measuring instrument, and they also depend on upper and lower bounds for thickness. However, it might be convenient to define the sample space as simply the positive real line

$$S = R^+ = \{x \mid x > 0\}$$

because a negative value for thickness cannot occur.

If it is known that all connectors will be between 10 and 11 millimeters thick, the sample space could be

$$S = \{x \mid 10 < x < 11\}$$

If the objective of the analysis is to consider only whether a particular part is low, medium, or high for thickness, the sample space might be taken to be the set of three outcomes:

$$S = \{low, medium, high\}$$

If the objective of the analysis is to consider only whether or not a particular part conforms to the manufacturing specifications, the sample space might be simplified to the set of two outcomes

$$S = \{yes, no\}$$

that indicate whether or not the part conforms.

It is useful to distinguish between two types of sample spaces.

Discrete and Continuous Sample Spaces

> A sample space is **discrete** if it consists of a finite or countable infinite set of outcomes. A sample space is **continuous** if it contains an interval (either finite or infinite) of real numbers.

In Example 2-1, the choice $S = R^+$ is an example of a continuous sample space, whereas $S = \{yes, no\}$ is a discrete sample space. As mentioned, the best choice of a sample space depends on the objectives of the study. As specific questions occur later in the book, appropriate sample spaces are discussed.

EXAMPLE 2-2
Manufacturing Specifications

If two connectors are selected and measured, the extension of the positive real line R is to take the sample space to be the positive quadrant of the plane:

$$S = R^+ \times R^+$$

If the objective of the analysis is to consider only whether or not the parts conform to the manufacturing specifications, either part may or may not conform. We abbreviate *yes* and *no* as *y* and *n*. If the ordered pair *yn* indicates that the first connector conforms and the second does not, the sample space can be represented by the four outcomes:

$$S = \{yy, yn, ny, nn\}$$

If we are only interested in the number of conforming parts in the sample, we might summarize the sample space as

$$S = \{0, 1, 2\}$$

As another example, consider an experiment in which the thickness is measured until a connector fails to meet the specifications. The sample space can be represented as

$$S = \{n, yn, yyn, yyyn, yyyyn, \text{ and so forth}\}$$

and this is an example of a discrete sample space that is countably infinite.

In random experiments in which items are selected from a batch, we will indicate whether or not a selected item is replaced before the next one is selected. For example, if the batch consists of three items $\{a, b, c\}$ and our experiment is to select two items without replacement, the sample space can be represented as

$$S_{\text{without}} = \{ab, ac, ba, bc, ca, cb\}$$

This description of the sample space maintains the order of the items selected so that the outcome ab and ba are separate elements in the sample space. A sample space with less detail only describes the two items selected $\{\{a, b\}, \{a, c\}, \{b, c\}\}$. This sample space is the possible subsets of two items. Sometimes the ordered outcomes are needed, but in other cases the simpler, unordered sample space is sufficient.

If items are replaced before the next one is selected, the sampling is referred to as with replacement. Then the possible ordered outcomes are

$$S_{\text{with}} = \{aa, ab, ac, ba, bb, bc, ca, cb, cc\}$$

The unordered description of the sample space is $\{\{a, a\}, \{a, b\}, \{a, c\}, \{b, b\}, \{b, c\}, \{c, c\}\}$. Sampling without replacement is more common for industrial applications.

Sometimes it is not necessary to specify the exact item selected, but only a property of the item. For example, suppose that there are 5 defective parts and 95 good parts in a batch. To study the quality of the batch, two are selected without replacement. Let g denote a good part and d denote a defective part. It might be sufficient to describe the sample space (ordered) in terms of quality of each part selected as

$$S = \{gg, gd, dg, dd\}$$

One must be cautious with this description of the sample space because there are many more pairs of items in which both are good than pairs in which both are defective. These differences must be accounted for when probabilities are computed later in this chapter. Still, this summary of the sample space will be convenient when conditional probabilities are used later in this chapter. Also, if there were only one defective part in the batch, there would be fewer possible outcomes

$$S = \{gg, gd, dg\}$$

because dd would be impossible. For sampling questions, sometimes the most important part of the solution is an appropriate description of the sample space.

2-1 SAMPLE SPACES AND EVENTS   21

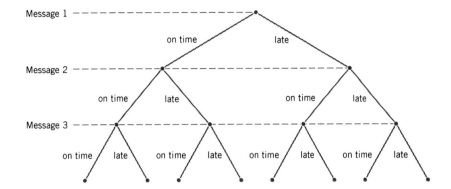

Figure 2-5 Tree diagram for three messages.

Sample spaces can also be described graphically with **tree diagrams.** When a sample space can be constructed in several steps or stages, we can represent each of the n_1 ways of completing the first step as a branch of a tree. Each of the ways of completing the second step can be represented as n_2 branches starting from the ends of the original branches, and so forth.

EXAMPLE 2-3   Each message in a digital communication system is classified as to whether it is received within the time specified by the system design. If three messages are classified, use a tree diagram to represent the sample space of possible outcomes.

Each message can either be received on time or late. The possible results for three messages can be displayed by eight branches in the tree diagram shown in Fig. 2-5.

EXAMPLE 2-4   An automobile manufacturer provides vehicles equipped with selected options. Each vehicle is ordered

With or without an automatic transmission   With one of three choices of a stereo system
With or without air-conditioning   With one of four exterior colors

If the sample space consists of the set of all possible vehicle types, what is the number of outcomes in the sample space? The sample space contains 48 outcomes. The tree diagram for the different types of vehicles is displayed in Fig. 2-6.

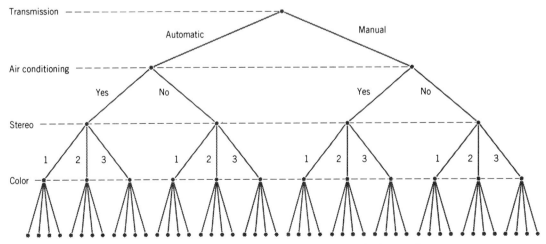

Figure 2-6 Tree diagram for different types of vehicles with 48 outcomes in the sample space.

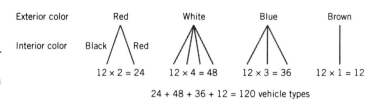

Figure 2-7 Tree diagram for different types of vehicles with interior colors.

EXAMPLE 2-5
Automobile Configurations

Consider an extension of the automobile manufacturer illustration in the previous example in which another vehicle option is the interior color. There are four choices of interior color: red, black, blue, or brown. However,

With a red exterior, only a black or red interior can be chosen.
With a white exterior, any interior color can be chosen.
With a blue exterior, only a black, red, or blue interior can be chosen.
With a brown exterior, only a brown interior can be chosen.

In Fig. 2-6, there are 12 vehicle types with each exterior color, but the number of interior color choices depends on the exterior color. As shown in Fig. 2-7, the tree diagram can be extended to show that there are 120 different vehicle types in the sample space.

2-1.3 Events

Often we are interested in a collection of related outcomes from a random experiment. Related outcomes can be described by subsets of the sample space and set operations can also be applied.

Event

> **An event is a subset of the sample space of a random experiment.**

We can also be interested in describing new events from combinations of existing events. Because events are subsets, we can use basic set operations such as unions, intersections, and complements to form other events of interest. Some of the basic set operations are summarized below in terms of events:

- The **union** of two events is the event that consists of all outcomes that are contained in either of the two events. We denote the union as $E_1 \cup E_2$.
- The **intersection** of two events is the event that consists of all outcomes that are contained in both of the two events. We denote the intersection as $E_1 \cap E_2$.
- The **complement** of an event in a sample space is the set of outcomes in the sample space that are not in the event. We denote the complement of the event E as E'. The notation E^C is also used in other literature to denote the complement.

EXAMPLE 2-6

Consider the sample space $S = \{yy, yn, ny, nn\}$ in Example 2-2. Suppose that the subset of outcomes for which at least one part conforms is denoted as E_1. Then,

$$E_1 = \{yy, yn, ny\}$$

The event in which both parts do not conform, denoted as E_2, contains only the single outcome, $E_2 = \{nn\}$. Other examples of events are $E_3 = \varnothing$, the null set, and $E_4 = S$, the sample space. If $E_5 = \{yn, ny, nn\}$,

$$E_1 \cup E_5 = S \qquad E_1 \cap E_5 = \{yn, ny\} \qquad E_1' = \{nn\}$$

EXAMPLE 2-7

Measurements of the thickness of a plastic connector might be modeled with the sample space $S = R^+$, the set of positive real numbers. Let

$$E_1 = \{x \mid 10 \le x < 12\} \quad \text{and} \quad E_2 = \{x \mid 11 < x < 15\}$$

Then,

$$E_1 \cup E_2 = \{x \mid 10 \le x < 15\} \quad \text{and} \quad E_1 \cap E_2 = \{x \mid 11 < x < 12\}$$

Also,

$$E_1' = \{x \mid x < 10 \text{ or } 12 \le x\} \quad \text{and} \quad E_1' \cap E_2 = \{x \mid 12 \le x < 15\}$$

EXAMPLE 2-8
Polycarbonate Plastic

Samples of polycarbonate plastic are analyzed for scratch and shock resistance. The results from 50 samples are summarized as follows:

		shock resistance	
		high	low
scratch resistance	high	40	4
	low	1	5

Let A denote the event that a sample has high shock resistance, and let B denote the event that a sample has high scratch resistance. Determine the number of samples in $A \cap B$, A', and $A \cup B$.

The event $A \cap B$ consists of the 40 samples for which scratch and shock resistances are high. The event A' consists of the 9 samples in which the shock resistance is low. The event $A \cup B$ consists of the 45 samples in which the shock resistance, scratch resistance, or both are high.

Diagrams are often used to portray relationships between sets, and these diagrams are also used to describe relationships between events. We can use **Venn diagrams** to represent a sample space and events in a sample space. For example, in Fig. 2-8(a) the sample space of the random experiment is represented as the points in the rectangle S. The events A and B are the subsets of points in the indicated regions. Figs. 2-8(b) to 2-8(d) illustrate additional joint events. Figure 2-9 illustrates two events with no common outcomes.

Mutually Exclusive Events

Two events, denoted as E_1 and E_2, such that

$$E_1 \cap E_2 = \varnothing$$

are said to be **mutually exclusive**.

24 CHAPTER 2 PROBABILITY

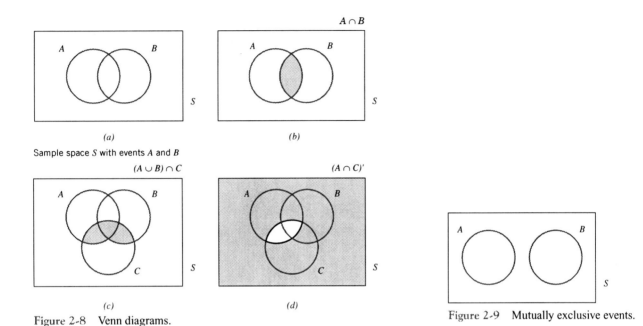

Figure 2-8 Venn diagrams.

Figure 2-9 Mutually exclusive events.

Additional results involving events are summarized below. The definition of the complement of an event implies that

$$(E')' = E$$

The distributive law for set operations implies that

$$(A \cup B) \cap C = (A \cap C) \cup (B \cap C), \quad \text{and} \quad (A \cap B) \cup C = (A \cup C) \cap (B \cup C)$$

DeMorgan's laws imply that

$$(A \cup B)' = A' \cap B' \quad \text{and} \quad (A \cap B)' = A' \cup B'$$

Also, remember that

$$A \cap B = B \cap A \quad \text{and} \quad A \cup B = B \cup A$$

2-1.4 Counting Techniques

In many of the examples in Chapter 2, it is easy to determine the number of outcomes in each event. In more complicated examples, determining the outcomes that comprise the sample space (or an event) becomes more difficult. Instead, counts of the numbers of outcomes in the sample space and various events are used to analyze the random experiments. These methods are referred to as counting techniques. Some simple rules can be used to simplify the calculations.

In Example 2-4, an automobile manufacturer provides vehicles equipped with selected options. Each vehicle is ordered

With or without an automatic transmission

With or without air conditioning

With one of three choices of a stereo system

With one of four exterior colors

The tree diagram in Fig. 2-6 describes the sample space of all possible vehicle types. The size of the sample space equals the number of branches in the last level of the tree and this quantity equals $2 \times 2 \times 3 \times 4 = 48$. This leads to the following useful result.

Multiplication Rule (for counting techniques)

If an operation can be described as a sequence of k steps, and

    if the number of ways of completing step 1 is n_1, and

    if the number of ways of completing step 2 is n_2 for each way of completing step 1, and

    if the number of ways of completing step 3 is n_3 for each way of completing step 2, and so forth,

the total number of ways of completing the operation is

$$n_1 \times n_2 \times \cdots \times n_k$$

EXAMPLE 2-9

In the design of a casing for a gear housing, we can use four different types of fasteners, three different bolt lengths, and three different bolt locations. From the multiplication rule, $4 \times 3 \times 3 = 36$ different designs are possible.

Permutations

Another useful calculation is the number of ordered sequences of the elements of a set. Consider a set of elements, such as $S = \{a, b, c\}$. A **permutation** of the elements is an ordered sequence of the elements. For example, *abc*, *acb*, *bac*, *bca*, *cab*, and *cba* are all of the permutations of the elements of S.

The number of **permutations** of n different elements is $n!$ where

$$n! = n \times (n-1) \times (n-2) \times \cdots \times 2 \times 1 \qquad (2\text{-}1)$$

This result follows from the multiplication rule. A permutation can be constructed by selecting the element to be placed in the first position of the sequence from the n elements, then selecting the element for the second position from the $n - 1$ remaining elements, then selecting the element for the third position from the remaining $n - 2$ elements, and so forth. Permutations such as these are sometimes referred to as linear permutations.

In some situations, we are interested in the number of arrangements of only some of the elements of a set. The following result also follows from the multiplication rule.

Permutations of Subsets

The number of permutations of subsets of r elements selected from a set of n different elements is

$$P_r^n = n \times (n-1) \times (n-2) \times \cdots \times (n-r+1) = \frac{n!}{(n-r)!} \qquad (2\text{-}2)$$

EXAMPLE 2-10
Printed Circuit Board

A printed circuit board has eight different locations in which a component can be placed. If four different components are to be placed on the board, how many different designs are possible?

Each design consists of selecting a location from the eight locations for the first component, a location from the remaining seven for the second component, a location from the remaining six for the third component, and a location from the remaining five for the fourth component. Therefore,

$$P_4^8 = 8 \times 7 \times 6 \times 5 = \frac{8!}{4!} = 1680 \text{ different designs are possible.}$$

Sometimes we are interested in counting the number of ordered sequences for objects that are not all different. The following result is a useful, general calculation.

Permutations of Similar Objects

The number of permutations of $n = n_1 + n_2 + \cdots + n_r$ objects of which n_1 are of one type, n_2 are of a second type, ..., and n_r are of an rth type is

$$\frac{n!}{n_1!\, n_2!\, n_3! \cdots n_r!} \qquad (2\text{-}3)$$

EXAMPLE 2-11
Machine Shop Schedule

Consider a machining operation in which a piece of sheet metal needs two identical diameter holes drilled and two identical size notches cut. We denote a drilling operation as d and a notching operation as n. In determining a schedule for a machine shop, we might be interested in the number of different possible sequences of the four operations. The number of possible sequences for two drilling operations and two notching operations is

$$\frac{4!}{2!\, 2!} = 6$$

The six sequences are easily summarized: *ddnn, dndn, dnnd, nddn, ndnd, nndd.*

EXAMPLE 2-12
Bar Codes

A part is labeled by printing with four thick lines, three medium lines, and two thin lines. If each ordering of the nine lines represents a different label, how many different labels can be generated by using this scheme?

From Equation 2-3, the number of possible part labels is

$$\frac{9!}{4!\, 3!\, 2!} = 1260$$

Combinations

Another counting problem of interest is the number of subsets of r elements that can be selected from a set of n elements. Here, order is not important. These are called **combinations.** Every subset of r elements can be indicated by listing the elements in the set and marking each element with a "*" if it is to be included in the subset. Therefore, each permutation of r *'s and $n - r$ blanks indicate a different subset and the number of these are obtained from Equation 2-3. For example, if the set is $S = \{a, b, c, d\}$ the subset $\{a, c\}$ can be indicated as

$$
\begin{array}{cccc}
a & b & c & d \\
* & & * &
\end{array}
$$

Combinations

> The number of combinations, subsets of size r that can be selected from a set of n elements, is denoted as $\binom{n}{r}$ or C_r^n and
>
> $$\binom{n}{r} = \frac{n!}{r!(n-r)!} \qquad (2\text{-}4)$$

EXAMPLE 2-13

A printed circuit board has eight different locations in which a component can be placed. If five identical components are to be placed on the board, how many different designs are possible?

Each design is a subset of the eight locations that are to contain the components. From Equation 2-4, the number of possible designs is

$$\frac{8!}{5!\,3!} = 56$$

The following example uses the multiplication rule in combination with Equation 2-4 to answer a more difficult, but common, question.

EXAMPLE 2-14
Sampling without Replacement

A bin of 50 manufactured parts contains three defective parts and 47 nondefective parts. A sample of six parts is selected from the 50 parts. Selected parts are not replaced. That is, each part can only be selected once and the sample is a subset of the 50 parts. How many different samples are there of size six that contain exactly two defective parts?

A subset containing exactly two defective parts can be formed by first choosing the two defective parts from the three defective parts. Using Equation 2-4, this step can be completed in

$$\binom{3}{2} = \frac{3!}{2!\,1!} = 3 \text{ different ways}$$

Then, the second step is to select the remaining four parts from the 47 acceptable parts in the bin. The second step can be completed in

$$\binom{47}{4} = \frac{47!}{4!\,43!} = 178{,}365 \text{ different ways}$$

Therefore, from the multiplication rule, the number of subsets of size six that contain exactly two defective items is

$$3 \times 178{,}365 = 535{,}095$$

As an additional computation, the total number of different subsets of size six is found to be

$$\binom{50}{6} = \frac{50!}{6!\,44!} = 15{,}890{,}700$$

When probability is discussed in this chapter, the probability of an event is determined as the ratio of the number of outcomes in the event to the number of outcomes in the sample space (for equally likely outcomes). Therefore, the probability that a sample contains exactly two defective parts is

$$\frac{535{,}095}{15{,}890{,}700} = 0.034$$

Note that this example illustrates a common distribution studied in Chapter 3 (hypergeometric distribution).

EXERCISES FOR SECTION 2-1

Provide a reasonable description of the sample space for each of the random experiments in Exercises 2-1 to 2-18. There can be more than one acceptable interpretation of each experiment. Describe any assumptions you make.

2-1. Each of three machined parts is classified as either above or below the target specification for the part.

2-2. Each of four transmitted bits is classified as either in error or not in error.

2-3. In the final inspection of electronic power supplies, three types of nonconformities might occur: functional, minor, or cosmetic. Power supplies that are defective are further classified as to type of nonconformity.

2-4. In the manufacturing of digital recording tape, electronic testing is used to record the number of bits in error in a 350-foot reel.

2-5. In the manufacturing of digital recording tape, each of 24 tracks is classified as containing or not containing one or more bits in error.

2-6. An ammeter that displays three digits is used to measure current in milliamperes.

2-7. A scale that displays two decimal places is used to measure material feeds in a chemical plant in tons.

2-8. The following two questions appear on an employee survey questionnaire. Each answer is chosen from the five-point scale 1 (never), 2, 3, 4, 5 (always).

Is the corporation willing to listen to and fairly evaluate new ideas?

How often are my coworkers important in my overall job performance?

2-9. The concentration of ozone to the nearest part per billion.

2-10. The time until a service transaction is requested of a computer to the nearest millisecond.

2-11. The pH reading of a water sample to the nearest tenth of a unit.

2-12. The voids in a ferrite slab are classified as small, medium, or large. The number of voids in each category is measured by an optical inspection of a sample.

2-13. The time of a chemical reaction is recorded to the nearest millisecond.

2-14. An order for an automobile can specify either an automatic or a standard transmission, either with or without air-conditioning, and any one of the four colors red, blue, black or white. Describe the set of possible orders for this experiment.

2-15. A sampled injection-molded part could have been produced in either one of two presses and in any one of the eight cavities in each press.

2-16. An order for a computer system can specify memory of 4, 8, or 12 gigabytes, and disk storage of 200, 300, or 400 gigabytes. Describe the set of possible orders.

2-17. Calls are repeatedly placed to a busy phone line until a connect is achieved.

2-18. In a magnetic storage device, three attempts are made to read data before an error recovery procedure that repositions the magnetic head is used. The error recovery procedure attempts three repositionings before an "abort" message is sent to the operator. Let

s denote the success of a read operation

f denote the failure of a read operation

F denote the failure of an error recovery procedure

S denote the success of an error recovery procedure

A denote an abort message sent to the operator.

Describe the sample space of this experiment with a tree diagram.

2-19. Three events are shown on the Venn diagram in the following figure:

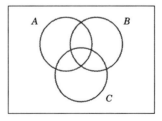

Reproduce the figure and shade the region that corresponds to each of the following events.

(a) A' (b) $A \cap B$ (c) $(A \cap B) \cup C$
(d) $(B \cup C)'$ (e) $(A \cap B)' \cup C$

2-20. Three events are shown on the Venn diagram in the following figure:

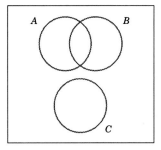

Reproduce the figure and shade the region that corresponds to each of the following events.
(a) A'
(b) $(A \cap B) \cup (A \cap B')$
(c) $(A \cap B) \cup C$
(d) $(B \cup C)'$
(e) $(A \cap B)' \cup C$

2-21. A digital scale is used that provides weights to the nearest gram.
(a) What is the sample space for this experiment?
Let A denote the event that a weight exceeds 11 grams, let B denote the event that a weight is less than or equal to 15 grams, and let C denote the event that a weight is greater than or equal to 8 grams and less than 12 grams.
Describe the following events.
(b) $A \cup B$
(c) $A \cap B$
(d) A'
(e) $A \cup B \cup C$
(f) $(A \cup C)'$
(g) $A \cap B \cap C$
(h) $B' \cap C$
(i) $A \cup (B \cap C)$

2-22. In an injection-molding operation, the length and width, denoted as X and Y, respectively, of each molded part are evaluated. Let

A denote the event of $48 < X < 52$ centimeters

B denote the event of $9 < Y < 11$ centimeters

C denote the event that a critical length meets customer requirements.

Construct a Venn diagram that includes these events. Shade the areas that represent the following:
(a) A
(b) $A \cap B$
(c) $A' \cup B$
(d) $A \cup B$
(e) If these events were mutually exclusive, how successful would this production operation be? Would the process produce parts with $X = 50$ centimeters and $Y = 10$ centimeters?

2-23. Four bits are transmitted over a digital communications channel. Each bit is either distorted or received without distortion. Let A_i denote the event that the ith bit is distorted, $i = 1, \ldots, 4$.
(a) Describe the sample space for this experiment.
(b) Are the A_i's mutually exclusive?

Describe the outcomes in each of the following events:
(c) A_1
(d) A_1'
(e) $A_1 \cap A_2 \cap A_3 \cap A_4$
(f) $(A_1 \cap A_2) \cup (A_3 \cap A_4)$

2-24. In light-dependent photosynthesis, light quality refers to the wavelengths of light that are important. The wavelength of a sample of photosynthetically active radiations (PAR) is measured to the nearest nanometer. The red range is 675–700 nm and the blue range is 450–500 nm. Let A denote the event that PAR occurs in the red range and let B denote the event that PAR occurs in the blue range. Describe the sample space and indicate each of the following events:
(a) A
(b) B
(c) $A \cap B$
(d) $A \cup B$

2-25. In control replication, cells are replicated over a period of two days. Not until mitosis is completed, can freshly-synthesized DNA be replicated again. Two control mechanisms have been identified—one positive and one negative. Suppose that a replication is observed in three cells. Let A denote the event that all cells are identified as positive and let B denote the event that all cells are negative. Describe the sample space graphically and display each of the following events:
(a) A
(b) B
(c) $A \cap B$
(d) $A \cup B$

2-26. Disks of polycarbonate plastic from a supplier are analyzed for scratch and shock resistance. The results from 100 disks are summarized below:

		shock resistance	
		high	low
scratch	high	70	9
resistance	low	16	5

Let A denote the event that a disk has high shock resistance, and let B denote the event that a disk has high scratch resistance. Determine the number of disks in $A \cap B$, A', and $A \cup B$.

2-27. Samples of a cast aluminum part are classified on the basis of surface finish (in microinches) and edge finish. The results of 100 parts are summarized as follows:

		edge finish	
		excellent	good
surface	excellent	80	2
finish	good	10	8

(a) Let A denote the event that a sample has excellent surface finish, and let B denote the event that a sample has excellent edge finish. Determine the number of samples in $A' \cap B$, B', and $A \cup B$.
(b) Assume that each of two samples is to be classified on the basis of surface finish, either excellent or good, edge finish, either excellent or good. Use a tree diagram to represent the possible outcomes of this experiment.

2-28. Samples of emissions from three suppliers are classified for conformance to air-quality specifications. The results from 100 samples are summarized as follows:

		conforms	
		yes	no
	1	22	8
supplier	2	25	5
	3	30	10

Let A denote the event that a sample is from supplier 1, and let B denote the event that a sample conforms to specifications. Determine the number of samples in $A' \cap B$, B', and $A \cup B$.

2-29. The rise time of a reactor is measured in minutes (and fractions of minutes). Let the sample space be positive, real numbers. Define the events A and B as follows:

$$A = \{x \mid x < 72.5\}$$

and

$$B = \{x \mid x > 52.5\}$$

Describe each of the following events:
(a) A' (b) B'
(c) $A \cap B$ (d) $A \cup B$

2-30. A sample of two items is selected without replacement from a batch. Describe the (ordered) sample space for each of the following batches:
(a) The batch contains the items $\{a, b, c, d\}$.
(b) The batch contains the items $\{a, b, c, d, e, f, g\}$.
(c) The batch contains 4 defective items and 20 good items.
(d) The batch contains 1 defective item and 20 good items.

2-31. A sample of two printed circuit boards is selected without replacement from a batch. Describe the (ordered) sample space for each of the following batches:
(a) The batch contains 90 boards that are not defective, 8 boards with minor defects, and 2 boards with major defects.
(b) The batch contains 90 boards that are not defective, 8 boards with minor defects, and 1 board with major defects.

2-32. Counts of the Web pages provided by each of two computer servers in a selected hour of the day are recorded. Let A denote the event that at least 10 pages are provided by server 1 and let B denote the event that at least 20 pages are provided by server 2.
(a) Describe the sample space for the numbers of pages for the two servers graphically.

Show each of the following events on the sample space graph:
(b) A (c) B
(d) $A \cap B$ (e) $A \cup B$

2-33. The rise time of a reactor is measured in minutes (and fractions of minutes). Let the sample space for the rise time of each batch be positive, real numbers. Consider the rise times of *two* batches. Let A denote the event that the rise time of batch 1 is less than 72.5 minutes, and let B denote the event that the rise time of batch 2 is greater than 52.5 minutes.

Describe the sample space for the rise time of two batches graphically and show each of the following events on a two-dimensional plot:
(a) A (b) B'
(c) $A \cap B$ (d) $A \cup B$

2-34. A wireless garage door opener has a code determined by the up or down setting of 12 switches. How many outcomes are in the sample space of possible codes?

2-35. An order for a personal digital assistant can specify any one of five memory sizes, any one of three types of displays, any one of four sizes of a hard disk, and can either include or not include a pen tablet. How many different systems can be ordered?

2-36. In a manufacturing operation, a part is produced by machining, polishing, and painting. If there are three machine tools, four polishing tools, and three painting tools, how many different routings (consisting of machining, followed by polishing, and followed by painting) for a part are possible?

2-37. New designs for a wastewater treatment tank have proposed three possible shapes, four possible sizes, three locations for input valves, and four locations for output valves. How many different product designs are possible?

2-38. A manufacturing process consists of 10 operations that can be completed in any order. How many different production sequences are possible?

2-39. A manufacturing operations consists of 10 operations. However, five machining operations must be completed before any of the remaining five assembly operations can begin. Within each set of five, operations can be completed in any order. How many different production sequences are possible?

2-40. In a sheet metal operation, three notches and four bends are required. If the operations can be done in any order, how many different ways of completing the manufacturing are possible?

2-41. A lot of 140 semiconductor chips is inspected by choosing a sample of five chips. Assume 10 of the chips do not conform to customer requirements.
(a) How many different samples are possible?
(b) How many samples of five contain exactly one nonconforming chip?
(c) How many samples of five contain at least one nonconforming chip?

2-42. In the layout of a printed circuit board for an electronic product, there are 12 different locations that can accommodate chips.
(a) If five different types of chips are to be placed on the board, how many different layouts are possible?

(b) If the five chips that are placed on the board are of the same type, how many different layouts are possible?

2-43. In the laboratory analysis of samples from a chemical process, five samples from the process are analyzed daily. In addition, a control sample is analyzed two times each day to check the calibration of the laboratory instruments.
(a) How many different sequences of process and control samples are possible each day? Assume that the five process samples are considered identical and that the two control samples are considered identical.
(b) How many different sequences of process and control samples are possible if we consider the five process samples to be different and the two control samples to be identical.
(c) For the same situation as part (b), how many sequences are possible if the first test of each day must be a control sample?

2-44. In the design of an electromechanical product, seven different components are to be stacked into a cylindrical casing that holds 12 components in a manner that minimizes the impact of shocks. One end of the casing is designated as the bottom and the other end is the top.
(a) How many different designs are possible?
(b) If the seven components are all identical, how many different designs are possible?
(c) If the seven components consist of three of one type of component and four of another type, how many different designs are possible? (more difficult)

2-45. The design of a communication system considered the following questions:
(a) How many three-digit phone prefixes that are used to represent a particular geographic area (such as an area code) can be created from the digits 0 through 9?
(b) As in part (a), how many three-digit phone prefixes are possible that do not start with 0 or 1, but contain 0 or 1 as the middle digit?
(c) How many three-digit phone prefixes are possible in which no digit appears more than once in each prefix?

2-46. A byte is a sequence of eight bits and each bit is either 0 or 1.
(a) How many different bytes are possible?
(b) If the first bit of a byte is a parity check, that is, the first byte is determined from the other seven bits, how many different bytes are possible?

2-47. In a chemical plant, 24 holding tanks are used for final product storage. Four tanks are selected at random and without replacement. Suppose that six of the tanks contain material in which the viscosity exceeds the customer requirements.
(a) What is the probability that exactly one tank in the sample contains high viscosity material?
(b) What is the probability that at least one tank in the sample contains high viscosity material?
(c) In addition to the six tanks with high viscosity levels, four different tanks contain material with high impurities. What is the probability that exactly one tank in the sample contains high viscosity material and exactly one tank in the sample contains material with high impurities?

2-48. Plastic parts produced by an injection-molding operation are checked for conformance to specifications. Each tool contains 12 cavities in which parts are produced, and these parts fall into a conveyor when the press opens. An inspector chooses 3 parts from among the 12 at random. Two cavities are affected by a temperature malfunction that results in parts that do not conform to specifications.
(a) What is the probability that the inspector finds exactly one nonconforming part?
(b) What is the probability that the inspector finds at least one nonconforming part?

2-49. A bin of 50 parts contains five that are defective. A sample of two is selected at random, without replacement.
(a) Determine the probability that both parts in the sample are defective by computing a conditional probability.
(b) Determine the answer to part (a) by using the subset approach that was described in this section.

2-2 INTERPRETATIONS OF PROBABILITY

2-2.1 Introduction

In this chapter, we introduce probability for **discrete sample spaces**—those with only a finite (or countably infinite) set of outcomes. The restriction to these sample spaces enables us to simplify the concepts and the presentation without excessive mathematics.

Probability is used to quantify the likelihood, or chance, that an outcome of a random experiment will occur. "The chance of rain today is 30%" is a statement that quantifies our feeling about the possibility of rain. The likelihood of an outcome is quantified by assigning a number from the interval [0, 1] to the outcome (or a percentage from 0 to 100%). Higher numbers indicate that the outcome is more likely than lower numbers. A 0 indicates an outcome will not occur. A probability of 1 indicates an outcome will occur with certainty.

32 CHAPTER 2 PROBABILITY

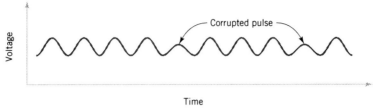

Figure 2-10 Relative frequency of corrupted pulses sent over a communication channel.

The probability of an outcome can be interpreted as our subjective probability, or **degree of belief,** that the outcome will occur. Different individuals will no doubt assign different probabilities to the same outcomes. Another interpretation of probability is based on the conceptual model of repeated replications of the random experiment. The probability of an outcome is interpreted as the limiting value of the proportion of times the outcome occurs in n repetitions of the random experiment as n increases beyond all bounds. For example, if we assign probability 0.2 to the outcome that there is a corrupted pulse in a digital signal, we might interpret this assignment as implying that, if we analyze many pulses, approximately 20% of them will be corrupted. This example provides a relative frequency interpretation of probability. The proportion, or relative frequency, of replications of the experiment that result in the outcome is 0.2. Probabilities are chosen so that the sum of the probabilities of all outcomes in an experiment add up to 1. This convention facilitates the relative frequency interpretation of probability. Figure 2-10 illustrates the concept of relative frequency.

Probabilities for a random experiment are often assigned on the basis of a reasonable model of the system under study. One approach is to base probability assignments on the simple concept of equally likely outcomes.

For example, suppose that we will select one laser diode **randomly** from a batch of 100. The sample space is the set of 100 diodes. *Randomly* implies that it is reasonable to assume that each diode in the batch has an equal chance of being selected. Because the sum of the probabilities must equal 1, the probability model for this experiment assigns probability of 0.01 to each of the 100 outcomes. We can interpret the probability by imagining many replications of the experiment. Each time we start with all 100 diodes and select one at random. The probability 0.01 assigned to a particular diode represents the proportion of replicates in which a particular diode is selected. When the model of **equally likely outcomes** is assumed, the probabilities are chosen to be equal.

Equally Likely Outcomes

Whenever a sample space consists of N possible outcomes that are equally likely, the probability of each outcome is $1/N$.

It is frequently necessary to assign probabilities to events that are composed of several outcomes from the sample space. This is straightforward for a discrete sample space.

EXAMPLE 2-15
Laser Diodes

Assume that 30% of the laser diodes in a batch of 100 meet the minimum power requirements of a specific customer. If a laser diode is selected randomly, that is, each laser diode is equally likely to be selected, our intuitive feeling is that the probability of meeting the customer's requirements is 0.30.

Let E denote the subset of 30 diodes that meet the customer's requirements. Because E contains 30 outcomes and each outcome has probability 0.01, we conclude that the probability of E is 0.3. The conclusion matches our intuition. Figure 2-11 illustrates this example.

Figure 2-11 Probability of the event E is the sum of the probabilities of the outcomes in E.

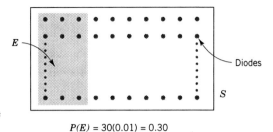

$P(E) = 30(0.01) = 0.30$

For a discrete sample space, the probability of an event can be defined by the reasoning used in the example above.

Probability of an Event

For a discrete sample space, the *probability of an event E*, denoted as $P(E)$, equals the sum of the probabilities of the outcomes in E.

EXAMPLE 2-16

A random experiment can result in one of the outcomes $\{a, b, c, d\}$ with probabilities 0.1, 0.3, 0.5, and 0.1, respectively. Let A denote the event $\{a, b\}$, B the event $\{b, c, d\}$, and C the event $\{d\}$. Then,

$$P(A) = 0.1 + 0.3 = 0.4$$
$$P(B) = 0.3 + 0.5 + 0.1 = 0.9$$
$$P(C) = 0.1$$

Also, $P(A') = 0.6$, $P(B') = 0.1$, and $P(C') = 0.9$. Furthermore, because $A \cap B = \{b\}$, $P(A \cap B) = 0.3$. Because $A \cup B = \{a, b, c, d\}$, $P(A \cup B) = 0.1 + 0.3 + 0.5 + 0.1 = 1$. Because $A \cap C$ is the null set, $P(A \cap C) = 0$.

EXAMPLE 2-17
Contamination Particles

A visual inspection of a location on wafers from a semiconductor manufacturing process resulted in the following table:

Number of Contamination Particles	Proportion of Wafers
0	0.40
1	0.20
2	0.15
3	0.10
4	0.05
5 or more	0.10

If one wafer is selected randomly from this process and the location is inspected, what is the probability that it contains no particles? If information were available for each wafer, we could define the sample space as the set of all wafers inspected and proceed as in the example with diodes. However, this level of detail is not needed in this case. We can consider the sample space to consist of the six categories that summarize the number of contamination particles on a wafer. Each category has probability equal to the proportion of wafers in the category. The event that there is no contamination particle in the inspected location on the wafer, denoted as E, can be considered to be comprised of the single outcome, namely, $E = \{0\}$. Therefore,

$$P(E) = 0.4$$

What is the probability that a wafer contains three or more particles in the inspected location? Let E denote the event that a wafer contains three or more particles in the inspected location. Then, E consists of the three outcomes {3, 4, 5 or more}. Therefore,

$$P(E) = 0.10 + 0.05 + 0.10 = 0.25$$

Often more than one item is selected from a batch without replacement when production is inspected. In this case, *randomly* selected implies that each possible subset of items is equally likely.

EXAMPLE 2-18 Suppose a batch contains six parts $\{a, b, c, d, e, f\}$ and two parts are selected randomly, without replacement. Suppose that part f is defective, but the others are good. What is the probability that part f appears in the sample?

The sample space consists of all possible (unordered) pairs selected without replacement. From Equation 2-4 or by enumeration there are 15 outcomes. Let E denote the event that part f is in the sample. Then E can be written as $E = \{\{a,f\}, \{b,f\}, \{c,f\}, \{d,f\}, \{e,f\}\}$. Because each outcome is equally likely, $P(E) = 5/15 = 1/3$.

2-2.2 Axioms of Probability

Now that the probability of an event has been defined, we can collect the assumptions that we have made concerning probabilities into a set of axioms that the probabilities in any random experiment must satisfy. The axioms ensure that the probabilities assigned in an experiment can be interpreted as relative frequencies and that the assignments are consistent with our intuitive understanding of relationships between relative frequencies. For example, if event A is contained in event B, we should have $P(A) \leq P(B)$. The **axioms do not determine probabilities**; the probabilities are assigned based on our knowledge of the system under study. However, the axioms enable us to easily calculate the probabilities of some events from knowledge of the probabilities of other events.

Axioms of Probability

Probability is a number that is assigned to each member of a collection of events from a random experiment that satisfies the following properties:

If S is the sample space and E is any event in a random experiment,

(1) $P(S) = 1$
(2) $0 \leq P(E) \leq 1$
(3) For two events E_1 and E_2 with $E_1 \cap E_2 = \emptyset$

$$P(E_1 \cup E_2) = P(E_1) + P(E_2)$$

The property that $0 \leq P(E) \leq 1$ is equivalent to the requirement that a relative frequency must be between 0 and 1. The property that $P(S) = 1$ is a consequence of the fact that an outcome from the sample space occurs on every trial of an experiment. Consequently, the relative frequency of S is 1. Property 3 implies that if the events E_1 and E_2 have no outcomes in common, the relative frequency of outcomes in $E_1 \cup E_2$ is the sum of the relative frequencies of the outcomes in E_1 and E_2.

These axioms imply the following results. The derivations are left as exercises at the end of this section. Now,

$$P(\emptyset) = 0$$

and for any event E,

$$P(E') = 1 - P(E)$$

For example, if the probability of the event E is 0.4, our interpretation of relative frequency implies that the probability of E' is 0.6. Furthermore, if the event E_1 is contained in the event E_2,

$$P(E_1) \leq P(E_2)$$

EXERCISES FOR SECTION 2-2

2-50. Each of the possible five outcomes of a random experiment is equally likely. The sample space is $\{a, b, c, d, e\}$. Let A denote the event $\{a, b\}$, and let B denote the event $\{c, d, e\}$. Determine the following:
(a) $P(A)$ (b) $P(B)$
(c) $P(A')$ (d) $P(A \cup B)$
(e) $P(A \cap B)$

2-51. The sample space of a random experiment is $\{a, b, c, d, e\}$ with probabilities 0.1, 0.1, 0.2, 0.4, and 0.2, respectively. Let A denote the event $\{a, b, c\}$, and let B denote the event $\{c, d, e\}$. Determine the following:
(a) $P(A)$ (b) $P(B)$
(c) $P(A')$ (d) $P(A \cup B)$
(e) $P(A \cap B)$

2-52. Orders for a computer are summarized by the optional features that are requested as follows:

	proportion of orders
no optional features	0.3
one optional feature	0.5
more than one optional feature	0.2

(a) What is the probability that an order requests at least one optional feature?
(b) What is the probability that an order does not request more than one optional feature?

2-53. If the last digit of a weight measurement is equally likely to be any of the digits 0 through 9,
(a) What is the probability that the last digit is 0?

(b) What is the probability that the last digit is greater than or equal to 5?

2-54. A part selected for testing is equally likely to have been produced on any one of six cutting tools.
(a) What is the sample space?
(b) What is the probability that the part is from tool 1?
(c) What is the probability that the part is from tool 3 or tool 5?
(d) What is the probability that the part is not from tool 4?

2-55. An injection-molded part is equally likely to be obtained from any one of the eight cavities on a mold.
(a) What is the sample space?
(b) What is the probability a part is from cavity 1 or 2?
(c) What is the probability that a part is neither from cavity 3 nor 4?

2-56. In an acid-base titration, a base or acid is gradually added to the other until they have completely neutralized each other. Because acids and bases are usually colorless (as are the water and salt produced in the neutralization reaction), pH is measured to monitor the reaction. Suppose that the equivalence point is reached after approximately 100 mL of a NaOH solution have been added (enough to react with all the acetic acid present) but that replicates are equally likely to indicate from 95 to 104 mL to the nearest mL. Assume that volumes are measured to the nearest mL and describe the sample space.
(a) What is the probability that equivalence is indicated at 100 mL?
(b) What is the probability that equivalence is indicated at less than 100 mL?
(c) What is the probability that equivalence is indicated between 98 and 102 mL (inclusive)?

2-57. In a NiCd battery, a fully charged cell is composed of Nickelic Hydroxide. Nickel is an element that has multiple oxidation states that is usually found in the following states:

nickel charge	proportions found
0	0.17
+2	0.35
+3	0.33
+4	0.15

(a) What is the probability that a cell has at least one of the positive nickel charged options?
(b) What is the probability that a cell is not composed of a positive nickel charge greater than +3?

2-58. A credit card contains 16 digits between 0 and 9. However, only 100 million numbers are valid. If a number is entered randomly, what is the probability that it is a valid number?

2-59. Suppose your vehicle is licensed in a state that issues license plates that consist of three digits (between 0 and 9) followed by three letters (between A and Z). If a license number is selected randomly, what is the probability that yours is the one selected?

2-60. A message can follow different paths through servers on a network. The sender's message can go to one of five servers for the first step, each of them can send to five servers at the second step, each of which can send to four servers at the third step, and then the message goes to the recipient's server.
(a) How many paths are possible?
(b) If all paths are equally likely, what is the probability that a message passes through the first of four servers at the third step?

2-61. Magnesium alkyls are used as homogenous catalysts in the production of linear low density polyethylene (LLDPE), which requires a finer magnesium powder to sustain a reaction. Redox reaction experiments using four different amounts of magnesium powder are performed. Each result may or may not be further reduced in a second step using three different magnesium powder amounts. Each of these results may or may not be further reduced in a third step using three different amounts of magnesium powder.
(a) How many experiments are possible?

(b) If all outcomes are equally likely, what is the probability that the best result is obtained from an experiment that uses all three steps?
(c) Does the result in the previous question change if five or six or seven different amounts are used in the first step? Explain.

2-62. Disks of polycarbonate plastic from a supplier are analyzed for scratch and shock resistance. The results from 100 disks are summarized as follows:

		shock resistance	
		high	low
scratch	high	70	9
resistance	low	16	5

Let A denote the event that a disk has high shock resistance, and let B denote the event that a disk has high scratch resistance. If a disk is selected at random, determine the following probabilities:
(a) $P(A)$ (b) $P(B)$
(c) $P(A')$ (d) $P(A \cap B)$
(e) $P(A \cup B)$ (f) $P(A' \cup B)$

2-63. Samples of emissions from three suppliers are classified for conformance to air-quality specifications. The results from 100 samples are summarized as follows:

		conforms	
		yes	no
	1	22	8
supplier	2	25	5
	3	30	10

Let A denote the event that a sample is from supplier 1, and let B denote the event that a sample conforms to specifications. If a sample is selected at random, determine the following probabilities:
(a) $P(A)$ (b) $P(B)$
(c) $P(A')$ (d) $P(A \cap B)$
(e) $P(A \cup B)$ (f) $P(A' \cup B)$

2-64. An article in the *Journal of Database Management* ["Experimental Study of a Self-Tuning Algorithm for DBMS Buffer Pools," 2005, Vol. 16, pp. 1–20)] provided the workload used in the TPC-C OLTP (Transaction Processing Performance Council's Version C On-Line Transaction Processing) benchmark, which simulates a typical order entry application.

Average Frequencies and Operations in TPC-C

Transaction	Frequency	Selects	Updates	Inserts	Deletes	Non-Unique Selects	Joins
New Order	43	23	11	12	0	0	0
Payment	44	4.2	3	1	0	0.6	0
Order Status	4	11.4	0	0	0	0.6	0
Delivery	5	130	120	0	10	0	0
Stock Level	4	0	0	0	0	0	1

The frequency of each type of transaction (in the second column) can be used as the percentage of each type of transaction. The average number of *selects* operations required for each type of transaction is shown. Let A denote the event of transactions with an average number of *selects* operations of 12 or fewer. Let B denote the event of transactions with an average number of *updates* operations of 12 or fewer. Calculate the following probabilities.

(a) $P(A)$ (b) $P(B)$ (c) $P(A \cap B)$
(d) $P(A \cap B')$ (e) $P(A \cup B)$

2-65. Use the axioms of probability to show the following:
(a) For any event E, $P(E') = 1 - P(E)$.
(b) $P(\emptyset) = 0$
(c) If A is contained in B, then $P(A) \leq P(B)$

2-3 ADDITION RULES

Joint events are generated by applying basic set operations to individual events. Unions of events, such as $A \cup B$; intersections of events, such as $A \cap B$; and complements of events, such as A', are commonly of interest. The probability of a joint event can often be determined from the probabilities of the individual events that comprise it. Basic set operations are also sometimes helpful in determining the probability of a joint event. In this section the focus is on unions of events.

EXAMPLE 2-19
Semiconductor Wafers

Table 2-1 lists the history of 940 wafers in a semiconductor manufacturing process. Suppose one wafer is selected at random. Let H denote the event that the wafer contains high levels of contamination. Then, $P(H) = 358/940$.

Let C denote the event that the wafer is in the center of a sputtering tool. Then, $P(C) = 626/940$. Also, $P(H \cap C)$ is the probability that the wafer is from the center of the sputtering tool and contains high levels of contamination. Therefore,

$$P(H \cap C) = 112/940$$

The event $H \cup C$ is the event that a wafer is from the center of the sputtering tool or contains high levels of contamination (or both). From the table, $P(H \cup C) = 872/940$. An alternative calculation of $P(H \cup C)$ can be obtained as follows. The 112 wafers that comprise the event $H \cap C$ are included once in the calculation of $P(H)$ and again in the calculation of $P(C)$. Therefore, $P(H \cup C)$ can be found to be

$$P(H \cup C) = P(H) + P(C) - P(H \cap C)$$
$$= 358/940 + 626/940 - 112/940 = 872/940$$

The preceding example illustrates that the probability of A or B is interpreted as $P(A \cup B)$ and that the following general **addition rule** applies.

Table 2-1 Wafers in Semiconductor Manufacturing Classified by Contamination and Location

Contamination	Location in Sputtering Tool		Total
	Center	Edge	
Low	514	68	582
High	112	246	358
Total	626	314	

Probability of a Union

$$P(A \cup B) = P(A) + P(B) - P(A \cap B) \qquad (2\text{-}5)$$

EXAMPLE 2-20
Semiconductor Wafers and Location

The wafers such as those described in Example 2-19 were further classified as either in the "center" or at the "edge" of the sputtering tool that was used in manufacturing, and by the degree of contamination. Table 2-2 shows the proportion of wafers in each category. What is the probability that a wafer was either at the edge or that it contains four or more particles? Let E_1 denote the event that a wafer contains four or more particles, and let E_2 denote the event that a wafer is at the edge.

The requested probability is $P(E_1 \cup E_2)$. Now, $P(E_1) = 0.15$ and $P(E_2) = 0.28$. Also, from the table, $P(E_1 \cap E_2) = 0.04$. Therefore, using Equation 2-1, we find that

$$P(E_1 \cup E_2) = 0.15 + 0.28 - 0.04 = 0.39$$

What is the probability that a wafer contains less than two particles or that it is both at the edge and contains more than four particles? Let E_1 denote the event that a wafer contains less than two particles, and let E_2 denote the event that a wafer is both from the edge and contains more than four particles. The requested probability is $P(E_1 \cup E_2)$. Now, $P(E_1) = 0.60$ and $P(E_2) = 0.03$. Also, E_1 and E_2 are mutually exclusive. Consequently, there are no wafers in the intersection and $P(E_1 \cap E_2) = 0$. Therefore,

$$P(E_1 \cup E_2) = 0.60 + 0.03 = 0.63$$

Recall that two events A and B are said to be mutually exclusive if $A \cap B = \varnothing$. Then, $P(A \cap B) = 0$, and the general result for the probability of $A \cup B$ simplifies to the third axiom of probability.

If A and B are mutually exclusive events,

$$P(A \cup B) = P(A) + P(B) \qquad (2\text{-}6)$$

Table 2-2 Wafers Classified by Contamination and Location

Number of Contamination Particles	Center	Edge	Totals
0	0.30	0.10	0.40
1	0.15	0.05	0.20
2	0.10	0.05	0.15
3	0.06	0.04	0.10
4	0.04	0.01	0.05
5 or more	0.07	0.03	0.10
Totals	0.72	0.28	1.00

Three or More Events

More complicated probabilities, such as $P(A \cup B \cup C)$, can be determined by repeated use of Equation 2-5 and by using some basic set operations. For example,

$$P(A \cup B \cup C) = P[(A \cup B) \cup C] = P(A \cup B) + P(C) - P[(A \cup B) \cap C]$$

Upon expanding $P(A \cup B)$ by Equation 2-5 and using the distributed rule for set operations to simplify $P[(A \cup B) \cap C]$, we obtain

$$\begin{aligned} P(A \cup B \cup C) &= P(A) + P(B) - P(A \cap B) + P(C) - P[(A \cap C) \cup (B \cap C)] \\ &= P(A) + P(B) - P(A \cap B) + P(C) \\ &\quad - [P(A \cap C) + P(B \cap C) - P(A \cap B \cap C)] \\ &= P(A) + P(B) + P(C) - P(A \cap B) - P(A \cap C) \\ &\quad - P(B \cap C) + P(A \cap B \cap C) \end{aligned}$$

We have developed a formula for the probability of the union of three events. Formulas can be developed for the probability of the union of any number of events, although the formulas become very complex. As a summary, for the case of three events

$$P(A \cup B \cup C) = P(A) + P(B) + P(C) - P(A \cap B) \\ - P(A \cap C) - P(B \cap C) + P(A \cap B \cap C) \quad (2\text{-}7)$$

Results for three or more events simplify considerably if the events are mutually exclusive. In general, a collection of events, $E_1, E_2, \ldots, E_k$, is said to be mutually exclusive if there is no overlap among any of them.

The Venn diagram for several mutually exclusive events is shown in Fig. 2-12. By generalizing the reasoning for the union of two events, the following result can be obtained:

Mutually Exclusive Events

A collection of events, $E_1, E_2, \ldots, E_k$, is said to be **mutually exclusive** if for all pairs,

$$E_i \cap E_j = \varnothing$$

For a collection of mutually exclusive events,

$$P(E_1 \cup E_2 \cup \ldots \cup E_k) = P(E_1) + P(E_2) + \ldots P(E_k) \quad (2\text{-}8)$$

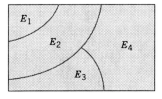

Figure 2-12 Venn diagram of four mutually exclusive events.

EXAMPLE 2-21
pH

A simple example of mutually exclusive events will be used quite frequently. Let X denote the pH of a sample. Consider the event that X is greater than 6.5 but less than or equal to 7.8. This probability is the sum of any collection of mutually exclusive events with union equal to the same range for X. One example is

$$P(6.5 < X \le 7.8) = P(6.5 < X \le 7.0) + P(7.0 < X \le 7.5) + P(7.5 < X \le 7.8)$$

Another example is

$$P(6.5 < X \le 7.8) = P(6.5 < X \le 6.6) + P(6.6 < X \le 7.1) \\ + P(7.1 < X \le 7.4) + P(7.4 < X \le 7.8)$$

The best choice depends on the particular probabilities available.

EXERCISES FOR SECTION 2-3

2-66. If $P(A) = 0.3$, $P(B) = 0.2$, and $P(A \cap B) = 0.1$, determine the following probabilities:
(a) $P(A')$
(b) $P(A \cup B)$
(c) $P(A' \cap B)$
(d) $P(A \cap B')$
(e) $P[(A \cup B)']$
(f) $P(A' \cup B)$

2-67. If A, B, and C are mutually exclusive events with $P(A) = 0.2, P(B) = 0.3,$ and $P(C) = 0.4$, determine the following probabilities:
(a) $P(A \cup B \cup C)$
(b) $P(A \cap B \cap C)$
(c) $P(A \cap B)$
(d) $P[(A \cup B) \cap C]$
(e) $P(A' \cap B' \cap C')$

2-68. In the article "ACL reconstruction using bone-patellar tendon-bone press-fit fixation: 10-year clinical results" in *Knee Surgery, Sports Traumatology, Arthroscopy* (2005, Vol. 13, pp. 248–255) the following causes for knee injuries were considered:

Activity	Percentage of Knee Injuries
Contact sport	46%
Noncontact sport	44%
Activity of daily living	9%
Riding motorcycle	1%

(a) What is the probability a knee injury resulted from a sport (contact or noncontact)?
(b) What is the probability a knee injury resulted from an activity other than a sport?

2-69. Disks of polycarbonate plastic from a supplier are analyzed for scratch and shock resistance. The results from 100 disks are summarized as follows:

		shock resistance	
		high	low
scratch resistance	high	70	9
	low	16	5

(a) If a disk is selected at random, what is the probability that its scratch resistance is high and its shock resistance is high?
(b) If a disk is selected at random, what is the probability that its scratch resistance is high or its shock resistance is high?
(c) Consider the event that a disk has high scratch resistance and the event that a disk has high shock resistance. Are these two events mutually exclusive?

2-70. Strands of copper wire from a manufacturer are analyzed for strength and conductivity. The results from 100 strands are as follows:

	strength	
	high	low
high conductivity	74	8
low conductivity	15	3

(a) If a strand is randomly selected, what is the probability that its conductivity is high and its strength is high?
(b) If a strand is randomly selected, what is the probability that its conductivity is low or the strength is low?
(c) Consider the event that a strand has low conductivity and the event that the strand has a low strength. Are these two events mutually exclusive?

2-71. The analysis of shafts for a compressor is summarized by conformance to specifications.

		roundness conforms	
		yes	no
surface finish conforms	yes	345	5
	no	12	8

(a) If a shaft is selected at random, what is the probability that the shaft conforms to surface finish requirements?
(b) What is the probability that the selected shaft conforms to surface finish requirements or to roundness requirements?

(c) What is the probability that the selected shaft either conforms to surface finish requirements or does not conform to roundness requirements?

(d) What is the probability that the selected shaft conforms to both surface finish and roundness requirements?

2-72. Cooking oil is produced in two main varieties: mono- and polyunsaturated. Two common sources of cooking oil are corn and canola. The following table shows the number of bottles of these oils at a supermarket:

		type of oil	
		canola	corn
type of	mono	7	13
unsaturation	poly	93	77

(a) If a bottle of oil is selected at random, what is the probability that it belongs to the polyunsaturated category?

(b) What is the probability that the chosen bottle is monounsaturated canola oil?

2-73. A manufacturer of front lights for automobiles tests lamps under a high humidity, high temperature environment using intensity and useful life as the responses of interest. The following table shows the performance of 130 lamps:

		useful life	
		satisfactory	unsatisfactory
intensity	satisfactory	117	3
	unsatisfactory	8	2

(a) Find the probability that a randomly selected lamp will yield unsatisfactory results under any criteria.

(b) The customers for these lamps demand 95% satisfactory results. Can the lamp manufacturer meet this demand?

2-74. A computer system uses passwords that are six characters and each character is one of the 26 letters (a–z) or 10 integers (0–9). Uppercase letters are not used. Let A denote the event that a password begins with a vowel (either a, e, i, o, u) and let B denote the event that a password ends with an even number (either 0, 2, 4, 6, or 8). Suppose a hacker selects a password at random. Determine the following probabilities:

(a) $P(A)$ (b) $P(B)$
(c) $P(A \cap B)$ (d) $P(A \cup B)$

2-4 CONDITIONAL PROBABILITY

Sometimes probabilities need to be reevaluated as additional information becomes available. A useful way to incorporate additional information into a probability model is to assume that the outcome that will be generated is a member of a given event. This event, say A, defines the conditions that the outcome is known to satisfy. Then probabilities can be revised to include this knowledge. The probability of an event B under the knowledge that the outcome will be in event A is denoted as

$$P(B|A)$$

and this is called the **conditional probability** of B **given** A.

A digital communication channel has an error rate of one bit per every thousand transmitted. Errors are rare, but when they occur, they tend to occur in bursts that affect many consecutive bits. If a single bit is transmitted, we might model the probability of an error as 1/1000. However, if the previous bit was in error, because of the bursts, we might believe that the probability that the next bit is in error is greater than 1/1000.

In a thin film manufacturing process, the proportion of parts that are not acceptable is 2%. However, the process is sensitive to contamination problems that can increase the rate of parts that are not acceptable. If we knew that during a particular shift there were problems with the filters used to control contamination, we would assess the probability of a part being unacceptable as higher than 2%.

In a manufacturing process, 10% of the parts contain visible surface flaws and 25% of the parts with surface flaws are (functionally) defective parts. However, only 5% of parts without surface flaws are defective parts. The probability of a defective part depends on our knowledge of the presence or absence of a surface flaw. Let D denote the event that a part is defective and let F denote the event that a part has a surface flaw. Then, we denote the probability of D given, or assuming, that a part has a surface flaw as $P(D|F)$. Because 25% of the parts with surface flaws are

42 CHAPTER 2 PROBABILITY

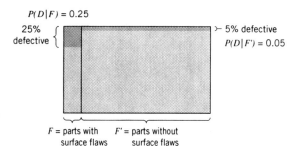

Figure 2-13 Conditional probabilities for parts with surface flaws.

defective, our conclusion can be stated as $P(D|F) = 0.25$. Furthermore, because F' denotes the event that a part does not have a surface flaw and because 5% of the parts without surface flaws are defective, we have that $P(D|F') = 0.05$. These results are shown graphically in Fig. 2-13.

EXAMPLE 2-22 Table 2-3 provides an example of 400 parts classified by surface flaws and as (functionally) defective. For this table the conditional probabilities match those discussed previously in this section. For example, of the parts with surface flaws (40 parts) the number defective is 10. Therefore,

$$P(D|F) = 10/40 = 0.25$$

and of the parts without surface flaws (360 parts) the number defective is 18. Therefore,

$$P(D|F') = 18/360 = 0.05$$

In Example 2-22 conditional probabilities were calculated directly. These probabilities can also be determined from the formal definition of conditional probability.

Conditional Probability

The **conditional probability** of an event B given an event A, denoted as $P(B|A)$, is

$$P(B|A) = P(A \cap B)/P(A) \qquad (2\text{-}9)$$

for $P(A) > 0$.

This definition can be understood in a special case in which all outcomes of a random experiment are equally likely. If there are n total outcomes,

$$P(A) = (\text{number of outcomes in } A)/n$$

Table 2-3 Parts Classified

		Surface Flaws		
		Yes (event F)	No	Total
Defective	Yes (event D)	10	18	28
	No	30	342	372
	Total	40	360	400

Also,

$$P(A \cap B) = (\text{number of outcomes in } A \cap B)/n$$

Consequently,

$$P(A \cap B)/P(A) = \frac{\text{number of outcomes in } A \cap B}{\text{number of outcomes in } A}$$

Therefore, $P(B|A)$ can be interpreted as the relative frequency of event B among the trials that produce an outcome in event A.

EXAMPLE 2-23
Surface Flaws

Again consider the 400 parts in Table 2-3. From this table

$$P(D|F) = P(D \cap F)/P(F) = \frac{10}{400} \Big/ \frac{40}{400} = \frac{10}{40}$$

Note that in this example all four of the following probabilities are different:

$$P(F) = 40/400 \qquad P(F|D) = 10/28$$
$$P(D) = 28/400 \qquad P(D|F) = 10/40$$

Here, $P(D)$ and $P(D|F)$ are probabilities of the same event, but they are computed under two different states of knowledge. Similarly, $P(F)$ and $P(F|D)$ are computed under two different states of knowledge.

The tree diagram in Fig. 2-14 can also be used to display conditional probabilities. The first branch is on surface flaw. Of the 40 parts with surface flaws, 10 are functionally defective and 30 are not. Therefore,

$$P(D|F) = 10/40 \qquad \text{and} \qquad P(D'|F) = 30/40$$

Of the 360 parts without surface flaws, 18 are functionally defective and 342 are not. Therefore,

$$P(D|F') = 18/360 \qquad \text{and} \qquad P(D'|F') = 342/360$$

Random Samples and Conditional Probability

Recall that to select one item randomly from a batch implies that each item is equally likely. If more than one item is selected, *randomly* implies that each element of the sample space is equally likely. For example, when sample spaces were presented earlier in this chapter, sampling with and without replacement were defined and illustrated for the simple case of a batch with three items $\{a, b, c\}$. If two items are selected randomly from this batch without replacement, each of the six outcomes in the ordered sample space

$$S_{\text{without}} = \{ab, ac, ba, bc, ca, cb\}$$

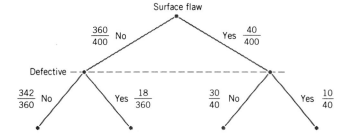

Figure 2-14 Tree diagram for parts classified

has probability 1/6. If the unordered sample space is used, each of the three outcomes in $\{\{a, b\}, \{a, c\}, \{b, c\}\}$ has probability 1/3.

When a sample is selected randomly from a large batch, it is usually easier to avoid enumeration of the sample space and calculate probabilities from conditional probabilities. For example, suppose that a batch contains 10 parts from tool 1 and 40 parts from tool 2. If two parts are selected randomly, without replacement, what is the conditional probability that a part from tool 2 is selected second given that a part from tool 1 is selected first? There are 50 possible parts to select first and 49 to select second. Therefore, the (ordered) sample space has $50 \times 49 = 2450$ outcomes. Let E_1 denote the event that the first part is from tool 1 and E_2 denote the event that the second part is from tool 2. Then

$$P(E_2|E_1) = P(E_1 \cap E_2)/P(E_1)$$

and a count of the number of outcomes in E_1 and the intersection is needed.

Although the answer can be determined from this start, this type of question can be answered more easily with the following result.

Random Samples

> To select randomly implies that at each step of the sample, the items that remain in the batch are equally likely to be selected.

If a part from tool 1 were selected with the first pick, 49 items would remain, 9 from tool 1 and 40 from tool 2, and they would be equally likely to be picked. Therefore, the probability that a part from tool 2 would be selected with the second pick given this first pick is

$$P(E_2|E_1) = 40/49$$

In this manner, other probabilities can also be simplified. For example, let the event E consist of the outcomes with the first selected part from tool 1 and the second part from tool 2. To determine the probability of E, consider each step. The probability that a part from tool 1 is selected with the first pick is $P(E_1) = 10/50$. The conditional probability that a part from tool 2 is selected with the second pick, given that a part from tool 1 is selected first is $P(E_2|E_1) = 40/49$. Therefore,

$$P(E) = P(E_2|E_1)P(E_1) = \frac{40}{49} \cdot \frac{10}{50} = \frac{8}{49}$$

Sometimes a partition of the question into successive picks is an easier method to solve the problem.

EXAMPLE 2-24 A day's production of 850 manufactured parts contains 50 parts that do not meet customer requirements. Two parts are selected randomly without replacement from the batch. What is the probability that the second part is defective given that the first part is defective?

Let A denote the event that the first part selected is defective, and let B denote the event that the second part selected is defective. The probability needed can be expressed as $P(B|A)$. If the first part is defective, prior to selecting the second part, the batch contains 849 parts, of which 49 are defective, therefore

$$P(B|A) = 49/849$$

2-4 CONDITIONAL PROBABILITY

EXAMPLE 2-25 Continuing the previous example, if three parts are selected at random, what is the probability that the first two are defective and the third is not defective? This event can be described in shorthand notation as simply $P(ddn)$. We have

$$P(ddn) = \frac{50}{850} \cdot \frac{49}{849} \cdot \frac{800}{848} = 0.0032$$

The third term is obtained as follows. After the first two parts are selected, there are 848 remaining. Of the remaining parts, 800 are not defective. In this example, it is easy to obtain the solution with a conditional probability for each selection.

EXERCISES FOR SECTION 2-4

2-75. Disks of polycarbonate plastic from a supplier are analyzed for scratch and shock resistance. The results from 100 disks are summarized as follows:

		shock resistance	
		high	low
scratch	high	70	9
resistance	low	16	5

Let A denote the event that a disk has high shock resistance, and let B denote the event that a disk has high scratch resistance. Determine the following probabilities:
(a) $P(A)$ (b) $P(B)$
(c) $P(A|B)$ (d) $P(B|A)$

2-76. Samples of skin experiencing desquamation are analyzed for both moisture and melanin content. The results from 100 skin samples are as follows:

		melanin content	
		high	low
moisture	high	13	7
content	low	48	32

Let A denote the event that a sample has low melanin content, and let B denote the event that a sample has high moisture content. Determine the following probabilities:
(a) $P(A)$ (b) $P(B)$
(c) $P(A|B)$ (d) $P(B|A)$

2-77. The analysis of results from a leaf transmutation experiment (turning a leaf into a petal) is summarized by type of transformation completed:

		total textural transformation	
		yes	no
total color	yes	243	26
transformation	no	13	18

(a) If a leaf completes the color transformation, what is the probability that it will complete the textural transformation?

(b) If a leaf does not complete the textural transformation, what is the probability it will complete the color transformation?

2-78. Samples of a cast aluminum part are classified on the basis of surface finish (in microinches) and length measurements. The results of 100 parts are summarized as follows:

		length	
		excellent	good
surface	excellent	80	2
finish	good	10	8

Let A denote the event that a sample has excellent surface finish, and let B denote the event that a sample has excellent length. Determine:
(a) $P(A)$ (b) $P(B)$
(c) $P(A|B)$ (d) $P(B|A)$
(e) If the selected part has excellent surface finish, what is the probability that the length is excellent?
(f) If the selected part has good length, what is the probability that the surface finish is excellent?

2-79. The following table summarizes the analysis of samples of galvanized steel for coating weight and surface roughness:

		coating weight	
		high	low
surface	high	12	16
roughness	low	88	34

(a) If the coating weight of a sample is high, what is the probability that the surface roughness is high?
(b) If the surface roughness of a sample is high, what is the probability that the coating weight is high?
(c) If the surface roughness of a sample is low, what is the probability that the coating weight is low?

2-80. Consider the data on wafer contamination and location in the sputtering tool shown in Table 2-2. Assume that one wafer is selected at random from this set. Let A denote the event that a wafer contains four or more particles, and let B denote the event that a wafer is from the center of the sputtering tool. Determine:
(a) $P(A)$ (b) $P(A|B)$

(c) $P(B)$ (d) $P(B|A)$
(e) $P(A \cap B)$ (f) $P(A \cup B)$

2-81. The following table summarizes the number of deceased beetles under autolysis (the destruction of a cell after its death by the action of its own enzymes) and putrefaction (decomposition of organic matter, especially protein, by microorganisms, resulting in production of foul-smelling matter):

		autolysis	
		high	low
putrefaction	high	14	59
	low	18	9

(a) If the autolysis of a sample is high, what is the probability that the putrefaction is low?
(b) If the putrefaction of a sample is high, what is the probability that the autolysis is high?
(c) If the putrefaction of a sample is low, what is the probability that the autolysis is low?

2-82. A maintenance firm has gathered the following information regarding the failure mechanisms for air conditioning systems:

		evidence of gas leaks	
		yes	no
evidence of electrical failure	yes	55	17
	no	32	3

The units without evidence of gas leaks or electrical failure showed other types of failure. If this is a representative sample of AC failure, find the probability
(a) That failure involves a gas leak
(b) That there is evidence of electrical failure given that there was a gas leak
(c) That there is evidence of a gas leak given that there is evidence of electrical failure

2-83. A lot of 100 semiconductor chips contains 20 that are defective. Two are selected randomly, without replacement, from the lot.
(a) What is the probability that the first one selected is defective?
(b) What is the probability that the second one selected is defective given that the first one was defective?
(c) What is the probability that both are defective?

(d) How does the answer to part (b) change if chips selected were replaced prior to the next selection?

2-84. A batch of 500 containers for frozen orange juice contains 5 that are defective. Two are selected, at random, without replacement from the batch.
(a) What is the probability that the second one selected is defective given that the first one was defective?
(b) What is the probability that both are defective?
(c) What is the probability that both are acceptable?

Three containers are selected, at random, without replacement, from the batch.
(d) What is the probability that the third one selected is defective given that the first and second ones selected were defective?
(e) What is the probability that the third one selected is defective given that the first one selected was defective and the second one selected was okay?
(f) What is the probability that all three are defective?

2-85. A batch of 350 samples of rejuvenated mitochondria contains 8 that are mutated (or defective). Two are selected, at random, without replacement from the batch.
(a) What is the probability that the second one selected is defective given that the first one was defective?
(b) What is the probability that both are defective?
(c) What is the probability that both are acceptable?

2-86. A computer system uses passwords that are exactly seven characters and each character is one of the 26 letters (a–z) or 10 integers (0–9). You maintain a password for this computer system. Let A denote the subset of passwords that begin with a vowel (either a, e, i, o, u) and let B denote the subset of passwords that end with an even number (either 0, 2, 4, 6, or 8).
(a) Suppose a hacker selects a password at random. What is the probability that your password is selected?
(b) Suppose a hacker knows your password is in event A and selects a password at random from this subset. What is the probability that your password is selected?
(c) Suppose a hacker knows your password is in A and B and selects a password at random from this subset. What is the probability that your password is selected?

2-87. If $P(A|B) = 1$, must $A = B$? Draw a Venn diagram to explain your answer.

2-88. Suppose A and B are mutually exclusive events. Construct a Venn diagram that contains the three events A, B, and C such that $P(A|C) = 1$ and $P(B|C) = 0$?

2-5 MULTIPLICATION AND TOTAL PROBABILITY RULES

2-5.1 Multiplication Rule

The probability of the intersection of two events is often needed. The conditional probability definition in Equation 2-9 can be rewritten to provide a formula known as the **multiplication rule** for probabilities.

Multiplication Rule

$$P(A \cap B) = P(B|A)P(A) = P(A|B)P(B) \qquad (2\text{-}10)$$

The last expression in Equation 2-10 is obtained by interchanging A and B.

EXAMPLE 2-26
Machining Stages

The probability that the first stage of a numerically controlled machining operation for high-rpm pistons meets specifications is 0.90. Failures are due to metal variations, fixture alignment, cutting blade condition, vibration, and ambient environmental conditions. Given that the first stage meets specifications, the probability that a second stage of machining meets specifications is 0.95. What is the probability that both stages meet specifications?

Let A and B denote the events that the first and second stages meet specifications, respectively. The probability requested is

$$P(A \cap B) = P(B|A)P(A) = 0.95(0.90) = 0.855$$

Although it is also true that $P(A \cap B) = P(A|B)P(B)$, the information provided in the problem does not match this second formulation.

2-5.2 Total Probability Rule

Sometimes the probability of an event is given under each of several conditions. With enough of these conditional probabilities, the probability of the event can be recovered. For example, suppose that in semiconductor manufacturing the probability is 0.10 that a chip that is subjected to high levels of contamination during manufacturing causes a product failure. The probability is 0.005 that a chip that is not subjected to high contamination levels during manufacturing causes a product failure. In a particular production run, 20% of the chips are subject to high levels of contamination. What is the probability that a product using one of these chips fails?

Clearly, the requested probability depends on whether or not the chip was exposed to high levels of contamination. For any event B, we can write B as the union of the part of B in A and the part of B in A'. That is,

$$B = (A \cap B) \cup (A' \cap B)$$

This result is shown in the Venn diagram in Fig. 2-15. Because A and A' are mutually exclusive, $A \cap B$ and $A' \cap B$ are mutually exclusive. Therefore, from the probability of the union of mutually exclusive events in Equation 2-6 and the Multiplication Rule in Equation 2-10, the following total probability rule is obtained.

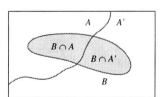

Figure 2-15 Partitioning an event into two mutually exclusive subsets.

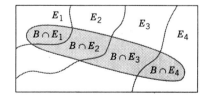

$B = (B \cap E_1) \cup (B \cap E_2) \cup (B \cap E_3) \cup (B \cap E_4)$

Figure 2-16 Partitioning an event into several mutually exclusive subsets.

48 CHAPTER 2 PROBABILITY

Total Probability Rule (two events)

For any events A and B,

$$P(B) = P(B \cap A) + P(B \cap A') = P(B|A)P(A) + P(B|A')P(A') \quad (2\text{-}11)$$

EXAMPLE 2-27
Semiconductor Contamination

Consider the contamination discussion at the start of this section. The information is summarized here

Probability of Failure	Level of Contamination	Probability of Level
0.1	High	0.2
0.005	Not High	0.8

Let F denote the event that the product fails, and let H denote the event that the chip is exposed to high levels of contamination. The requested probability is $P(F)$, and the information provided can be represented as

$$P(F|H) = 0.10 \quad \text{and} \quad P(F|H') = 0.005$$
$$P(H) = 0.20 \quad \text{and} \quad P(H') = 0.80$$

From Equation 2-11,

$$P(F) = 0.10(0.20) + 0.005(0.80) = 0.024$$

which can be interpreted as just the weighted average of the two probabilities of failure.

The reasoning used to develop Equation 2-11 can be applied more generally. Because $A \cup A' = S$ we know $(A \cap B) \cup (A' \cap B)$ equals B and because $A \cap A' = \phi$ we know $A \cap B$ and $A' \cap B$ are mutually exclusive. In general, a collection of sets $E_1, E_2, \ldots, E_k$ such that $E_1 \cup E_2 \cup \ldots \cup E_k = S$ is said to be **exhaustive**. A graphical display of partitioning an event B among a collection of mutually exclusive and exhaustive events is shown in Fig. 2-16.

Total Probability Rule (multiple events)

Assume $E_1, E_2, \ldots, E_k$ are k mutually exclusive and exhaustive sets. Then

$$P(B) = P(B \cap E_1) + P(B \cap E_2) + \cdots + P(B \cap E_k)$$
$$= P(B|E_1)P(E_1) + P(B|E_2)P(E_2) + \cdots + P(B|E_k)P(E_k) \quad (2\text{-}12)$$

EXAMPLE 2-28
Semiconductor Failures

Continuing with semiconductor manufacturing, assume the following probabilities for product failure subject to levels of contamination in manufacturing:

Probability of Failure	Level of Contamination
0.10	High
0.01	Medium
0.001	Low

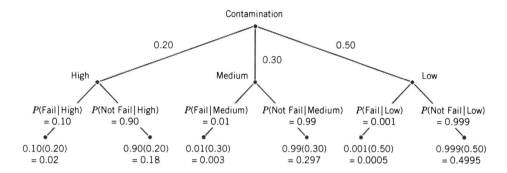

Figure 2-17 Tree diagram for Example 2-28.

In a particular production run, 20% of the chips are subjected to high levels of contamination, 30% to medium levels of contamination, and 50% to low levels of contamination. What is the probability that a product using one of these chips fails? Let

H denote the event that a chip is exposed to high levels of contamination

M denote the event that a chip is exposed to medium levels of contamination

L denote the event that a chip is exposed to low levels of contamination

Then,

$$P(F) = P(F|H)P(H) + P(F|M)P(M) + P(F|L)P(L)$$
$$= 0.10(0.20) + 0.01(0.30) + 0.001(0.50) = 0.0235$$

The calculations are conveniently organized with the tree diagram in Fig. 2-17.

EXERCISES FOR SECTION 2-5

2-89. Suppose that $P(A|B) = 0.4$ and $P(B) = 0.5$. Determine the following:
(a) $P(A \cap B)$
(b) $P(A' \cap B)$

2-90. Suppose that $P(A|B) = 0.2$, $P(A|B') = 0.3$, and $P(B) = 0.8$. What is $P(A)$?

2-91. The probability is 1% that an electrical connector that is kept dry fails during the warranty period of a portable computer. If the connector is ever wet, the probability of a failure during the warranty period is 5%. If 90% of the connectors are kept dry and 10% are wet, what proportion of connectors fail during the warranty period?

2-92. Suppose 2% of cotton fabric rolls and 3% of nylon fabric rolls contain flaws. Of the rolls used by a manufacturer, 70% are cotton and 30% are nylon. What is the probability that a randomly selected roll used by the manufacturer contains flaws?

2-93. The edge roughness of slit paper products increases as knife blades wear. Only 1% of products slit with new blades have rough edges, 3% of products slit with blades of average sharpness exhibit roughness, and 5% of products slit with worn blades exhibit roughness. If 25% of the blades in manufacturing are new, 60% are of average sharpness, and 15% are worn, what is the proportion of products that exhibit edge roughness?

2-94. In the 2004 presidential election, exit polls from the critical state of Ohio provided the following results:

total	Bush, 2004	Kerry, 2004
no college degree (62%)	50%	50%
college graduate (38%)	53%	46%

What is the probability a randomly selected respondent voted for Bush?

2-95. Computer keyboard failures are due to faulty electrical connects (12%) or mechanical defects (88%). Mechanical defects are related to loose keys (27%) or improper assembly (73%). Electrical connect defects are caused by defective wires (35%), improper connections (13%), or poorly welded wires (52%).
(a) Find the probability that a failure is due to loose keys.
(b) Find the probability that a failure is due to improperly connected or poorly welded wires.

2-96. Heart failures are due to either natural occurrences (87%) or outside factors (13%). Outside factors are related to induced substances (73%) or foreign objects (27%). Natural occurrences are caused by arterial blockage (56%), disease (27%), and infection (e.g., staph infection) (17%).
(a) Determine the probability that a failure is due to induced substance.
(b) Determine the probability that a failure is due to disease or infection.

2-97. A batch of 25 injection-molded parts contains 5 that have suffered excessive shrinkage.
(a) If two parts are selected at random, and without replacement, what is the probability that the second part selected is one with excessive shrinkage?
(b) If three parts are selected at random, and without replacement, what is the probability that the third part selected is one with excessive shrinkage?

2-98. A lot of 100 semiconductor chips contains 20 that are defective.
(a) Two are selected, at random, without replacement, from the lot. Determine the probability that the second chip selected is defective.
(b) Three are selected, at random, without replacement, from the lot. Determine the probability that all are defective.

2-99. An article in the *British Medical Journal* "Comparison of treatment of renal calculi by operative surgery, percutaneous nephrolithotomy, and extracorporeal shock wave lithotripsy," (1986, Vol. 82, pp. 879–892) provided the following discussion of success rates in kidney stone removals. Open surgery had a success rate of 78% (273/350) while a newer method, percutaneous nephrolithotomy (PN), had a success rate of 83% (289/350). This newer method looked better, but the results changed when stone diameter was considered. For stones with diameters less than two centimeters, 93% (81/87) of cases of open surgery were successful compared with only 83% (234/270) of cases of PN. For stones greater than or equal to two centimeters, the success rates were 73% (192/263) and 69% (55/80) for open surgery and PN, respectively. Open surgery is better for both stone sizes, but less successful in total. In 1951, E. H. Simpson pointed out this apparent contradiction (known as **Simpson's Paradox**) but the hazard still persists today. Explain how open surgery can be better for both stone sizes but worse in total.

2-6 INDEPENDENCE

In some cases, the conditional probability of $P(B|A)$ might equal $P(B)$. In this special case, knowledge that the outcome of the experiment is in event A does not affect the probability that the outcome is in event B.

EXAMPLE 2-29
Sampling with Replacement

Suppose a day's production of 850 manufactured parts contains 50 parts that do not meet customer requirements. Suppose two parts are selected from the batch, but the first part is replaced before the second part is selected. What is the probability that the second part is defective (denoted as B) given that the first part is defective (denoted as A)? The probability needed can be expressed as $P(B|A)$.

Because the first part is replaced prior to selecting the second part, the batch still contains 850 parts, of which 50 are defective. Therefore, the probability of B does not depend on whether or not the first part was defective. That is,

$$P(B|A) = 50/850$$

Also, the probability that both parts are defective is

$$P(A \cap B) = P(B|A)P(A) = \left(\frac{50}{850}\right) \cdot \left(\frac{50}{850}\right) = 0.0035$$

EXAMPLE 2-30
Flaws and Functions

The information in Table 2-3 related surface flaws to functionally defective parts. In that case, we determined that $P(D|F) = 10/40 = 0.25$ and $P(D) = 28/400 = 0.07$. Suppose that the situation is different and follows Table 2-4. Then,

$$P(D|F) = 2/40 = 0.05 \quad \text{and} \quad P(D) = 20/400 = 0.05$$

Table 2-4 Parts Classified

		Surface Flaws		
		Yes (event F)	No	Total
Defective	Yes (event D)	2	18	20
	No	38	342	380
	Total	40	360	400

That is, the probability that the part is defective does not depend on whether it has surface flaws. Also,

$$P(F|D) = 2/20 = 0.10 \quad \text{and} \quad P(F) = 40/400 = 0.10$$

so the probability of a surface flaw does not depend on whether the part is defective. Furthermore, the definition of conditional probability implies that

$$P(F \cap D) = P(D|F)P(F)$$

but in the special case of this problem

$$P(F \cap D) = P(D)P(F) = \frac{2}{40} \cdot \frac{2}{20} = \frac{1}{200}$$

The preceding example illustrates the following conclusions. In the special case that $P(B|A) = P(B)$, we obtain

$$P(A \cap B) = P(B|A)P(A) = P(B)P(A)$$

and

$$P(A|B) = \frac{P(A \cap B)}{P(B)} = \frac{P(A)P(B)}{P(B)} = P(A)$$

These conclusions lead to an important definition.

Independence (two events)

Two events are independent if any one of the following equivalent statements is true:

(1) $P(A|B) = P(A)$
(2) $P(B|A) = P(B)$
(3) $P(A \cap B) = P(A)P(B)$ (2-13)

It is left as a mind-expanding exercise to show that independence implies related results such as

$$P(A' \cap B') = P(A')P(B').$$

The concept of independence is an important relationship between events and is used throughout this text. A mutually exclusive relationship between two events is based only on the outcomes that comprise the events. However, an independence relationship depends on the probability model used for the random experiment. Often, independence is assumed to be part of the random experiment that describes the physical system under study.

EXAMPLE 2-31

A day's production of 850 manufactured parts contains 50 parts that do not meet customer requirements. Two parts are selected at random, without replacement, from the batch. Let A denote the event that the first part is defective, and let B denote the event that the second part is defective.

We suspect that these two events are not independent because knowledge that the first part is defective suggests that it is less likely that the second part selected is defective. Indeed, $P(B|A) = 49/849$. Now, what is $P(B)$? Finding the unconditional $P(B)$ is somewhat difficult because the possible values of the first selection need to be considered:

$$P(B) = P(B|A)P(A) + P(B|A')P(A')$$
$$= (49/849)(50/850) + (50/849)(800/850)$$
$$= 50/850$$

Interestingly, $P(B)$, the unconditional probability that the second part selected is defective, without any knowledge of the first part, is the same as the probability that the first part selected is defective. Yet, our goal is to assess independence. Because $P(B|A)$ does not equal $P(B)$, the two events are not independent, as we suspected.

When considering three or more events, we can extend the definition of independence with the following general result.

Independence (multiple events)

The events $E_1, E_2, \ldots, E_n$ are independent if and only if for any subset of these events $E_{i_1}, E_{i_2}, \ldots, E_{i_k}$,

$$P(E_{i_1} \cap E_{i_2} \cap \cdots \cap E_{i_k}) = P(E_{i_1}) \times P(E_{i_2}) \times \cdots \times P(E_{i_k}) \quad (2\text{-}14)$$

This definition is typically used to calculate the probability that several events occur assuming that they are independent and the individual event probabilities are known. The knowledge that the events are independent usually comes from a fundamental understanding of the random experiment.

EXAMPLE 2-32
Series Circuit

The following circuit operates only if there is a path of functional devices from left to right. The probability that each device functions is shown on the graph. Assume that devices fail independently. What is the probability that the circuit operates?

Let L and R denote the events that the left and right devices operate, respectively. There is only a path if both operate. The probability the circuit operates is

$$P(L \text{ and } R) = P(L \cap R) = P(L)P(R) = 0.80(0.90) = 0.72$$

EXAMPLE 2-33

Assume that the probability that a wafer contains a large particle of contamination is 0.01 and that the wafers are independent; that is, the probability that a wafer contains a large particle is not dependent on the characteristics of any of the other wafers. If 15 wafers are analyzed, what is the probability that no large particles are found?

Let E_i denote the event that the ith wafer contains no large particles, $i = 1, 2, \ldots, 15$. Then, $P(E_i) = 0.99$. The probability requested can be represented as $P(E_1 \cap E_2 \cap \cdots \cap E_{15})$. From the independence assumption and Equation 2-14,

$$P(E_1 \cap E_2 \cap \cdots \cap E_{15}) = P(E_1) \times P(E_2) \times \cdots \times P(E_{15}) = 0.99^{15} = 0.86$$

EXAMPLE 2-34
Parallel Circuit

The following circuit operates only if there is a path of functional devices from left to right. The probability that each device functions is shown on the graph. Assume that devices fail independently. What is the probability that the circuit operates?

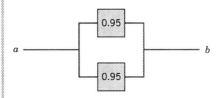

Let T and B denote the events that the top and bottom devices operate, respectively. There is a path if at least one device operates. The probability that the circuit operates is

$$P(T \text{ or } B) = 1 - P[(T \text{ or } B)'] = 1 - P(T' \text{ and } B')$$

A simple formula for the solution can be derived from the complements T' and B'. From the independence assumption,

$$P(T' \text{ and } B') = P(T')P(B') = (1 - 0.95)^2 = 0.05^2$$

so

$$P(T \text{ or } B) = 1 - 0.05^2 = 0.9975$$

EXAMPLE 2-35
Advanced Circuit

The following circuit operates only if there is a path of functional devices from left to right. The probability that each device functions is shown on the graph. Assume that devices fail independently. What is the probability that the circuit operates?

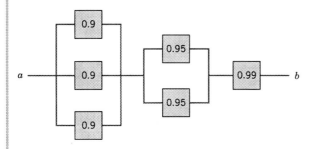

The solution can be obtained from a partition of the graph into three columns. The probability that there is a path of functional devices only through the three units on the left can be determined from the independence in a manner similar to the previous example. It is

$$1 - 0.1^3$$

Similarly, the probability that there is a path of functional devices only through the two units in the middle is

$$1 - 0.05^2$$

The probability that there is a path of functional devices only through the one unit on the right is simply the probability that the device functions, namely, 0.99. Therefore, with the independence assumption used again, the solution is

$$(1 - 0.1^3)(1 - 0.05^2)(0.99) = 0.987$$

EXERCISES FOR SECTION 2-6

2-100. If $P(A|B) = 0.4$, $P(B) = 0.8$, and $P(A) = 0.5$, are the events A and B independent?

2-101. If $P(A|B) = 0.3$, $P(B) = 0.8$, and $P(A) = 0.3$, are the events B and the complement of A independent?

2-102. If $P(A) = 0.2$, $P(B) = 0.2$, and A and B are mutually exclusive, are they independent?

2-103. A batch of 500 containers for frozen orange juice contains 5 that are defective. Two are selected, at random, without replacement, from the batch. Let A and B denote the events that the first and second container selected is defective, respectively.
(a) Are A and B independent events?
(b) If the sampling were done with replacement, would A and B be independent?

2-104. Disks of polycarbonate plastic from a supplier are analyzed for scratch and shock resistance. The results from 100 disks are summarized as follows:

		shock resistance	
		high	low
scratch	high	70	9
resistance	low	16	5

Let A denote the event that a disk has high shock resistance, and let B denote the event that a disk has high scratch resistance. Are events A and B independent?

2-105. Samples of emissions from three suppliers are classified for conformance to air-quality specifications. The results from 100 samples are summarized as follows:

		conforms	
		yes	no
	1	22	8
supplier	2	25	5
	3	30	10

Let A denote the event that a sample is from supplier 1, and let B denote the event that a sample conforms to specifications.
(a) Are events A and B independent?
(b) Determine $P(B|A)$.

2-106. Redundant Array of Inexpensive Discs (RAID) is a technology that uses multiple hard drives to increase the speed of data transfer and provide instant data backup. Suppose that the probability of any hard drive failing in a day is 0.001 and the drive failures are independent.
(a) Suppose you implement a RAID 0 scheme that uses two hard drives each containing a mirror image of the other. What is the probability of data loss? Assume that data loss occurs if both drives fail within the same day.
(b) Suppose you implement a RAID 1 scheme that splits the data over two hard drives. What is the probability of data loss? Assume that data loss occurs if at least one drive fails within the same day.

2-107. The probability that a lab specimen contains high levels of contamination is 0.10. Five samples are checked, and the samples are independent.
(a) What is the probability that none contains high levels of contamination?
(b) What is the probability that exactly one contains high levels of contamination?
(c) What is the probability that at least one contains high levels of contamination?

2-108. In a test of a printed circuit board using a random test pattern, an array of 10 bits is equally likely to be 0 or 1. Assume the bits are independent.
(a) What is the probability that all bits are 1s?
(b) What is the probability that all bits are 0s?
(c) What is the probability that exactly five bits are 1s and five bits are 0s?

2-109. Six tissues are extracted from an ivy plant infested by spider mites. The plant in infested in 20% of its area. A sample is chosen from randomly selected areas on the ivy plant. Assume that the samples are independent.
(a) What is the probability that four successive samples show the signs of infestation?
(b) What is the probability that three out of four successive samples show the signs of infestation?

2-110. A player of a video game is confronted with a series of four opponents and an 80% probability of defeating each opponent. Assume that the results from opponents are independent

(and that when the player is defeated by an opponent the game ends).
(a) What is the probability that a player defeats all four opponents in a game?
(b) What is the probability that a player defeats at least two opponents in a game?
(c) If the game is played three times, what is the probability that the player defeats all four opponents at least once?

2-111. In an acid-base titration, a base or acid is gradually added to the other until they have completely neutralized each other. Since acids and bases are usually colorless (as are the water and salt produced in the neutralization reaction), pH is measured to monitor the reaction. Suppose that the equivalence point is reached after approximately 100 mL of a NaOH solution have been added (enough to react with all the acetic acid present) but that replicates are equally likely to indicate from 95 to 104 mL, measured to the nearest mL. Assume that two technicians each conduct titrations independently.
(a) What is the probability that both technicians obtain equivalence at 100 mL?
(b) What is the probability that both technicians obtain equivalence between 98 and 104 mL (inclusive)?
(c) What is the probability that the average volume at equivalence from the technicians is 100 mL?

2-112. A credit card contains 16 digits. It also contains a month and year of expiration. Suppose there are one million users of a credit card with unique card numbers. A hacker randomly selects a 16 digit credit card number.
(a) What is the probability that it belongs to a user?
(b) Suppose a hacker has a 25% chance of correctly guessing the year your card expires and randomly selects one of the 12 months. What is the probability that the hacker correctly selects the month and year of expiration?

2-113. Eight cavities in an injection-molding tool produce plastic connectors that fall into a common stream. A sample is chosen every several minutes. Assume that the samples are independent.
(a) What is the probability that five successive samples were all produced in cavity one of the mold?
(b) What is the probability that five successive samples were all produced in the same cavity of the mold?
(c) What is the probability that four out of five successive samples were produced in cavity one of the mold?

2-114. The following circuit operates if and only if there is a path of functional devices from left to right. The probability that each device functions is as shown. Assume that the probability that a device is functional does not depend on whether or not other devices are functional. What is the probability that the circuit operates?

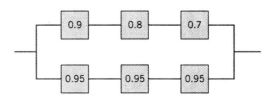

2-115. The following circuit operates if and only if there is a path of functional devices from left to right. The probability each device functions is as shown. Assume that the probability that a device functions does not depend on whether or not other devices are functional. What is the probability that the circuit operates?

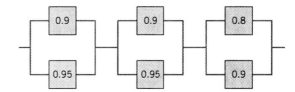

2-7 BAYES' THEOREM

The examples in this chapter indicate that information is often presented in terms of conditional probabilities. These conditional probabilities commonly provide the probability of an event (such as failure) given a condition (such as high or low contamination). But after a random experiment generates an outcome, we are naturally interested in the probability that a condition was present (high contamination) given an outcome (a semiconductor failure). Thomas Bayes addressed this essential question in the 1700s and developed the fundamental result known as Bayes' theorem. Do not let the simplicity of the mathematics conceal the importance. There is extensive interest in such probabilities in modern statistical analysis.

From the definition of conditional probability,

$$P(A \cap B) = P(A|B)P(B) = P(B \cap A) = P(B|A)P(A)$$

56 CHAPTER 2 PROBABILITY

Now considering the second and last terms in the expression above, we can write

$$P(A|B) = \frac{P(B|A)P(A)}{P(B)} \quad \text{for} \quad P(B) > 0 \tag{2-15}$$

This is a useful result that enables us to solve for $P(A|B)$ in terms of $P(B|A)$.

EXAMPLE 2-36 Reconsider Example 2-27. The conditional probability a high level of contamination was present when a failure occurred is to be determined. The information from Example 2-27 is summarized here.

Probability of Failure	Level of Contamination	Probability of Level
0.1	High	0.2
0.005	Not High	0.8

The probability of $P(H|F)$ is determined from

$$P(H|F) = \frac{P(F|H)P(H)}{P(F)} = \frac{0.10(0.20)}{0.024} = 0.83$$

The value of $P(F)$ in the denominator of our solution was found from $P(F) = P(F|H)P(H) + P(F|H')P(H')$.

In general, if $P(B)$ in the denominator of Equation 2-15 is written using the Total Probability Rule in Equation 2-12, we obtain the following general result, which is known as **Bayes' Theorem**.

Bayes' Theorem

If $E_1, E_2, \ldots, E_k$ are k mutually exclusive and exhaustive events and B is any event,

$$P(E_1|B) = \frac{P(B|E_1)P(E_1)}{P(B|E_1)P(E_1) + P(B|E_2)P(E_2) + \cdots + P(B|E_k)P(E_k)} \tag{2-16}$$

for $P(B) > 0$

Notice that the numerator always equals one of the terms in the sum in the denominator.

EXAMPLE 2-37
Medical
Diagnostic

Because a new medical procedure has been shown to be effective in the early detection of an illness, a medical screening of the population is proposed. The probability that the test correctly identifies someone with the illness as positive is 0.99, and the probability that the test correctly identifies someone without the illness as negative is 0.95. The incidence of the illness in the general population is 0.0001. You take the test, and the result is positive. What is the probability that you have the illness?

Let D denote the event that you have the illness, and let S denote the event that the test signals positive. The probability requested can be denoted as $P(D|S)$. The probability that the test correctly signals someone without the illness as negative is 0.95. Consequently, the probability of a positive test without the illness is

$$P(S|D') = 0.05$$

From Bayes' Theorem,

$$P(D|S) = P(S|D)P(D)/[P(S|D)P(D) + P(S|D')P(D')]$$
$$= 0.99(0.0001)/[0.99(0.0001) + 0.05(1 - 0.0001)]$$
$$= 1/506 = 0.002$$

That is, the probability of you having the illness given a positive result from the test is only 0.002. Surprisingly, even though the test is effective, in the sense that $P(S|D)$ is high and $P(S|D')$ is low, because the incidence of the illness in the general population is low, the chances are quite small that you actually have the disease even if the test is positive.

EXAMPLE 2-38
Bayesian Network

Bayesian networks are used on the Web sites of high-technology manufacturers to allow customers to quickly diagnose problems with products. An oversimplified example is presented here. A printer manufacturer obtained the following probabilities from a database of test results. Printer failures are associated with three types of problems: hardware, software, and other (such as connectors), with probabilities 0.1, 0.6, and 0.3, respectively. The probability of a printer failure given a hardware problem is 0.9, given a software problem is 0.2, and given any other type of problem is 0.5. If a customer enters the manufacturer's Web site to diagnose a printer failure, what is the most likely cause of the problem?

Let the events H, S, and O denote a hardware, software, or other problem, respectively, and let F denote a printer failure. The most likely cause of the problem is the one that corresponds to the largest of $P(H|F)$, $P(S|F)$, and $P(O|F)$. In Bayes' Theorem the denominator is

$$P(F) = P(F|H)P(H) + P(F|S)P(S) + P(F|O)P(O) = 0.9(0.1) + 0.2(0.6) + 0.5(0.3) = 0.36$$

Then

$$P(H|F) = P(F|H)P(H)/P(F) = 0.9(0.1)/0.36 = 0.250$$
$$P(S|F) = P(F|S)P(S)/P(F) = 0.2(0.6)/0.36 = 0.333$$
$$P(O|F) = P(F|O)P(O)/P(F) = 0.5(0.3)/0.36 = 0.417$$

Notice that $P(H|F) + P(S|F) + P(O|F) = 1$ because one of the three types of problems is responsible for the failure. Because $P(O|F)$ is largest, the most likely cause of the problem is in the *other* category. A Web site dialog to diagnose the problem quickly should start with a check into that type of problem.

EXERCISES FOR SECTION 2-7

2-116. Suppose that $P(A|B) = 0.7$, $P(A) = 0.5$, and $P(B) = 0.2$. Determine $P(B|A)$.

2-117. Suppose that $P(A|B) = 0.4$, $P(A|B') = 0.2$, $P(B) = 0.8$. Determine $P(B|A)$.

2-118. Software to detect fraud in consumer phone cards tracks the number of metropolitan areas where calls originate each day. It is found that 1% of the legitimate users originate calls from two or more metropolitan areas in a single day. However, 30% of fraudulent users originate calls from two or more metropolitan areas in a single day. The proportion of fraudulent users is 0.01%. If the same user originates calls from two or more metropolitan areas in a single day, what is the probability that the user is fraudulent?

2-119. A new process of more accurately detecting anaerobic respiration in cells is being tested. The new process is important due to its high accuracy, its lack of extensive experimentation, and the fact that it could be used to identify five different categories of organisms: obligate anaerobes, facultative anaerobes, aerotolerant, microaerophiles, and nanaerobes instead of using a single test for each category. The process claims that it can identify obligate anaerobes with 97.8% accuracy, facultative anaerobes with 98.1% accuracy, aerotolerant with 95.9% accuracy, microaerophiles with 96.5% accuracy, and nanaerobes with 99.2% accuracy. If any category is not present, the process does not signal. Samples are prepared for the calibration of the process and 31% of them contain obligate anaerobes, 27% contain

facultative anaerobes, 21% contain microaerophiles, 13% contain nanaerobes, and 8% contain aerotolerant. A test sample is selected randomly.
(a) What is the probability that the process will signal?
(b) If the test signals, what is the probability that microaerophiles are present?

2-120. In the 2004 presidential election, exit polls from the critical state of Ohio provided the following results:

	Bush	Kerry
no college degree (62%)	50%	50%
college graduate (38%)	53%	46%

If a randomly selected respondent voted for Bush, what is the probability that the person has a college degree?

2-121. Customers are used to evaluate preliminary product designs. In the past, 95% of highly successful products received good reviews, 60% of moderately successful products received good reviews, and 10% of poor products received good reviews. In addition, 40% of products have been highly successful, 35% have been moderately successful, and 25% have been poor products.
(a) What is the probability that a product attains a good review?
(b) If a new design attains a good review, what is the probability that it will be a highly successful product?
(c) If a product does not attain a good review, what is the probability that it will be a highly successful product?

2-122. An inspector working for a manufacturing company has a 99% chance of correctly identifying defective items and a 0.5% chance of incorrectly classifying a good item as defective. The company has evidence that its line produces 0.9% of nonconforming items.
(a) What is the probability that an item selected for inspection is classified as defective?
(b) If an item selected at random is classified as nondefective, what is the probability that it is indeed good?

2-123. A new analytical method to detect pollutants in water is being tested. This new method of chemical analysis is important because, if adopted, it could be used to detect three different contaminants—organic pollutants, volatile solvents, and chlorinated compounds—instead of having to use a single test for each pollutant. The makers of the test claim that it can detect high levels of organic pollutants with 99.7% accuracy, volatile solvents with 99.95% accuracy, and chlorinated compounds with 89.7% accuracy. If a pollutant is not present, the test does not signal. Samples are prepared for the calibration of the test and 60% of them are contaminated with organic pollutants, 27% with volatile solvents, and 13% with traces of chlorinated compounds. A test sample is selected randomly.
(a) What is the probability that the test will signal?
(b) If the test signals, what is the probability that chlorinated compounds are present?

2-8 RANDOM VARIABLES

We often summarize the outcome from a random experiment by a simple number. In many of the examples of random experiments that we have considered, the sample space has been a description of possible outcomes. In some cases, descriptions of outcomes are sufficient, but in other cases, it is useful to associate a number with each outcome in the sample space. Because the particular outcome of the experiment is not known in advance, the resulting value of our variable is not known in advance. For this reason, the variable that associates a number with the outcome of a random experiment is referred to as a random variable.

Random Variable

A **random variable** is a function that assigns a real number to each outcome in the sample space of a random experiment.

Notation is used to distinguish between a random variable and the real number.

A random variable is denoted by an uppercase letter such as X. After an experiment is conducted, the measured value of the random variable is denoted by a lowercase letter such as $x = 70$ milliamperes.

Sometimes a measurement (such as current in a copper wire or length of a machined part) can assume any value in an interval of real numbers (at least theoretically). Then arbitrary precision in the measurement is possible. Of course, in practice, we might round off to the nearest tenth or hundredth of a unit. The random variable that represents this measurement is said to be a continuous random variable. The range of the random variable includes all values in an interval of real numbers; that is, the range can be thought of as a continuum.

In other experiments, we might record a count such as the number of transmitted bits that are received in error. Then the measurement is limited to integers. Or we might record that a proportion such as 0.0042 of the 10,000 transmitted bits were received in error. Then the measurement is fractional, but it is still limited to discrete points on the real line. Whenever the measurement is limited to discrete points on the real line, the random variable is said to be a discrete random variable.

Discrete and Continuous Random Variables

> A **discrete** random variable is a random variable with a finite (or countably infinite) range.
> A **continuous** random variable is a random variable with an interval (either finite or infinite) of real numbers for its range.

In some cases, the random variable X is actually discrete but, because the range of possible values is so large, it might be more convenient to analyze X as a continuous random variable. For example, suppose that current measurements are read from a digital instrument that displays the current to the nearest one-hundredth of a milliampere. Because the possible measurements are limited, the random variable is discrete. However, it might be a more convenient, simple approximation to assume that the current measurements are values of a continuous random variable.

Examples of Random Variables

> Examples of **continuous** random variables:
>   electrical current, length, pressure, temperature, time, voltage, weight
>
> Examples of **discrete** random variables:
>   number of scratches on a surface, proportion of defective parts among 1000 tested, number of transmitted bits received in error

EXERCISES FOR SECTION 2-8

2-124. Decide whether a discrete or continuous random variable is the best model for each of the following variables:
(a) The time until a projectile returns to earth.
(b) The number of times a transistor in a computer memory changes state in one operation.
(c) The volume of gasoline that is lost to evaporation during the filling of a gas tank.
(d) The outside diameter of a machined shaft.
(e) The number of cracks exceeding one-half inch in 10 miles of an interstate highway.
(f) The weight of an injection-molded plastic part.
(g) The number of molecules in a sample of gas.
(h) The concentration of output from a reactor.
(i) The current in an electronic circuit.

Supplemental Exercises

2-125. Samples of laboratory glass are in small, light packaging or heavy, large packaging. Suppose that 2 and 1% of the sample shipped in small and large packages, respectively, break during transit. If 60% of the samples are shipped in large packages and 40% are shipped in small packages, what proportion of samples break during shipment?

2-126. A sample of three calculators is selected from a manufacturing line, and each calculator is classified as either defective or acceptable. Let A, B, and C denote the events that the first, second, and third calculators respectively, are defective.

(a) Describe the sample space for this experiment with a tree diagram.

Use the tree diagram to describe each of the following events:
(b) A (c) B
(d) $A \cap B$ (e) $B \cup C$

2-127. Samples of a cast aluminum part are classified on the basis of surface finish (in microinches) and edge finish. The results of 100 parts are summarized as follows:

		edge finish	
		excellent	good
surface	excellent	80	2
finish	good	10	8

Let A denote the event that a sample has excellent surface finish, and let B denote the event that a sample has excellent edge finish. If a part is selected at random, determine the following probabilities:

(a) $P(A)$ (b) $P(B)$
(c) $P(A')$ (d) $P(A \cap B)$
(e) $P(A \cup B)$ (f) $P(A' \cup B)$

2-128. Shafts are classified in terms of the machine tool that was used for manufacturing the shaft and conformance to surface finish and roundness.

2-129. If A, B, and C are mutually exclusive events, is it possible for $P(A) = 0.3$, $P(B) = 0.4$, and $P(C) = 0.5$? Why or why not?

Tool 1		roundness conforms	
		yes	no
surface finish	yes	200	1
conforms	no	4	2
Tool 2		roundness conforms	
		yes	no
surface finish	yes	145	4
conforms	no	8	6

(a) If a shaft is selected at random, what is the probability that the shaft conforms to surface finish requirements or to roundness requirements or is from Tool 1?
(b) If a shaft is selected at random, what is the probability that the shaft conforms to surface finish requirements or does not conform to roundness requirements or is from Tool 2?
(c) If a shaft is selected at random, what is the probability that the shaft conforms to both surface finish and roundness requirements or the shaft is from Tool 2?

(d) If a shaft is selected at random, what is the probability that the shaft conforms to surface finish requirements or the shaft is from Tool 2?

2-130. The analysis of shafts for a compressor is summarized by conformance to specifications:

		roundness conforms	
		yes	no
surface finish	yes	345	5
conforms	no	12	8

(a) If we know that a shaft conforms to roundness requirements, what is the probability that it conforms to surface finish requirements?
(b) If we know that a shaft does not conform to roundness requirements, what is the probability that it conforms to surface finish requirements?

2-131. A researcher receives 100 containers of oxygen. Of those containers, twenty of them have oxygen that is not ionized and the rest are ionized. Two samples are randomly selected, without replacement, from the lot.

(a) What is the probability that the first one selected is not ionized?
(b) What is the probability that the second one selected is not ionized given that the first one was ionized?
(c) What is the probability that both are ionized?
(d) How does the answer in part (b) change if samples selected were replaced prior to the next selection?

2-132. A lot contains 15 castings from a local supplier and 25 castings from a supplier in the next state. Two castings are selected randomly, without replacement, from the lot of 40. Let A be the event that the first casting selected is from the local supplier, and let B denote the event that the second casting is selected from the local supplier. Determine:

(a) $P(A)$ (b) $P(B|A)$
(c) $P(A \cap B)$ (d) $P(A \cup B)$

Suppose three castings are selected at random, without replacement, from the lot of 40. In addition to the definitions of events A and B, let C denote the event that the third casting selected is from the local supplier. Determine:

(e) $P(A \cap B \cap C)$
(f) $P(A \cap B \cap C')$

2-133. In the manufacturing of a chemical adhesive, 3% of all batches have raw materials from two different lots. This occurs when holding tanks are replenished and the remaining portion of a lot is insufficient to fill the tanks.

Only 5% of batches with material from a single lot require reprocessing. However, the viscosity of batches consisting of two or more lots of material is more difficult to control, and 40% of such batches require additional processing to achieve the required viscosity.

Let A denote the event that a batch is formed from two different lots, and let B denote the event that a lot

requires additional processing. Determine the following probabilities:
(a) $P(A)$ (b) $P(A')$
(c) $P(B|A)$ (d) $P(B|A')$
(e) $P(A \cap B)$ (f) $P(A \cap B')$
(g) $P(B)$

2-134. Incoming calls to a customer service center are classified as complaints (75% of call) or requests for information (25% of calls). Of the complaints, 40% deal with computer equipment that does not respond and 57% deal with incomplete software installation; and in the remaining 3% of complaints the user has improperly followed the installation instructions. The requests for information are evenly divided on technical questions (50%) and requests to purchase more products (50%).
(a) What is the probability that an incoming call to the customer service center will be from a customer who has not followed installation instructions properly?
(b) Find the probability that an incoming call is a request for purchasing more products.

2-135. A congested computer network has a 0.002 probability of losing a data packet and packet losses are independent events. A lost packet must be resent.
(a) What is the probability that an e-mail message with 100 packets will need any resent?
(b) What is the probability that an e-mail message with 3 packets will need exactly one to be resent?
(c) If 10 e-mail messages are sent, each with 100 packets, what is the probability that at least one message will need some packets to be resent?

2-136. Samples of a cast aluminum part are classified on the basis of surface finish (in microinches) and length measurements. The results of 100 parts are summarized as follows:

		length	
		excellent	good
surface	excellent	80	2
finish	good	10	8

Let A denote the event that a sample has excellent surface finish, and let B denote the event that a sample has excellent length. Are events A and B independent?

2-137. An optical storage device uses an error recovery procedure that requires an immediate satisfactory readback of any written data. If the readback is not successful after three writing operations, that sector of the disk is eliminated as unacceptable for data storage. On an acceptable portion of the disk, the probability of a satisfactory readback is 0.98. Assume the readbacks are independent. What is the probability that an acceptable portion of the disk is eliminated as unacceptable for data storage?

2-138. Semiconductor lasers used in optical storage products require higher power levels for write operations than for read operations. High-power-level operations lower the useful life of the laser.

Lasers in products used for backup of higher speed magnetic disks primarily write, and the probability that the useful life exceeds five years is 0.95. Lasers that are in products that are used for main storage spend approximately an equal amount of time reading and writing, and the probability that the useful life exceeds five years is 0.995. Now, 25% of the products from a manufacturer are used for backup and 75% of the products are used for main storage.

Let A denote the event that a laser's useful life exceeds five years, and let B denote the event that a laser is in a product that is used for backup.

Use a tree diagram to determine the following:
(a) $P(B)$ (b) $P(A|B)$
(c) $P(A|B')$ (d) $P(A \cap B)$
(e) $P(A \cap B')$ (f) $P(A)$
(g) What is the probability that the useful life of a laser exceeds five years?
(h) What is the probability that a laser that failed before five years came from a product used for backup?

2-139. Energy released from three cells breaks the molecular bond and converts ATP (Adenosine Tri-Phosphate) into ADP (adenosine diphosphate). Storage of ATP in muscle cells (even for an athlete) can only sustain maximal muscle power for less than five seconds (a short dash). Three systems are used to replenish ATP: phosphagen system; glycogen-lactic acid system (anaerobic); and aerobic respiration, but the first is only useful for less than 10 seconds, and even the second system provides less than two minutes of ATP. An endurance athlete needs to perform below the anaerobic threshold to sustain energy for extended periods. A sample of 100 individuals is described by the energy system used in exercise at different intensity levels.

	primarily aerobic	
period	yes	no
1	50	7
2	13	30

Let A denote the event that an individual is in period 2, and let B denote the event that the energy is primarily aerobic. Determine the number of individuals in
(a) $A' \cap B$ (b) B' (c) $A \cup B$

2-140. A sample preparation for a chemical measurement is completed correctly by 25% of the lab technicians, completed with a minor error by 70%, and completed with a major error by 5%.
(a) If a technician is selected randomly to complete the preparation, what is the probability it is completed without error?
(b) What is the probability that it is completed with either a minor or a major error?

2-141. In circuit testing of printed circuit boards, each board either fails or does not fail the test. A board that fails the test is then checked further to determine which one of five defect types is the primary failure mode. Represent the sample space for this experiment.

2-142. The data from 200 machined parts are summarized as follows:

edge condition	depth of bore	
	above target	below target
coarse	15	10
moderate	25	20
smooth	50	80

(a) What is the probability that a part selected has a moderate edge condition and a below-target bore depth?
(b) What is the probability that a part selected has a moderate edge condition or a below-target bore depth?
(c) What is the probability that a part selected does not have a moderate edge condition or does not have a below-target bore depth?

2-143. Computers in a shipment of 100 units contain a portable hard drive, CD RW drive, or both according to the following table:

	portable hard drive	
CD RW	yes	no
yes	15	80
no	4	1

Let A denote the events that a computer has a portable hard drive and let B denote the event that a computer has a CD RW drive. If one computer is selected randomly, compute
(a) $P(A)$ (b) $P(A \cap B)$ (c) $P(A \cup B)$
(d) $P(A' \cap B)$ (e) $P(A|B)$

2-144. The probability that a customer's order is not shipped on time is 0.05. A particular customer places three orders, and the orders are placed far enough apart in time that they can be considered to be independent events.
(a) What is the probability that all are shipped on time?
(b) What is the probability that exactly one is not shipped on time?
(c) What is the probability that two or more orders are not shipped on time?

2-145. Let E_1, E_2, and E_3 denote the samples that conform to a percentage of solids specification, a molecular weight specification, and a color specification, respectively. A total of 240 samples are classified by the E_1, E_2, and E_3 specifications, where *yes* indicates that the sample conforms.

E_3 yes

		E_2		
		yes	no	Total
E_1	yes	200	1	201
	no	5	4	9
Total		205	5	210

E_3 no

		E_2		
		yes	no	
E_1	yes	20	4	24
	no	6	0	6
Total		26	4	30

(a) Are E_1, E_2, and E_3 mutually exclusive events?
(b) Are E'_1, E'_2, and E'_3 mutually exclusive events?
(c) What is $P(E'_1 \text{ or } E'_2 \text{ or } E'_3)$?
(d) What is the probability that a sample conforms to all three specifications?
(e) What is the probability that a sample conforms to the E_1 or E_3 specification?
(f) What is the probability that a sample conforms to the E_1 or E_2 or E_3 specification?

2-146. Transactions to a computer database are either new items or changes to previous items. The addition of an item can be completed in less than 100 milliseconds 90% of the time, but only 20% of changes to a previous item can be completed in less than this time. If 30% of transactions are changes, what is the probability that a transaction can be completed in less than 100 milliseconds?

2-147. A steel plate contains 20 bolts. Assume that 5 bolts are not torqued to the proper limit. Four bolts are selected at random, without replacement, to be checked for torque.
(a) What is the probability that all four of the selected bolts are torqued to the proper limit?
(b) What is the probability that at least one of the selected bolts is not torqued to the proper limit?

2-148. The following circuit operates if and only if there is a path of functional devices from left to right. Assume devices fail independently and that the probability of *failure* of each device is as shown. What is the probability that the circuit operates?

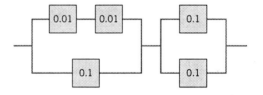

2-149. The probability of getting through by telephone to buy concert tickets is 0.92. For the same event, the probability of accessing the vendor's Web site is 0.95. Assume that these two ways to buy tickets are independent. What is the probability that someone who tries to buy tickets through the Internet and by phone will obtain tickets?

2-150. The British government has stepped up its information campaign regarding foot and mouth disease by mailing brochures to farmers around the country. It is estimated that 99% of Scottish farmers who receive the brochure possess enough information to deal with an outbreak of the disease, but only 90% of those without the brochure can deal with an outbreak. After the first three months of mailing, 95% of the farmers in Scotland received the informative brochure. Compute the probability that a randomly selected farmer will have enough information to deal effectively with an outbreak of the disease.

2-151. In an automated filling operation, the probability of an incorrect fill when the process is operated at a low speed is 0.001. When the process is operated at a high speed, the probability of an incorrect fill is 0.01. Assume that 30% of the containers are filled when the process is operated at a high speed and the remainder are filled when the process is operated at a low speed.
(a) What is the probability of an incorrectly filled container?
(b) If an incorrectly filled container is found, what is the probability that it was filled during the high-speed operation?

2-152. An encryption-decryption system consists of three elements: encode, transmit, and decode. A faulty encode occurs in 0.5% of the messages processed, transmission errors occur in 1% of the messages, and a decode error occurs in 0.1% of the messages. Assume the errors are independent.
(a) What is the probability of a completely defect-free message?
(b) What is the probability of a message that has either an encode or a decode error?

2-153. It is known that two defective copies of a commercial software program were erroneously sent to a shipping lot that has now a total of 75 copies of the program. A sample of copies will be selected from the lot without replacement.
(a) If three copies of the software are inspected, determine the probability that exactly one of the defective copies will be found.
(b) If three copies of the software are inspected, determine the probability that both defective copies will be found.
(c) If 73 copies are inspected, determine the probability that both copies will be found. Hint: Work with the copies that remain in the lot.

2-154. A robotic insertion tool contains 10 primary components. The probability that any component fails during the warranty period is 0.01. Assume that the components fail independently and that the tool fails if any component fails. What is the probability that the tool fails during the warranty period?

2-155. An e-mail message can travel through one of two server routes. The probability of transmission error in each of the servers and the proportion of messages that travel each route are shown in the following table. Assume that the servers are independent.

	percentage of messages	probability of error			
		server 1	server 2	server 3	server 4
route 1	30	0.01	0.015		
route 2	70			0.02	0.003

(a) What is the probability that a message will arrive without error?
(b) If a message arrives in error, what is the probability it was sent through route 1?

2-156. A machine tool is idle 15% of the time. You request immediate use of the tool on five different occasions during the year. Assume that your requests represent independent events.
(a) What is the probability that the tool is idle at the time of all of your requests?
(b) What is the probability that the machine is idle at the time of exactly four of your requests?
(c) What is the probability that the tool is idle at the time of at least three of your requests?

2-157. A lot of 50 spacing washers contains 30 washers that are thicker than the target dimension. Suppose that three washers are selected at random, without replacement, from the lot.
(a) What is the probability that all three washers are thicker than the target?
(b) What is the probability that the third washer selected is thicker than the target if the first two washers selected are thinner than the target?
(c) What is the probability that the third washer selected is thicker than the target?

2-158. Continuation of Exercise 2-157. Washers are selected from the lot at random, without replacement.
(a) What is the minimum number of washers that need to be selected so that the probability that all the washers are thinner than the target is less than 0.10?
(b) What is the minimum number of washers that need to be selected so that the probability that one or more washers are thicker than the target is at least 0.90?

2-159. The following table lists the history of 940 orders for features in an entry-level computer product.

		extra memory	
		no	yes
optional high-	no	514	68
speed processor	yes	112	246

Let A be the event that an order requests the optional high-speed processor, and let B be the event that an order requests extra memory. Determine the following probabilities:
(a) $P(A \cup B)$ (b) $P(A \cap B)$
(c) $P(A' \cup B)$ (d) $P(A' \cap B')$
(e) What is the probability that an order requests an optional high-speed processor given that the order requests extra memory?
(f) What is the probability that an order requests extra memory given that the order requests an optional high-speed processor?

2-160. The alignment between the magnetic tape and head in a magnetic tape storage system affects the performance of the system. Suppose that 10% of the read operations are

degraded by skewed alignments, 5% of the read operations are degraded by off-center alignments, and the remaining read operations are properly aligned. The probability of a read error is 0.01 from a skewed alignment, 0.02 from an off-center alignment, and 0.001 from a proper alignment.
(a) What is the probability of a read error?
(b) If a read error occurs, what is the probability that it is due to a skewed alignment?

2-161. The following circuit operates if and only if there is a path of functional devices from left to right. Assume that devices fail independently and that the probability of *failure* of each device is as shown. What is the probability that the circuit does not operate?

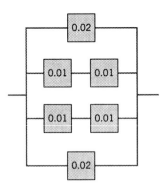

2-162. A company that tracks the use of its web site determined that the more pages a visitor views, the more likely the visitor is to provide contact information. Use the following tables to answer the questions:

Number of pages viewed:	1	2	3	4 or more
Percentage of visitors:	40	30	20	10
Percentage of visitors in each page-view category that provide contact information:	10	10	20	40

(a) What is the probability that a visitor to the web site provides contact information?
(b) If a visitor provides contact information, what is the probability that the visitor viewed four or more pages?

2-163. An article in *Genome Research* "An Assessment of Gene Prediction Accuracy in Large DNA Sequences," (2000, Vol. 10, pp. 1631–1642), considered the accuracy of commercial software to predict nucleotides in gene sequences. The following table shows the number of sequences for which the programs produced predictions and the number of nucleotides correctly predicted (computed globally from the total number of prediction successes and failures on all sequences).

	Number of sequences	Proportion
GenScan	177	0.93
Blastx default	175	0.91
Blastx topcomboN	174	0.97
Blastx 2 stages	175	0.90
GeneWise	177	0.98
Procrustes	177	0.93

Assume the prediction successes and failures are independent between the programs.
(a) What is the probability that all programs predict a nucleotide correctly?
(b) What is the probability that all programs predict a nucleotide incorrectly?
(c) What is the probability that at least one Blastx program predicts a nucleotide correctly?

2-164. A batch contains 36 bacteria cells. Assume that 12 of the cells are not capable of cellular replication. Six cells are selected at random, without replacement, to be checked for replication.
(a) What is the probability that all six cells of the selected cells are able to replicate?
(b) What is the probability that at least one of the selected cells is not capable of replication?

2-165. A computer system uses passwords that are exactly seven characters and each character is one of the 26 letters (a–z) or 10 integers (0–9). Uppercase letters are not used.
(a) How many passwords are possible?
(b) If a password consists of exactly six letters and one number, how many passwords are possible?
(c) If a password consists of five letters followed by two numbers, how many passwords are possible?

2-166. Natural red hair consists of two genes. People with red hair have two dominant genes, two regressive genes, or one dominant and one regressive gene. A group of 1000 people was categorized as follows:

	Gene 2		
Gene 1	Dominant	Regressive	Other
Dominant	5	25	30
Regressive	7	63	35
Other	20	15	800

Let A denote the event that a person has a dominant red-hair gene and let B denote the event that a person has a regressive red hair gene. If a person is selected at random from this group, compute the following.
(a) $P(A)$ (b) $P(A \cap B)$
(c) $P(A \cup B)$ (d) $P(A' \cap B)$ (e) $P(A \mid B)$
(f) Probability that the selected person has red hair

2-167. Two suppliers each supplied 2000 parts and these were evaluated for conformance to specifications. One part type was of greater complexity than the other. The proportion of nonconforming parts of each type are shown in the table.

Supplier	Simple component	Complex assembly	Total
1 Nonconforming	2	10	12
Total	1000	1000	2000
2 Nonconforming	4	6	10
Total	1600	400	2000

One part is selected at random from each supplier. For each supplier, separately calculate the following probabilities:
(a) What is the probability a part conforms to specifications?
(b) What is the probability a part conforms to specifications given it is a complex assembly?
(c) What is the probability a part conforms to specifications given it is a simple component?
(d) Compare your answers for each supplier in part (a) to those in parts (b) and (c) and explain any unusual results.

MIND-EXPANDING EXERCISES

2-168. The alignment between the magnetic tape and head in a magnetic tape storage system affects the performance of the system. Suppose that 10% of the read operations are degraded by skewed alignments, 5% by off-center alignments, 1% by both skewness and off-center, and the remaining read operations are properly aligned. The probability of a read error is 0.01 from a skewed alignment, 0.02 from an off-center alignment, 0.06 from both conditions, and 0.001 from a proper alignment. What is the probability of a read error?

2-169. Suppose that a lot of washers is large enough that it can be assumed that the sampling is done with replacement. Assume that 60% of the washers exceed the target thickness.
(a) What is the minimum number of washers that need to be selected so that the probability that none are thicker than the target is less than 0.10?
(b) What is the minimum number of washers that need to be selected so that the probability that one or more washers are thicker than the target is at least 0.90?

2-170. A biotechnology manufacturing firm can produce diagnostic test kits at a cost of $20. Each kit for which there is a demand in the week of production can be sold for $100. However, the half-life of components in the kit requires the kit to be scrapped if it is not sold in the week of production. The cost of scrapping the kit is $5. The weekly demand is summarized as follows:

weekly demand

Number of units	0	50	100	200
Probability of demand	0.05	0.4	0.3	0.25

How many kits should be produced each week to maximize the mean earnings of the firm?

2-171. Assume the following characteristics of the inspection process in Exercise 2-147. If an operator checks a bolt, the probability that an incorrectly torqued bolt is identified is 0.95. If a checked bolt is correctly torqued, the operator's conclusion is always correct. What is the probability that at least one bolt in the sample of four is identified as being incorrectly torqued?

2-172. If the events A and B are independent, show that A' and B' are independent.

2-173. Suppose that a table of part counts is generalized as follows:

		conforms	
		yes	no
supplier	1	ka	kb
	2	a	b

where a, b, and k are positive integers. Let A denote the event that a part is from supplier 1 and let B denote the event that a part conforms to specifications. Show that A and B are independent events.

This exercise illustrates the result that whenever the rows of a table (with r rows and c columns) are proportional, an event defined by a row category and an event defined by a column category are independent.

IMPORTANT TERMS AND CONCEPTS

Addition rule
Axioms of probability
Bayes' theorem
Combination
Conditional probability
Equally likely outcomes
Event
Independence
Multiplication rule
Mutually exclusive events
Outcome
Permutation
Probability
Random experiment
Random variables—discrete and continuous
Sample spaces—discrete and continuous
Simpson's paradox
Total probability rule
Tree Diagram
Venn Diagram
With or without replacement

3 Discrete Random Variables and Probability Distributions

CHAPTER OUTLINE

3-1 DISCRETE RANDOM VARIABLES

3-2 PROBABILITY DISTRIBUTIONS AND PROBABILITY MASS FUNCTIONS

3-3 CUMULATIVE DISTRIBUTION FUNCTIONS

3-4 MEAN AND VARIANCE OF A DISCRETE RANDOM VARIABLE

3-5 DISCRETE UNIFORM DISTRIBUTION

3-6 BINOMIAL DISTRIBUTION

3-7 GEOMETRIC AND NEGATIVE BINOMIAL DISTRIBUTIONS

    3-7.1 Geometric Distribution

    3-7.2 Negative Binomial Distribution

3-8 HYPERGEOMETRIC DISTRIBUTION

3-9 POISSON DISTRIBUTION

LEARNING OBJECTIVES

After careful study of this chapter you should be able to do the following:
1. Determine probabilities from probability mass functions and the reverse
2. Determine probabilities from cumulative distribution functions and cumulative distribution functions from probability mass functions, and the reverse
3. Calculate means and variances for discrete random variables
4. Understand the assumptions for each of the discrete probability distributions presented
5. Select an appropriate discrete probability distribution to calculate probabilities in specific applications
6. Calculate probabilities, determine means and variances for each of the discrete probability distributions presented

3-1 DISCRETE RANDOM VARIABLES

Many physical systems can be modeled by the same or similar random experiments and random variables. The distribution of the random variables involved in each of these common systems can be analyzed, and the results of that analysis can be used in different applications and examples. In this chapter, we present the analysis of several random experiments and **discrete random variables** that frequently arise in applications. We often omit a discussion of the underlying sample space of the random experiment and directly describe the distribution of a particular random variable.

EXAMPLE 3-1
Voice Lines

A voice communication system for a business contains 48 external lines. At a particular time, the system is observed, and some of the lines are being used. Let the random variable X denote the number of lines in use. Then, X can assume any of the integer values 0 through 48. When the system is observed, if 10 lines are in use, $x = 10$.

EXAMPLE 3-2

In a semiconductor manufacturing process, two wafers from a lot are tested. Each wafer is classified as *pass* or *fail*. Assume that the probability that a wafer passes the test is 0.8 and that wafers are independent. The sample space for the experiment and associated probabilities are shown in Table 3-1. For example, because of the independence, the probability of the outcome that the first wafer tested passes and the second wafer tested fails, denoted as *pf*, is

$$P(pf) = 0.8(0.2) = 0.16$$

The random variable X is defined to be equal to the number of wafers that pass. The last column of the table shows the values of X that are assigned to each outcome in the experiment.

EXAMPLE 3-3

Define the random variable X to be the number of contamination particles on a wafer in semiconductor manufacturing. Although wafers possess a number of characteristics, the random variable X summarizes the wafer only in terms of the number of particles.

The possible values of X are integers from zero up to some large value that represents the maximum number of particles that can be found on one of the wafers. If this maximum number is very large, we might simply assume that the range of X is the set of integers from zero to infinity.

Note that more than one random variable can be defined on a sample space. In Example 3-3, we might define the random variable Y to be the number of chips from a wafer that fail the final test.

Table 3-1 Wafer Tests

Outcome			
Wafer 1	Wafer 2	Probability	x
Pass	Pass	0.64	2
Fail	Pass	0.16	1
Pass	Fail	0.16	1
Fail	Fail	0.04	0

EXERCISES FOR SECTION 3-1

For each of the following exercises, determine the range (possible values) of the random variable.

3-1. The random variable is the number of nonconforming solder connections on a printed circuit board with 1000 connections.

3-2. In a voice communication system with 50 lines, the random variable is the number of lines in use at a particular time.

3-3. An electronic scale that displays weights to the nearest pound is used to weigh packages. The display shows only five digits. Any weight greater than the display can indicate is shown as 99999. The random variable is the displayed weight.

3-4. A batch of 500 machined parts contains 10 that do not conform to customer requirements. The random variable is the number of parts in a sample of 5 parts that do not conform to customer requirements.

3-5. A batch of 500 machined parts contains 10 that do not conform to customer requirements. Parts are selected successively, without replacement, until a nonconforming part is obtained. The random variable is the number of parts selected.

3-6. The random variable is the moisture content of a lot of raw material, measured to the nearest percentage point.

3-7. The random variable is the number of surface flaws in a large coil of galvanized steel.

3-8. The random variable is the number of computer clock cycles required to complete a selected arithmetic calculation.

3-9. An order for an automobile can select the base model or add any number of 15 options. The random variable is the number of options selected in an order.

3-10. Wood paneling can be ordered in thicknesses of 1/8, 1/4, or 3/8 inch. The random variable is the total thickness of paneling in two orders.

3-11. A group of 10,000 people are tested for a gene called Ifi202 that has been found to increase the risk for lupus. The random variable is the number of people who carry the gene.

3-12. In an acid-base titration, the milliliters of base that are needed to reach equivalence are measured to the nearest milliliter between 0.1 and 0.15 liters (inclusive).

3-13. The number of mutations in a nucleotide sequence of length 40,000 in a DNA strand after exposure to radiation is measured. Each nucleotide may be mutated.

3-2 PROBABILITY DISTRIBUTIONS AND PROBABILITY MASS FUNCTIONS

Random variables are so important in random experiments that sometimes we essentially ignore the original sample space of the experiment and focus on the probability distribution of the random variable. For example, in Example 3-1, our analysis might focus exclusively on the integers $\{0, 1, \ldots, 48\}$ in the range of X. In Example 3-2, we might summarize the random experiment in terms of the three possible values of X, namely $\{0, 1, 2\}$. In this manner, a random variable can simplify the description and analysis of a random experiment.

The **probability distribution** of a random variable X is a description of the probabilities associated with the possible values of X. For a discrete random variable, the distribution is often specified by just a list of the possible values along with the probability of each. In some cases, it is convenient to express the probability in terms of a formula.

EXAMPLE 3-4
Digital Channel

There is a chance that a bit transmitted through a digital transmission channel is received in error. Let X equal the number of bits in error in the next four bits transmitted. The possible values for X are $\{0, 1, 2, 3, 4\}$. Based on a model for the errors that is presented in the following section, probabilities for these values will be determined. Suppose that the probabilities are

$$P(X = 0) = 0.6561 \quad P(X = 1) = 0.2916 \quad P(X = 2) = 0.0486$$

$$P(X = 3) = 0.0036 \quad P(X = 4) = 0.0001$$

The probability distribution of X is specified by the possible values along with the probability of each. A graphical description of the probability distribution of X is shown in Fig. 3-1.

Suppose a loading on a long, thin beam places mass only at discrete points. See Fig. 3-2. The loading can be described by a function that specifies the mass at each of the discrete

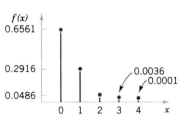

Figure 3-1 Probability distribution for bits in error.

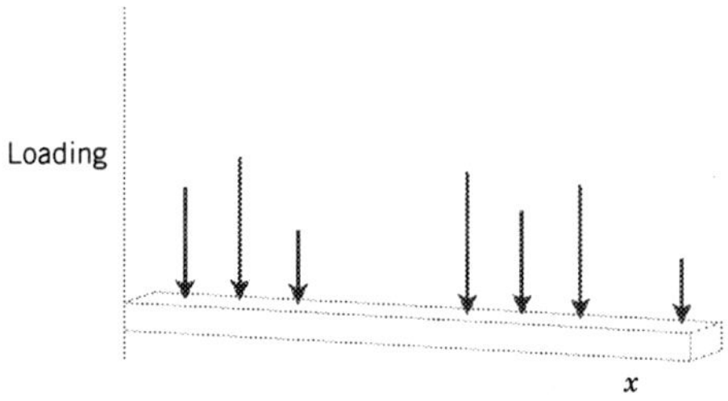

Figure 3-2 Loadings at discrete points on a long, thin beam.

points. Similarly, for a discrete random variable X, its distribution can be described by a function that specifies the probability at each of the possible discrete values for X.

Probability Mass Function

For a discrete random variable X with possible values $x_1, x_2, \ldots, x_n$, a **probability mass function** is a function such that

(1) $f(x_i) \geq 0$

(2) $\sum_{i=1}^{n} f(x_i) = 1$

(3) $f(x_i) = P(X = x_i)$ (3-1)

For example, in Example 3-4, $f(0) = 0.6561, f(1) = 0.2916, f(2) = 0.0486, f(3) = 0.0036$, and $f(4) = 0.0001$. Check that the sum of the probabilities in Example 3-4 is 1.

EXAMPLE 3-5 Wafer Contamination

Let the random variable X denote the number of semiconductor wafers that need to be analyzed in order to detect a large particle of contamination. Assume that the probability that a wafer contains a large particle is 0.01 and that the wafers are independent. Determine the probability distribution of X.

Let p denote a wafer in which a large particle is present, and let a denote a wafer in which it is absent. The sample space of the experiment is infinite, and it can be represented as all possible sequences that start with a string of a's and end with p. That is,

$$s = \{p, ap, aap, aaap, aaaap, aaaaap, \text{ and so forth}\}$$

Consider a few special cases. We have $P(X = 1) = P(p) = 0.01$. Also, using the independence assumption

$$P(X = 2) = P(ap) = 0.99(0.01) = 0.0099$$

A general formula is

$$P(X = x) = P(\underbrace{aa \ldots ap}_{(x-1)a\text{'s}}) = 0.99^{x-1}(0.01), \quad \text{for } x = 1, 2, 3, \ldots$$

Describing the probabilities associated with X in terms of this formula is the simplest method of describing the distribution of X in this example. Clearly $f(x) \geq 0$. The fact that the sum of the probabilities is 1 is left as an exercise. This is an example of a geometric random variable, and details are provided later in this chapter.

EXERCISES FOR SECTION 3-2

3-14. The sample space of a random experiment is $\{a, b, c, d, e, f\}$, and each outcome is equally likely. A random variable is defined as follows:

outcome	a	b	c	d	e	f
x	0	0	1.5	1.5	2	3

Determine the probability mass function of X. Use the probability mass function to determine the following probabilities:

(a) $P(X = 1.5)$ (b) $P(0.5 < X < 2.7)$
(c) $P(X > 3)$ (d) $P(0 \leq X < 2)$
(e) $P(X = 0$ or $X = 2)$

For Exercises 3-15 to 3-18 verify that the following functions are probability mass functions, and determine the requested probabilities.

3-15.

x	-2	-1	0	1	2
f(x)	1/8	2/8	2/8	2/8	1/8

(a) $P(X \leq 2)$ (b) $P(X > -2)$
(c) $P(-1 \leq X \leq 1)$ (d) $P(X \leq -1$ or $X = 2)$

3-16. $f(x) = (8/7)(1/2)^x$, $x = 1, 2, 3$
(a) $P(X \leq 1)$ (b) $P(X > 1)$
(c) $P(2 < X < 6)$ (d) $P(X \leq 1$ or $X > 1)$

3-17. $f(x) = \dfrac{2x + 1}{25}$, $x = 0, 1, 2, 3, 4$
(a) $P(X = 4)$ (b) $P(X \leq 1)$
(c) $P(2 \leq X < 4)$ (d) $P(X > -10)$

3-18. $f(x) = (3/4)(1/4)^x$, $x = 0, 1, 2, \ldots$
(a) $P(X = 2)$ (b) $P(X \leq 2)$
(c) $P(X > 2)$ (d) $P(X \geq 1)$

3-19. An article in *Knee Surgery, Sports Traumatology, Arthroscopy*, "Arthroscopic meniscal repair with an absorbable screw: results and surgical technique," (2005, Vol. 13, pp. 273–279) cites a success rate more than 90% for meniscal tears with a rim width of less than 3 mm, but only a 67% success rate for tears from 3–6 mm. If you are unlucky enough to suffer a meniscal tear of less than 3 mm on your left knee, and one of width 3–6 mm on your right knee, what is the probability mass function of the number of successful surgeries? Assume the surgeries are independent.

3-20. An optical inspection system is to distinguish among different part types. The probability of a correct classification of any part is 0.98. Suppose that three parts are inspected and that the classifications are independent. Let the random variable X denote the number of parts that are correctly classified. Determine the probability mass function of X.

3-21. In a semiconductor manufacturing process, three wafers from a lot are tested. Each wafer is classified as *pass* or *fail*. Assume that the probability that a wafer passes the test is 0.8 and that wafers are independent. Determine the probability mass function of the number of wafers from a lot that pass the test.

3-22. The space shuttle flight control system called PASS (Primary Avionics Software Set) uses four independent computers working in parallel. At each critical step, the computers "vote" to determine the appropriate step. The probability that a computer will ask for a roll to the left when a roll to the right is appropriate is 0.0001. Let X denote the number of computers that vote for a left roll when a right roll is appropriate. What is the probability mass function of X?

3-23. A disk drive manufacturer estimates that in five years a storage device with 1 terabyte of capacity will sell with probability 0.5, a storage device with 500 gigabytes capacity will sell with a probability 0.3, and a storage device with 100 gigabytes capacity will sell with probability 0.2. The revenue associated with the sales in that year are estimated to be $50 million, $25 million, and $10 million, respectively. Let X be the revenue of storage devices during that year. Determine the probability mass function of X.

3-24. Marketing estimates that a new instrument for the analysis of soil samples will be very successful, moderately successful, or unsuccessful, with probabilities 0.3, 0.6, and 0.1, respectively. The yearly revenue associated with a very successful, moderately successful, or unsuccessful product is $10 million, $5 million, and $1 million, respectively. Let the random variable X denote the yearly revenue of the product. Determine the probability mass function of X.

3-25. The distributor of a machine for cytogenics has developed a new model. The company estimates that when it is introduced into the market, it will be very successful with a probability 0.6, moderately successful with a probability 0.3, and not successful with probability 0.1. The estimated yearly profit associated with the model being very successful is $15 million and being moderately successful is $5 million; not successful would result in a loss of $500,000. Let X be the yearly profit of the new model. Determine the probability mass function of X.

3-26. An assembly consists of two mechanical components. Suppose that the probabilities that the first and second components meet specifications are 0.95 and 0.98. Assume that the components are independent. Determine the probability mass function of the number of components in the assembly that meet specifications.

3-27. An assembly consists of three mechanical components. Suppose that the probabilities that the first, second, and third components meet specifications are 0.95, 0.98, and 0.99. Assume that the components are independent. Determine the probability mass function of the number of components in the assembly that meet specifications.

72   CHAPTER 3 DISCRETE RANDOM VARIABLES AND PROBABILITY DISTRIBUTIONS

3-3 CUMULATIVE DISTRIBUTION FUNCTIONS

EXAMPLE 3-6
Digital Channel

In Example 3-4, we might be interested in the probability of three or fewer bits being in error. This question can be expressed as $P(X \leq 3)$.

The event that $\{X \leq 3\}$ is the union of the events $\{X = 0\}, \{X = 1\}, \{X = 2\}$, and $\{X = 3\}$. Clearly, these three events are mutually exclusive. Therefore,

$$P(X \leq 3) = P(X = 0) + P(X = 1) + P(X = 2) + P(X = 3)$$
$$= 0.6561 + 0.2916 + 0.0486 + 0.0036 = 0.9999$$

This approach can also be used to determine

$$P(X = 3) = P(X \leq 3) - P(X \leq 2) = 0.0036$$

Example 3-6 shows that it is sometimes useful to be able to provide **cumulative probabilities** such as $P(X \leq x)$ and that such probabilities can be used to find the probability mass function of a random variable. Therefore, using cumulative probabilities is an alternate method of describing the probability distribution of a random variable.

In general, for any discrete random variable with possible values $x_1, x_2, \ldots, x_n$, the events $\{X = x_1\}, \{X = x_2\}, \ldots, \{X = x_n\}$ are mutually exclusive. Therefore, $P(X \leq x) = \sum_{x_i \leq x} f(x_i)$.

Cumulative Distribution Function

The **cumulative distribution function** of a discrete random variable X, denoted as $F(x)$, is

$$F(x) = P(X \leq x) = \sum_{x_i \leq x} f(x_i)$$

For a discrete random variable X, $F(x)$ satisfies the following properties.

(1) $F(x) = P(X \leq x) = \sum_{x_i \leq x} f(x_i)$
(2) $0 \leq F(x) \leq 1$
(3) If $x \leq y$, then $F(x) \leq F(y)$ (3-2)

Like a probability mass function, a cumulative distribution function provides probabilities. Notice that even if the random variable X can only assume integer values, the cumulative distribution function can be defined at noninteger values. In Example 3-6, $F(1.5) = P(X \leq 1.5) = P\{X = 0\} + P(X = 1) = 0.6561 + 0.2916 = 0.9477$. Properties (1) and (2) of a cumulative distribution function follow from the definition. Property (3) follows from the fact that if $x \leq y$, the event that $\{X \leq x\}$ is contained in the event $\{X \leq y\}$.

The next example shows how the cumulative distribution function can be used to determine the probability mass function of a discrete random variable.

EXAMPLE 3-7
Cumulative Distribution Function

Determine the probability mass function of X from the following cumulative distribution function:

$$F(x) = \begin{cases} 0 & x < -2 \\ 0.2 & -2 \leq x < 0 \\ 0.7 & 0 \leq x < 2 \\ 1 & 2 \leq x \end{cases}$$

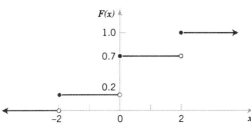

Figure 3-3 Cumulative distribution function for Example 3-7.

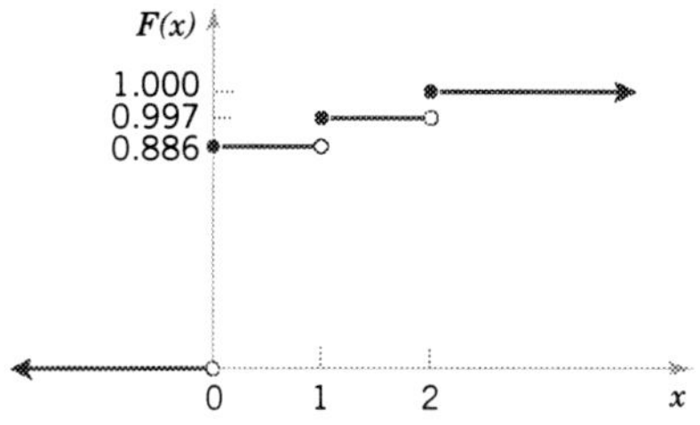

Figure 3-4 Cumulative distribution function for Example 3-8.

Figure 3-3 displays a plot of $F(x)$. From the plot, the only points that receive nonzero probability are -2, 0, and 2. The probability mass function at each point is the change in the cumulative distribution function at the point. Therefore,

$$f(-2) = 0.2 - 0 = 0.2 \qquad f(0) = 0.7 - 0.2 = 0.5 \qquad f(2) = 1.0 - 0.7 = 0.3$$

EXAMPLE 3-8
Sampling without Replacement

Suppose that a day's production of 850 manufactured parts contains 50 parts that do not conform to customer requirements. Two parts are selected at random, without replacement, from the batch. Let the random variable X equal the number of nonconforming parts in the sample. What is the cumulative distribution function of X?

The question can be answered by first finding the probability mass function of X.

$$P(X = 0) = \frac{800}{850} \cdot \frac{799}{849} = 0.886$$

$$P(X = 1) = 2 \cdot \frac{800}{850} \cdot \frac{50}{849} = 0.111$$

$$P(X = 2) = \frac{50}{850} \cdot \frac{49}{849} = 0.003$$

Therefore,

$$F(0) = P(X \leq 0) = 0.886$$

$$F(1) = P(X \leq 1) = 0.886 + 0.111 = 0.997$$

$$F(2) = P(X \leq 2) = 1$$

The cumulative distribution function for this example is graphed in Fig. 3-4. Note that $F(x)$ is defined for all x from $-\infty < x < \infty$ and not only for 0, 1, and 2.

EXERCISES FOR SECTION 3-3

3-28. Determine the cumulative distribution function of the random variable in Exercise 3-14.

3-29. Determine the cumulative distribution function for the random variable in Exercise 3-15; also determine the following probabilities:
(a) $P(X \leq 1.25)$ (b) $P(X \leq 2.2)$
(c) $P(-1.1 < X \leq 1)$ (d) $P(X > 0)$

3-30. Determine the cumulative distribution function for the random variable in Exercise 3-16; also determine the following probabilities:
(a) $P(X < 1.5)$ (b) $P(X \leq 3)$
(c) $P(X > 2)$ (d) $P(1 < X \leq 2)$

3-31. Determine the cumulative distribution function for the random variable in Exercise 3-21.

3-32. Determine the cumulative distribution function for the random variable in Exercise 3-22.

3-33. Determine the cumulative distribution function for the random variable in Exercise 3-23.

3-34. Determine the cumulative distribution function for the variable in Exercise 3-24.

Verify that the following functions are cumulative distribution functions, and determine the probability mass function and the requested probabilities.

3-35.
$$F(x) = \begin{cases} 0 & x < 1 \\ 0.5 & 1 \leq x < 3 \\ 1 & 3 \leq x \end{cases}$$
(a) $P(X \leq 3)$ (b) $P(X \leq 2)$
(c) $P(1 \leq X \leq 2)$ (d) $P(X > 2)$

3-36. Errors in an experimental transmission channel are found when the transmission is checked by a certifier that detects missing pulses. The number of errors found in an eight-bit byte is a random variable with the following distribution:

$$F(x) = \begin{cases} 0 & x < 1 \\ 0.7 & 1 \leq x < 4 \\ 0.9 & 4 \leq x < 7 \\ 1 & 7 \leq x \end{cases}$$

Determine each of the following probabilities:
(a) $P(X \leq 4)$ (b) $P(X > 7)$
(c) $P(X \leq 5)$ (d) $P(X > 4)$
(e) $P(X \leq 2)$

3-37.
$$F(x) = \begin{cases} 0 & x < -10 \\ 0.25 & -10 \leq x < 30 \\ 0.75 & 30 \leq x < 50 \\ 1 & 50 \leq x \end{cases}$$
(a) $P(X \leq 50)$ (b) $P(X \leq 40)$
(c) $P(40 \leq X \leq 60)$ (d) $P(X < 0)$
(e) $P(0 \leq X < 10)$ (f) $P(-10 < X < 10)$

3-38. The thickness of wood paneling (in inches) that a customer orders is a random variable with the following cumulative distribution function:

$$F(x) = \begin{cases} 0 & x < 1/8 \\ 0.2 & 1/8 \leq x < 1/4 \\ 0.9 & 1/4 \leq x < 3/8 \\ 1 & 3/8 \leq x \end{cases}$$

Determine the following probabilities:
(a) $P(X \leq 1/18)$ (b) $P(X \leq 1/4)$
(c) $P(X \leq 5/16)$ (d) $P(X > 1/4)$
(e) $P(X \leq 1/2)$

3-4 MEAN AND VARIANCE OF A DISCRETE RANDOM VARIABLE

Two numbers are often used to summarize a probability distribution for a random variable X. The mean is a measure of the center or middle of the probability distribution, and the variance is a measure of the dispersion, or variability in the distribution. These two measures do not uniquely identify a probability distribution. That is, two different distributions can have the same mean and variance. Still, these measures are simple, useful summaries of the probability distribution of X.

Mean and Variance

The mean or expected value of the discrete random variable X, denoted as μ or $E(X)$, is

$$\mu = E(X) = \sum_x x f(x) \qquad (3\text{-}3)$$

The variance of X, denoted as σ^2 or $V(X)$, is

$$\sigma^2 = V(X) = E(X - \mu)^2 = \sum_x (x - \mu)^2 f(x) = \sum_x x^2 f(x) - \mu^2$$

The standard deviation of X is $\sigma = \sqrt{\sigma^2}$.

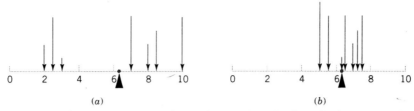

(a)   (b)

Figure 3-5 A probability distribution can be viewed as a loading with the mean equal to the balance point. Parts (a) and (b) illustrate equal means, but Part (a) illustrates a larger variance.

The mean of a discrete random variable X is a weighted average of the possible values of X, with weights equal to the probabilities. If $f(x)$ is the probability mass function of a loading on a long, thin beam, $E(X)$ is the point at which the beam balances. Consequently, $E(X)$ describes the "center" of the distribution of X in a manner similar to the balance point of a loading. See Fig. 3-5.

The variance of a random variable X is a measure of dispersion or scatter in the possible values for X. The variance of X uses weight $f(x)$ as the multiplier of each possible squared deviation $(x - \mu)^2$. Figure 3-5 illustrates probability distributions with equal means but different variances. Properties of summations and the definition of μ can be used to show the equality of the formulas for variance.

$$V(X) = \sum_x (x - \mu)^2 f(x) = \sum_x x^2 f(x) - 2\mu \sum_x x f(x) + \mu^2 \sum_x f(x)$$

$$= \sum_x x^2 f(x) - 2\mu^2 + \mu^2 = \sum_x x^2 f(x) - \mu^2$$

Either formula for $V(x)$ can be used. Figure 3-6 illustrates that two probability distributions can differ even though they have identical means and variances.

EXAMPLE 3-9
Digital Channel

In Example 3-4, there is a chance that a bit transmitted through a digital transmission channel is received in error. Let X equal the number of bits in error in the next four bits transmitted. The possible values for X are $\{0, 1, 2, 3, 4\}$. Based on a model for the errors that is presented in the following section, probabilities for these values will be determined. Suppose that the probabilities are

$P(X = 0) = 0.6561$ $P(X = 2) = 0.0486$ $P(X = 4) = 0.0001$
$P(X = 1) = 0.2916$ $P(X = 3) = 0.0036$

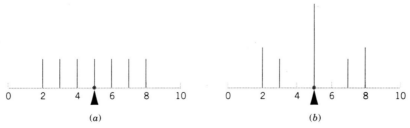

(a)   (b)

Figure 3-6 The probability distributions illustrated in Parts (a) and (b) differ even though they have equal means and equal variances.

Now

$$\mu = E(X) = 0f(0) + 1f(1) + 2f(2) + 3f(3) + 4f(4)$$
$$= 0(0.6561) + 1(0.2916) + 2(0.0486) + 3(0.0036) + 4(0.0001)$$
$$= 0.4$$

Although X never assumes the value 0.4, the weighted average of the possible values is 0.4.
To calculate $V(X)$, a table is convenient.

x	$x - 0.4$	$(x - 0.4)^2$	$f(x)$	$f(x)(x - 0.4)^2$
0	−0.4	0.16	0.6561	0.104976
1	0.6	0.36	0.2916	0.104976
2	1.6	2.56	0.0486	0.124416
3	2.6	6.76	0.0036	0.024336
4	3.6	12.96	0.0001	0.001296

$$V(X) = \sigma^2 = \sum_{i=1}^{5} f(x_i)(x_i - 0.4)^2 = 0.36$$

The alternative formula for variance could also be used to obtain the same result.

EXAMPLE 3-10
Marketing

Two new product designs are to be compared on the basis of revenue potential. Marketing feels that the revenue from design A can be predicted quite accurately to be $3 million. The revenue potential of design B is more difficult to assess. Marketing concludes that there is a probability of 0.3 that the revenue from design B will be $7 million, but there is a 0.7 probability that the revenue will be only $2 million. Which design do you prefer?

Let X denote the revenue from design A. Because there is no uncertainty in the revenue from design A, we can model the distribution of the random variable X as $3 million with probability 1. Therefore, $E(X) = \$3$ million.

Let Y denote the revenue from design B. The expected value of Y in millions of dollars is

$$E(Y) = \$7(0.3) + \$2(0.7) = \$3.5$$

Because $E(Y)$ exceeds $E(X)$, we might prefer design B. However, the variability of the result from design B is larger. That is,

$$\sigma^2 = (7 - 3.5)^2(0.3) + (2 - 3.5)^2(0.7)$$
$$= 5.25 \text{ millions of dollars squared}$$

Because the units of the variables in this example are millions of dollars, and because the variance of a random variable squares the deviations from the mean, the units of σ^2 are millions of dollars squared. These units make interpretation difficult.

Because the units of standard deviation are the same as the units of the random variable, the standard deviation σ is easier to interpret. In this example, we can summarize our results as "the average deviation of Y from its mean is $2.29 million."

EXAMPLE 3-11
Messages

The number of messages sent per hour over a computer network has the following distribution:

x = number of messages	10	11	12	13	14	15
f(x)	0.08	0.15	0.30	0.20	0.20	0.07

Determine the mean and standard deviation of the number of messages sent per hour.

$$E(X) = 10(0.08) + 11(0.15) + \cdots + 15(0.07) = 12.5$$
$$V(X) = 10^2(0.08) + 11^2(0.15) + \cdots + 15^2(0.07) - 12.5^2 = 1.85$$
$$\sigma = \sqrt{V(X)} = \sqrt{1.85} = 1.36$$

The variance of a random variable X can be considered to be the expected value of a specific function of X, namely, $h(X) = (X - \mu)^2$. In general, the expected value of any function $h(X)$ of a discrete random variable is defined in a similar manner.

Expected Value of a Function of a Discrete Random Variable

If X is a discrete random variable with probability mass function $f(x)$,

$$E[h(X)] = \sum_x h(x) f(x) \tag{3-4}$$

EXAMPLE 3-12
Digital Channel

In Example 3-9, X is the number of bits in error in the next four bits transmitted. What is the expected value of the square of the number of bits in error? Now, $h(X) = X^2$. Therefore,

$$E[h(X)] = 0^2 \times 0.6561 + 1^2 \times 0.2916 + 2^2 \times 0.0486$$
$$+ 3^2 \times 0.0036 + 4^2 \times 0.0001 = 0.52$$

In the previous example, the expected value of X^2 does not equal $E(X)$ squared. However, in the special case that $h(X) = aX + b$ for any constants a and b, $E[h(X)] = aE(X) + b$. This can be shown from the properties of sums in the definition in Equation 3-4.

EXERCISES FOR SECTION 3-4

3-39. If the range of X is the set $\{0, 1, 2, 3, 4\}$ and $P(X = x) = 0.2$ determine the mean and variance of the random variable.

3-40. Determine the mean and variance of the random variable in Exercise 3-14.

3-41. Determine the mean and variance of the random variable in Exercise 3-15.

3-42. Determine the mean and variance of the random variable in Exercise 3-16.

3-43. Determine the mean and variance of the random variable in Exercise 3-17.

3-44. Determine the mean and variance of the random variable in Exercise 3-18.

3-45. Determine the mean and variance of the random variable in Exercise 3-19.

3-46. Determine the mean and variance of the random variable in Exercise 3-20.

3-47. The range of the random variable X is $[0, 1, 2, 3, x]$, where x is unknown. If each value is equally likely and the mean of X is 6, determine x.

3-48. In a NiCd battery, a fully charged cell is composed of Nickelic Hydroxide. Nickel is an element that has multiple

78 CHAPTER 3 DISCRETE RANDOM VARIABLES AND PROBABILITY DISTRIBUTIONS

oxidation states. Assume the following proportions of the states:

Nickel Charge	Proportions Found
0	0.17
+2	0.35
+3	0.33
+4	0.15

(a) Determine the cumulative distribution function of nickel charge.
(b) Determine the mean and variance of the nickel charge.

3-49. The space shuttle flight control system called PASS (Primary Avionics Software Set) uses four independent computers working in parallel. At each critical step, the computers "vote" to determine the appropriate step. The probability that a computer will ask for a roll to the left when a roll to the right is appropriate is 0.0001. Let X denote the number of computers that vote for a left roll when a right roll is appropriate. What is the mean and variance of X?

3-50. Trees are subjected to different levels of carbon dioxide atmosphere with 6% of the trees in a minimal growth condition at 350 parts per million (ppm), 10% at 450 ppm (slow growth), 47% at 550 ppm (moderate growth), and 37% at 650 ppm (rapid growth). What is the mean and standard deviation of the carbon dioxide atmosphere (in ppm) for these trees in ppm?

3-51. An article in the *Journal of Database Management* "Experimental Study of a Self-Tuning Algorithm for DBMS Buffer Pools," (2005, Vol. 16, pp. 1–20) provided the workload used in the TPC-C OLTP (Transaction Processing Performance Council's Version C On-Line Transaction Processing) benchmark, which simulates a typical order entry application.

The frequency of each type of transaction (in the second column) can be used as the percentage of each type of transaction. The average number of *selects* operations required for each type transaction is shown.

(a) Determine the mean and standard deviation of the number of *selects* operations for a transaction from the distribution of types shown in the table.
(b) Determine the mean and standard deviation of the total number of operations (*selects*, *updates*, . . . , and *joins*) for a transaction from the distribution of types shown in the table.

Average Frequencies and Operations in TPC-C

Transaction	Frequency	Selects	Updates	Inserts	Deletes	Non-Unique Selects	Joins
New Order	43	23	11	12	0	0	0
Payment	44	4.2	3	1	0	0.6	0
Order Status	4	11.4	0	0	0	0.6	0
Delivery	5	130	120	0	10	0	0
Stock Level	4	0	0	0	0	0	1

3-5 DISCRETE UNIFORM DISTRIBUTION

The simplest discrete random variable is one that assumes only a finite number of possible values, each with equal probability. A random variable X that assumes each of the values $x_1, x_2, \ldots, x_n$, with equal probability $1/n$, is frequently of interest.

Discrete Uniform Distribution

A random variable X has a discrete uniform distribution if each of the n values in its range, say, $x_1, x_2, \ldots, x_n$, has equal probability. Then,

$$f(x_i) = 1/n \qquad (3\text{-}5)$$

Figure 3-7 Probability mass function for a discrete uniform random variable.

EXAMPLE 3-13

The first digit of a part's serial number is equally likely to be any one of the digits 0 through 9. If one part is selected from a large batch and X is the first digit of the serial number, X has a discrete uniform distribution with probability 0.1 for each value in $R = \{0, 1, 2, \ldots, 9\}$. That is,

$$f(x) = 0.1$$

for each value in R. The probability mass function of X is shown in Fig. 3-7.

Suppose the range of the discrete random variable X is the consecutive integers $a, a + 1, a + 2, \ldots, b$, for $a \leq b$. The range of X contains $b - a + 1$ values each with probability $1/(b - a + 1)$. Now,

$$\mu = \sum_{k=a}^{b} k \left(\frac{1}{b - a + 1} \right)$$

The algebraic identity $\sum_{k=a}^{b} k = \dfrac{b(b + 1) - (a - 1)a}{2}$ can be used to simplify the result to $\mu = (b + a)/2$. The derivation of the variance is left as an exercise.

Mean and Variance

Suppose X is a discrete uniform random variable on the consecutive integers $a, a + 1, a + 2, \ldots, b$, for $a \leq b$. The mean of X is

$$\mu = E(X) = \frac{b + a}{2}$$

The variance of X is

$$\sigma^2 = \frac{(b - a + 1)^2 - 1}{12} \qquad (3\text{-}6)$$

EXAMPLE 3-14
Number of Voice Lines

As in Example 3-1, let the random variable X denote the number of the 48 voice lines that are in use at a particular time. Assume that X is a discrete uniform random variable with a range of 0 to 48. Then,

$$E(X) = (48 + 0)/2 = 24$$

and

$$\sigma = \{[(48 - 0 + 1)^2 - 1]/12\}^{1/2} = 14.14$$

Equation 3-6 is more useful than it might first appear. If all the values in the range of a random variable X are multiplied by a constant (without changing any probabilities), the mean and standard deviation of X are multiplied by the constant. You are asked to verify this result in an exercise. Because the variance of a random variable is the square of the standard deviation, the variance of X is multiplied by the constant squared. More general results of this type are discussed in Chapter 5.

EXAMPLE 3-15
Proportion of
Voice Lines

Let the random variable Y denote the proportion of the 48 voice lines that are in use at a particular time, and X denotes the number of lines that are in use at a particular time. Then, $Y = X/48$. Therefore,

$$E(Y) = E(X)/48 = 0.5$$

and

$$V(Y) = V(X)/48^2 = 0.087$$

EXERCISES FOR SECTION 3-5

3-52. Let the random variable X have a discrete uniform distribution on the integers $0 \leq x \leq 100$. Determine the mean and variance of X.

3-53. Let the random variable X have a discrete uniform distribution on the integers $1 \leq x \leq 3$. Determine the mean and variance of X.

3-54. Thickness measurements of a coating process are made to the nearest hundredth of a millimeter. The thickness measurements are uniformly distributed with values 0.15, 0.16, 0.17, 0.18, and 0.19. Determine the mean and variance of the coating thickness for this process.

3-55. Product codes of 2, 3, or 4 letters are equally likely. What is the mean and standard deviation of the number of letters in 100 codes?

3-56. The lengths of plate glass parts are measured to the nearest tenth of a millimeter. The lengths are uniformly distributed, with values at every tenth of a millimeter starting at 590.0 and continuing through 590.9. Determine the mean and variance of lengths.

3-57. Assume that the wavelengths of photosynthetically active radiations (PAR) are uniformly distributed at integer nanometers in the red spectrum from 675 to 700 nm.

(a) What is the mean and variance of the wavelength distribution for this radiation?
(b) If wavelengths are uniformly distributed at integer nanometers from 75 to 100 nanometers, how does the mean and variance of the wavelength distribution compare to the previous part? Explain.

3-58. The probability of an operator entering alphanumeric data incorrectly into a field in a database is equally likely. The random variable X is the number of fields on a data entry form with an error. The data entry form has 28 fields. Is X a discrete uniform random variable? Why or why not.

3-59. Suppose that X has a discrete uniform distribution on the integers 0 through 9. Determine the mean, variance, and standard deviation of the random variable $Y = 5X$ and compare to the corresponding results for X.

3-60. Show that for a discrete uniform random variable X, if each of the values in the range of X is multiplied by the constant c, the effect is to multiply the mean of X by c and the variance of X by c^2. That is, show that $E(cX) = cE(X)$ and $V(cX) = c^2V(X)$.

3-6 BINOMIAL DISTRIBUTION

Consider the following random experiments and random variables:

1. Flip a coin 10 times. Let X = number of heads obtained.
2. A worn machine tool produces 1% defective parts. Let X = number of defective parts in the next 25 parts produced.
3. Each sample of air has a 10% chance of containing a particular rare molecule. Let X = the number of air samples that contain the rare molecule in the next 18 samples analyzed.

4. Of all bits transmitted through a digital transmission channel, 10% are received in error. Let $X =$ the number of bits in error in the next five bits transmitted.
5. A multiple choice test contains 10 questions, each with four choices, and you guess at each question. Let $X =$ the number of questions answered correctly.
6. In the next 20 births at a hospital, let $X =$ the number of female births.
7. Of all patients suffering a particular illness, 35% experience improvement from a particular medication. In the next 100 patients administered the medication, let $X =$ the number of patients who experience improvement.

These examples illustrate that a general probability model that includes these experiments as particular cases would be very useful.

Each of these random experiments can be thought of as consisting of a series of repeated, random trials: 10 flips of the coin in experiment 1, the production of 25 parts in experiment 2, and so forth. The random variable in each case is a count of the number of trials that meet a specified criterion. The outcome from each trial either meets the criterion that X counts or it does not; consequently, each trial can be summarized as resulting in either a success or a failure. For example, in the multiple choice experiment, for each question, only the choice that is correct is considered a success. Choosing any one of the three incorrect choices results in the trial being summarized as a failure.

The terms *success* and *failure* are just labels. We can just as well use A and B or 0 or 1. Unfortunately, the usual labels can sometimes be misleading. In experiment 2, because X counts defective parts, the production of a defective part is called a success.

A trial with only two possible outcomes is used so frequently as a building block of a random experiment that it is called a **Bernoulli trial**. It is usually assumed that the trials that constitute the random experiment are **independent**. This implies that the outcome from one trial has no effect on the outcome to be obtained from any other trial. Furthermore, it is often reasonable to assume that the **probability of a success in each trial is constant.** In the multiple choice experiment, if the test taker has no knowledge of the material and just guesses at each question, we might assume that the probability of a correct answer is 1/4 *for each question.*

EXAMPLE 3-16
Digital Channel

The chance that a bit transmitted through a digital transmission channel is received in error is 0.1. Also, assume that the transmission trials are independent. Let $X =$ the number of bits in error in the next four bits transmitted. Determine $P(X = 2)$.

Let the letter E denote a bit in error, and let the letter O denote that the bit is okay, that is, received without error. We can represent the outcomes of this experiment as a list of four letters that indicate the bits that are in error and those that are okay. For example, the outcome $OEOE$ indicates that the second and fourth bits are in error and the other two bits are okay. The corresponding values for x are

Outcome	x	Outcome	x
OOOO	0	EOOO	1
OOOE	1	EOOE	2
OOEO	1	EOEO	2
OOEE	2	EOEE	3
OEOO	1	EEOO	2
OEOE	2	EEOE	3
OEEO	2	EEEO	3
OEEE	3	EEEE	4

The event that $X = 2$ consists of the six outcomes:

$$\{EEOO, EOEO, EOOE, OEEO, OEOE, OOEE\}$$

Using the assumption that the trials are independent, the probability of $\{EEOO\}$ is

$$P(EEOO) = P(E)P(E)P(O)P(O) = (0.1)^2(0.9)^2 = 0.0081$$

Also, any one of the six mutually exclusive outcomes for which $X = 2$ has the same probability of occurring. Therefore,

$$P(X = 2) = 6(0.0081) = 0.0486$$

In general,

$$P(X = x) = \text{(number of outcomes that result in } x \text{ errors) times } (0.1)^x(0.9)^{4-x}$$

To complete a general probability formula, only an expression for the number of outcomes that contain x errors is needed. An outcome that contains x errors can be constructed by partitioning the four trials (letters) in the outcome into two groups. One group is of size x and contains the errors, and the other group is of size $n - x$ and consists of the trials that are okay. The number of ways of partitioning four objects into two groups, one of which is of size x, is $\binom{4}{x} = \dfrac{4!}{x!(4-x)!}$. Therefore, in this example

$$P(X = x) = \binom{4}{x}(0.1)^x(0.9)^{4-x}$$

Notice that $\binom{4}{2} = 4!/[2!\,2!] = 6$, as found above. The probability mass function of X was shown in Example 3-4 and Fig. 3-1.

The previous example motivates the following result.

Binomial Distribution

A random experiment consists of n Bernoulli trials such that

(1) The trials are independent

(2) Each trial results in only two possible outcomes, labeled as "success" and "failure"

(3) The probability of a success in each trial, denoted as p, remains constant

The random variable X that equals the number of trials that result in a success has a **binomial random variable** with parameters $0 < p < 1$ and $n = 1, 2, \ldots$. The probability mass function of X is

$$f(x) = \binom{n}{x} p^x(1-p)^{n-x} \quad x = 0, 1, \ldots, n \quad (3\text{-}7)$$

As in Example 3-16, $\binom{n}{x}$ equals the total number of different sequences of trials that contain x successes and $n - x$ failures. The total number of different sequences that contain x successes and $n - x$ failures times the probability of each sequence equals $P(X = x)$.

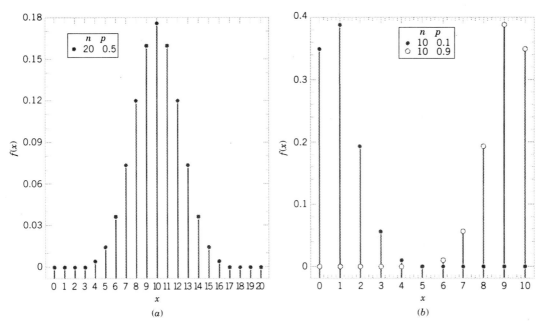

Figure 3-8 Binomial distributions for selected values of n and p.

The probability expression above is a very useful formula that can be applied in a number of examples. The name of the distribution is obtained from the *binomial expansion*. For constants a and b, the binomial expansion is

$$(a + b)^n = \sum_{k=0}^{n} \binom{n}{k} a^k b^{n-k}$$

Let p denote the probability of success on a single trial. Then, by using the binomial expansion with $a = p$ and $b = 1 - p$, we see that the sum of the probabilities for a binomial random variable is 1. Furthermore, because each trial in the experiment is classified into two outcomes, {success, failure}, the distribution is called a "bi"-nomial. A more general distribution, which includes the binomial as a special case, is the multinomial distribution and this is presented in Chapter 5.

Examples of binomial distributions are shown in Fig. 3-8. For a fixed n, the distribution becomes more symmetric as p increases from 0 to 0.5 or decreases from 1 to 0.5. For a fixed p, the distribution becomes more symmetric as n increases.

EXAMPLE 3-17 Several examples using the binomial coefficient $\binom{n}{x}$ follow.

$$\binom{10}{3} = 10!/[3!\,7!] = (10 \cdot 9 \cdot 8)/(3 \cdot 2) = 120$$

$$\binom{15}{10} = 15!/[10!\,5!] = (15 \cdot 14 \cdot 13 \cdot 12 \cdot 11)/(5 \cdot 4 \cdot 3 \cdot 2) = 3003$$

$$\binom{100}{4} = 100!/[4!\,96!] = (100 \cdot 99 \cdot 98 \cdot 97)/(4 \cdot 3 \cdot 2) = 3{,}921{,}225$$

Also recall that $0! = 1$.

EXAMPLE 3-18
Organic Pollution

Each sample of water has a 10% chance of containing a particular organic pollutant. Assume that the samples are independent with regard to the presence of the pollutant. Find the probability that in the next 18 samples, exactly 2 contain the pollutant.

Let $X =$ the number of samples that contain the pollutant in the next 18 samples analyzed. Then X is a binomial random variable with $p = 0.1$ and $n = 18$. Therefore,

$$P(X = 2) = \binom{18}{2}(0.1)^2(0.9)^{16}$$

Now $\binom{18}{2} = 18!/[2!\,16!] = 18(17)/2 = 153$. Therefore,

$$P(X = 2) = 153(0.1)^2(0.9)^{16} = 0.284$$

Determine the probability that at least four samples contain the pollutant. The requested probability is

$$P(X \geq 4) = \sum_{x=4}^{18} \binom{18}{x}(0.1)^x(0.9)^{18-x}$$

However, it is easier to use the complementary event,

$$P(X \geq 4) = 1 - P(X < 4) = 1 - \sum_{x=0}^{3} \binom{18}{x}(0.1)^x(0.9)^{18-x}$$

$$= 1 - [0.150 + 0.300 + 0.284 + 0.168] = 0.098$$

Determine the probability that $3 \leq X < 7$. Now

$$P(3 \leq X < 7) = \sum_{x=3}^{6} \binom{18}{x}(0.1)^x(0.9)^{18-x}$$

$$= 0.168 + 0.070 + 0.022 + 0.005$$

$$= 0.265$$

A table of cumulative binomial probabilities is provided in Appendix A and this can simplify some calculations. For example, the binomial distribution in Example 3-16 has $p = 0.1$ and $n = 4$. A probability such as $P(X = 2)$ can be calculated from the table as

$$P(X = 2) = P(X \leq 2) - P(X \leq 1) = 0.9963 - 0.9477 = 0.0486$$

and this agrees with the result obtained previously.

The mean and variance of a binomial random variable can be obtained from an analysis of the independent trials that comprise the binomial experiment. Define new random variables

$$X_i = \begin{cases} 1 \text{ if } i\text{th trial is a success} \\ 0 \text{ otherwise} \end{cases}$$

for $i = 1, 2, \ldots, n$. Then

$$X = X_1 + X_2 + \cdots + X_n$$

Also, it is easy to derive the mean and variance of each X_i as

$$E(X_i) = 1p + 0(1-p) = p$$

and

$$V(X_i) = (1-p)^2 p + (0-p)^2(1-p) = p(1-p)$$

Sums of random variables are discussed in Chapter 5 and there the intuitively reasonable result that

$$E(X) = E(X_1) + E(X_2) + \cdots + E(X_n)$$

is derived. Furthermore, for the independent trials of a binomial experiment it is also shown in Chapter 5 that

$$V(X) = V(X_1) + V(X_2) + \cdots + V(X_n)$$

Because $E(X_i) = p$ and $V(X_i) = p(1-p)$, we obtain the solution $E(X) = np$ and $V(X) = np(1-p)$.

Mean and Variance

If X is a binomial random variable with parameters p and n,

$$\mu = E(X) = np \quad \text{and} \quad \sigma^2 = V(X) = np(1-p) \tag{3-8}$$

EXAMPLE 3-19 For the number of transmitted bits received in error in Example 3-16, $n = 4$ and $p = 0.1$, so

$$E(X) = 4(0.1) = 0.4 \quad \text{and} \quad V(X) = 4(0.1)(0.9) = 0.36$$

and these results match those obtained from a direct calculation in Example 3-9.

EXERCISES FOR SECTION 3-6

3-61. For each scenario described below, state whether or not the binomial distribution is a reasonable model for the random variable and why. State any assumptions you make.

(a) A production process produces thousands of temperature transducers. Let X denote the number of nonconforming transducers in a sample of size 30 selected at random from the process.

(b) From a batch of 50 temperature transducers, a sample of size 30 is selected without replacement. Let X denote the number of nonconforming transducers in the sample.

(c) Four identical electronic components are wired to a controller that can switch from a failed component to one of the remaining spares. Let X denote the number of components that have failed after a specified period of operation.

(d) Let X denote the number of accidents that occur along the federal highways in Arizona during a one-month period.

(e) Let X denote the number of correct answers by a student taking a multiple choice exam in which a student can eliminate some of the choices as being incorrect in some questions and all of the incorrect choices in other questions.

(f) Defects occur randomly over the surface of a semiconductor chip. However, only 80% of defects can be found by testing. A sample of 40 chips with one defect each is tested. Let X denote the number of chips in which the test finds a defect.

(g) Reconsider the situation in part (f). Now, suppose the sample of 40 chips consists of chips with 1 and with 0 defects.

(h) A filling operation attempts to fill detergent packages to the advertised weight. Let X denote the number of detergent packages that are underfilled.

(i) Errors in a digital communication channel occur in bursts that affect several consecutive bits. Let X denote the number of bits in error in a transmission of 100,000 bits.

(j) Let X denote the number of surface flaws in a large coil of galvanized steel.

3-62. Let X be a binomial random variable with $p = 0.2$ and $n = 20$. Use the binomial table in Appendix A to determine the following probabilities.
(a) $P(X \leq 3)$ (b) $P(X > 10)$
(c) $P(X = 6)$ (d) $P(6 \leq X \leq 11)$

3-63. Let X be a binomial random variable with $p = 0.1$ and $n = 10$. Calculate the following probabilities from the binomial probability mass function and also from the binomial table in Appendix A and compare results.
(a) $P(X \leq 2)$ (b) $P(X > 8)$
(c) $P(X = 4)$ (d) $P(5 \leq X \leq 7)$

3-64. The random variable X has a binomial distribution with $n = 10$ and $p = 0.5$. Determine the following probabilities:
(a) $P(X = 5)$ (b) $P(X \leq 2)$
(c) $P(X \geq 9)$ (d) $P(3 \leq X < 5)$

3-65. The random variable X has a binomial distribution with $n = 10$ and $p = 0.01$. Determine the following probabilities.
(a) $P(X = 5)$ (b) $P(X \leq 2)$
(c) $P(X \geq 9)$ (d) $P(3 \leq X < 5)$

3-66. The random variable X has a binomial distribution with $n = 10$ and $p = 0.5$. Sketch the probability mass function of X.
(a) What value of X is most likely?
(b) What value(s) of X is(are) least likely?

3-67. Sketch the probability mass function of a binomial distribution with $n = 10$ and $p = 0.01$ and comment on the shape of the distribution.
(a) What value of X is most likely?
(b) What value of X is least likely?

3-68. Determine the cumulative distribution function of a binomial random variable with $n = 3$ and $p = 1/2$.

3-69. Determine the cumulative distribution function of a binomial random variable with $n = 3$ and $p = 1/4$.

3-70. An electronic product contains 40 integrated circuits. The probability that any integrated circuit is defective is 0.01, and the integrated circuits are independent. The product operates only if there are no defective integrated circuits. What is the probability that the product operates?

3-71. The phone lines to an airline reservation system are occupied 40% of the time. Assume that the events that the lines are occupied on successive calls are independent. Assume that 10 calls are placed to the airline.
(a) What is the probability that for exactly three calls the lines are occupied?
(b) What is the probability that for at least one call the lines are not occupied?
(c) What is the expected number of calls in which the lines are all occupied?

3-72. A multiple choice test contains 25 questions, each with four answers. Assume a student just guesses on each question.
(a) What is the probability that the student answers more than 20 questions correctly?
(b) What is the probability the student answers less than 5 questions correctly?

3-73. A particularly long traffic light on your morning commute is green 20% of the time that you approach it. Assume that each morning represents an independent trial.
(a) Over five mornings, what is the probability that the light is green on exactly one day?
(b) Over 20 mornings, what is the probability that the light is green on exactly four days?
(c) Over 20 mornings, what is the probability that the light is green on more than four days?

3-74. Samples of rejuvenated mitochondria are mutated (defective) in 1% of cases. Suppose 15 samples are studied, and they can be considered to be independent for mutation. Determine the following probabilities. The binomial table in Appendix A can help.
(a) No samples are mutated.
(b) At most one sample is mutated.
(c) More than half the samples are mutated.

3-75. An article in *Information Security Technical Report*, "Malicious software—past, present and future," (2004, Vol. 9, pp. 6–18) provided the following data on the top ten malicious software instances for 2002. The clear leader in the number of registered incidences for the year 2002 was the Internet worm "Klez," and it is still one of the most widespread threats. This virus was first detected on 26 October 2001, and it has held the top spot among malicious software for the longest period in the history of virology.

Place	Name	% Instances
1	I-Worm.Klez	61.22%
2	I-Worm.Lentin	20.52%
3	I-Worm.Tanatos	2.09%
4	I-Worm.BadtransII	1.31%
5	Macro.Word97.Thus	1.19%
6	I-Worm.Hybris	0.60%
7	I-Worm.Bridex	0.32%
8	I-Worm.Magistr	0.30%
9	Win95.CIH	0.27%
10	I-Worm.Sircam	0.24%

The 10 most widespread malicious programs for 2002 *(Source—Kaspersky Labs)*.

Suppose that 20 malicious software instances are reported. Assume that the malicious sources can be assumed to be independent.
(a) What is the probability at least one instance is "Klez"?
(b) What is the probability that three or more instances are "Klez"?
(c) What is the mean and standard deviation of the number of "Klez" instances among the 20 reported?

3-76. Heart failure is due to either natural occurrences (87%) or outside factors (13%). Outside factors are related to induced substances or foreign objects. Natural occurrences are caused by arterial blockage, disease, and infection. Suppose that 20 patients will visit an emergency room with heart failure. Assume that causes of heart failure between individuals are independent.
(a) What is the probability that three individuals have conditions caused by outside factors?
(b) What is the probability that three or more individuals have conditions caused by outside factors?
(c) What is the mean and standard deviation of the number of individuals with conditions caused by outside factors?

3-77. A computer system uses passwords that are exactly six characters and each character is one of the 26 letters (a–z) or 10 integers (0–9). Suppose there are 10000 users on the system with unique passwords. A hacker randomly selects (with replacement) one billion passwords from the potential set in the milliseconds before security software closes the unauthorized access.
(a) What is the distribution of the number of user passwords selected by the hacker?
(b) What is the probability that no user passwords are selected?
(c) What is the mean and variance of the number of user passwords selected?

3-78. **A statistical process control chart example.** Samples of 20 parts from a metal punching process are selected every hour. Typically, 1% of the parts require rework. Let X denote the number of parts in the sample of 20 that require rework. A process problem is suspected if X exceeds its mean by more than three standard deviations.
(a) If the percentage of parts that require rework remains at 1%, what is the probability that X exceeds its mean by more than three standard deviations?
(b) If the rework percentage increases to 4%, what is the probability that X exceeds 1?
(c) If the rework percentage increases to 4%, what is the probability that X exceeds 1 in at least one of the next five hours of samples?

3-79. Because not all airline passengers show up for their reserved seat, an airline sells 125 tickets for a flight that holds only 120 passengers. The probability that a passenger does not show up is 0.10, and the passengers behave independently.
(a) What is the probability that every passenger who shows up can take the flight?
(b) What is the probability that the flight departs with empty seats?

3-80. This exercise illustrates that poor quality can affect schedules and costs. A manufacturing process has 100 customer orders to fill. Each order requires one component part that is purchased from a supplier. However, typically, 2% of the components are identified as defective, and the components can be assumed to be independent.
(a) If the manufacturer stocks 100 components, what is the probability that the 100 orders can be filled without reordering components?
(b) If the manufacturer stocks 102 components, what is the probability that the 100 orders can be filled without reordering components?
(c) If the manufacturer stocks 105 components, what is the probability that the 100 orders can be filled without reordering components?

3-7 GEOMETRIC AND NEGATIVE BINOMIAL DISTRIBUTIONS

3-7.1 Geometric Distribution

Consider a random experiment that is closely related to the one used in the definition of a binomial distribution. Again, assume a series of Bernoulli trials (independent trials with constant probability p of a success on each trial). However, instead of a fixed number of trials, trials are conducted until a success is obtained. Let the random variable X denote the number of trials until the first success. In Example 3-5, successive wafers are analyzed until a large particle is detected. Then, X is the number of wafers analyzed. In the transmission of bits, X might be the number of bits transmitted until an error occurs.

EXAMPLE 3-20
Digital Channel

The probability that a bit transmitted through a digital transmission channel is received in error is 0.1. Assume the transmissions are independent events, and let the random variable X denote the number of bits transmitted *until* the first error.

Then, $P(X = 5)$ is the probability that the first four bits are transmitted correctly and the fifth bit is in error. This event can be denoted as $\{OOOOE\}$, where O denotes an okay bit. Because the trials are independent and the probability of a correct transmission is 0.9,

$$P(X = 5) = P(OOOOE) = 0.9^4 0.1 = 0.066$$

Note that there is some probability that X will equal any integer value. Also, if the first trial is a success, $X = 1$. Therefore, the range of X is $\{1, 2, 3, \ldots\}$, that is, all positive integers.

Geometric Distribution

In a series of Bernoulli trials (independent trials with constant probability p of a success), let the random variable X denote the number of trials until the first success. Then X is a geometric random variable with parameter $0 < p < 1$ and

$$f(x) = (1 - p)^{x-1} p \qquad x = 1, 2, \ldots \qquad (3\text{-}9)$$

Examples of the probability mass functions for geometric random variables are shown in Fig. 3-9. Note that the height of the line at x is $(1 - p)$ times the height of the line at $x - 1$. That is, the probabilities decrease in a geometric progression. The distribution acquires its name from this result.

EXAMPLE 3-21   The probability that a wafer contains a large particle of contamination is 0.01. If it is assumed that the wafers are independent, what is the probability that exactly 125 wafers need to be analyzed before a large particle is detected?

Let X denote the number of samples analyzed until a large particle is detected. Then X is a geometric random variable with $p = 0.01$. The requested probability is

$$P(X = 125) = (0.99)^{124} 0.01 = 0.0029$$

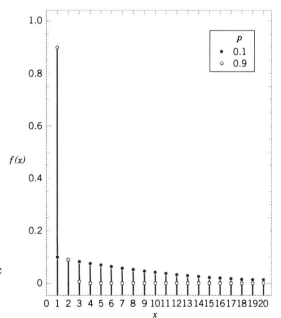

Figure 3-9 Geometric distributions for selected values of the parameter p.

3-7 GEOMETRIC AND NEGATIVE BINOMIAL DISTRIBUTIONS

The mean of a geometric random variable is

$$\mu = \sum_{k=1}^{\infty} kp(1-p)^{k-1} = p\sum_{k=1}^{\infty} kq^{k-1}$$

where $q = p - 1$. The right-hand side of the previous equation is recognized to be the partial derivative with respect to q of

$$p\sum_{k=1}^{\infty} q^k = \frac{pq}{1-q}$$

where the last equality is obtained from the known sum of a geometric series. Therefore,

$$\mu = \frac{\partial}{\partial q}\left[\frac{pq}{1-q}\right] = \frac{p}{(1-q)^2} = \frac{p}{p^2} = \frac{1}{p}$$

and the mean is derived. To obtain the variance of a geometric random variable, we can first derive $E(X^2)$ by a similar approach. This can be obtained from partial second derivatives with respect to q. Then the formula $V(X) = E(X^2) - (EX)^2$ is applied. The details are a bit more work and this is left as a mind-expanding exercise.

Mean and Variance

If X is a geometric random variable with parameter p,

$$\mu = E(X) = 1/p \quad \text{and} \quad \sigma^2 = V(X) = (1-p)/p^2 \quad (3\text{-}10)$$

EXAMPLE 3-22 Consider the transmission of bits in Example 3-20. Here, $p = 0.1$. The mean number of transmissions until the first error is $1/0.1 = 10$. The standard deviation of the number of transmissions before the first error is

$$\sigma = [(1 - 0.1)/0.1^2]^{1/2} = 9.49$$

Lack of Memory Property

A geometric random variable has been defined as the number of trials until the first success. However, because the trials are independent, the count of the number of trials until the next success can be started at any trial without changing the probability distribution of the random variable. For example, in the transmission of bits, if 100 bits are transmitted, the probability that the first error, after bit 100, occurs on bit 106 is the probability that the next six outcomes are *OOOOOE*. This probability is $(0.9)^5(0.1) = 0.059$, which is identical to the probability that the initial error occurs on bit 6.

The implication of using a geometric model is that the system presumably will not wear out. The probability of an error remains constant for all transmissions. In this sense, the geometric distribution is said to lack any memory. The lack of memory property will be discussed again in the context of an exponential random variable in Chapter 4.

EXAMPLE 3-23
Lack of Memory

In Example 3-20, the probability that a bit is transmitted in error is equal to 0.1. Suppose 50 bits have been transmitted. The mean number of bits until the next error is $1/0.1 = 10$—the same result as the mean number of bits until the first error.

3-7.2 Negative Binomial Distribution

A generalization of a geometric distribution in which the random variable is the number of Bernoulli trials required to obtain r successes results in the **negative binomial distribution**.

EXAMPLE 3-24
Digital Channel

As in Example 3-20, suppose the probability that a bit transmitted through a digital transmission channel is received in error is 0.1. Assume the transmissions are independent events, and let the random variable X denote the number of bits transmitted until the *fourth* error.

Then, X has a negative binomial distribution with $r = 4$. Probabilities involving X can be found as follows. The $P(X = 10)$ is the probability that exactly three errors occur in the first nine trials and then trial 10 results in the fourth error. The probability that exactly three errors occur in the first nine trials is determined from the binomial distribution to be

$$\binom{9}{3}(0.1)^3(0.9)^6$$

Because the trials are independent, the probability that exactly three errors occur in the first 9 trials and trial 10 results in the fourth error is the product of the probabilities of these two events, namely,

$$\binom{9}{3}(0.1)^3(0.9)^6(0.1) = \binom{9}{3}(0.1)^4(0.9)^6$$

The previous result can be generalized as follows.

Negative Binomial Distribution

In a series of Bernoulli trials (independent trials with constant probability p of a success), let the random variable X denote the number of trials until r successes occur. Then X is a negative binomial random variable with parameters $0 < p < 1$ and $r = 1, 2, 3, \ldots,$ and

$$f(x) = \binom{x-1}{r-1}(1-p)^{x-r}p^r \qquad x = r, r+1, r+2, \ldots. \qquad (3\text{-}11)$$

Because at least r trials are required to obtain r successes, the range of X is from r to ∞. In the special case that $r = 1$, a negative binomial random variable is a geometric random variable. Selected negative binomial distributions are illustrated in Fig. 3-10.

The lack of memory property of a geometric random variable implies the following. Let X denote the total number of trials required to obtain r successes. Let X_1 denote the number of trials required to obtain the first success, let X_2 denote the number of extra trials required to obtain the second success, let X_3 denote the number of extra trials to obtain the third success, and so forth. Then, the total number of trials required to obtain r successes is $X = X_1 + X_2 + \cdots + X_r$. Because of the lack of memory property, each of the random variables $X_1, X_2, \ldots, X_r$ has a geometric distribution with the same value of p. Consequently, a

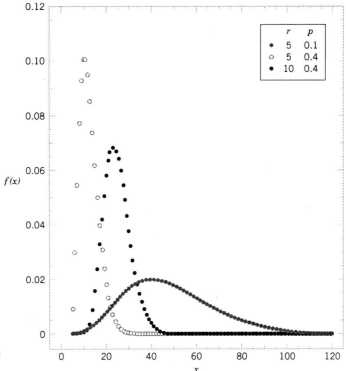

Figure 3-10 Negative binomial distributions for selected values of the parameters r and p.

negative binomial random variable can be interpreted as the sum of r geometric random variables. This concept is illustrated in Fig. 3-11.

Recall that a binomial random variable is a count of the number of successes in n Bernoulli trials. That is, the number of trials is predetermined, and the number of successes is random. A negative binomial random variable is a count of the number of trials required to obtain r successes. That is, the number of successes is predetermined, and the number of trials is random. In this sense, a negative binomial random variable can be considered the opposite, or negative, of a binomial random variable.

The description of a negative binomial random variable as a sum of geometric random variables leads to the following results for the mean and variance. Sums of random variables are studied in Chapter 5.

Mean and Variance

If X is a negative binomial random variable with parameters p and r,

$$\mu = E(X) = r/p \quad \text{and} \quad \sigma^2 = V(X) = r(1-p)/p^2 \qquad (3\text{-}12)$$

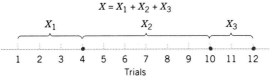

Figure 3-11 Negative binomial random variable represented as a sum of geometric random variables.

EXAMPLE 3-25
Web Servers

A Web site contains three identical computer servers. Only one is used to operate the site, and the other two are spares that can be activated in case the primary system fails. The probability of a failure in the primary computer (or any activated spare system) from a request for service is 0.0005. Assuming that each request represents an independent trial, what is the mean number of requests until failure of all three servers?

Let X denote the number of requests until all three servers fail, and let X_1, X_2, and X_3 denote the number of requests before a failure of the first, second, and third servers used, respectively. Now, $X = X_1 + X_2 + X_3$. Also, the requests are assumed to comprise independent trials with constant probability of failure $p = 0.0005$. Furthermore, a spare server is not affected by the number of requests before it is activated. Therefore, X has a negative binomial distribution with $p = 0.0005$ and $r = 3$. Consequently,

$$E(X) = 3/0.0005 = 6000 \text{ requests}$$

What is the probability that all three servers fail within five requests? The probability is $P(X \leq 5)$ and

$$P(X \leq 5) = P(X = 3) + P(X = 4) + P(X = 5)$$
$$= 0.0005^3 + \binom{3}{2}0.0005^3(0.9995) + \binom{4}{2}0.0005^3(0.9995)^2$$
$$= 1.25 \times 10^{-10} + 3.75 \times 10^{-10} + 7.49 \times 10^{-10}$$
$$= 1.249 \times 10^{-9}$$

EXERCISES FOR SECTION 3-7

3-81. Suppose the random variable X has a geometric distribution with $p = 0.5$. Determine the following probabilities:
(a) $P(X = 1)$ (b) $P(X = 4)$
(c) $P(X = 8)$ (d) $P(X \leq 2)$
(e) $P(X > 2)$

3-82. Suppose the random variable X has a geometric distribution with a mean of 2.5. Determine the following probabilities:
(a) $P(X = 1)$ (b) $P(X = 4)$
(c) $P(X = 5)$ (d) $P(X \leq 3)$
(e) $P(X > 3)$

3-83. Consider a sequence of independent Bernoulli trials with $p = 0.2$.
(a) What is the expected number of trials to obtain the first success?
(b) After the eighth success occurs, what is the expected number of trials to obtain the ninth success?

3-84. Suppose that X is a negative binomial random variable with $p = 0.2$ and $r = 4$. Determine the following:
(a) $E(X)$ (b) $P(X = 20)$
(c) $P(X = 19)$ (d) $P(X = 21)$
(e) The most likely value for X

3-85. The probability of a successful optical alignment in the assembly of an optical data storage product is 0.8. Assume the trials are independent.
(a) What is the probability that the first successful alignment requires exactly four trials?
(b) What is the probability that the first successful alignment requires at most four trials?
(c) What is the probability that the first successful alignment requires at least four trials?

3-86. In a clinical study, volunteers are tested for a gene that has been found to increase the risk for a disease. The probability that a person carries the gene is 0.1.
(a) What is the probability 4 or more people will have to be tested before 2 with the gene are detected?
(b) How many people are expected to be tested before 2 with the gene are detected?

3-87. Assume that each of your calls to a popular radio station has a probability of 0.02 of connecting, that is, of not obtaining a busy signal. Assume that your calls are independent.
(a) What is the probability that your first call that connects is your tenth call?
(b) What is the probability that it requires more than five calls for you to connect?
(c) What is the mean number of calls needed to connect?

3-88. A player of a video game is confronted with a series of opponents and has an 80% probability of defeating each one. Success with any opponent is independent of previous encounters. The player continues to contest opponents until defeated.
(a) What is the probability mass function of the number of opponents contested in a game?
(b) What is the probability that a player defeats at least two opponents in a game?

(c) What is the expected number of opponents contested in a game?
(d) What is the probability that a player contests four or more opponents in a game?
(e) What is the expected number of game plays until a player contests four or more opponents?

3-89. Heart failure is due to either natural occurrences (87%) or outside factors (13%). Outside factors are related to induced substances or foreign objects. Natural occurrences are caused by arterial blockage, disease, and infection. Assume that causes of heart failure between individuals are independent.
(a) What is the probability that the first patient with heart failure that enters the emergency room has the condition due to outside factors?
(b) What is the probability that third patient with heart failure that enters the emergency room is the first one due to outside factors?
(c) What is the mean number of heart failure patients with the condition due to natural causes that enter the emergency room before the first patient with heart failure from outside factors?

3-90. A computer system uses passwords constructed from the 26 letters (a–z) or 10 integers (0–9). Suppose there are 10000 users on the system with unique passwords. A hacker randomly selects (with replacement) passwords from the potential set.
(a) There are 9900 users with unique six-character passwords on the system and the hacker randomly selects six-character passwords. What is the mean and standard deviation of the number of attempts before the hacker selects a user password?
(b) Suppose there are 100 users with unique three-character passwords on the system and the hacker randomly selects three-character passwords. What is the mean and standard deviation of the number of attempts before the hacker selects a user password?
(c) Comment on the security differences between six- and three-character passwords.

3-91. A trading company has eight computers that it uses to trade on the New York Stock Exchange (NYSE). The probability of a computer failing in a day is 0.005, and the computers fail independently. Computers are repaired in the evening and each day is an independent trial.
(a) What is the probability that all eight computers fail in a day?
(b) What is the mean number of days until a specific computer fails?

(c) What is the mean number of days until all eight computers fail in the same day?

3-92. Assume that 20 parts are checked each hour and that X denotes the number of parts in the sample of 20 that require rework. Parts are assumed to be independent with respect to rework.
(a) If the percentage of parts that require rework remains at 1%, what is the probability that hour 10 is the first sample at which X exceeds 1?
(b) If the rework percentage increases to 4%, what is the probability that hour 10 is the first sample at which X exceeds 1?
(c) If the rework percentage increases to 4%, what is the expected number of hours until X exceeds 1?

3-93. A fault-tolerant system that processes transactions for a financial services firm uses three separate computers. If the operating computer fails, one of the two spares can be immediately switched online. After the second computer fails, the last computer can be immediately switched online. Assume that the probability of a failure during any transaction is 10^{-8} and that the transactions can be considered to be independent events.
(a) What is the mean number of transactions before all computers have failed?
(b) What is the variance of the number of transactions before all computers have failed?

3-94. In the process of meiosis, a single parent diploid cell goes through eight different phases. However, only 60% of the processes pass the first six phases and only 40% pass all eight. Assume the results from each phase are independent.
(a) If the probability of a successful pass of each one of the first six phases is constant, what is the probability of a successful pass of a single one of these phases?
(b) If the probability of a successful pass of each one of the last two phases is constant, what is the probability of a successful pass of a single one of these phases?

3-95. Show that the probability density function of a negative binomial random variable equals the probability density function of a geometric random variable when $r = 1$. Show that the formulas for the mean and variance of a negative binomial random variable equal the corresponding results for a geometric random variable when $r = 1$.

3-96. Derive the expression for the variance of a geometric random variable with parameter p. (Formulas for infinite series are required.)

3-8 HYPERGEOMETRIC DISTRIBUTION

In Example 3-8, a day's production of 850 manufactured parts contains 50 parts that do not conform to customer requirements. Two parts are selected at random, without replacement from the day's production. That is, selected units are not replaced before the next selection is made. Let A and B denote the events that the first and second parts are nonconforming, respectively. In Chapter 2, we found $P(B|A) = 49/849$ and $P(A) = 50/850$. Consequently, knowledge that the first part is nonconforming suggests that it is less likely that the second part selected is nonconforming.

This experiment is fundamentally different from the examples based on the binomial distribution. In this experiment, the trials are not independent. Note that, in the unusual case that each unit selected is replaced before the next selection, the trials are independent and there is a constant probability of a nonconforming part on each trial. Then, the number of nonconforming parts in the sample is a binomial random variable.

Let X equal the number of nonconforming parts in the sample. Then

$$P(X = 0) = P(\text{both parts conform}) = (800/850)(799/849) = 0.886$$

$P(X = 1) = P(\text{first part selected conforms and the second part selected does not, or the first part selected does not and the second part selected conforms})$

$$= (800/850)(50/849) + (50/850)(800/849) = 0.111$$

$$P(X = 2) = P(\text{both parts do not conform}) = (50/850)(49/849) = 0.003$$

As in this example, samples are often selected without replacement. Although probabilities can be determined by the reasoning used in the example above, a general formula for computing probabilities when samples are selected without replacement is quite useful. The counting rules presented in Chapter 2 can be used to justify the formula given below.

Hypergeometric Distribution

> A set of N objects contains
> $\quad K$ objects classified as successes
> $\quad N - K$ objects classified as failures
>
> A sample of size n objects is selected randomly (without replacement) from the N objects, where $K \leq N$ and $n \leq N$.
>
> Let the random variable X denote the number of successes in the sample. Then X is a **hypergeometric random variable** and
>
> $$f(x) = \frac{\binom{K}{x}\binom{N-K}{n-x}}{\binom{N}{n}} \qquad x = \max\{0, n + K - N\} \text{ to } \min\{K, n\} \qquad (3\text{-}13)$$

The expression $\min\{K, n\}$ is used in the definition of the range of X because the maximum number of successes that can occur in the sample is the smaller of the sample size, n, and the number of successes available, K. Also, if $n + K > N$, at least $n + K - N$ successes must occur in the sample. Selected hypergeometric distributions are illustrated in Fig. 3-12.

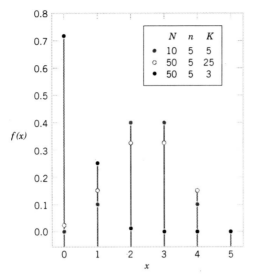

Figure 3-12 Hypergeometric distributions for selected values of parameters N, K, and n.

EXAMPLE 3-26
Sampling without Replacement

The example at the start of this section can be reanalyzed by using the general expression in the definition of a hypergeometric random variable. That is,

$$P(X = 0) = \frac{\binom{50}{0}\binom{800}{2}}{\binom{850}{2}} = \frac{319600}{360825} = 0.886$$

$$P(X = 1) = \frac{\binom{50}{1}\binom{800}{1}}{\binom{850}{2}} = \frac{40000}{360825} = 0.111$$

$$P(X = 2) = \frac{\binom{50}{2}\binom{800}{0}}{\binom{850}{2}} = \frac{1225}{360825} = 0.003$$

EXAMPLE 3-27
Parts from Suppliers

A batch of parts contains 100 parts from a local supplier of tubing and 200 parts from a supplier of tubing in the next state. If four parts are selected randomly and without replacement, what is the probability they are all from the local supplier?

Let X equal the number of parts in the sample from the local supplier. Then, X has a hypergeometric distribution and the requested probability is $P(X = 4)$. Consequently,

$$P(X = 4) = \frac{\binom{100}{4}\binom{200}{0}}{\binom{300}{4}} = 0.0119$$

What is the probability that two or more parts in the sample are from the local supplier?

$$P(X \geq 2) = \frac{\binom{100}{2}\binom{200}{2}}{\binom{300}{4}} + \frac{\binom{100}{3}\binom{200}{1}}{\binom{300}{4}} + \frac{\binom{100}{4}\binom{200}{0}}{\binom{300}{4}}$$

$$= 0.298 + 0.098 + 0.0119 = 0.408$$

What is the probability that at least one part in the sample is from the local supplier?

$$P(X \geq 1) = 1 - P(X = 0) = 1 - \frac{\binom{100}{0}\binom{200}{4}}{\binom{300}{4}} = 0.804$$

The mean and variance of a hypergeometric random variable can be determined from the trials that comprise the experiment. However, the trials are not independent, and so the calculations are more difficult than for a binomial distribution. The results are stated as follows.

Mean and Variance

If X is a hypergeometric random variable with parameters N, K, and n, then

$$\mu = E(X) = np \quad \text{and} \quad \sigma^2 = V(X) = np(1-p)\left(\frac{N-n}{N-1}\right) \quad (3\text{-}14)$$

where $p = K/N$.

Here p is interpreted as the proportion of successes in the set of N objects.

EXAMPLE 3-28 In the previous example, the sample size is 4. The random variable X is the number of parts in the sample from the local supplier. Then, $p = 100/300 = 1/3$. Therefore,

$$E(X) = 4(100/300) = 1.33$$

and

$$V(X) = 4(1/3)(2/3)[(300-4)/299] = 0.88$$

For a hypergeometric random variable, $E(X)$ is similar to the mean of a binomial random variable. Also, $V(X)$ differs from the result for a binomial random variable only by the term shown below.

Finite Population Correction Factor

The term in the variance of a hypergeometric random variable

$$\frac{N-n}{N-1} \quad (3\text{-}15)$$

is called the **finite population correction factor**.

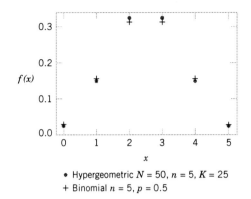

Figure 3-13 Comparison of hypergeometric and binomial distributions.

	0	1	2	3	4	5
Hypergeometric probability	0.025	0.149	0.326	0.326	0.149	0.025
Binomial probability	0.031	0.156	0.312	0.312	0.156	0.031

Sampling with replacement is equivalent to sampling from an infinite set because the proportion of success remains constant for every trial in the experiment. As mentioned previously, if sampling were done with replacement, X would be a binomial random variable and its variance would be $np(1 - p)$. Consequently, the finite population correction represents the correction to the binomial variance that results because the sampling is without replacement from the finite set of size N.

If n is small relative to N, the correction is small and the hypergeometric distribution is similar to the binomial. In this case, a binomial distribution can effectively approximate the distribution of the number of units of a specified type in the sample. A case is illustrated in Fig. 3-13.

EXAMPLE 3-29
Customer Sample

A listing of customer accounts at a large corporation contains 1000 customers. Of these, 700 have purchased at least one of the corporation's products in the last three months. To evaluate a new product design, 50 customers are sampled at random from the corporate listing. What is the probability that more than 45 of the sampled customers have purchased from the corporation in the last three months?

The sampling is without replacement. However, because the sample size of 50 is small relative to the number of customer accounts, 1000, the probability of selecting a customer who has purchased from the corporation in the last three months remains approximately constant as the customers are chosen.

For example, let A denote the event that the first customer selected has purchased from the corporation in the last three months, and let B denote the event that the second customer selected has purchased from the corporation in the last three months. Then, $P(A) = 700/1000 = 0.7$ and $P(B|A) = 699/999 = 0.6997$. That is, the trials are approximately independent.

Let X denote the number of customers in the sample who have purchased from the corporation in the last three months. Then, X is a hypergeometric random variable with $N = 1000$, $n = 50$, and $K = 700$. Consequently, $p = K/N = 0.7$. The requested probability is $P(X > 45)$. Because the sample size is small relative to the batch size, the distribution of X can be approximated as binomial with $n = 50$ and $p = 0.7$. Using the binomial approximation to the distribution of X results in

$$P(X > 45) = \sum_{x=46}^{50} \binom{50}{x} 0.7^x (1 - 0.7)^{50-x} = 0.00017$$

The probability from the hypergeometric distribution is 0.00013, but this requires computer software. The result agrees well with the binomial approximation.

EXERCISES FOR SECTION 3-8

3-97. Suppose X has a hypergeometric distribution with $N = 100, n = 4$, and $K = 20$. Determine the following:
(a) $P(X = 1)$
(b) $P(X = 6)$
(c) $P(X = 4)$
(d) Determine the mean and variance of X.

3-98. Suppose X has a hypergeometric distribution with $N = 20, n = 4$, and $K = 4$. Determine the following:
(a) $P(X = 1)$
(b) $P(X = 4)$
(c) $P(X \le 2)$
(d) Determine the mean and variance of X.

3-99. Suppose X has a hypergeometric distribution with $N = 10, n = 3$, and $K = 4$. Sketch the probability mass function of X. Determine the cumulative distribution function for X.

3-100. A batch contains 36 bacteria cells and 12 of the cells are not capable of cellular replication. Suppose you examine three bacteria cells selected at random, without replacement.
(a) What is the probability mass function of the number of cells in the sample that can replicate?
(b) What is the mean and variance of the number of cells in the sample that can replicate?
(c) What is the probability that at least one of the selected cells cannot replicate?

3-101. A company employs 800 men under the age of 55. Suppose that 30% carry a marker on the male chromosome that indicates an increased risk for high blood pressure.
(a) If 10 men in the company are tested for the marker in this chromosome, what is the probability that exactly 1 man has the marker?
(b) If 10 men in the company are tested for the marker in this chromosome, what is the probability that more than 1 has the marker?

3-102. Printed circuit cards are placed in a functional test after being populated with semiconductor chips. A lot contains 140 cards, and 20 are selected without replacement for functional testing.
(a) If 20 cards are defective, what is the probability that at least 1 defective card is in the sample?
(b) If 5 cards are defective, what is the probability that at least 1 defective card appears in the sample?

3-103. The analysis of results from a leaf transmutation experiment (turning a leaf into a petal) is summarized by type of transformation completed:

		Total Textural Transformation	
		Yes	No
Total Color	Yes	243	26
Transformation	No	13	18

A naturalist randomly selects three leaves from this set, without replacement. Determine the following probabilities.
(a) Exactly one has undergone both types of transformations.
(b) At least one has undergone both transformations.
(c) Exactly one has undergone one but not both transformations.
(d) At least one has undergone at least one transformation.

3-104. A state runs a lottery in which 6 numbers are randomly selected from 40, without replacement. A player chooses 6 numbers before the state's sample is selected.
(a) What is the probability that the 6 numbers chosen by a player match all 6 numbers in the state's sample?
(b) What is the probability that 5 of the 6 numbers chosen by a player appear in the state's sample?
(c) What is the probability that 4 of the 6 numbers chosen by a player appear in the state's sample?
(d) If a player enters one lottery each week, what is the expected number of weeks until a player matches all 6 numbers in the state's sample?

3-105. Magnetic tape is slit into half-inch widths that are wound into cartridges. A slitter assembly contains 48 blades. Five blades are selected at random and evaluated each day for sharpness. If any dull blade is found, the assembly is replaced with a newly sharpened set of blades.
(a) If 10 of the blades in an assembly are dull, what is the probability that the assembly is replaced the first day it is evaluated?
(b) If 10 of the blades in an assembly are dull, what is the probability that the assembly is not replaced until the third day of evaluation? [*Hint:* Assume the daily decisions are independent, and use the geometric distribution.]
(c) Suppose on the first day of evaluation, two of the blades are dull, on the second day of evaluation six are dull, and on the third day of evaluation, ten are dull. What is the probability that the assembly is not replaced until the third day of evaluation? [*Hint:* Assume the daily decisions are independent. However, the probability of replacement changes every day.]

3-106.
(a) Calculate the finite population corrections for Exercises 3-97 and 3-98. For which exercise should the binomial approximation to the distribution of X be better?
(b) For Exercise 3-97, calculate $P(X = 1)$ and $P(X = 4)$ assuming that X has a binomial distribution and compare these results to results derived from the hypergeometric distribution.
(c) For Exercise 3-98, calculate $P(X = 1)$ and $P(X = 4)$ assuming that X has a binomial distribution and compare these results to the results derived from the hypergeometric distribution.
(d) Use the binomial approximation to the hypergeometric distribution to approximate the probabilities in Exercise 3-102. What is the finite population correction in this exercise?

3-9 POISSON DISTRIBUTION

We introduce the Poisson distribution with an example.

EXAMPLE 3-30 Consider the transmission of n bits over a digital communication channel. Let the random variable X equal the number of bits in error. When the probability that a bit is in error is constant and the transmissions are independent, X has a binomial distribution. Let p denote the probability that a bit is in error. Let $\lambda = pn$. Then, $E(x) = pn = \lambda$ and

$$P(X = x) = \binom{n}{x} p^x (1-p)^{n-x} = \binom{n}{x}\left(\frac{\lambda}{n}\right)^x \left(1 - \frac{\lambda}{n}\right)^{n-x}$$

Now, suppose that the number of bits transmitted increases and the probability of an error decreases exactly enough that pn remains equal to a constant. That is, n increases and p decreases accordingly, such that $E(X) = \lambda$ remains constant. Then, with some work, it can be shown that

$$\binom{n}{x}\left(\frac{1}{n}\right)^x \to \frac{1}{x!} \qquad \left(1 - \frac{\lambda}{n}\right)^{-x} \to 1 \qquad \left(1 - \frac{\lambda}{n}\right)^n \to e^{-\lambda}$$

so that

$$\lim_{n \to \infty} P(X = x) = \frac{e^{-\lambda} \lambda^x}{x!}, \qquad x = 0, 1, 2, \ldots$$

Also, because the number of bits transmitted tends to infinity, the number of errors can equal any non-negative integer. Therefore, the range of X is the integers from zero to infinity.

The distribution obtained as the limit in the above example is more useful than the derivation above implies. The following example illustrates the broader applicability.

EXAMPLE 3-31
Wire Flaws

Flaws occur at random along the length of a thin copper wire. Let X denote the random variable that counts the number of flaws in a length of L millimeters of wire and suppose that the average number of flaws in L millimeters is λ.

The probability distribution of X can be found by reasoning in a manner similar to the previous example. Partition the length of wire into n subintervals of small length, say, 1 micrometer each. If the subinterval chosen is small enough, the probability that more than one flaw occurs in the subinterval is negligible. Furthermore, we can interpret the assumption that flaws occur at random to imply that every subinterval has the same probability of containing a flaw, say, p. Finally, if we assume that the probability that a subinterval contains a flaw is independent of other subintervals, we can model the distribution of X as approximately a binomial random variable. Because

$$E(X) = \lambda = np$$

we obtain

$$p = \lambda/n$$

That is, the probability that a subinterval contains a flaw is λ/n. With small enough subintervals, n is very large and p is very small. Therefore, the distribution of X is obtained as in the previous example.

Example 3-31 can be generalized to include a broad array of random experiments. The interval that was partitioned was a length of wire. However, the same reasoning can be applied

100   CHAPTER 3 DISCRETE RANDOM VARIABLES AND PROBABILITY DISTRIBUTIONS

to any interval, including an interval of time, an area, or a volume. For example, counts of (1) particles of contamination in semiconductor manufacturing, (2) flaws in rolls of textiles, (3) calls to a telephone exchange, (4) power outages, and (5) atomic particles emitted from a specimen have all been successfully modeled by the probability mass function in the following definition.

Poisson Distribution

Given an interval of real numbers, assume events occur at random throughout the interval. If the interval can be partitioned into subintervals of small enough length such that

(1) the probability of more than one event in a subinterval is zero,
(2) the probability of one event in a subinterval is the same for all subintervals and proportional to the length of the subinterval, and
(3) the event in each subinterval is independent of other subintervals, the random experiment is called a **Poisson process**.

The random variable X that equals the number of events in the interval is a **Poisson random variable** with parameter $0 < \lambda$, and the probability mass function of X is

$$f(x) = \frac{e^{-\lambda}\lambda^x}{x!} \qquad x = 0, 1, 2, \ldots \qquad (3\text{-}16)$$

The sum of the probabilities is one because

$$\sum_{k=0}^{\infty} \frac{e^{-\lambda}\lambda^k}{k!} = e^{-\lambda} \sum_{k=0}^{\infty} \frac{\lambda^k}{k!}$$

and the summation on the right-hand side of the previous equation is recognized to be Taylor's expansion of e^x evaluated at λ. Therefore, the summation equals e^λ and the right-hand side equals $e^{-\lambda}e^\lambda = 1$.

Historically, the term *process* has been used to suggest the observation of a system over time. In our example with the copper wire, we showed that the Poisson distribution could also apply to intervals such as lengths. Figure 3-14 provides graphs of selected Poisson distributions.

It is important to **use consistent units** in the calculation of probabilities, means, and variances involving Poisson random variables. The following example illustrates unit conversions. For example, if the

average number of flaws per millimeter of wire is 3.4, then the

average number of flaws in 10 millimeters of wire is 34, and the

average number of flaws in 100 millimeters of wire is 340.

If a Poisson random variable represents the number of events in some interval, the mean of the random variable must equal the expected number of events in the same length of interval.

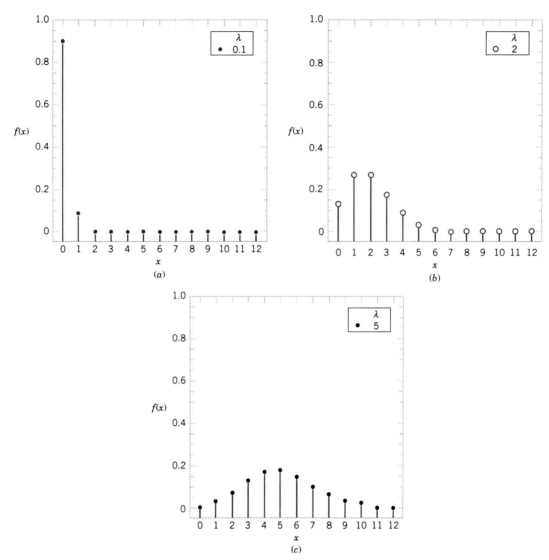

Figure 3-14 Poisson distributions for selected values of the parameters.

EXAMPLE 3-32
Calculations for
Wire Flaws

For the case of the thin copper wire, suppose that the number of flaws follows a Poisson distribution with a mean of 2.3 flaws per millimeter. Determine the probability of exactly 2 flaws in 1 millimeter of wire.

Let X denote the number of flaws in 1 millimeter of wire. Then, $E(X) = 2.3$ flaws and

$$P(X = 2) = \frac{e^{-2.3}2.3^2}{2!} = 0.265$$

Determine the probability of 10 flaws in 5 millimeters of wire. Let X denote the number of flaws in 5 millimeters of wire. Then, X has a Poisson distribution with

$$E(X) = 5 \text{ mm} \times 2.3 \text{ flaws/mm} = 11.5 \text{ flaws}$$

Therefore,

$$P(X = 10) = e^{-11.5} \frac{11.5^{10}}{10!} = 0.113$$

Determine the probability of at least 1 flaw in 2 millimeters of wire. Let X denote the number of flaws in 2 millimeters of wire. Then, X has a Poisson distribution with

$$E(X) = 2 \text{ mm} \times 2.3 \text{ flaws/mm} = 4.6 \text{ flaws}$$

Therefore,

$$P(X \geq 1) = 1 - P(X = 0) = 1 - e^{-4.6} = 0.9899$$

**EXAMPLE 3-33
CDs**

Contamination is a problem in the manufacture of optical storage disks (CDs). The number of particles of contamination that occur on an optical disk has a Poisson distribution, and the average number of particles per centimeter squared of media surface is 0.1. The area of a disk under study is 100 squared centimeters. Find the probability that 12 particles occur in the area of a disk under study.

Let X denote the number of particles in the area of a disk under study. Because the mean number of particles is 0.1 particles per cm^2

$$E(X) = 100 \text{ cm}^2 \times 0.1 \text{ particles/cm}^2 = 10 \text{ particles}$$

Therefore,

$$P(X = 12) = \frac{e^{-10} 10^{12}}{12!} = 0.095$$

The probability that zero particles occur in the area of the disk under study is

$$P(X = 0) = e^{-10} = 4.54 \times 10^{-5}$$

Determine the probability that 12 or fewer particles occur in the area of the disk under study. The probability is

$$P(X \leq 12) = P(X = 0) + P(X = 1) + \cdots + P(X = 12) = \sum_{i=0}^{12} \frac{e^{-10} 10^i}{i!}$$

Because this sum is tedious to compute, many computer programs calculate cumulative Poisson probabilities. From one such program, $P(X \leq 12) = 0.792$.

The mean of a Poisson random variable is

$$\mu = \sum_{k=1}^{\infty} \frac{k e^{-\lambda} \lambda^k}{k!} = \lambda \sum_{k=1}^{\infty} \frac{e^{-\lambda} \lambda^{k-1}}{(k-1)!}$$

where the summation can start at $k = 1$ because the $k = 0$ term is zero. If a change of **variable** $j = k - 1$ is used, the summation on the right-hand side of the previous equation is **recognized** to be the sum of the probabilities of a Poisson random variable and this equals one. Therefore, the previous equation simplifies to $\mu = \lambda$ and the mean is derived.

To obtain the variance of a Poisson random variable we can start with $E(X^2)$ and this equals

$$E(X^2) = \sum_{k=1}^{\infty} \frac{k^2 e^{-\lambda} \lambda^k}{k!} = \lambda \sum_{k=1}^{\infty} \frac{k e^{-\lambda} \lambda^{k-1}}{(k-1)!}$$

Write $k = (k-1) + 1$ to obtain

$$E(X^2) = \lambda \sum_{k=1}^{\infty} \frac{(k-1) e^{-\lambda} \lambda^{k-1}}{(k-1)!} + \lambda \sum_{k=1}^{\infty} \frac{e^{-\lambda} \lambda^{k-1}}{(k-1)!}$$

The summation in the first term on the right-hand side of the previous equation is recognized to be the mean of X and this equals λ so that the first term is λ^2. The summation in the second term on the right-hand side is recognized to be the sum of the probabilities and this equals one. Therefore, the previous equation simplifies to $E(X^2) = \lambda^2 + \lambda$. Because the $V(X) = E(X^2) - (EX)^2$ we have

$$V(X) = \lambda^2 + \lambda - \lambda^2 = \lambda$$

and the variance is derived.

Mean and Variance

If X is a Poisson random variable with parameter λ, then

$$\mu = E(X) = \lambda \quad \text{and} \quad \sigma^2 = V(X) = \lambda \quad \quad (3\text{-}17)$$

The mean and variance of a Poisson random variable are equal. For example, if particle counts follow a Poisson distribution with a mean of 25 particles per square centimeter, the variance is also 25 and the standard deviation of the counts is 5 per square centimeter. Consequently, information on the variability is very easily obtained. Conversely, if the variance of count data is much greater than the mean of the same data, the Poisson distribution is not a good model for the distribution of the random variable.

EXERCISES FOR SECTION 3-9

3-107. Suppose X has a Poisson distribution with a mean of 4. Determine the following probabilities:
(a) $P(X = 0)$ (b) $P(X \leq 2)$
(c) $P(X = 4)$ (d) $P(X = 8)$

3-108. Suppose X has a Poisson distribution with a mean of 0.4. Determine the following probabilities:
(a) $P(X = 0)$ (b) $P(X \leq 2)$
(c) $P(X = 4)$ (d) $P(X = 8)$

3-109. Suppose that the number of customers that enter a bank in an hour is a Poisson random variable, and suppose that $P(X = 0) = 0.05$. Determine the mean and variance of X.

3-110. The number of telephone calls that arrive at a phone exchange is often modeled as a Poisson random variable. Assume that on the average there are 10 calls per hour.

(a) What is the probability that there are exactly 5 calls in one hour?
(b) What is the probability that there are 3 or fewer calls in one hour?
(c) What is the probability that there are exactly 15 calls in two hours?
(d) What is the probability that there are exactly 5 calls in 30 minutes?

3-111. Astronomers treat the number of stars in a given volume of space as a Poisson random variable. The density in the Milky Way Galaxy in the vicinity of our solar system is one star per 16 cubic light years.
(a) What is the probability of two or more stars in 16 cubic light years?

(b) How many cubic light years of space must be studied so that the probability of one or more stars exceeds 0.95?

3-112. Data from www.centralhudsonlabs determined the mean number of insect fragments in 225 gram chocolate bars was 14.4, but three brands had insect contamination more than twice the average. See the U.S. Food and Drug Administration–Center for Food Safety and Applied Nutrition for Defect Action Levels for food products. Assume the number of fragments (contaminants) follows a Poisson distribution.
(a) If you consume a 225 gram bar from a brand at the mean contamination level, what is the probability of no insect contaminants?
(b) Suppose you consume a bar that is one-fifth the size tested (45 grams) from a brand at the mean contamination level. What is the probability of no insect contaminants?
(c) If you consume seven 28.35 gram (one ounce) bars this week from a brand at the mean contamination level, what is the probability that you consume one or more insect fragments in more than one bar?
(d) Is the probability of a test result more than twice the mean of 14.4 unusual, or can it be considered typical variation? Explain.

3-113. In 1898 L. J. Bortkiewicz published a book entitled *The Law of Small Numbers*. He used data collected over 20 years to show that the number of soldiers killed by horse kicks each year in each corps in the Prussian cavalry followed a Poisson distribution with a mean of 0.61.
(a) What is the probability of more than one death in a corps in a year?
(b) What is the probability of no deaths in a corps over five years?

3-114. The number of flaws in bolts of cloth in textile manufacturing is assumed to be Poisson distributed with a mean of 0.1 flaw per square meter.
(a) What is the probability that there are two flaws in 1 square meter of cloth?
(b) What is the probability that there is one flaw in 10 square meters of cloth?
(c) What is the probability that there are no flaws in 20 square meters of cloth?
(d) What is the probability that there are at least two flaws in 10 square meters of cloth?

3-115. When a computer disk manufacturer tests a disk, it writes to the disk and then tests it using a certifier. The certifier counts the number of missing pulses or errors. The number of errors on a test area on a disk has a Poisson distribution with $\lambda = 0.2$.
(a) What is the expected number of errors per test area?
(b) What percentage of test areas have two or fewer errors?

3-116. The number of cracks in a section of interstate highway that are significant enough to require repair is assumed to follow a Poisson distribution with a mean of two cracks per mile.
(a) What is the probability that there are no cracks that require repair in 5 miles of highway?

(b) What is the probability that at least one crack requires repair in 1/2 mile of highway?
(c) If the number of cracks is related to the vehicle load on the highway and some sections of the highway have a heavy load of vehicles whereas other sections carry a light load, how do you feel about the assumption of a Poisson distribution for the number of cracks that require repair?

3-117. The number of surface flaws in plastic panels used in the interior of automobiles has a Poisson distribution with a mean of 0.05 flaws per square foot of plastic panel. Assume an automobile interior contains 10 square feet of plastic panel.
(a) What is the probability that there are no surface flaws in an auto's interior?
(b) If 10 cars are sold to a rental company, what is the probability that none of the 10 cars has any surface flaws?
(c) If 10 cars are sold to a rental company, what is the probability that at most one car has any surface flaws?

3-118. The number of failures of a testing instrument from contamination particles on the product is a Poisson random variable with a mean of 0.02 failure per hour.
(a) What is the probability that the instrument does not fail in an 8-hour shift?
(b) What is the probability of at least one failure in a 24-hour day?

Supplemental Exercises

3-119. Let the random variable X be equally likely to assume any of the values 1/8, 1/4, or 3/8. Determine the mean and variance of X.

3-120. Let X denote the number of bits received in error in a digital communication channel, and assume that X is a binomial random variable with $p = 0.001$. If 1000 bits are transmitted, determine the following:
(a) $P(X = 1)$ (b) $P(X \geq 1)$
(c) $P(X \leq 2)$ (d) mean and variance of X

3-121. Batches that consist of 50 coil springs from a production process are checked for conformance to customer requirements. The mean number of nonconforming coil springs in a batch is 5. Assume that the number of nonconforming springs in a batch, denoted as X, is a binomial random variable.
(a) What are n and p?
(b) What is $P(X \leq 2)$?
(c) What is $P(X \geq 49)$?

3-122. An automated egg carton loader has a 1% probability of cracking an egg, and a customer will complain if more than one egg per dozen is cracked. Assume each egg load is an independent event.
(a) What is the distribution of cracked eggs per dozen? Include parameter values.
(b) What is the probability that a carton of a dozen eggs results in a complaint?
(c) What is the mean and standard deviation of the number of cracked eggs in a carton of one dozen?

3-123. A total of 12 cells are replicated. Freshly-synthesized DNA cannot be replicated again until mitosis is completed. Two control mechanisms have been identified—one positive and one negative—that are used with equal probability. Assume that each cell independently uses a control mechanism. Determine the following probabilities.
(a) All cells use a positive control mechanism.
(b) Exactly half the cells use a positive control mechanism.
(c) More than four, but less than seven cells use a positive control mechanism.

3-124. A congested computer network has a 1% chance of losing a data packet and packet losses are independent events. An e-mail message requires 100 packets.
(a) What is the distribution of data packets that must be resent? Include the parameter values.
(b) What is the probability that at least one packet must be resent?
(c) What is the probability that two or more packets must be resent?
(d) What is the mean and standard deviation of the number of packets that must be resent?
(e) If there are 10 messages and each contains 100 packets, what is the probability that at least one message requires two or more packets be resent?

3-125. A particularly long traffic light on your morning commute is green 20% of the time that you approach it. Assume that each morning represents an independent trial.
(a) What is the probability that the first morning that the light is green is the fourth morning that you approach it?
(b) What is the probability that the light is not green for 10 consecutive mornings?

3-126. The probability is 0.6 that a calibration of a transducer in an electronic instrument conforms to specifications for the measurement system. Assume the calibration attempts are independent. What is the probability that at most three calibration attempts are required to meet the specifications for the measurement system?

3-127. An electronic scale in an automated filling operation stops the manufacturing line after three underweight packages are detected. Suppose that the probability of an underweight package is 0.001 and each fill is independent.
(a) What is the mean number of fills before the line is stopped?
(b) What is the standard deviation of the number of fills before the line is stopped?

3-128. The probability that an eagle kills a jackrabbit in a day of hunting is 10%. Assume that results are independent between days.
(a) What is the distribution of the number of days until a successful jackrabbit hunt?
(b) What is the probability that the eagle must wait 5 days for its first successful hunt?
(c) What is the expected number of days until a successful hunt?
(d) If the eagle can survive up to 10 days without food (it requires a successful hunt on the tenth day), what is the probability the eagle is still alive 10 days from now?

3-129. Traffic flow is traditionally modeled as a Poisson distribution. A traffic engineer monitors the traffic flowing through an intersection with an average of 6 cars per minute. To set the timing of a traffic signal the following probabilities are used.
(a) What is the probability of no cars through the intersection within 30 seconds?
(b) What is the probability of three or more cars through the intersection within 30 seconds?
(c) Calculate the minimum number of cars through the intersection so that the probability of this number or fewer cars in 30 seconds is at least 90%.
(d) If the variance of the number of cars through the intersection per minute is 20, is the Poisson distribution appropriate? Explain.

3-130. A shipment of chemicals arrives in 15 totes. Three of the totes are selected at random, without replacement, for an inspection of purity. If two of the totes do not conform to purity requirements, what is the probability that at least one of the nonconforming totes is selected in the sample?

3-131. The probability that your call to a service line is answered in less than 30 seconds is 0.75. Assume that your calls are independent.
(a) If you call 10 times, what is the probability that exactly 9 of your calls are answered within 30 seconds?
(b) If you call 20 times, what is the probability that at least 16 calls are answered in less than 30 seconds?
(c) If you call 20 times, what is the mean number of calls that are answered in less than 30 seconds?

3-132. Continuation of Exercise 3-131.
(a) What is the probability that you must call four times to obtain the first answer in less than 30 seconds?
(b) What is the mean number of calls until you are answered in less than 30 seconds?

3-133. Continuation of Exercise 3-131.
(a) What is the probability that you must call six times in order for two of your calls to be answered in less than 30 seconds?
(b) What is the mean number of calls to obtain two answers in less than 30 seconds?

3-134. The number of messages sent to a computer bulletin board is a Poisson random variable with a mean of 5 messages per hour.
(a) What is the probability that 5 messages are received in 1 hour?
(b) What is the probability that 10 messages are received in 1.5 hours?
(c) What is the probability that less than two messages are received in one-half hour?

3-135. A Web site is operated by four identical computer servers. Only one is used to operate the site; the others are spares that can be activated in case the active server fails. The probability that a request to the Web site generates a failure in the active server is 0.0001. Assume that each request is an independent trial. What is the mean time until failure of all four computers?

3-136. The number of errors in a textbook follows a Poisson distribution with a mean of 0.01 error per page. What is the probability that there are three or less errors in 100 pages?

3-137. The probability that an individual recovers from an illness in a one-week time period without treatment is 0.1. Suppose that 20 independent individuals suffering from this illness are treated with a drug and 4 recover in a one-week time period. If the drug has no effect, what is the probability that 4 or more people recover in a one-week time period?

3-138. Patient response to a generic drug to control pain is scored on a 5-point scale, where a 5 indicates complete relief. Historically the distribution of scores is

1	2	3	4	5
0.05	0.1	0.2	0.25	0.4

Two patients, assumed to be independent, are each scored.
(a) What is the probability mass function of the total score?
(b) What is the probability mass function of the average score?

3-139. In a manufacturing process that laminates several ceramic layers, 1% of the assemblies are defective. Assume that the assemblies are independent.
(a) What is the mean number of assemblies that need to be checked to obtain five defective assemblies?
(b) What is the standard deviation of the number of assemblies that need to be checked to obtain five defective assemblies?

3-140. Continuation of Exercise 3-139. Determine the minimum number of assemblies that need to be checked so that the probability of at least one defective assembly exceeds 0.95.

3-141. Determine the constant c so that the following function is a probability mass function: $f(x) = cx$ for $x = 1, 2, 3, 4$.

3-142. A manufacturer of a consumer electronics product expects 2% of units to fail during the warranty period. A sample of 500 independent units is tracked for warranty performance.
(a) What is the probability that none fails during the warranty period?
(b) What is the expected number of failures during the warranty period?
(c) What is the probability that more than two units fail during the warranty period?

3-143. Messages that arrive at a service center for an information systems manufacturer have been classified on the basis of the number of keywords (used to help route messages) and the type of message, either email or voice. Also, 70% of the messages arrive via email and the rest are voice.

number of keywords	0	1	2	3	4
email	0.1	0.1	0.2	0.4	0.2
voice	0.3	0.4	0.2	0.1	0

Determine the probability mass function of the number of keywords in a message.

3-144. The random variable X has the following probability distribution:

x	2	3	5	8
probability	0.2	0.4	0.3	0.1

Determine the following:
(a) $P(X \leq 3)$ (b) $P(X > 2.5)$
(c) $P(2.7 < X < 5.1)$ (d) $E(X)$
(e) $V(X)$

3-145. Determine the probability mass function for the random variable with the following cumulative distribution function:

$$F(x) = \begin{cases} 0 & x < 2 \\ 0.2 & 2 \leq x < 5.7 \\ 0.5 & 5.7 \leq x < 6.5 \\ 0.8 & 6.5 \leq x < 8.5 \\ 1 & 8.5 \leq x \end{cases}$$

3-146. Each main bearing cap in an engine contains four bolts. The bolts are selected at random, without replacement, from a parts bin that contains 30 bolts from one supplier and 70 bolts from another.
(a) What is the probability that a main bearing cap contains all bolts from the same supplier?
(b) What is the probability that exactly three bolts are from the same supplier?

3-147. Assume the number of errors along a magnetic recording surface is a Poisson random variable with a mean of one error every 10^5 bits. A sector of data consists of 4096 eight-bit bytes.
(a) What is the probability of more than one error in a sector?
(b) What is the mean number of sectors until an error is found?

3-148. An installation technician for a specialized communication system is dispatched to a city only when three or more orders have been placed. Suppose orders follow a Poisson distribution with a mean of 0.25 per week for a city

with a population of 100,000 and suppose your city contains a population of 800,000.
(a) What is the probability that a technician is required after a one-week period?
(b) If you are the first one in the city to place an order, what is the probability that you have to wait more than two weeks from the time you place your order until a technician is dispatched?

3-149. From 500 customers, a major appliance manufacturer will randomly select a sample without replacement. The company estimates that 25% of the customers will provide useful data. If this estimate is correct, what is the probability mass function of the number of customers that will provide useful data?
(a) Assume that the company samples 5 customers.
(b) Assume that the company samples 10 customers.

3-150. It is suspected that some of the totes containing chemicals purchased from a supplier exceed the moisture content target. Samples from 30 totes are to be tested for moisture content. Assume that the totes are independent. Determine the proportion of totes from the supplier that must exceed the moisture content target so that the probability is 0.90 that at least one tote in the sample of 30 fails the test.

3-151. Messages arrive to a computer server according to a Poisson distribution with a mean rate of 10 per hour. Determine the length of an interval of time such that the probability that no messages arrive during this interval is 0.90.

3-152. Flaws occur in the interior of plastic used for automobiles according to a Poisson distribution with a mean of 0.02 flaw per panel.
(a) If 50 panels are inspected, what is the probability that there are no flaws?
(b) What is the expected number of panels that need to be inspected before a flaw is found?
(c) If 50 panels are inspected, what is the probability that the number of panels that have one or more flaws is less than or equal to 2?

MIND-EXPANDING EXERCISES

3-153. Derive the convergence results used to obtain a Poisson distribution as the limit of a binomial distribution.

3-154. Show that the function $f(x)$ in Example 3-5 satisfies the properties of a probability mass function by summing the infinite series.

3-155. Derive the formula for the mean and standard deviation of a discrete uniform random variable over the range of integers $a, a + 1, \ldots, b$.

3-156. A company performs inspection on shipments from suppliers in order to defect nonconforming products. Assume a lot contains 1000 items and 1% are nonconforming. What sample size is needed so that the probability of choosing at least one nonconforming item in the sample is at least 0.90? Assume the binomial approximation to the hypergeometric distribution is adequate.

3-157. A company performs inspection on shipments from suppliers in order to detect nonconforming products. The company's policy is to use a sample size that is always 10% of the lot size. Comment on the effectiveness of this policy as a general rule for all sizes of lots.

3-158. Surface flaws in automobile exterior panels follow a Poisson distribution with a mean of 0.1 flaw per panel. If 100 panels are checked, what is the probability that fewer than five panels have any flaws?

3-159. A large bakery can produce rolls in lots of either 0, 1000, 2000, or 3000 per day. The production cost per item is $0.10. The demand varies randomly according to the following distribution:

demand for rolls	0	1000	2000	3000
probability of demand	0.3	0.2	0.3	0.2

Every roll for which there is a demand is sold for $0.30. Every roll for which there is no demand is sold in a secondary market for $0.05. How many rolls should the bakery produce each day to maximize the mean profit?

3-160. A manufacturer stocks components obtained from a supplier. Suppose that 2% of the components are defective and that the defective components occur independently. How many components must the manufacturer have in stock so that the probability that 100 orders can be completed without reordering components is at least 0.95?

IMPORTANT TERMS AND CONCEPTS

- Bernoulli trial
- Binomial distribution
- Cumulative probability distribution function-discrete random variable
- Discrete uniform distribution
- Expected value of a function of a random variable
- Finite population correction factor
- Geometric distribution
- Hypergeometric distribution
- Lack of memory property-discrete random variable
- Mean-discrete random variable
- Mean-function of a discrete random variable
- Negative binomial distribution
- Poisson distribution
- Poisson process
- Probability distribution-discrete random variable
- Probability mass function
- Standard deviation-discrete random variable
- Variance-discrete random variable

Continuous Random Variables and Probability Distributions

CHAPTER OUTLINE

4-1 CONTINUOUS RANDOM VARIABLES

4-2 PROBABILITY DISTRIBUTIONS AND PROBABILITY DENSITY FUNCTIONS

4-3 CUMULATIVE DISTRIBUTION FUNCTIONS

4-4 MEAN AND VARIANCE OF A CONTINUOUS RANDOM VARIABLE

4-5 CONTINUOUS UNIFORM DISTRIBUTION

4-6 NORMAL DISTRIBUTION

4-7 NORMAL APPROXIMATION TO THE BINOMIAL AND POISSON DISTRIBUTIONS

4-8 EXPONENTIAL DISTRIBUTION

4-9 ERLANG AND GAMMA DISTRIBUTIONS

4-10 WEIBULL DISTRIBUTION

4-11 LOGNORMAL DISTRIBUTION

LEARNING OBJECTIVES

After careful study of this chapter you should be able to do the following:
1. Determine probabilities from probability density functions
2. Determine probabilities from cumulative distribution functions and cumulative distribution functions from probability density functions, and the reverse
3. Calculate means and variances for continuous random variables
4. Understand the assumptions for each of the continuous probability distributions presented
5. Select an appropriate continuous probability distribution to calculate probabilities in specific applications
6. Calculate probabilities, determine means and variances for each of the continuous probability distributions presented
7. Standardize normal random variables
8. Use the table for the cumulative distribution function of a standard normal distribution to calculate probabilities
9. Approximate probabilities for some binomial and Poisson distributions

4-1 CONTINUOUS RANDOM VARIABLES

Previously, we discussed the measurement of the current in a thin copper wire. We noted that the results might differ slightly in day-to-day replications because of small variations in variables that are not controlled in our experiment—changes in ambient temperatures, small impurities in the chemical composition of the wire, current source drifts, and so forth.

Another example is the selection of one part from a day's production and very accurately measuring a dimensional length. In practice, there can be small variations in the actual measured lengths due to many causes, such as vibrations, temperature fluctuations, operator differences, calibrations, cutting tool wear, bearing wear, and raw material changes. Even the measurement procedure can produce variations in the final results.

In these types of experiments, the measurement of interest—current in a copper wire experiment, length of a machined part—can be represented by a random variable. It is reasonable to model the range of possible values of the random variable by an interval (finite or infinite) of real numbers. For example, for the length of a machined part, our model enables the measurement from the experiment to result in any value within an interval of real numbers. Because the range is any value in an interval, the model provides for any precision in length measurements. However, because the number of possible values of the random variable X is uncountably infinite, X has a distinctly different distribution from the discrete random variables studied previously. The range of X includes all values in an interval of real numbers; that is, the range of X can be thought of as a continuum.

A number of continuous distributions frequently arise in applications. These distributions are described, and example computations of probabilities, means, and variances are provided in the remaining sections of this chapter.

4-2 PROBABILITY DISTRIBUTIONS AND PROBABILITY DENSITY FUNCTIONS

Density functions are commonly used in engineering to describe physical systems. For example, consider the density of a loading on a long, thin beam as shown in Fig. 4-1. For any point x along the beam, the density can be described by a function (in grams/cm). Intervals with large loadings correspond to large values for the function. The total loading between points a and b is determined as the integral of the density function from a to b. This integral is the area under the density function over this interval, and it can be loosely interpreted as the sum of all the loadings over this interval.

Similarly, a **probability density function** $f(x)$ can be used to describe the probability distribution of a **continuous random variable** X. If an interval is likely to contain a value for X, its probability is large and it corresponds to large values for $f(x)$. The probability that X is between a and b is determined as the integral of $f(x)$ from a to b. See Fig. 4-2.

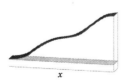

Figure 4-1 Density function of a loading on a long, thin beam.

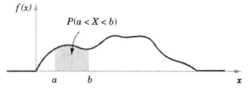

Figure 4-2 Probability determined from the area under $f(x)$.

Probability Density Function

For a continuous random variable X, a **probability density function** is a function such that

(1) $f(x) \geq 0$

(2) $\int_{-\infty}^{\infty} f(x)\, dx = 1$

(3) $P(a \leq X \leq b) = \int_{a}^{b} f(x)\, dx =$ area under $f(x)$ from a to b

for any a and b (4-1)

A probability density function provides a simple description of the probabilities associated with a random variable. As long as $f(x)$ is nonnegative and $\int_{-\infty}^{\infty} f(x)\, dx = 1$, $0 \leq P(a < X < b) \leq 1$ so that the probabilities are properly restricted. A probability density function is zero for x values that cannot occur and it is assumed to be zero wherever it is not specifically defined.

A histogram is an approximation to a probability density function. See Fig. 4-3. For each interval of the histogram, the area of the bar equals the relative frequency (proportion) of the measurements in the interval. The relative frequency is an estimate of the probability that a measurement falls in the interval. Similarly, the area under $f(x)$ over any interval equals the true probability that a measurement falls in the interval.

The important point is that $f(x)$ **is used to calculate an area** that represents the probability that X assumes a value in $[a, b]$. For the current measurement example, the probability that X results in [14 mA, 15 mA] is the integral of the probability density function of X over this interval. The probability that X results in [14.5 mA, 14.6 mA] is the integral of the same function, $f(x)$, over the smaller interval. By appropriate choice of the shape of $f(x)$, we can represent the probabilities associated with any continuous random variable X. The shape of $f(x)$ determines how the probability that X assumes a value in [14.5 mA, 14.6 mA] compares to the probability of any other interval of equal or different length.

For the density function of a loading on a long thin beam, because every point has zero width, the loading at any point is zero. Similarly, for a continuous random variable X and *any* value x.

$$P(X = x) = 0$$

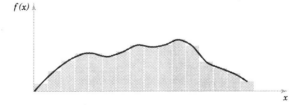

Figure 4-3 Histogram approximates a probability density function.

Based on this result, it might appear that our model of a continuous random variable is useless. However, in practice, when a particular current measurement is observed, such as 14.47 milliamperes, this result can be interpreted as the rounded value of a current measurement that is actually in a range such as $14.465 \le x \le 14.475$. Therefore, the probability that the rounded value 14.47 is observed as the value for X is the probability that X assumes a value in the interval [14.465, 14.475], which is not zero. Similarly, because each point has zero probability, one need not distinguish between inequalities such as $<$ or $\le$ for continuous random variables.

If X is a continuous random variable, for any x_1 and x_2,

$$P(x_1 \le X \le x_2) = P(x_1 < X \le x_2) = P(x_1 \le X < x_2) = P(x_1 < X < x_2) \quad (4\text{-}2)$$

EXAMPLE 4-1
Electric Current

Let the continuous random variable X denote the current measured in a thin copper wire in milliamperes. Assume that the range of X is [0, 20 mA], and assume that the probability density function of X is $f(x) = 0.05$ for $0 \le x \le 20$. What is the probability that a current measurement is less than 10 milliamperes?

The probability density function is shown in Fig. 4-4. It is assumed that $f(x) = 0$ wherever it is not specifically defined. The probability requested is indicated by the shaded area in Fig. 4-4.

$$P(X < 10) = \int_0^{10} f(x)\,dx = \int_0^{10} 0.05\,dx = 0.5$$

As another example,

$$P(5 < X < 20) = \int_5^{20} f(x)\,dx = 0.75$$

EXAMPLE 4-2
Hole Diameter

Let the continuous random variable X denote the diameter of a hole drilled in a sheet metal component. The target diameter is 12.5 millimeters. Most random disturbances to the process result in larger diameters. Historical data show that the distribution of X can be modeled by a probability density function $f(x) = 20e^{-20(x-12.5)}, x \ge 12.5$.

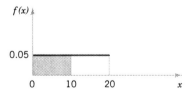

Figure 4-4 Probability density function for Example 4-1.

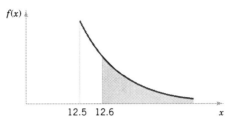

Figure 4-5 Probability density function for Example 4-2.

If a part with a diameter larger than 12.60 millimeters is scrapped, what proportion of parts is scrapped? The density function and the requested probability are shown in Fig. 4-5. A part is scrapped if $X > 12.60$. Now,

$$P(X > 12.60) = \int_{12.6}^{\infty} f(x)\, dx = \int_{12.6}^{\infty} 20 e^{-20(x-12.5)}\, dx = -e^{-20(x-12.5)} \Big|_{12.6}^{\infty} = 0.135$$

What proportion of parts is between 12.5 and 12.6 millimeters? Now,

$$P(12.5 < X < 12.6) = \int_{12.5}^{12.6} f(x)\, dx = -e^{-20(x-12.5)} \Big|_{12.5}^{12.6} = 0.865$$

Because the total area under $f(x)$ equals 1, we can also calculate $P(12.5 < X < 12.6) = 1 - P(X > 12.6) = 1 - 0.135 = 0.865$.

EXERCISES FOR SECTION 4-2

4-1. Suppose that $f(x) = e^{-x}$ for $0 < x$. Determine the following probabilities:
(a) $P(1 < X)$ (b) $P(1 < X < 2.5)$
(c) $P(X = 3)$ (d) $P(X < 4)$
(e) $P(3 \le X)$

4-2. Suppose that $f(x) = e^{-x}$ for $0 < x$.
(a) Determine x such that $P(x < X) = 0.10$.
(b) Determine x such that $P(X \le x) = 0.10$.

4-3. Suppose that $f(x) = x/8$ for $3 < x < 5$. Determine the following probabilities:
(a) $P(X < 4)$ (b) $P(X > 3.5)$
(c) $P(4 < X < 5)$ (d) $P(X < 4.5)$
(e) $P(X < 3.5 \text{ or } X > 4.5)$

4-4. Suppose that $f(x) = e^{-(x-4)}$ for $4 < x$. Determine the following probabilities:
(a) $P(1 < X)$ (b) $P(2 \le X < 5)$
(c) $P(5 < X)$ (d) $P(8 < X < 12)$
(e) Determine x such that $P(X < x) = 0.90$.

4-5. Suppose that $f(x) = 1.5x^2$ for $-1 < x < 1$. Determine the following probabilities:
(a) $P(0 < X)$ (b) $P(0.5 < X)$
(c) $P(-0.5 \le X \le 0.5)$ (d) $P(X < -2)$

(e) $P(X < 0 \text{ or } X > -0.5)$
(f) Determine x such that $P(x < X) = 0.05$.

4-6. The probability density function of the time to failure of an electronic component in a copier (in hours) is $f(x) = e^{-x/1000}/1000$ for $x > 0$. Determine the probability that
(a) A component lasts more than 3000 hours before failure.
(b) A component fails in the interval from 1000 to 2000 hours.
(c) A component fails before 1000 hours.
(d) Determine the number of hours at which 10% of all components have failed.

4-7. The probability density function of the net weight in pounds of a packaged chemical herbicide is $f(x) = 2.0$ for $49.75 < x < 50.25$ pounds.
(a) Determine the probability that a package weighs more than 50 pounds.
(b) How much chemical is contained in 90% of all packages?

4-8. The probability density function of the length of a hinge for fastening a door is $f(x) = 1.25$ for $74.6 < x < 75.4$ millimeters. Determine the following:
(a) $P(X < 74.8)$
(b) $P(X < 74.8 \text{ or } X > 75.2)$

(c) If the specifications for this process are from 74.7 to 75.3 millimeters, what proportion of hinges meets specifications?

4-9. The probability density function of the length of a metal rod is $f(x) = 2$ for $2.3 < x < 2.8$ meters.
(a) If the specifications for this process are from 2.25 to 2.75 meters, what proportion of the bars fail to meet the specifications?
(b) Assume that the probability density function is $f(x) = 2$ for an interval of length 0.5 meters. Over what value should the density be centered to achieve the greatest proportion of bars within specifications?

4-10. If X is a continuous random variable, argue that $P(x_1 \leq X \leq x_2) = P(x_1 < X \leq x_2) = P(x_1 \leq X < x_2) = P(x_1 < X < x_2)$.

4-3 CUMULATIVE DISTRIBUTION FUNCTIONS

An alternative method to describe the distribution of a discrete random variable can also be used for continuous random variables.

Cumulative Distribution Function

The **cumulative distribution function** of a continuous random variable X is

$$F(x) = P(X \leq x) = \int_{-\infty}^{x} f(u)\, du \qquad (4\text{-}3)$$

for $-\infty < x < \infty$.

Extending the definition of $f(x)$ to the entire real line enables us to define the cumulative distribution function for all real numbers. The following example illustrates the definition.

EXAMPLE 4-3
Electric Current

For the copper current measurement in Example 4-1, the cumulative distribution function of the random variable X consists of three expressions. If $x < 0$, $f(x) = 0$. Therefore,

$$F(x) = 0, \quad \text{for} \quad x < 0$$

and

$$F(x) = \int_{0}^{x} f(u)\, du = 0.05x, \quad \text{for} \quad 0 \leq x < 20$$

Finally,

$$F(x) = \int_{0}^{x} f(u)\, du = 1, \quad \text{for} \quad 20 \leq x$$

Therefore,

$$F(x) = \begin{cases} 0 & x < 0 \\ 0.05x & 0 \leq x < 20 \\ 1 & 20 \leq x \end{cases}$$

The plot of $F(x)$ is shown in Fig. 4-6.

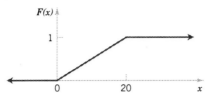
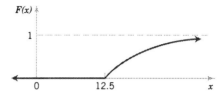

Figure 4-6 Cumulative distribution function for Example 4-3.

Figure 4-7 Cumulative distribution function for Example 4-4.

Notice that in the definition of $F(x)$ any $<$ can be changed to $\leq$ and vice versa. That is, $F(x)$ can be defined as either $0.05x$ or 0 at the end-point $x = 0$, and $F(x)$ can be defined as either $0.05x$ or 1 at the end-point $x = 20$. In other words, $F(x)$ is a continuous function. For a discrete random variable, $F(x)$ is not a continuous function. Sometimes, a continuous random variable is defined as one that has a continuous cumulative distribution function.

EXAMPLE 4-4
Hole Diameter

For the drilling operation in Example 4-2, $F(x)$ consists of two expressions.

$$F(x) = 0 \quad \text{for} \quad x < 12.5$$

and for $12.5 \leq x$

$$F(x) = \int_{12.5}^{x} 20 e^{-20(u-12.5)} \, du$$
$$= 1 - e^{-20(x-12.5)}$$

Therefore,

$$F(x) = \begin{cases} 0 & x < 12.5 \\ 1 - e^{-20(x-12.5)} & 12.5 \leq x \end{cases}$$

Figure 4-7 displays a graph of $F(x)$.

The probability density function of a continuous random variable can be determined from the cumulative distribution function by differentiating. Recall that the fundamental theorem of calculus states that

$$\frac{d}{dx} \int_{-\infty}^{x} f(u) \, du = f(x)$$

Then, given $F(x)$

$$f(x) = \frac{dF(x)}{dx}$$

as long as the derivative exists.

EXAMPLE 4-5
Reaction Time

The time until a chemical reaction is complete (in milliseconds) is approximated by the cumulative distribution function

$$F(x) = \begin{cases} 0 & x < 0 \\ 1 - e^{-0.01x} & 0 \le x \end{cases}$$

Determine the probability density function of X. What proportion of reactions is complete within 200 milliseconds? Using the result that the probability density function is the derivative of the $F(x)$, we obtain

$$f(x) = \begin{cases} 0 & x < 0 \\ 0.01 e^{-0.01x} & 0 \le x \end{cases}$$

The probability that a reaction completes within 200 milliseconds is

$$P(X < 200) = F(200) = 1 - e^{-2} = 0.8647.$$

EXERCISES FOR SECTION 4-3

4-11. Suppose the cumulative distribution function of the random variable X is

$$F(x) = \begin{cases} 0 & x < 0 \\ 0.2x & 0 \le x < 5 \\ 1 & 5 \le x \end{cases}$$

Determine the following:
(a) $P(X < 2.8)$ (b) $P(X > 1.5)$
(c) $P(X < -2)$ (d) $P(X > 6)$

4-12. Suppose the cumulative distribution function of the random variable X is

$$F(x) = \begin{cases} 0 & x < -2 \\ 0.25x + 0.5 & -2 \le x < 2 \\ 1 & 2 \le x \end{cases}$$

Determine the following:
(a) $P(X < 1.8)$ (b) $P(X > -1.5)$
(c) $P(X < -2)$ (d) $P(-1 < X < 1)$

4-13. The gap width is an important property of a magnetic recording head. In coded units, if the width is a continuous random variable over the range from $0 < x < 2$ with $f(x) = 0.5x$, determine the cumulative distribution function of the gap width.

4-14. The probability density function of the time customers arrive at a terminal (in minutes after 8:00 A.M.) is $f(x) = e^{-x/10}/10$ for $0 < x$. Determine the probability that
(a) The first customer arrives by 9:00 A.M.
(b) The first customer arrives between 8:15 A.M. and 8:30 A.M.
(c) Two or more customers arrive before 8:40 A.M. among five that arrive at the terminal. Assume customer arrivals are independent.
(d) Determine the cumulative distribution function and use the cumulative distribution function to determine the probability that the first customer arrives between 8:15 A.M. and 8:30 A.M.

4-15. Determine the cumulative distribution function for the distribution in Exercise 4-1.

4-16. Determine the cumulative distribution function for the distribution in Exercise 4-3.

4-17. Determine the cumulative distribution function for the distribution in Exercise 4-4.

4-18. Determine the cumulative distribution function for the distribution in Exercise 4-6. Use the cumulative distribution function to determine the probability that a component lasts more than 3000 hours before failure.

4-19. Determine the cumulative distribution function for the distribution in Exercise 4-9. Use the cumulative distribution function to determine the probability that a length exceeds 2.7 meters.

Determine the probability density function for each of the following cumulative distribution functions.

4-20. $F(x) = 1 - e^{-2x} \quad x > 0$

4-21.

$$F(x) = \begin{cases} 0 & x < 0 \\ 0.2x & 0 \le x < 4 \\ 0.04x + 0.64 & 4 \le x < 9 \\ 1 & 9 \le x \end{cases}$$

4-22.

$$F(x) = \begin{cases} 0 & x < -2 \\ 0.25x + 0.5 & -2 \le x < 1 \\ 0.5x + 0.25 & 1 \le x < 1.5 \\ 1 & 1.5 \le x \end{cases}$$

4-4 MEAN AND VARIANCE OF A CONTINUOUS RANDOM VARIABLE

The mean and variance can also be defined for a continuous random variable. Integration replaces summation in the discrete definitions. If a probability density function is viewed as a loading on a beam as in Fig. 4-1, the mean is the balance point.

Mean and Variance

Suppose X is a continuous random variable with probability density function $f(x)$. The mean or expected value of X, denoted as μ or $E(X)$, is

$$\mu = E(X) = \int_{-\infty}^{\infty} x f(x)\, dx \qquad (4\text{-}4)$$

The variance of X, denoted as $V(X)$ or σ^2, is

$$\sigma^2 = V(X) = \int_{-\infty}^{\infty} (x - \mu)^2 f(x)\, dx = \int_{-\infty}^{\infty} x^2 f(x)\, dx - \mu^2$$

The standard deviation of X is $\sigma = \sqrt{\sigma^2}$.

The equivalence of the two formulas for variance can be derived as one, as was done for discrete random variables.

EXAMPLE 4-6 Electric Current

For the copper current measurement in Example 4-1, the mean of X is

$$E(X) = \int_0^{20} x f(x)\, dx = 0.05 x^2 / 2 \Big|_0^{20} = 10$$

The variance of X is

$$V(X) = \int_0^{20} (x - 10)^2 f(x)\, dx = 0.05(x - 10)^3/3 \Big|_0^{20} = 33.33$$

The expected value of a function $h(X)$ of a continuous random variable is also defined in a straightforward manner.

Expected Value of a Function of a Continuous Random Variable

If X is a continuous random variable with probability density function $f(x)$,

$$E[h(X)] = \int_{-\infty}^{\infty} h(x) f(x)\, dx \qquad (4\text{-}5)$$

EXAMPLE 4-7

In Example 4-1, X is the current measured in milliamperes. What is the expected value of the squared current? Now, $h(X) = X^2$. Therefore,

$$E[h(X)] = \int_{-\infty}^{\infty} x^2 f(x)\, dx = \int_0^{20} 0.05 x^2\, dx = 0.05 \left. \frac{x^3}{3} \right|_0^{20} = 133.33$$

In the previous example, the expected value of X^2 does not equal $E(X)$ squared. However, in the special case that $h(X) = aX + b$ for any constants a and b, $E[h(X)] = aE(X) + b$. This can be shown from the properties of integrals.

EXAMPLE 4-8
Hole Diameter

For the drilling operation in Example 4-2, the mean of X is

$$E(X) = \int_{12.5}^{\infty} x f(x)\, dx = \int_{12.5}^{\infty} x\, 20 e^{-20(x-12.5)}\, dx$$

Integration by parts can be used to show that

$$E(X) = -x e^{-20(x-12.5)} - \left. \frac{e^{-20(x-12.5)}}{20} \right|_{12.5}^{\infty} = 12.5 + 0.05 = 12.55$$

The variance of X is

$$V(X) = \int_{12.5}^{\infty} (x - 12.55)^2 f(x)\, dx$$

Although more difficult, integration by parts can be used two times to show that $V(X) = 0.0025$.

EXERCISES FOR SECTION 4-4

4-23. Suppose $f(x) = 0.25$ for $0 < x < 4$. Determine the mean and variance of X.

4-24. Suppose $f(x) = 0.125x$ for $0 < x < 4$. Determine the mean and variance of X.

4-25. Suppose $f(x) = 1.5x^2$ for $-1 < x < 1$. Determine the mean and variance of X.

4-26. Suppose that $f(x) = x/8$ for $3 < x < 5$. Determine the mean and variance for x.

4-27. Suppose that contamination particle size (in micrometers) can be modeled as $f(x) = 2x^{-3}$ for $1 < x$. Determine the mean of X.

4-28. Suppose the probability density function of the length of computer cables is $f(x) = 0.1$ from 1200 to 1210 millimeters.
(a) Determine the mean and standard deviation of the cable length.
(b) If the length specifications are $1195 < x < 1205$ millimeters, what proportion of cables are within specifications?

4-29. The thickness of a conductive coating in micrometers has a density function of $600 x^{-2}$ for $100\ \mu m < x < 120\ \mu m$.

(a) Determine the mean and variance of the coating thickness.
(b) If the coating costs \$0.50 per micrometer of thickness on each part, what is the average cost of the coating per part?

4-30. The probability density function of the weight of packages delivered by a post office is $f(x) = 70/(69 x^2)$ for $1 < x < 70$ pounds.
(a) Determine the mean and variance of weight.
(b) If the shipping cost is \$2.50 per pound, what is the average shipping cost of a package?
(c) Determine the probability that the weight of a package exceeds 50 pounds.

4-31. Integration by parts is required. The probability density function for the diameter of a drilled hole in millimeters is $10 e^{-10(x-5)}$ for $x > 5$ mm. Although the target diameter is 5 millimeters, vibrations, tool wear, and other nuisances produce diameters larger than 5 millimeters.
(a) Determine the mean and variance of the diameter of the holes.
(b) Determine the probability that a diameter exceeds 5.1 millimeters.

4-5 CONTINUOUS UNIFORM DISTRIBUTION

The simplest continuous distribution is analogous to its discrete counterpart.

Continuous Uniform Distribution

A continuous random variable X with probability density function

$$f(x) = 1/(b - a), \quad a \leq x \leq b \tag{4-6}$$

is a continuous uniform random variable.

The probability density function of a continuous uniform random variable is shown in Fig. 4-8. The mean of the continuous uniform random variable X is

$$E(X) = \int_a^b \frac{x}{b-a} dx = \left. \frac{0.5x^2}{b-a} \right|_a^b = \frac{(a+b)}{2}$$

The variance of X is

$$V(X) = \int_a^b \frac{\left(x - \left(\frac{a+b}{2}\right)\right)^2}{b-a} dx = \left. \frac{\left(x - \frac{a+b}{2}\right)^3}{3(b-a)} \right|_a^b = \frac{(b-a)^2}{12}$$

These results are summarized as follows.

Mean and Variance

If X is a continuous uniform random variable over $a \leq x \leq b$,

$$\mu = E(X) = \frac{(a+b)}{2} \quad \text{and} \quad \sigma^2 = V(X) = \frac{(b-a)^2}{12} \tag{4-7}$$

EXAMPLE 4-9
Uniform Current

Let the continuous random variable X denote the current measured in a thin copper wire in milliamperes. Assume that the range of X is [0, 20 mA], and assume that the probability density function of X is $f(x) = 0.05, 0 \leq x \leq 20$.

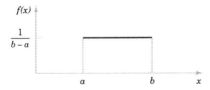

Figure 4-8 Continuous uniform probability density function.

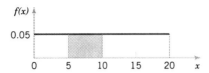

Figure 4-9 Probability for Example 4-9.

What is the probability that a measurement of current is between 5 and 10 milliamperes? The requested probability is shown as the shaded area in Fig. 4-9.

$$P(5 < X < 10) = \int_5^{10} f(x)\,dx$$

$$= 5(0.05) = 0.25$$

The mean and variance formulas can be applied with $a = 0$ and $b = 20$. Therefore,

$$E(X) = 10 \text{ mA} \quad \text{and} \quad V(X) = 20^2/12 = 33.33 \text{ mA}^2$$

Consequently, the standard deviation of X is 5.77 mA.

The cumulative distribution function of a continuous uniform random variable is obtained by integration. If $a < x < b$,

$$F(x) = \int_a^x 1/(b-a)\,du = x/(b-a) - a/(b-a)$$

Therefore, the complete description of the cumulative distribution function of a continuous uniform random variable is

$$F(x) = \begin{cases} 0 & x < a \\ (x-a)/(b-a) & a \le x < b \\ 1 & b \le x \end{cases}$$

An example of $F(x)$ for a continuous uniform random variable is shown in Fig. 4-6.

EXERCISES FOR SECTION 4-5

4-32. Suppose X has a continuous uniform distribution over the interval $[1.5, 5.5]$.
(a) Determine the mean, variance, and standard deviation of X.
(b) What is $P(X < 2.5)$?
(c) Determine the cumulative distribution function.

4-33. Suppose X has a continuous uniform distribution over the interval $[-1, 1]$.
(a) Determine the mean, variance, and standard deviation of X.
(b) Determine the value for x such that $P(-x < X < x) = 0.90$.
(c) Determine the cumulative distribution function.

4-34. The net weight in pounds of a packaged chemical herbicide is uniform for $49.75 < x < 50.25$ pounds.

(a) Determine the mean and variance of the weight of packages.
(b) Determine the cumulative distribution function of the weight of packages.
(c) Determine $P(X < 50.1)$.

4-35. The thickness of a flange on an aircraft component is uniformly distributed between 0.95 and 1.05 millimeters.
(a) Determine the cumulative distribution function of flange thickness.
(b) Determine the proportion of flanges that exceeds 1.02 millimeters.
(c) What thickness is exceeded by 90% of the flanges?
(d) Determine the mean and variance of flange thickness.

4-36. Suppose the time it takes a data collection operator to fill out an electronic form for a database is uniformly between 1.5 and 2.2 minutes.
(a) What is the mean and variance of the time it takes an operator to fill out the form?
(b) What is the probability that it will take less than two minutes to fill out the form?
(c) Determine the cumulative distribution function of the time it takes to fill out the form.

4-37. The thickness of photoresist applied to wafers in semiconductor manufacturing at a particular location on the wafer is uniformly distributed between 0.2050 and 0.2150 micrometers.
(a) Determine the cumulative distribution function of photoresist thickness.
(b) Determine the proportion of wafers that exceeds 0.2125 micrometers in photoresist thickness.
(c) What thickness is exceeded by 10% of the wafers?
(d) Determine the mean and variance of photoresist thickness.

4-38. An adult can lose or gain two pounds of water in the course of a day. Assume that the changes in water weight is uniformly distributed between minus two and plus two pounds in a day. What is the standard deviation of your weight over a day?

4-39. A dolphin show is scheduled to start at 9:00 A.M., 9:30 A.M., and 10:00 A.M. Once the show starts, the gate will be closed. A visitor will arrive at the gate at a time uniformly distributed between 8:30 A.M. and 10:00 A.M. Determine
(a) The cumulative distribution function of the time (in minutes) between arrival and 8:30 A.M.
(b) The mean and variance of the distribution in the previous part.
(c) The probability that a visitor waits less than 10 minutes for a show.
(d) The probability that a visitor waits more than 20 minutes for a show.

4-40. Measurement error that is continuous and uniformly distributed from -3 to $+3$ millivolts is added to the true voltage of a circuit. Then the measurement is rounded to the nearest millivolt so that it becomes discrete. Suppose that the true voltage is 250 millivolts.
(a) What is the probability mass function of the measured voltage?
(b) What is the mean and variance of the measured voltage?

4-6 NORMAL DISTRIBUTION

Undoubtedly, the most widely used model for the distribution of a random variable is a **normal distribution**. Whenever a random experiment is replicated, the random variable that equals the average (or total) result over the replicates tends to have a normal distribution as the number of replicates becomes large. De Moivre presented this fundamental result, known as the **central limit theorem**, in 1733. Unfortunately, his work was lost for some time, and Gauss independently developed a normal distribution nearly 100 years later. Although De Moivre was later credited with the derivation, a normal distribution is also referred to as a **Gaussian distribution**.

When do we average (or total) results? Almost always. For example, an automotive engineer may plan a study to average pull-off force measurements from several connectors. If we assume that each measurement results from a replicate of a random experiment, the normal distribution can be used to make approximate conclusions about this average. These conclusions are the primary topics in the subsequent chapters of this book.

Furthermore, sometimes the central limit theorem is less obvious. For example, assume that the deviation (or error) in the length of a machined part is the sum of a large number of infinitesimal effects, such as temperature and humidity drifts, vibrations, cutting angle variations, cutting tool wear, bearing wear, rotational speed variations, mounting and fixture variations, variations in numerous raw material characteristics, and variation in levels of contamination. If the component errors are independent and equally likely to be positive or negative, the total error can be shown to have an approximate normal distribution. Furthermore, the normal distribution arises in the study of numerous basic physical phenomena. For example, the physicist Maxwell developed a normal distribution from simple assumptions regarding the velocities of molecules.

The theoretical basis of a normal distribution is mentioned to justify the somewhat complex form of the probability density function. Our objective now is to calculate probabilities for a normal random variable. The central limit theorem will be stated more carefully later.

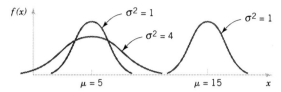

Figure 4-10 Normal probability density functions for selected values of the parameters μ and σ^2.

Random variables with different means and variances can be modeled by normal probability density functions with appropriate choices of the center and width of the curve. The value of $E(X) = \mu$ determines the center of the probability density function and the value of $V(X) = \sigma^2$ determines the width. Figure 4-10 illustrates several normal probability density functions with selected values of μ and σ^2. Each has the characteristic symmetric bell-shaped curve, but the centers and dispersions differ. The following definition provides the formula for normal probability density functions.

Normal Distribution

A random variable X with probability density function

$$f(x) = \frac{1}{\sqrt{2\pi}\sigma} e^{\frac{-(x-\mu)^2}{2\sigma^2}} \qquad -\infty < x < \infty \qquad (4\text{-}8)$$

is a **normal random variable** with parameters μ, where $-\infty < \mu < \infty$, and $\sigma > 0$. Also,

$$E(X) = \mu \quad \text{and} \quad V(X) = \sigma^2 \qquad (4\text{-}9)$$

and the notation $N(\mu, \sigma^2)$ is used to denote the distribution. The mean and variance of X are shown to equal μ and σ^2, respectively, at the end of this Section 5-6.

EXAMPLE 4-10   Assume that the current measurements in a strip of wire follow a normal distribution with a mean of 10 milliamperes and a variance of 4 (milliamperes)^2. What is the probability that a measurement exceeds 13 milliamperes?

Let X denote the current in milliamperes. The requested probability can be represented as $P(X > 13)$. This probability is shown as the shaded area under the normal probability density function in Fig. 4-11. Unfortunately, there is no closed-form expression for the integral of a normal probability density function, and probabilities based on the normal distribution are typically found numerically or from a table (that we will later introduce).

Some useful results concerning a normal distribution are summarized below and in Fig. 4-12. For any normal random variable,

$$P(\mu - \sigma < X < \mu + \sigma) = 0.6827$$
$$P(\mu - 2\sigma < X < \mu + 2\sigma) = 0.9545$$
$$P(\mu - 3\sigma < X < \mu + 3\sigma) = 0.9973$$

Also, from the symmetry of $f(x)$, $P(X > \mu) = P(X < \mu) = 0.5$. Because $f(x)$ is positive for all x, this model assigns some probability to each interval of the real line. However, the

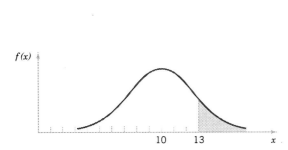

Figure 4-11 Probability that $X > 13$ for a normal random variable with $\mu = 10$ and $\sigma^2 = 4$.

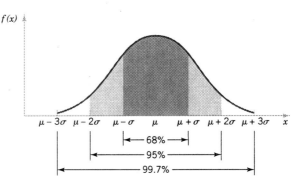

Figure 4-12 Probabilities associated with a normal distribution.

probability density function decreases as x moves farther from μ. Consequently, the probability that a measurement falls far from μ is small, and at some distance from μ the probability of an interval can be approximated as zero.

The area under a normal probability density function beyond 3σ from the mean is quite small. This fact is convenient for quick, rough sketches of a normal probability density function. The sketches help us determine probabilities. Because more than 0.9973 of the probability of a normal distribution is within the interval $(\mu - 3\sigma, \mu + 3\sigma)$, 6σ is often referred to as the **width** of a normal distribution. Advanced integration methods can be used to show that the area under the normal probability density function from $-\infty < x < \infty$ is 1.

Standard Normal Random Variable

A normal random variable with

$$\mu = 0 \quad \text{and} \quad \sigma^2 = 1$$

is called a standard normal random variable and is denoted as Z.

The cumulative distribution function of a standard normal random variable is denoted as

$$\Phi(z) = P(Z \leq z)$$

Appendix Table III provides cumulative probabilities for a standard normal random variable. Cumulative distribution functions for normal random variables are also widely available in computer packages. They can be used in the same manner as Appendix Table III to obtain probabilities for these random variables. The use of Table III is illustrated by the following example.

EXAMPLE 4-11
Standard Normal Distribution

Assume Z is a standard normal random variable. Appendix Table III provides probabilities of the form $\Phi(z) = P(Z \leq z)$. The use of Table II to find $P(Z \leq 1.5)$ is illustrated in Fig. 4-13. Read down the z column to the row that equals 1.5. The probability is read from the adjacent column, labeled 0.00, to be 0.93319.

The column headings refer to the hundredth's digit of the value of z in $P(Z \leq z)$. For example, $P(Z \leq 1.53)$ is found by reading down the z column to the row 1.5 and then selecting the probability from the column labeled 0.03 to be 0.93699.

124 CHAPTER 4 CONTINUOUS RANDOM VARIABLES AND PROBABILITY DISTRIBUTIONS

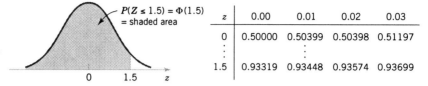

Figure 4-13 Standard normal probability density function.

Probabilities that are not of the form $P(Z \leq z)$ are found by using the basic rules of probability and the symmetry of the normal distribution along with Appendix Table III. The following examples illustrate the method.

EXAMPLE 4-12 The following calculations are shown pictorially in Fig. 4-14. In practice, a probability is often rounded to one or two significant digits.

(1) $P(Z > 1.26) = 1 - P(Z \leq 1.26) = 1 - 0.89616 = 0.10384$

(2) $P(Z < -0.86) = 0.19490$.

(3) $P(Z > -1.37) = P(Z < 1.37) = 0.91465$

(4) $P(-1.25 < Z < 0.37)$. This probability can be found from the difference of two areas, $P(Z < 0.37) - P(Z < -1.25)$. Now,

$$P(Z < 0.37) = 0.64431 \quad \text{and} \quad P(Z < -1.25) = 0.10565$$

Therefore,

$$P(-1.25 < Z < 0.37) = 0.64431 - 0.10565 = 0.53866$$

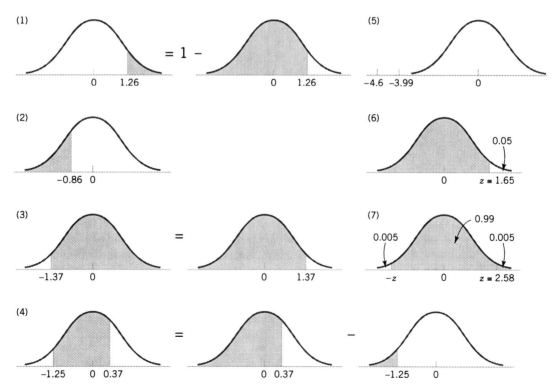

Figure 4-14 Graphical displays for standard normal distributions.

(5) $P(Z \leq -4.6)$ cannot be found exactly from Appendix Table II. However, the last entry in the table can be used to find that $P(Z \leq -3.99) = 0.00003$. Because $P(Z \leq -4.6) < P(Z \leq -3.99)$, $P(Z \leq -4.6)$ is nearly zero.

(6) Find the value z such that $P(Z > z) = 0.05$. This probability expression can be written as $P(Z \leq z) = 0.95$. Now, Table II is used in reverse. We search through the probabilities to find the value that corresponds to 0.95. The solution is illustrated in Fig. 4-14. We do not find 0.95 exactly; the nearest value is 0.95053, corresponding to $z = 1.65$.

(7) Find the value of z such that $P(-z < Z < z) = 0.99$. Because of the symmetry of the normal distribution, if the area of the shaded region in Fig. 4-14(7) is to equal 0.99, the area in each tail of the distribution must equal 0.005. Therefore, the value for z corresponds to a probability of 0.995 in Table II. The nearest probability in Table II is 0.99506, when $z = 2.58$.

The preceding examples show how to calculate probabilities for standard normal random variables. To use the same approach for an arbitrary normal random variable would require a separate table for every possible pair of values for μ and σ. Fortunately, all normal probability distributions are related algebraically, and Appendix Table III can be used to find the probabilities associated with an arbitrary normal random variable by first using a simple transformation.

Standardizing a Normal Random Variable

If X is a normal random variable with $E(X) = \mu$ and $V(X) = \sigma^2$, the random variable

$$Z = \frac{X - \mu}{\sigma} \qquad (4\text{-}10)$$

is a normal random variable with $E(Z) = 0$ and $V(Z) = 1$. That is, Z is a standard normal random variable.

Creating a new random variable by this transformation is referred to as *standardizing*. The random variable Z represents the distance of X from its mean in terms of standard deviations. It is the key step to calculate a probability for an arbitrary normal random variable.

EXAMPLE 4-13
Normally Distributed Current

Suppose the current measurements in a strip of wire are assumed to follow a normal distribution with a mean of 10 milliamperes and a variance of 4 (milliamperes)^2. What is the probability that a measurement will exceed 13 milliamperes?

Let X denote the current in milliamperes. The requested probability can be represented as $P(X > 13)$. Let $Z = (X - 10)/2$. The relationship between the several values of X and the transformed values of Z are shown in Fig. 4-15. We note that $X > 13$ corresponds to $Z > 1.5$. Therefore, from Appendix Table II,

$$P(X > 13) = P(Z > 1.5) = 1 - P(Z \leq 1.5) = 1 - 0.93319 = 0.06681$$

Rather than using Fig. 4-15, the probability can be found from the inequality $X > 13$. That is,

$$P(X > 13) = P\left(\frac{(X - 10)}{2} > \frac{(13 - 10)}{2}\right) = P(Z > 1.5) = 0.06681$$

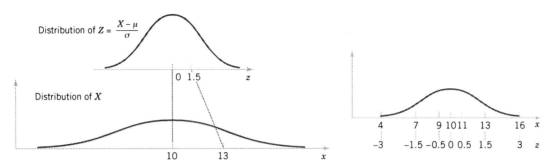

Figure 4-15 Standardizing a normal random variable.

In the preceding example, the value 13 is transformed to 1.5 by standardizing, and 1.5 is often referred to as the *z*-value associated with a probability. The following summarizes the calculation of probabilities derived from normal random variables.

Standardizing to Calculate a Probability

Suppose X is a normal random variable with mean μ and variance σ^2. Then,

$$P(X \leq x) = P\left(\frac{X - \mu}{\sigma} \leq \frac{x - \mu}{\sigma}\right) = P(Z \leq z) \qquad (4\text{-}11)$$

where Z is a standard normal random variable, and $z = \dfrac{(x - \mu)}{\sigma}$ is the *z*-value obtained by standardizing X.

The probability is obtained by using Appendix Table II with $z = (x - \mu)/\sigma$.

EXAMPLE 4-14
Normally Distributed Current

Continuing the previous example, what is the probability that a current measurement is between 9 and 11 milliamperes? From Fig. 4-15, or by proceeding algebraically, we have

$$P(9 < X < 11) = P((9 - 10)/2 < (X - 10)/2 < (11 - 10)/2)$$
$$= P(-0.5 < Z < 0.5) = P(Z < 0.5) - P(Z < -0.5)$$
$$= 0.69146 - 0.30854 = 0.38292$$

Determine the value for which the probability that a current measurement is below this value is 0.98. The requested value is shown graphically in Fig. 4-16. We need the value of x such that $P(X < x) = 0.98$. By standardizing, this probability expression can be written as

$$P(X < x) = P((X - 10)/2 < (x - 10)/2)$$
$$= P(Z < (x - 10)/2)$$
$$= 0.98$$

Appendix Table III is used to find the z-value such that $P(Z < z) = 0.98$. The nearest probability from Table III results in

$$P(Z < 2.05) = 0.97982$$

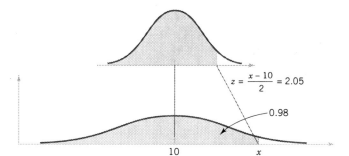

Figure 4-16 Determining the value of x to meet a specified probability.

Therefore, $(x - 10)/2 = 2.05$, and the standardizing transformation is used in reverse to solve for x. The result is

$$x = 2(2.05) + 10 = 14.1 \text{ milliamperes}$$

EXAMPLE 4-15
Signal Detection

Assume that in the detection of a digital signal the background noise follows a normal distribution with a mean of 0 volt and standard deviation of 0.45 volt. The system assumes a digital 1 has been transmitted when the voltage exceeds 0.9. What is the probability of detecting a digital 1 when none was sent?

Let the random variable N denote the voltage of noise. The requested probability is

$$P(N > 0.9) = P\left(\frac{N}{0.45} > \frac{0.9}{0.45}\right) = P(Z > 2) = 1 - 0.97725 = 0.02275$$

This probability can be described as the probability of a false detection.

Determine symmetric bounds about 0 that include 99% of all noise readings. The question requires us to find x such that $P(-x < N < x) = 0.99$. A graph is shown in Fig. 4-17. Now,

$$P(-x < N < x) = P(-x/0.45 < N/0.45 < x/0.45)$$
$$= P(-x/0.45 < Z < x/0.45) = 0.99$$

From Appendix Table III

$$P(-2.58 < Z < 2.58) = 0.99$$

Therefore,

$$x/0.45 = 2.58$$

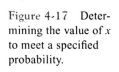

Figure 4-17 Determining the value of x to meet a specified probability.

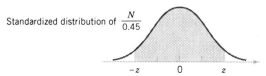

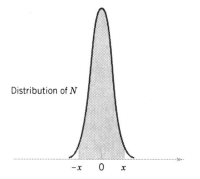

Figure 4-18 Distribution for Example 4-16.

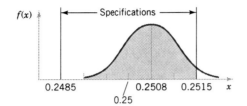

and

$$x = 2.58(0.45) = 1.16$$

Suppose that when a digital 1 signal is transmitted the mean of the noise distribution shifts to 1.8 volts. What is the probability that a digital 1 is not detected? Let the random variable S denote the voltage when a digital 1 is transmitted. Then,

$$P(S < 0.9) = P\left(\frac{S - 1.8}{0.45} < \frac{0.9 - 1.8}{0.45}\right) = P(Z < -2) = 0.02275$$

This probability can be interpreted as the probability of a missed signal.

EXAMPLE 4-16
Shaft Diameter

The diameter of a shaft in an optical storage drive is normally distributed with mean 0.2508 inch and standard deviation 0.0005 inch. The specifications on the shaft are 0.2500 ± 0.0015 inch. What proportion of shafts conforms to specifications?

Let X denote the shaft diameter in inches. The requested probability is shown in Fig. 4-18 and

$$P(0.2485 < X < 0.2515) = P\left(\frac{0.2485 - 0.2508}{0.0005} < Z < \frac{0.2515 - 0.2508}{0.0005}\right)$$
$$= P(-4.6 < Z < 1.4) = P(Z < 1.4) - P(Z < -4.6)$$
$$= 0.91924 - 0.0000 = 0.91924$$

Most of the nonconforming shafts are too large, because the process mean is located very near to the upper specification limit. If the process is centered so that the process mean is equal to the target value of 0.2500,

$$P(0.2485 < X < 0.2515) = P\left(\frac{0.2485 - 0.2500}{0.0005} < Z < \frac{0.2515 - 0.2500}{0.0005}\right)$$
$$= P(-3 < Z < 3)$$
$$= P(Z < 3) - P(Z < -3)$$
$$= 0.99865 - 0.00135$$
$$= 0.9973$$

By recentering the process, the yield is increased to approximately 99.73%.

EXERCISES FOR SECTION 4-6

4-41. Use Appendix Table III to determine the following probabilities for the standard normal random variable Z:
(a) $P(Z < 1.32)$ (b) $P(Z < 3.0)$
(c) $P(Z > 1.45)$ (d) $P(Z > -2.15)$
(e) $P(-2.34 < Z < 1.76)$

4-42. Use Appendix Table III to determine the following probabilities for the standard normal random variable Z:
(a) $P(-1 < Z < 1)$ (b) $P(-2 < Z < 2)$
(c) $P(-3 < Z < 3)$ (d) $P(Z > 3)$
(e) $P(0 < Z < 1)$

4-43. Assume Z has a standard normal distribution. Use Appendix Table III to determine the value for z that solves each of the following:
(a) $P(Z < z) = 0.9$
(b) $P(Z < z) = 0.5$
(c) $P(Z > z) = 0.1$
(d) $P(Z > z) = 0.9$
(e) $P(-1.24 < Z < z) = 0.8$

4-44. Assume Z has a standard normal distribution. Use Appendix Table III to determine the value for z that solves each of the following:
(a) $P(-z < Z < z) = 0.95$
(b) $P(-z < Z < z) = 0.99$
(c) $P(-z < Z < z) = 0.68$
(d) $P(-z < Z < z) = 0.9973$

4-45. Assume X is normally distributed with a mean of 10 and a standard deviation of 2. Determine the following:
(a) $P(X < 13)$
(b) $P(X > 9)$
(c) $P(6 < X < 14)$
(d) $P(2 < X < 4)$
(e) $P(-2 < X < 8)$

4-46. Assume X is normally distributed with a mean of 10 and a standard deviation of 2. Determine the value for x that solves each of the following:
(a) $P(X > x) = 0.5$
(b) $P(X > x) = 0.95$
(c) $P(x < X < 10) = 0.2$
(d) $P(-x < X - 10 < x) = 0.95$
(e) $P(-x < X - 10 < x) = 0.99$

4-47. Assume X is normally distributed with a mean of 5 and a standard deviation of 4. Determine the following:
(a) $P(X < 11)$
(b) $P(X > 0)$
(c) $P(3 < X < 7)$
(d) $P(-2 < X < 9)$
(e) $P(2 < X < 8)$

4-48. Assume X is normally distributed with a mean of 5 and a standard deviation of 4. Determine the value for x that solves each of the following:
(a) $P(X > x) = 0.5$
(b) $P(X > x) = 0.95$
(c) $P(x < X < 9) = 0.2$
(d) $P(3 < X < x) = 0.95$
(e) $P(-x < X - 5 < x) = 0.99$

4-49. The compressive strength of samples of cement can be modeled by a normal distribution with a mean of 6000 kilograms per square centimeter and a standard deviation of 100 kilograms per square centimeter.
(a) What is the probability that a sample's strength is less than 6250 Kg/cm^2?
(b) What is the probability that a sample's strength is between 5800 and 5900 Kg/cm^2?
(c) What strength is exceeded by 95% of the samples?

4-50. The time until recharge for a battery in a laptop computer under common conditions is normally distributed with mean of 260 minutes and a standard deviation of 50 minutes.
(a) What is the probability that a battery lasts more than four hours?
(b) What are the quartiles (the 25% and 75% values) of battery life?
(c) What value of life in minutes is exceeded with 95% probability?

4-51. An article in *Knee Surgery Sports Traumatol Arthrosc*, "Effect of provider volume on resource utilization for surgical procedures," (2005, Vol. 13, pp. 273–279) showed a mean time of 129 minutes and a standard deviation of 14 minutes for ACL reconstruction surgery at high-volume hospitals (with more than 300 such surgeries per year).
(a) What is the probability that your ACL surgery at a high-volume hospital requires a time more than two standard deviations above the mean?
(b) What is the probability that your ACL surgery at a high-volume hospital is completed in less than 100 minutes?
(c) The probability of a completed ACL surgery at a high-volume hospital is equal to 95% at what time?
(d) If your surgery requires 199 minutes, what do you conclude about the volume of such surgeries at your hospital? Explain.

4-52. Cholesterol is a fatty substance that is an important part of the outer lining (membrane) of cells in the body of animals. Its normal range for an adult is 120–240 mg/dl. The Food and Nutrition Institute of the Philippines found that the total cholesterol level for Filipino adults has a mean of 159.2 mg/dl and 84.1% of adults have a cholesterol level below 200 mg/dl (http://www.fnri.dost.gov.ph/facts/p3lipnat.html). Suppose that the total cholesterol level is normally distributed.
(a) Determine the standard deviation of this distribution.
(b) What are the quartiles (the 25% and 75% values) of this distribution?
(c) What is the value of the cholesterol level that exceeds 90% of the population?
(d) An adult is at moderate risk if cholesterol level is more than one but less than two standard deviations above the mean. What percentage of the population is at moderate risk according to this criterion?
(e) An adult is thought to be at high risk if his cholesterol level is more than two standard deviations above the mean. What percentage of the population is at high risk?
(f) An adult has low risk if cholesterol level is one standard deviation or more below the mean. What percentage of the population is at low risk?

4-53. The line width for semiconductor manufacturing is assumed to be normally distributed with a mean of 0.5 micrometer and a standard deviation of 0.05 micrometer.
(a) What is the probability that a line width is greater than 0.62 micrometer?
(b) What is the probability that a line width is between 0.47 and 0.63 micrometer?
(c) The line width of 90% of samples is below what value?

4-54. The fill volume of an automated filling machine used for filling cans of carbonated beverage is normally distributed with a mean of 12.4 fluid ounces and a standard deviation of 0.1 fluid ounce.
(a) What is the probability a fill volume is less than 12 fluid ounces?
(b) If all cans less than 12.1 or greater than 12.6 ounces are scrapped, what proportion of cans is scrapped?

(c) Determine specifications that are symmetric about the mean that include 99% of all cans.

4-55. In the previous exercise, suppose that the mean of the filling operation can be adjusted easily, but the standard deviation remains at 0.1 ounce.
(a) At what value should the mean be set so that 99.9% of all cans exceed 12 ounces?
(b) At what value should the mean be set so that 99.9% of all cans exceed 12 ounces if the standard deviation can be reduced to 0.05 fluid ounce?

4-56. The reaction time of a driver to visual stimulus is normally distributed with a mean of 0.4 seconds and a standard deviation of 0.05 seconds.
(a) What is the probability that a reaction requires more than 0.5 seconds?
(b) What is the probability that a reaction requires between 0.4 and 0.5 seconds?
(c) What is the reaction time that is exceeded 90% of the time?

4-57. The speed of a file transfer from a server on campus to a personal computer at a student's home on a weekday evening is normally distributed with a mean of 60 kilobits per second and a standard deviation of 4 kilobits per second.
(a) What is the probability that the file will transfer at a speed of 70 kilobits per second or more?
(b) What is the probability that the file will transfer at a speed of less than 58 kilobits per second?
(c) If the file is 1 megabyte, what is the average time it will take to transfer the file? (Assume eight bits per byte.)

4-58. The average height of a woman aged 20–74 years is 64 inches in 2002 with an increase of approximately one inch from 1960 (http://usgovinfo.about.com/od/healthcare). Suppose the height of a woman is normally distributed with a standard deviation of 2 inches.
(a) What is the probability a randomly selected woman in this population is between 58 inches and 70 inches?
(b) What are the quartiles of this distribution?
(c) Determine the height that is symmetric about the mean that includes 90% of this population.
(d) What is the probability that five women selected at random from this population all exceed 68 inches?

4-59. In an accelerator center, an experiment needs a 1.41 cm thick aluminum cylinder (http://puhep1.princeton.edu/mumu/target/Solenoid_Coil.pdf). Suppose that the thickness of a cylinder has a normal distribution with a mean 1.41 cm and a standard deviation of 0.01 cm.
(a) What is the probability that a thickness is greater than 1.42 cm?
(b) What thickness is exceeded by 95% of the samples?
(c) If the specifications require that the thickness is between 1.39 cm and 1.43 cm, what proportion of the samples meet specifications?

4-60. The demand for water use in Phoenix in 2003 hit a high of about 442 million gallons per day on June 27, 2003 (http://phoenix.gov/WATER/wtrfacts.html). Water use in the summer is normally distributed with a mean of 310 million gallons per day and a standard deviation of 45 million gallons per day. City reservoirs have a combined storage capacity of nearly 350 million gallons.
(a) What is the probability that a day requires more water than is stored in city reservoirs?
(b) What reservoir capacity is needed so that the probability it is exceeded is 1%?
(c) What amount of water use is exceeded with 95% probability?
(d) Water is provided to approximately 1.4 million people. What is the mean daily consumption per person at which the probability that the demand exceeds the current reservoir capacity is 1%? Assume that the standard deviation of demand remains the same.

4-61. The life of a semiconductor laser at a constant power is normally distributed with a mean of 7000 hours and a standard deviation of 600 hours.
(a) What is the probability that a laser fails before 5000 hours?
(b) What is the life in hours that 95% of the lasers exceed?
(c) If three lasers are used in a product and they are assumed to fail independently, what is the probability that all three are still operating after 7000 hours?

4-62. The diameter of the dot produced by a printer is normally distributed with a mean diameter of 0.002 inch and a standard deviation of 0.0004 inch.
(a) What is the probability that the diameter of a dot exceeds 0.0026 inch?
(b) What is the probability that a diameter is between 0.0014 and 0.0026 inch?
(c) What standard deviation of diameters is needed so that the probability in part (b) is 0.995?

4-63. The weight of a sophisticated running shoe is normally distributed with a mean of 12 ounces and a standard deviation of 0.5 ounce.
(a) What is the probability that a shoe weighs more than 13 ounces?
(b) What must the standard deviation of weight be in order for the company to state that 99.9% of its shoes are less than 13 ounces?
(c) If the standard deviation remains at 0.5 ounce, what must the mean weight be in order for the company to state that 99.9% of its shoes are less than 13 ounces?

4-64. Measurement error that is normally distributed with a mean of zero and a standard deviation of 0.5 gram is added to the true weight of a sample. Then the measurement is rounded to the nearest gram. Suppose that the true weight of a sample is 165.5 grams.
(a) What is the probability that the rounded result is 167 grams?
(b) What is the probability that the rounded result is 167 grams or greater?

4-7 NORMAL APPROXIMATION TO THE BINOMIAL AND POISSON DISTRIBUTIONS

We began our section on the normal distribution with the central limit theorem and the normal distribution as an approximation to a random variable with a large number of trials. Consequently, it should not be a surprise to learn that the normal distribution can be used to approximate binomial probabilities for cases in which n is large. The following example illustrates that for many physical systems the binomial model is appropriate with an extremely large value for n. In these cases, it is difficult to calculate probabilities by using the binomial distribution. Fortunately, the normal approximation is most effective in these cases. An illustration is provided in Fig. 4-19. The area of each bar equals the binomial probability of x. Notice that the area of bars can be approximated by areas under the normal density function.

From Fig. 4-19 it can be seen that a probability such as $P(3 \leq X \leq 7)$ is better approximated by the area under the normal curve from 2.5 to 7.5. This observation provides a method to improve the approximation of binomial probabilities. Because a continuous normal distribution is used to approximate a discrete binomial distribution, the modification is referred to as a continuity correction.

EXAMPLE 4-17 In a digital communication channel, assume that the number of bits received in error can be modeled by a binomial random variable, and assume that the probability that a bit is received in error is 1×10^{-5}. If 16 million bits are transmitted, what is the probability that 150 or fewer errors occur?

Let the random variable X denote the number of errors. Then X is a binomial random variable and

$$P(X \leq 150) = \sum_{x=0}^{150} \binom{16{,}000{,}000}{x} (10^{-5})^x (1 - 10^{-5})^{16{,}000{,}000 - x}$$

Clearly, the probability in Example 4-17 is difficult to compute. Fortunately, the normal distribution can be used to provide an excellent approximation in this example.

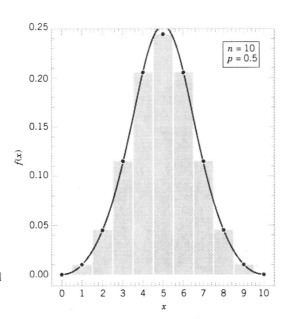

Figure 4-19 Normal approximation to the binomial distribution.

Normal Approximation to the Binomial Distribution

If X is a binomial random variable with parameters n and p

$$Z = \frac{X - np}{\sqrt{np(1-p)}} \qquad (4\text{-}12)$$

is approximately a standard normal random variable. To approximate a binomial probability with a normal distribution a **continuity correction** is applied as follows

$$P(X \leq x) = P(X \leq x + 0.5) \cong P\left(Z \leq \frac{x + 0.5 - np}{\sqrt{np(1-p)}}\right)$$

and

$$P(x \leq X) = P(x - 0.5 \leq X) \cong P\left(\frac{x - 0.5 - np}{\sqrt{np(1-p)}} \leq Z\right)$$

The approximation is good for $np > 5$ and $n(1-p) > 5$.

Recall that for a binomial variable X, $E(X) = np$ and $V(X) = np(1-p)$. Consequently, the expression in Equation 4-12 is nothing more than the formula for standardizing the random variable X. Probabilities involving X can be approximated by using a standard normal distribution. The approximation is good when n is large relative to p.

A way to remember the approximation is to write the probability in terms of $\leq$ or $\geq$ and then add or subtract the 0.5 correction factor to make the probability greater.

EXAMPLE 4-18

The digital communication problem in the previous example is solved as follows:

$$P(X \leq 150) = P(X \leq 150.5) = P\left(\frac{X - 160}{\sqrt{160(1 - 10^{-5})}} \leq \frac{150.5 - 160}{\sqrt{160(1 - 10^{-5})}}\right)$$

$$\cong P(Z \leq -0.75) = 0.227$$

Because $np = (16 \times 10^6)(1 \times 10^{-5}) = 160$ and $n(1-p)$ is much larger, the approximation is expected to work well in this case.

EXAMPLE 4-19
Normal Approximation to Binomial

Again consider the transmission of bits in Example 4-18. To judge how well the normal approximation works, assume only $n = 50$ bits are to be transmitted and that the probability of an error is $p = 0.1$. The exact probability that 2 or less errors occur is

$$P(X \leq 2) = \binom{50}{0} 0.9^{50} + \binom{50}{1} 0.1(0.9^{49}) + \binom{50}{2} 0.1^2(0.9^{48}) = 0.112$$

Based on the normal approximation

$$P(X \leq 2) = P\left(\frac{X - 5}{\sqrt{50(0.1)(0.9)}} < \frac{2.5 - 5}{\sqrt{50(0.1)(0.9)}}\right) = P(Z < -1.18) = 0.119$$

Even for a sample as small as 50 bits, the normal approximation is reasonable.

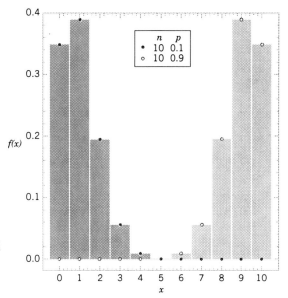

Figure 4-20 Binomial distribution is not symmetrical if p is near 0 or 1.

As another example, $P(8 < X) = P(9 \leq X)$ and this is better approximated as

$$P(9 \leq X) = P(8.5 \leq X) \approx P\left(\frac{8.5 - 5}{2.12} \leq Z\right) = P(1.65 \leq Z) = 0.05$$

We can even approximate $P(X = 5) = P(5 \leq X \leq 5)$ as

$$P(5 \leq X \leq 5) \approx P\left(\frac{5 - 0.5 - 5}{2.12} \leq Z \leq \frac{5 + 0.5 - 5}{2.12}\right) = P(-0.24 \leq Z \leq 0.24) = 0.19$$

and this compares well with the exact answer of 0.1849.

The correction factor is used to improve the approximation. However, if np or $n(1 - p)$ is small, the binomial distribution is quite skewed and the symmetric normal distribution is not a good approximation. Two cases are illustrated in Fig. 4-20.

Recall that the binomial distribution is a satisfactory approximation to the hypergeometric distribution when n, the sample size, is small relative to N, the size of the population from which the sample is selected. A rule of thumb is that the binomial approximation is effective if $n/N < 0.1$. Recall that for a hypergeometric distribution p is defined as $p = K/N$. That is, p is interpreted as the number of successes in the population. Therefore, the normal distribution can provide an effective approximation of hypergeometric probabilities when $n/N < 0.1$, $np > 5$ and $n(1 - p) > 5$. Figure 4-21 provides a summary of these guidelines.

Recall that the Poisson distribution was developed as the limit of a binomial distribution as the number of trials increased to infinity. Consequently, it should not be surprising to find that the normal distribution can also be used to approximate probabilities of a Poisson random variable.

hypergometric distribution	$\approx$ $\frac{n}{N} < 0.1$	binomial distribution	$\approx$ $np > 5$ $n(1 - p) > 5$	normal distribution

Figure 4-21 Conditions for approximating hypergeometric and binomial probabilities.

Normal Approximation to the Poisson Distribution

If X is a Poisson random variable with $E(X) = \lambda$ and $V(X) = \lambda$,

$$Z = \frac{X - \lambda}{\sqrt{\lambda}} \qquad (4\text{-}13)$$

is approximately a standard normal random variable. The approximation is good for

$$\lambda > 5$$

EXAMPLE 4-20
Normal Approximation to Poisson

Assume that the number of asbestos particles in a squared meter of dust on a surface follows a Poisson distribution with a mean of 1000. If a squared meter of dust is analyzed, what is the probability that 950 or fewer particles are found?

This probability can be expressed exactly as

$$P(X \leq 950) = \sum_{x=0}^{950} \frac{e^{-1000} 1000^x}{x!}$$

The computational difficulty is clear. The probability can be approximated as

$$P(X \leq x) = P\left(Z \leq \frac{950 - 1000}{\sqrt{1000}}\right) = P(Z \leq -1.58) = 0.057$$

EXERCISES FOR SECTION 4-7

4-65. Suppose that X is a binomial random variable with $n = 200$ and $p = 0.4$.
(a) Approximate the probability that X is less than or equal to 70.
(b) Approximate the probability that X is greater than 70 and less than 90.
(c) Approximate the probability $X = 80$.

4-66. Suppose that X is a Poisson random variable with $\lambda = 6$.
(a) Compute the exact probability that X is less than 4.
(b) Approximate the probability that X is less than 4 and compare to the result in part (a).
(c) Approximate the probability that $8 < X < 12$.

4-67. Suppose that X has a Poisson distribution with a mean of 64. Approximate the following probabilities:
(a) $P(X > 72)$
(b) $P(X < 64)$
(c) $P(60 < X \leq 68)$

4-68. The manufacturing of semiconductor chips produces 2% defective chips. Assume the chips are independent and that a lot contains 1000 chips.
(a) Approximate the probability that more than 25 chips are defective.
(b) Approximate the probability that between 20 and 30 chips are defective.

4-69. There were 49.7 million people with some type of long-lasting condition or disability living in the United States in 2000. This represented 19.3 percent of the majority of civilians aged five and over (http://factfinder.census.gov). A sample of 1000 persons is selected at random.
(a) Approximate the probability that more than 200 persons in the sample have a disability.
(b) Approximate the probability that between 180 and 300 people in the sample have a disability.

4-70. Phoenix water is provided to approximately 1.4 million people, who are served through more than 362,000 accounts (http://phoenix.gov/WATER/wtrfacts.html). All accounts are metered and billed monthly. The probability that an account has an error in a month is 0.001, and accounts can be assumed to be independent.
(a) What is the mean and standard deviation of the number of account errors each month?
(b) Approximate the probability of fewer than 350 errors in a month.
(c) Approximate a value so that the probability that the number of errors exceeding this value is 0.05.
(d) Approximate the probability of more than 400 errors per month in the next two months. Assume that results between months are independent.

4-71. An electronic office product contains 5000 electronic components. Assume that the probability that each component operates without failure during the useful life of the product is 0.999, and assume that the components fail independently. Approximate the probability that 10 or more of the original 5000 components fail during the useful life of the product.

4-72. A corporate Web site contains errors on 50 of 1000 pages. If 100 pages are sampled randomly, without replacement, approximate the probability that at least 1 of the pages in error are in the sample.

4-73. Suppose that the number of asbestos particles in a sample of 1 squared centimeter of dust is a Poisson random variable with a mean of 1000. What is the probability that 10 squared centimeters of dust contains more than 10,000 particles?

4-74. A high-volume printer produces minor print-quality errors on a test pattern of 1000 pages of text according to a Poisson distribution with a mean of 0.4 per page.

(a) Why are the number of errors on each page independent random variables?
(b) What is the mean number of pages with errors (one or more)?
(c) Approximate the probability that more than 350 pages contain errors (one or more).

4-75. Hits to a high-volume Web site are assumed to follow a Poisson distribution with a mean of 10,000 per day. Approximate each of the following:
(a) The probability of more than 20,000 hits in a day
(b) The probability of less than 9900 hits in a day
(c) The value such that the probability that the number of hits in a day exceed the value is 0.01
(d) Approximate the expected number of days in a year (365 days) that exceed 10,200 hits.
(e) Approximate the probability that over a year (365 days) more than 15 days each have more than 10,200 hits.

4-8 EXPONENTIAL DISTRIBUTION

The discussion of the Poisson distribution defined a random variable to be the number of flaws along a length of copper wire. The distance between flaws is another random variable that is often of interest. Let the random variable X denote the length from any starting point on the wire until a flaw is detected.

As you might expect, the distribution of X can be obtained from knowledge of the distribution of the number of flaws. The key to the relationship is the following concept. The distance to the first flaw exceeds 3 millimeters if and only if there are no flaws within a length of 3 millimeters—simple, but sufficient for an analysis of the distribution of X.

In general, let the random variable N denote the number of flaws in x millimeters of wire. If the mean number of flaws is λ per millimeter, N has a Poisson distribution with mean λx. We assume that the wire is longer than the value of x. Now,

$$P(X > x) = P(N = 0) = \frac{e^{-\lambda x}(\lambda x)^0}{0!} = e^{-\lambda x}$$

Therefore,

$$F(x) = P(X \le x) = 1 - e^{-\lambda x}, \quad x \ge 0$$

is the cumulative distribution function of X. By differentiating $F(x)$, the probability density function of X is calculated to be

$$f(x) = \lambda e^{-\lambda x}, \quad x \ge 0$$

The derivation of the distribution of X depends only on the assumption that the flaws in the wire follow a Poisson process. Also, the starting point for measuring X doesn't matter because the probability of the number of flaws in an interval of a Poisson process depends only on the length of the interval, not on the location. For any Poisson process, the following general result applies.

136   CHAPTER 4 CONTINUOUS RANDOM VARIABLES AND PROBABILITY DISTRIBUTIONS

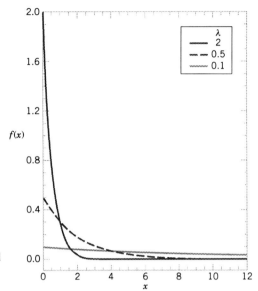

Figure 4-22 Probability density function of exponential random variables for selected values of λ.

Exponential Distribution

The random variable X that equals the distance between successive events of a Poisson process with mean $\lambda > 0$ is an **exponential random variable** with parameter λ. The probability density function of X is

$$f(x) = \lambda e^{-\lambda x} \quad \text{for} \quad 0 \leq x < \infty \tag{4-14}$$

The exponential distribution obtains its name from the exponential function in the probability density function. Plots of the exponential distribution for selected values of λ are shown in Fig. 4-22. For any value of λ, the exponential distribution is quite skewed. The following results are easily obtained and are left as an exercise.

Mean and Variance

If the random variable X has an exponential distribution with parameter λ,

$$\mu = E(X) = \frac{1}{\lambda} \quad \text{and} \quad \sigma^2 = V(X) = \frac{1}{\lambda^2} \tag{4-15}$$

It is important to **use consistent units** in the calculation of probabilities, means, and variances involving exponential random variables. The following example illustrates unit conversions.

EXAMPLE 4-21
Computer Usage

In a large corporate computer network, user log-ons to the system can be modeled as a Poisson process with a mean of 25 log-ons per hour. What is the probability that there are no log-ons in an interval of 6 minutes?

Let X denote the time in hours from the start of the interval until the first log-on. Then, X has an exponential distribution with $\lambda = 25$ log-ons per hour. We are interested in the probability that X

Figure 4-23 Probability for the exponential distribution in Example 4-21.

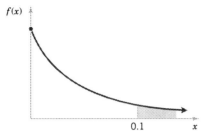

exceeds 6 minutes. Because λ is given in log-ons per hour, we express all time units in hours. That is, 6 minutes = 0.1 hour. The probability requested is shown as the shaded area under the probability density function in Fig. 4-23. Therefore,

$$P(X > 0.1) = \int_{0.1}^{\infty} 25e^{-25x}\, dx = e^{-25(0.1)} = 0.082$$

Also, the cumulative distribution function can be used to obtain the same result as follows:

$$P(X > 0.1) = 1 - F(0.1) = e^{-25(0.1)}$$

An identical answer is obtained by expressing the mean number of log-ons as 0.417 log-ons per minute and computing the probability that the time until the next log-on exceeds 6 minutes. Try it.

What is the probability that the time until the next log-on is between 2 and 3 minutes? Upon converting all units to hours,

$$P(0.033 < X < 0.05) = \int_{0.033}^{0.05} 25e^{-25x}\, dx = -e^{-25x}\Big|_{0.033}^{0.05} = 0.152$$

An alternative solution is

$$P(0.033 < X < 0.05) = F(0.05) - F(0.033) = 0.152$$

Determine the interval of time such that the probability that no log-on occurs in the interval is 0.90. The question asks for the length of time x such that $P(X > x) = 0.90$. Now,

$$P(X > x) = e^{-25x} = 0.90$$

Take the (natural) log of both sides to obtain $-25x = \ln(0.90) = -0.1054$. Therefore,

$$x = 0.00421 \text{ hour} = 0.25 \text{ minute}$$

Furthermore, the mean time until the next log-on is

$$\mu = 1/25 = 0.04 \text{ hour} = 2.4 \text{ minutes}$$

The standard deviation of the time until the next log-on is

$$\sigma = 1/25 \text{ hours} = 2.4 \text{ minutes}$$

138 CHAPTER 4 CONTINUOUS RANDOM VARIABLES AND PROBABILITY DISTRIBUTIONS

In the previous example, the probability that there are no log-ons in a 6-minute interval is 0.082 regardless of the starting time of the interval. A Poisson process assumes that events occur uniformly throughout the interval of observation; that is, there is no clustering of events. If the log-ons are well modeled by a Poisson process, the probability that the first log-on after noon occurs after 12:06 P.M. is the same as the probability that the first log-on after 3:00 P.M. occurs after 3:06 P.M. And if someone logs on at 2:22 P.M., the probability the next log-on occurs after 2:28 P.M. is still 0.082.

Our starting point for observing the system does not matter. However, if there are high-use periods during the day, such as right after 8:00 A.M., followed by a period of low use, a Poisson process is not an appropriate model for log-ons and the distribution is not appropriate for computing probabilities. It might be reasonable to model each of the high- and low-use periods by a separate Poisson process, employing a larger value for λ during the high-use periods and a smaller value otherwise. Then, an exponential distribution with the corresponding value of λ can be used to calculate log-on probabilities for the high- and low-use periods.

Lack of Memory Property

An even more interesting property of an exponential random variable is concerned with conditional probabilities.

EXAMPLE 4-22
Lack of Memory Property

Let X denote the time between detections of a particle with a Geiger counter and assume that X has an exponential distribution with $E(X) = 1.4$ minutes. The probability that we detect a particle within 30 seconds of starting the counter is

$$P(X < 0.5 \text{ minute}) = F(0.5) = 1 - e^{-0.5/1.4} = 0.30$$

In this calculation, all units are converted to minutes. Now, suppose we turn on the Geiger counter and wait 3 minutes without detecting a particle. What is the probability that a particle is detected in the next 30 seconds?

Because we have already been waiting for 3 minutes, we feel that we are "due." That is, the probability of a detection in the next 30 seconds should be greater than 0.3. However, for an exponential distribution, this is not true. The requested probability can be expressed as the conditional probability that $P(X < 3.5 | X > 3)$. From the definition of conditional probability,

$$P(X < 3.5 | X > 3) = P(3 < X < 3.5)/P(X > 3)$$

where

$$P(3 < X < 3.5) = F(3.5) - F(3) = [1 - e^{-3.5/1.4}] - [1 - e^{-3/1.4}] = 0.035$$

and

$$P(X > 3) = 1 - F(3) = e^{-3/1.4} = 0.117$$

Therefore,

$$P(X < 3.5 | X > 3) = 0.035/0.117 = 0.30$$

After waiting for 3 minutes without a detection, the probability of a detection in the next 30 seconds is the same as the probability of a detection in the 30 seconds immediately after starting the counter. The fact that you have waited 3 minutes without a detection does not change the probability of a detection in the next 30 seconds.

Figure 4-24 Lack of memory property of an exponential distribution.

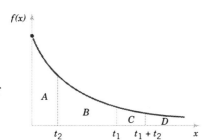

Example 4-22 illustrates the **lack of memory property** of an exponential random variable and a general statement of the property follows. In fact, the exponential distribution is the only continuous distribution with this property.

Lack of Memory Property

For an exponential random variable X,

$$P(X < t_1 + t_2 | X > t_1) = P(X < t_2) \qquad (4\text{-}16)$$

Figure 4-24 graphically illustrates the lack of memory property. The area of region A divided by the total area under the probability density function ($A + B + C + D = 1$) equals $P(X < t_2)$. The area of region C divided by the area $C + D$ equals $P(X < t_1 + t_2 | X > t_1)$. The lack of memory property implies that the proportion of the total area that is in A equals the proportion of the area in C and D that is in C. The mathematical verification of the lack of memory property is left as a mind-expanding exercise.

The lack of memory property is not that surprising when you consider the development of a Poisson process. In that development, we assumed that an interval could be partitioned into small intervals that were independent. These subintervals are similar to independent Bernoulli trials that comprise a binomial process; knowledge of previous results does not affect the probabilities of events in future subintervals. An exponential random variable is the continuous analog of a geometric random variable, and they share a similar lack of memory property.

The exponential distribution is often used in reliability studies as the model for the time until failure of a device. For example, the lifetime of a semiconductor chip might be modeled as an exponential random variable with a mean of 40,000 hours. The lack of memory property of the exponential distribution implies that the device does not wear out. That is, regardless of how long the device has been operating, the probability of a failure in the next 1000 hours is the same as the probability of a failure in the first 1000 hours of operation. The lifetime L of a device with failures caused by random shocks might be appropriately modeled as an exponential random variable. However, the lifetime L of a device that suffers slow mechanical wear, such as bearing wear, is better modeled by a distribution such that $P(L < t + \Delta t | L > t)$ increases with t. Distributions such as the Weibull distribution are often used, in practice, to model the failure time of this type of device. The Weibull distribution is presented in a later section.

EXERCISES FOR SECTION 4-8

4-76. Suppose X has an exponential distribution with $\lambda = 2$. Determine the following:
(a) $P(X \le 0)$ (b) $P(X \ge 2)$
(c) $P(X \le 1)$ (d) $P(1 < X < 2)$
(e) Find the value of x such that $P(X < x) = 0.05$.

4-77. Suppose X has an exponential distribution with mean equal to 10. Determine the following:
(a) $P(X > 10)$ (b) $P(X > 20)$
(c) $P(X < 30)$
(d) Find the value of x such that $P(X < x) = 0.95$.

4-78. Suppose X has an exponential distribution with a mean of 10. Determine the following:
(a) $P(X < 5)$
(b) $P(X < 15 \mid X > 10)$
(c) Compare the results in parts (a) and (b) and comment on the role of the lack of memory property.

4-79. Suppose the counts recorded by a Geiger counter follow a Poisson process with an average of two counts per minute.
(a) What is the probability that there are no counts in a 30-second interval?
(b) What is the probability that the first count occurs in less than 10 seconds?
(c) What is the probability that the first count occurs between 1 and 2 minutes after start-up?

4-80. Suppose that the log-ons to a computer network follow a Poisson process with an average of 3 counts per minute.
(a) What is the mean time between counts?
(b) What is the standard deviation of the time between counts?
(c) Determine x such that the probability that at least one count occurs before time x minutes is 0.95.

4-81. The time between calls to a plumbing supply business is exponentially distributed with a mean time between calls of 15 minutes.
(a) What is the probability that there are no calls within a 30-minute interval?
(b) What is the probability that at least one call arrives within a 10-minute interval?
(c) What is the probability that the first call arrives within 5 and 10 minutes after opening?
(d) Determine the length of an interval of time such that the probability of at least one call in the interval is 0.90.

4-82. The life of automobile voltage regulators has an exponential distribution with a mean life of six years. You purchase an automobile that is six years old, with a working voltage regulator, and plan to own it for six years.
(a) What is the probability that the voltage regulator fails during your ownership?
(b) If your regulator fails after you own the automobile three years and it is replaced, what is the mean time until the next failure?

4-83. Suppose that the time to failure (in hours) of fans in a personal computer can be modeled by an exponential distribution with $\lambda = 0.0003$.
(a) What proportion of the fans will last at least 10,000 hours?
(b) What proportion of the fans will last at most 7000 hours?

4-84. The time between the arrival of electronic messages at your computer is exponentially distributed with a mean of two hours.
(a) What is the probability that you do not receive a message during a two-hour period?
(b) If you have not had a message in the last four hours, what is the probability that you do not receive a message in the next two hours?
(c) What is the expected time between your fifth and sixth messages?

4-85. The time between arrivals of taxis at a busy intersection is exponentially distributed with a mean of 10 minutes.
(a) What is the probability that you wait longer than one hour for a taxi?
(b) Suppose you have already been waiting for one hour for a taxi, what is the probability that one arrives within the next 10 minutes?
(c) Determine x such that the probability that you wait more than x minutes is 0.10.
(d) Determine x such that the probability that you wait less than x minutes is 0.90.
(e) Determine x such that the probability that you wait less than x minutes is 0.50.

4-86. The number of stork sightings on a route in South Carolina follows a Poisson process with a mean of 2.3 per year.
(a) What is the mean time between sightings?
(b) What is the probability that there are no sightings within three months (0.25 years)?
(c) What is the probability that the time until the first sighting exceeds six months?
(d) What is the probability of no sighting within three years?

4-87. According to results from the analysis of chocolate bars in Chapter 3, the mean number of insect fragments was 14.4 in 225 grams. Assume the number of fragments follow a Poisson distribution.
(a) What is the mean number of grams of chocolate until a fragment is detected?
(b) What is the probability that there are no fragments in a 28.35 gram (one ounce) chocolate bar?
(c) Suppose you consume seven one-ounce (28.35 grams) bars this week. What is the probability of no insect fragments?

4-88. The distance between major cracks in a highway follows an exponential distribution with a mean of 5 miles.
(a) What is the probability that there are no major cracks in a 10-mile stretch of the highway?
(b) What is the probability that there are two major cracks in a 10-mile stretch of the highway?
(c) What is the standard deviation of the distance between major cracks?
(d) What is the probability that the first major crack occurs between 12 and 15 miles of the start of inspection?
(e) What is the probability that there are no major cracks in two separate 5-mile stretches of the highway?
(f) Given that there are no cracks in the first 5 miles inspected, what is the probability that there are no major cracks in the next 10 miles inspected?

4-89. The lifetime of a mechanical assembly in a vibration test is exponentially distributed with a mean of 400 hours.
(a) What is the probability that an assembly on test fails in less than 100 hours?

(b) What is the probability that an assembly operates for more than 500 hours before failure?
(c) If an assembly has been on test for 400 hours without a failure, what is the probability of a failure in the next 100 hours?
(d) If 10 assemblies are tested, what is the probability that at least one fails in less than 100 hours? Assume that the assemblies fail independently.
(e) If 10 assemblies are tested, what is the probability that all have failed by 800 hours? Assume the assemblies fail independently.

4-90. The time between arrivals of small aircraft at a county airport is exponentially distributed with a mean of one hour.
(a) What is the probability that more than three aircraft arrive within an hour?
(b) If 30 separate one-hour intervals are chosen, what is the probability that no interval contains more than three arrivals?
(c) Determine the length of an interval of time (in hours) such that the probability that no arrivals occur during the interval is 0.10.

4-91. The time between calls to a corporate office is exponentially distributed with a mean of 10 minutes.
(a) What is the probability that there are more than three calls in one-half hour?
(b) What is the probability that there are no calls within one-half hour?
(c) Determine x such that the probability that there are no calls within x hours is 0.01.
(d) What is the probability that there are no calls within a two-hour interval?
(e) If four nonoverlapping one-half hour intervals are selected, what is the probability that none of these intervals contains any call?
(f) Explain the relationship between the results in part (a) and (b).

4-92. Assume that the flaws along a magnetic tape follow a Poisson distribution with a mean of 0.2 flaw per meter. Let X denote the distance between two successive flaws.
(a) What is the mean of X?
(b) What is the probability that there are no flaws in 10 consecutive meters of tape?
(c) Does your answer to part (b) change if the 10 meters are not consecutive?
(d) How many meters of tape need to be inspected so that the probability that at least one flaw is found is 90%?
(e) What is the probability that the first time the distance between two flaws exceeds 8 meters is at the fifth flaw?
(f) What is the mean number of flaws before a distance between two flaws exceeds 8 meters?

4-93. If the random variable X has an exponential distribution with mean θ, determine the following:
(a) $P(X > \theta)$ (b) $P(X > 2\theta)$
(c) $P(X > 3\theta)$
(d) How do the results depend on θ?

4-94. Derive the formula for the mean and variance of an exponential random variable.

4-9 ERLANG AND GAMMA DISTRIBUTIONS

An exponential random variable describes the length until the first count is obtained in a Poisson process. A generalization of the exponential distribution is the length until r counts occur in a Poisson process. Consider the following example.

EXAMPLE 4-23
Processor Failure

The failures of the central processor units of large computer systems are often modeled as a Poisson process. Typically, failures are not caused by components wearing out, but by more random failures of the large number of semiconductor circuits in the units. Assume that the units that fail are immediately repaired, and assume that the mean number of failures per hour is 0.0001. Let X denote the time until four failures occur in a system. Determine the probability that X exceeds 40,000 hours.

Let the random variable N denote the number of failures in 40,000 hours of operation. The time until four failures occur exceeds 40,000 hours if and only if the number of failures in 40,000 hours is three or less. Therefore,

$$P(X > 40{,}000) = P(N \le 3)$$

The assumption that the failures follow a Poisson process implies that N has a Poisson distribution with

$$E(N) = 40{,}000(0.0001) = 4 \text{ failures per 40,000 hours}$$

Therefore,

$$P(X > 40{,}000) = P(N \le 3) = \sum_{k=0}^{3} \frac{e^{-4} 4^k}{k!} = 0.433$$

The previous example can be generalized to show that if X is the time until the rth event in a Poisson process then

$$P(X > x) = \sum_{k=0}^{r-1} \frac{e^{-\lambda x}(\lambda x)^k}{k!}$$

Because $P(X > x) = 1 - F(x)$, the probability density function of X equals the negative of the derivative of the right-hand side of the previous equation. After extensive algebraic simplification, the probability density function of X can be shown to equal

$$f(x) = \frac{\lambda^r x^{r-1} e^{-\lambda x}}{(r-1)!}$$

for $x > 0$ and $r = 1, 2, \ldots$. This probability density function defines an **Erlang distribution**. Clearly, an Erlang random variable with $r = 1$ is an exponential random variable.

It is convenient to generalize the Erlang distribution to allow r to assume any nonnegative value. Then the Erlang and some other common distributions become special cases of this generalized distribution. To accomplish this step, the factorial function $(r - 1)!$ has to be generalized to apply to any nonnegative value or r, but the generalized function should still equal $(r - 1)!$ when r is a positive integer.

Gamma Function

The gamma function is

$$\Gamma(r) = \int_0^\infty x^{r-1} e^{-x} \, dx, \quad \text{for } r > 0 \qquad (4\text{-}17)$$

It can be shown that the integral in the definition of $\Gamma(r)$ is finite. Furthermore, by using integration by parts it can be shown that

$$\Gamma(r) = (r-1)\Gamma(r-1)$$

This result is left as an exercise. Therefore, if r is a positive integer (as in the Erlang distribution),

$$\Gamma(r) = (r-1)!$$

Also, $\Gamma(1) = 0! = 1$ and it can be shown that $\Gamma(1/2) = \pi^{1/2}$. The gamma function can be interpreted as a generalization to noninteger values of r of the term $(r - 1)!$ that is used in the Erlang probability density function. Now the Erlang distribution can be generalized.

Gamma Distribution

The random variable X with probability density function

$$f(x) = \frac{\lambda^r x^{r-1} e^{-\lambda x}}{\Gamma(r)}, \quad \text{for } x > 0 \qquad (4\text{-}18)$$

has a gamma random variable with parameters $\lambda > 0$ and $r > 0$. If r is an integer, X has an Erlang distribution.

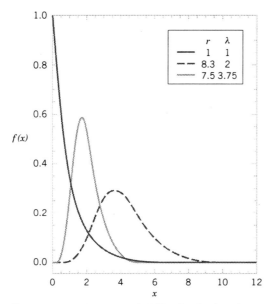

Figure 4-25 Gamma probability density functions for selected values of λ and r.

The parameters λ and r are often called the **scale** and **shape** parameters, respectively. However, one should check the definitions used in software packages. For example, Minitab defines the scale parameter as $1/\lambda$. Sketches of the gamma distribution for several values of λ and r are shown in Fig. 4-25. Many different shapes can be generated from changes to the parameters. Also, the change of variable $u = \lambda x$ and the definition of the gamma function can be used to show that the density integrates to one.

Recall that for an exponential distribution with parameter λ the mean and variance are $1/\lambda$ and $1/\lambda^2$, respectively. An Erlang random variable is the time until the rth event in a Poisson process and the time between events are independent. Therefore, it is plausible that the mean and variance of a gamma random variable multiply the exponential results by r. This motivates the following conclusions. Repeated integration by parts can be used to derive these, but the details are lengthy and omitted.

Mean and Variance

If X is a gamma random variable with parameters λ and r,

$$\mu = E(X) = r/\lambda \quad \text{and} \quad \sigma^2 = V(X) = r/\lambda^2 \qquad (4\text{-}19)$$

EXAMPLE 4-24

The time to prepare a micro-array slide for high-throughput genomics is a Poisson process with a mean of two hours per slide. What is the probability that 10 slides require more than 25 hours to prepare?

Let X denote the time to prepare 10 slides. Because of the assumption of a Poisson process, X has a gamma distribution with $\lambda = 1/2$, $r = 10$, and the requested probability is $P(X > 25)$. The probability can be obtained from software that provides cumulative Poisson probabilities or gamma probabilities.

For the cumulative Poisson probabilities we use the method in Example 4-23 to obtain

$$P(X > 25) = \sum_{k=0}^{9} \frac{e^{-12.5}(12.5)^k}{k!}$$

In Minitab we set the mean = 12.5 and the input = 9 to obtain $P(X > 25) = 0.2014$.

As a check, we use the gamma cumulative probability function in Minitab. Set the shape parameter to 10, the scale parameter to 2 because Minitab uses the inverse of our definition, and the input to 25. The probability computed is $P(X \leq 25) = 0.7986$ and when this is subtracted from one we match with the previous result that $P(X > 25) = 0.2014$.

What are the mean and standard deviation of the time to prepare 10 slides?

The mean time is

$$E(X) = r/\lambda = 10/0.5 = 20$$

The variance of time is

$$V(X) = r/\lambda^2 = 10/0.5^2 = 40$$

so that the standard deviation is $40^{1/2} = 6.32$ hours.

The slides will be completed by what length of time with probability equal to 0.95?

The question asks for x such that

$$P(X \leq x) = 0.95$$

where X is gamma with $\lambda = 0.5$ and $r = 10$. In Minitab, we use the gamma inverse cumulative probability function and set the shape parameter to 10, the scale parameter to 2 because Minitab uses the inverse of our definition, and the probability to 0.95. The solution computed is

$$P(X \leq 31.41) = 0.95$$

Furthermore, the **chi-squared distribution** is a special case of the gamma distribution in which $\lambda = 1/2$ and r equals one of the values $1/2, 1, 3/2, 2, \ldots$. This distribution is used extensively in interval estimation and tests of hypotheses that are discussed in subsequent chapters. The chi-squared distribution is discussed in Chapter 7.

EXERCISES FOR SECTION 4-9

4-95. Use the properties of the gamma function to evaluate the following:
(a) $\Gamma(6)$ (b) $\Gamma(5/2)$
(c) $\Gamma(9/2)$

4-96. Given the probability density function $f(x) = 0.01^3 x^2 e^{-0.01x}/\Gamma(3)$, determine the mean and variance of the distribution.

4-97. Calls to a telephone system follow a Poisson distribution with a mean of five calls per minute.
(a) What is the name applied to the distribution and parameter values of the time until the tenth call?
(b) What is the mean time until the tenth call?
(c) What is the mean time between the ninth and tenth calls?
(d) What is the probability that exactly four calls occur within one minute?
(e) If 10 separate one-minute intervals are chosen, what is the probability that all intervals contain more than two calls?

4-98. Raw materials are studied for contamination. Suppose that the number of particles of contamination per pound of material is a Poisson random variable with a mean of 0.01 particle per pound.
(a) What is the expected number of pounds of material required to obtain 15 particles of contamination?
(b) What is the standard deviation of the pounds of materials required to obtain 15 particles of contamination?

4-99. The time between failures of a laser in a cytogenics machine is exponentially distributed with a mean of 25,000 hours.
(a) What is the expected time until the second failure?
(b) What is the probability that the time until the third failure exceeds 50,000 hours?

4-100. In a data communication system, several messages that arrive at a node are bundled into a packet before they are transmitted over the network. Assume the messages arrive at the node according to a Poisson process with

$\tau = 30$ messages per minute. Five messages are used to form a packet.
(a) What is the mean time until a packet is formed, that is, until five messages arrived at the node?
(b) What is the standard deviation of the time until a packet is formed?
(c) What is the probability that a packet is formed in less than 10 seconds?
(d) What is the probability that a packet is formed in less than 5 seconds?

4-101. Errors caused by contamination on optical disks occur at the rate of one error every 10^5 bits. Assume the errors follow a Poisson distribution.
(a) What is the mean number of bits until five errors occur?
(b) What is the standard deviation of the number of bits until five errors occur?
(c) The error-correcting code might be ineffective if there are three or more errors within 10^5 bits. What is the probability of this event?

4-102. Calls to the help line of a large computer distributor follow a Poisson distribution with a mean of 20 calls per minute.
(a) What is the mean time until the one-hundredth call?
(b) What is the mean time between call numbers 50 and 80?
(c) What is the probability that three or more calls occur within 15 seconds?

4-103. The time between arrivals of customers at an automatic teller machine is an exponential random variable with a mean of 5 minutes.
(a) What is the probability that more than three customers arrive in 10 minutes?
(b) What is the probability that the time until the fifth customer arrives is less than 15 minutes?

4-104. Use integration by parts to show that $\Gamma(r) = (r - 1) \Gamma(r - 1)$.

4-105. Show that the gamma density function $f(x, \lambda, r)$ integrates to 1.

4-106. Use the result for the gamma distribution to determine the mean and variance of a chi-square distribution with $r = 7/2$.

4-10 WEIBULL DISTRIBUTION

As mentioned previously, the Weibull distribution is often used to model the time until failure of many different physical systems. The parameters in the distribution provide a great deal of flexibility to model systems in which the number of failures increases with time (bearing wear), decreases with time (some semiconductors), or remains constant with time (failures caused by external shocks to the system).

Weibull Distribution

The random variable X with probability density function

$$f(x) = \frac{\beta}{\delta}\left(\frac{x}{\delta}\right)^{\beta-1} \exp\left[-\left(\frac{x}{\delta}\right)^{\beta}\right], \quad \text{for } x > 0 \quad (4\text{-}20)$$

is a Weibull random variable with scale parameter $\delta > 0$ and shape parameter $\beta > 0$.

The flexibility of the Weibull distribution is illustrated by the graphs of selected probability density functions in Fig. 4-26. By inspecting the probability density function, it is seen that when $\beta = 1$, the Weibull distribution is identical to the exponential distribution. Also, the **Raleigh distribution** is a special case when the shape parameter is 2.

The cumulative distribution function is often used to compute probabilities. The following result can be obtained.

Cumulative Distribution Function

If X has a Weibull distribution with parameters δ and β, then the cumulative distribution function of X is

$$F(x) = 1 - e^{-\left(\frac{x}{\delta}\right)^{\beta}} \quad (4\text{-}21)$$

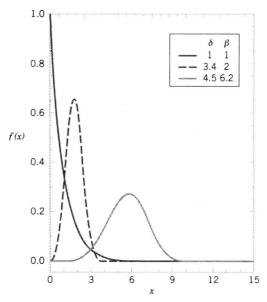

Figure 4-26 Weibull probability density functions for selected values of δ and β.

Also, the following result can be obtained.

Mean and Variance

If X has a Weibull distribution with parameters δ and β,

$$\mu = E(X) = \delta\Gamma\left(1 + \frac{1}{\beta}\right) \quad \text{and} \quad \sigma^2 = V(X) = \delta^2\Gamma\left(1 + \frac{2}{\beta}\right) - \delta^2\left[\Gamma\left(1 + \frac{1}{\beta}\right)\right]^2$$

(4-21)

EXAMPLE 4-25
Bearing Wear

The time to failure (in hours) of a bearing in a mechanical shaft is satisfactorily modeled as a Weibull random variable with $\beta = 1/2$, and $\delta = 5000$ hours. Determine the mean time until failure.

From the expression for the mean,

$$E(X) = 5000\Gamma[1 + (1/0.5)] = 5000\Gamma[3] = 5000 \times 2! = 10{,}000 \text{ hours}$$

Determine the probability that a bearing lasts at least 6000 hours. Now

$$P(x > 6000) = 1 - F(6000) = \exp\left[-\left(\frac{6000}{5000}\right)^{1/2}\right] = e^{-1.095} = 0.334$$

Consequently, only 33.4% of all bearings last at least 6000 hours.

EXERCISES FOR SECTION 4-10

4-107. Suppose that X has a Weibull distribution with $\beta = 0.2$ and $\delta = 100$ hours. Determine the mean and variance of X.

4-108. Suppose that X has a Weibull distribution with $\beta = 0.2$ and $\delta = 100$ hours. Determine the following:
(a) $P(X < 10{,}000)$ (b) $P(X > 5000)$

4-109. If X is a Weibull random variable with $\beta = 1$, and $\delta = 1000$, what is another name for the distribution of X and what is the mean of X?

4-110. Assume that the life of a roller bearing follows a Weibull distribution with parameters $\beta = 2$ and $\delta = 10,000$ hours.
(a) Determine the probability that a bearing lasts at least 8000 hours.
(b) Determine the mean time until failure of a bearing.
(c) If 10 bearings are in use and failures occur independently, what is the probability that all 10 bearings last at least 8000 hours?

4-111. The life (in hours) of a computer processing unit (CPU) is modeled by a Weibull distribution with parameters $\beta = 3$ and $\delta = 900$ hours.
(a) Determine the mean life of the CPU.
(b) Determine the variance of the life of the CPU.
(c) What is the probability that the CPU fails before 500 hours?

4-112. Assume the life of a packaged magnetic disk exposed to corrosive gases has a Weibull distribution with $\beta = 0.5$ and the mean life is 600 hours.
(a) Determine the probability that a packaged disk lasts at least 500 hours.
(b) Determine the probability that a packaged disk fails before 400 hours.

4-113. The life (in hours) of a magnetic resonance imaging machine (MRI) is modeled by a Weibull distribution with parameters $\beta = 2$ and $\delta = 500$ hours.
(a) Determine the mean life of the MRI.
(b) Determine the variance of the life of the MRI.
(c) What is the probability that the MRI fails before 250 hours?

4-114. An article in the *Journal of the Indian Geophysical Union* titled "Weibull and gamma distributions for wave parameter predictions," (2005, Vol. 9, pp. 55–64) used the Weibull distribution to model ocean wave heights. Assume that the mean wave height at the observation station is 2.5 m and the shape parameter equals 2. Determine the standard deviation of wave height.

4-115. An article in the *Journal of Geophysical Research*, "Spatial and temporal distributions of U.S. of winds and wind power at 80 m derived from measurements," (2003, vol. 108, pp. 10–1: 10–20) considered wind speed at stations throughout the U.S. A Weibull distribution can be used to model the distribution of wind speeds at a given location. Every location is characterized by a particular shape and scale parameter. For a station at Amarillo, Texas, the mean wind speed at 80 m (the hub height of large wind turbines) in 2000 was 10.3 m/s with a standard deviation of 4.9 m/s. Determine the shape and scale parameters of a Weibull distribution with these properties.

4-11 LOGNORMAL DISTRIBUTION

Variables in a system sometimes follow an exponential relationship as $x = \exp(w)$. If the exponent is a random variable, say W, $X = \exp(W)$ is a random variable and the distribution of X is of interest. An important special case occurs when W has a normal distribution. In that case, the distribution of X is called a **lognormal distribution**. The name follows from the transformation $\ln(X) = W$. That is, the natural logarithm of X is normally distributed.

Probabilities for X are obtained from the transformation to W, but we need to recognize that the range of X is $(0, \infty)$. Suppose that W is normally distributed with mean θ and variance ω^2; then the cumulative distribution function for X is

$$F(x) = P[X \leq x] = P[\exp(W) \leq x] = P[W \leq \ln(x)]$$
$$= P\left[Z \leq \frac{\ln(x) - \theta}{\omega}\right] = \Phi\left[\frac{\ln(x) - \theta}{\omega}\right]$$

for $x > 0$, where Z is a standard normal random variable. Therefore, Appendix Table III can be used to determine the probability. Also, $F(x) = 0$, for $x \leq 0$.

The probability density function of X can be obtained from the derivative of $F(x)$. This derivative is applied to the last term in the expression for $F(x)$, the integral of the standard normal density function. Furthermore, from the probability density function, the mean and variance of X can be derived. The details are omitted, but a summary of results follows.

Lognormal Distribution

Let W have a normal distribution with mean θ and variance ω^2; then $X = \exp(W)$ is a **lognormal random variable** with probability density function

$$f(x) = \frac{1}{x\omega\sqrt{2\pi}} \exp\left[-\frac{(\ln x - \theta)^2}{2\omega^2}\right] \quad 0 < x < \infty$$

The mean and variance of X are

$$E(X) = e^{\theta + \omega^2/2} \quad \text{and} \quad V(X) = e^{2\theta + \omega^2}(e^{\omega^2} - 1) \quad (4\text{-}22)$$

The parameters of a lognormal distribution are θ and ω^2, but care is needed to interpret that these are the mean and variance of the normal random variable W. The mean and variance of X are the functions of these parameters shown in (4-22). Figure 4-27 illustrates lognormal distributions for selected values of the parameters.

The lifetime of a product that degrades over time is often modeled by a lognormal random variable. For example, this is a common distribution for the lifetime of a semiconductor laser. A Weibull distribution can also be used in this type of application, and with an appropriate choice for parameters, it can approximate a selected lognormal distribution. However, a lognormal distribution is derived from a simple exponential function of a normal random variable, so it is easy to understand and easy to evaluate probabilities.

EXAMPLE 4-26
Semiconductor Laser

The lifetime of a semiconductor laser has a lognormal distribution with $\theta = 10$ hours and $\omega = 1.5$ hours. What is the probability the lifetime exceeds 10,000 hours?

From the cumulative distribution function for X

$$P(X > 10{,}000) = 1 - P[\exp(W) \leq 10{,}000] = 1 - P[W \leq \ln(10{,}000)]$$

$$= 1 - \Phi\left(\frac{\ln(10{,}000) - 10}{1.5}\right) = 1 - \Phi(-0.52) = 1 - 0.30 = 0.70$$

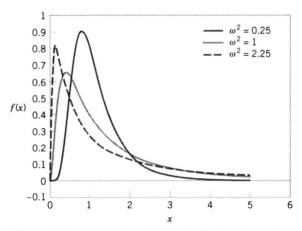

Figure 4-27 Lognormal probability density functions with $\theta = 0$ for selected values of ω^2.

What lifetime is exceeded by 99% of lasers? The question is to determine x such that $P(X > x) = 0.99$. Therefore,

$$P(X > x) = P[\exp(W) > x] = P[W > \ln(x)] = 1 - \Phi\left(\frac{\ln(x) - 10}{1.5}\right) = 0.99$$

From Appendix Table III, $1 - \Phi(z) = 0.99$ when $z = -2.33$. Therefore,

$$\frac{\ln(x) - 10}{1.5} = -2.33 \quad \text{and} \quad x = \exp(6.505) = 668.48 \text{ hours.}$$

Determine the mean and standard deviation of lifetime. Now,

$$E(X) = e^{\theta + \omega^2/2} = \exp(10 + 1.125) = 67{,}846.3$$
$$V(X) = e^{2\theta + \omega^2}(e^{\omega^2} - 1) = \exp(20 + 2.25)[\exp(2.25) - 1] = 39{,}070{,}059{,}886.6$$

so the standard deviation of X is 197,661.5 hours. Notice that the standard deviation of lifetime is large relative to the mean.

EXERCISES FOR SECTION 4-11

4-116. Suppose that X has a lognormal distribution with parameters $\theta = 5$ and $\omega^2 = 9$. Determine the following:
(a) $P(X < 13{,}300)$
(b) The value for x such that $P(X \leq x) = 0.95$
(c) The mean and variance of X

4-117. Suppose that X has a lognormal distribution with parameters $\theta = -2$ and $\omega^2 = 9$. Determine the following:
(a) $P(500 < X < 1000)$
(b) The value for x such that $P(X < x) = 0.1$
(c) The mean and variance of X

4-118. Suppose that X has a lognormal distribution with parameters $\theta = 2$ and $\omega^2 = 4$. Determine the following:
(a) $P(X < 500)$
(b) The conditional probability that $X < 1500$ given that $X > 1000$
(c) What does the difference between the probabilities in parts (a) and (b) imply about lifetimes of lognormal random variables?

4-119. The length of time (in seconds) that a user views a page on a Web site before moving to another page is a lognormal random variable with parameters $\theta = 0.5$ and $\omega^2 = 1$.
(a) What is the probability that a page is viewed for more than 10 seconds?
(b) By what length of time have 50% of the users moved to another page?
(c) What is the mean and standard deviation of the time until a user moves from the page?

4-120. Suppose that X has a lognormal distribution and that the mean and variance of X are 100 and 85,000, respectively. Determine the parameters θ and ω^2 of the lognormal distribution. (*Hint:* define $x = \exp(\theta)$ and $y = \exp(\omega^2)$ and write two equations in terms of x and y.)

4-121. The lifetime of a semiconductor laser has a lognormal distribution, and it is known that the mean and standard deviation of lifetime are 10,000 and 20,000, respectively.
(a) Calculate the parameters of the lognormal distribution
(b) Determine the probability that a lifetime exceeds 10,000 hours
(c) Determine the lifetime that is exceeded by 90% of lasers

4-122. An article in *Health and Population: Perspectives and Issues* (2000, Vol. 23, pp. 28–36) used the lognormal distribution to model blood pressure in humans. The mean systolic blood pressure (SBP) in males age 17 was 120.87 mm Hg. If the coefficient of variation (100% × standard deviation/mean) is 9%, what are the parameter values of the lognormal distribution?

4-123. Derive the probability density function of a lognormal random variable from the derivative of the cumulative distribution function.

Supplemental Exercises

4-124. The probability density function of the time it takes a hematology cell counter to complete a test on a blood sample is $f(x) = 0.04$ for $50 < x < 75$ seconds.
(a) What percentage of tests require more than 70 seconds to complete.
(b) What percentage of tests require less than one minute to complete.
(c) Determine the mean and variance of the time to complete a test on a sample.

4-125. The tensile strength of paper is modeled by a normal distribution with a mean of 35 pounds per square inch and a standard deviation of 2 pounds per square inch.
(a) What is the probability that the strength of a sample is less than 40 lb/in^2?

(b) If the specifications require the tensile strength to exceed 30 lb/in², what proportion of the samples is scrapped?

4-126. The time it takes a cell to divide (called mitosis) is normally distributed with an average time of one hour and a standard deviation of 5 minutes.
(a) What is the probability that a cell divides in less than 45 minutes?
(b) What is the probability that it takes a cell more than 65 minutes to divide?
(c) By what time have approximately 99% of all cells completed mitosis?

4-127. The length of an injection-molded plastic case that holds magnetic tape is normally distributed with a length of 90.2 millimeters and a standard deviation of 0.1 millimeter.
(a) What is the probability that a part is longer than 90.3 millimeters or shorter than 89.7 millimeters?
(b) What should the process mean be set at to obtain the greatest number of parts between 89.7 and 90.3 millimeters?
(c) If parts that are not between 89.7 and 90.3 millimeters are scrapped, what is the yield for the process mean that you selected in part (b)?

Assume that the process is centered so that the mean is 90 millimeters and the standard deviation is 0.1 millimeter. Suppose that 10 cases are measured, and they are assumed to be independent.
(d) What is the probability that all 10 cases are between 89.7 and 90.3 millimeters?
(e) What is the expected number of the 10 cases that are between 89.7 and 90.3 millimeters?

4-128. The sick-leave time of employees in a firm in a month is normally distributed with a mean of 100 hours and a standard deviation of 20 hours.
(a) What is the probability that the sick-leave time for next month will be between 50 and 80 hours?
(b) How much time should be budgeted for sick leave if the budgeted amount should be exceeded with a probability of only 10%?

4-129. The percentage of people exposed to a bacteria who become ill is 20%. Assume that people are independent. Assume that 1000 people are exposed to the bacteria. Approximate each of the following:
(a) The probability that more than 225 become ill
(b) The probability that between 175 and 225 become ill
(c) The value such that the probability that the number of people that become ill exceeds the value is 0.01

4-130. The time to failure (in hours) for a laser in a cytometry machine is modeled by an exponential distribution with $\lambda = 0.00004$.
(a) What is the probability that the laser will last at least 20,000 hours?
(b) What is the probability that the laser will last at most 30,000 hours?
(c) What is the probability that the laser will last between 20,000 and 30,000 hours?

4-131. When a bus service reduces fares, a particular trip from New York City to Albany, New York, is very popular. A small bus can carry four passengers. The time between calls for tickets is exponentially distributed with a mean of 30 minutes. Assume that each call orders one ticket. What is the probability that the bus is filled in less than 3 hours from the time of the fare reduction?

4-132. The time between process problems in a manufacturing line is exponentially distributed with a mean of 30 days.
(a) What is the expected time until the fourth problem?
(b) What is the probability that the time until the fourth problem exceeds 120 days?

4-133. The life of a recirculating pump follows a Weibull distribution with parameters $\beta = 2$, and $\delta = 700$ hours.
(a) Determine the mean life of a pump.
(b) Determine the variance of the life of a pump.
(c) What is the probability that a pump will last longer than its mean?

4-134. The size of silver particles in a photographic emulsion is known to have a log normal distribution with a mean of 0.001 mm and a standard deviation of 0.002 mm.
(a) Determine the parameter values for the lognormal distribution.
(b) What is the probability of a particle size greater than 0.005 mm?

4-135. Suppose that $f(x) = 0.5x - 1$ for $2 < x < 4$. Determine the following:
(a) $P(X < 2.5)$
(b) $P(X > 3)$
(c) $P(2.5 < X < 3.5)$
(d) Determine the cumulative distribution function of the random variable.
(e) Determine the mean and variance of the random variable.

4-136. The time between calls is exponentially distributed with a mean time between calls of 10 minutes.
(a) What is the probability that the time until the first call is less than 5 minutes?
(b) What is the probability that the time until the first call is between 5 and 15 minutes?
(c) Determine the length of an interval of time such that the probability of at least one call in the interval is 0.90.
(d) If there has not been a call in 10 minutes, what is the probability that the time until the next call is less than 5 minutes?
(e) What is the probability that there are no calls in the intervals from 10:00 to 10:05, from 11:30 to 11:35, and from 2:00 to 2:05?
(f) What is the probability that the time until the third call is greater than 30 minutes?
(g) What is the mean time until the fifth call?

4-137. The CPU of a personal computer has a lifetime that is exponentially distributed with a mean lifetime of six years. You have owned this CPU for three years.
(a) What is the probability that the CPU fails in the next three years?
(b) Assume that your corporation has owned 10 CPUs for three years, and assume that the CPUs fail independently. What is the probability that at least one fails within the next three years?

4-138. Suppose that X has a lognormal distribution with parameters $\theta = 0$ and $\omega^2 = 4$. Determine the following:
(a) $P(10 < X < 50)$
(b) The value for x such that $P(X < x) = 0.05$
(c) The mean and variance of X

4-139. Suppose that X has a lognormal distribution and that the mean and variance of X are 50 and 4000, respectively. Determine the following:
(a) The parameters θ and ω^2 of the lognormal distribution
(b) The probability that X is less than 150

4-140. Asbestos fibers in a dust sample are identified by an electron microscope after sample preparation. Suppose that the number of fibers is a Poisson random variable and the mean number of fibers per squared centimeter of surface dust is 100. A sample of 800 square centimeters of dust is analyzed. Assume a particular grid cell under the microscope represents 1/160,000 of the sample.
(a) What is the probability that at least one fiber is visible in the grid cell?
(b) What is the mean of the number of grid cells that need to be viewed to observe 10 that contain fibers?
(c) What is the standard deviation of the number of grid cells that need to be viewed to observe 10 that contain fibers?

4-141. Without an automated irrigation system, the height of plants two weeks after germination is normally distributed with a mean of 2.5 centimeters and a standard deviation of 0.5 centimeters.
(a) What is the probability that a plant's height is greater than 2.25 centimeters?
(b) What is the probability that a plant's height is between 2.0 and 3.0 centimeters?
(c) What height is exceeded by 90% of the plants?

4-142. Continuation of Exercise 4-141. With an automated irrigation system, a plant grows to a height of 3.5 centimeters two weeks after germination.
(a) What is the probability of obtaining a plant of this height or greater from the distribution of heights in Exercise 4-135.
(b) Do you think the automated irrigation system increases the plant height at two weeks after germination?

4-143. The thickness of a laminated covering for a wood surface is normally distributed with a mean of 5 millimeters and a standard deviation of 0.2 millimeter.
(a) What is the probability that a covering thickness is greater than 5.5 millimeters?
(b) If the specifications require the thickness to be between 4.5 and 5.5 millimeters, what proportion of coverings do not meet specifications?
(c) The covering thickness of 95% of samples is below what value?

4-144. The diameter of the dot produced by a printer is normally distributed with a mean diameter of 0.002 inch. Suppose that the specifications require the dot diameter to be between 0.0014 and 0.0026 inch. If the probability that a dot meets specifications is to be 0.9973, what standard deviation is needed?

4-145. Continuation of Exercise 4-144. Assume that the standard deviation of the size of a dot is 0.0004 inch. If the probability that a dot meets specifications is to be 0.9973, what specifications are needed? Assume that the specifications are to be chosen symmetrically around the mean of 0.002.

4-146. The life of a semiconductor laser at a constant power is normally distributed with a mean of 7000 hours and a standard deviation of 600 hours.
(a) What is the probability that a laser fails before 5,800 hours?
(b) What is the life in hours that 90% of the lasers exceed?
(c) What should the mean life equal in order for 99% of the lasers to exceed 10,000 hours before failure?
(d) A product contains three lasers, and the product fails if any of the lasers fails. Assume the lasers fail independently. What should the mean life equal in order for 99% of the products to exceed 10,000 hours before failure?

4-147. Continuation of Exercise 4-146. Rework parts (a) and (b). Assume that the lifetime is an exponential random variable with the same mean.

4-148. Continuation of Exercise 4-146. Rework parts (a) and (b). Assume that the lifetime is a lognormal random variable with the same mean and standard deviation.

4-149. A square inch of carpeting contains 50 carpet fibers. The probability of a damaged fiber is 0.0001. Assume the damaged fibers occur independently.
(a) Approximate the probability of one or more damaged fibers in 1 square yard of carpeting.
(b) Approximate the probability of four or more damaged fibers in 1 square yard of carpeting.

4-150. An airline makes 200 reservations for a flight that holds 185 passengers. The probability that a passenger arrives for the flight is 0.9 and the passengers are assumed to be independent.
(a) Approximate the probability that all the passengers that arrive can be seated.
(b) Approximate the probability that there are empty seats.
(c) Approximate the number of reservations that the airline should make so that the probability that everyone who arrives can be seated is 0.95. [*Hint:* Successively try values for the number of reservations.]

MIND-EXPANDING EXERCISES

4-151. The steps in this exercise lead to the probability density function of an Erlang random variable X with parameters λ and r, $f(x) = \lambda^r x^{r-1} e^{-\lambda x}/(r-1)!$, $x > 0$, $r = 1, 2, \ldots$.
(a) Use the Poisson distribution to express $P(X > x)$.
(b) Use the result from part (a) to determine the cumulative distribution function of X.
(c) Differentiate the cumulative distribution function in part (b) and simplify to obtain the probability density function of X.

4-152. A bearing assembly contains 10 bearings. The bearing diameters are assumed to be independent and normally distributed with a mean of 1.5 millimeters and a standard deviation of 0.025 millimeter. What is the probability that the maximum diameter bearing in the assembly exceeds 1.6 millimeters?

4-153. Let the random variable X denote a measurement from a manufactured product. Suppose the target value for the measurement is m. For example, X could denote a dimensional length, and the target might be 10 millimeters. The **quality loss** of the process producing the product is defined to be the expected value of $\$k(X - m)^2$, where k is a constant that relates a deviation from target to a loss measured in dollars.
(a) Suppose X is a continuous random variable with $E(X) = m$ and $V(X) = \sigma^2$. What is the quality loss of the process?
(b) Suppose X is a continuous random variable with $E(X) = \mu$ and $V(X) = \sigma^2$. What is the quality loss of the process?

4-154. The lifetime of an electronic amplifier is modeled as an exponential random variable. If 10% of the amplifiers have a mean of 20,000 hours and the remaining amplifiers have a mean of 50,000 hours, what proportion of the amplifiers fail before 60,000 hours?

4-155. **Lack of Memory Property.** Show that for an exponential random variable X, $P(X < t_1 + t_2 \mid X > t_1) = P(X < t_2)$

4-156. A process is said to be of **six-sigma quality** if the process mean is at least six standard deviations from the nearest specification. Assume a normally distributed measurement.
(a) If a process mean is centered between upper and lower specifications at a distance of six standard deviations from each, what is the probability that a product does not meet specifications? Using the result that 0.000001 equals one part per million, express the answer in parts per million.
(b) Because it is difficult to maintain a process mean centered between the specifications, the probability of a product not meeting specifications is often calculated after assuming the process shifts. If the process mean positioned as in part (a) shifts upward by 1.5 standard deviations, what is the probability that a product does not meet specifications? Express the answer in parts per million.
(c) Rework part (a). Assume that the process mean is at a distance of three standard deviations.
(d) Rework part (b). Assume that the process mean is at a distance of three standard deviations and then shifts upward by 1.5 standard deviations.
(e) Compare the results in parts (b) and (d) and comment.

IMPORTANT TERMS AND CONCEPTS

Chi-squared distribution
Continuity correction
Continuous uniform distribution
Cumulative probability distribution function-continuous random variable
Erlang distribution
Exponential distribution
Gamma distribution
Lack of memory property-continuous random variable
Lognormal distribution
Mean-continuous random variable
Mean-function of a continuous random variable
Normal approximation to binomial and Poisson probabilities
Normal distribution
Probability density function
Probability distribution-continuous random variable
Standard deviation-continuous random variable
Standardizing
Standard normal distribution
Variance-continuous random variable
Weibull distribution

188 CHAPTER 5 JOINT PROBABILITY DISTRIBUTIONS

fact that the integral of a normal probability density function for a single variable is 1.]

5-53. If X and Y have a bivariate normal distribution with joint probability density $f_{XY}(x, y; \sigma_X, \sigma_Y, \mu_X, \mu_Y, \rho)$, show that the marginal probability distribution of X is normal with mean μ_X and standard deviation σ_X. [*Hint:* Complete the square in the exponent and use the fact that the integral of a normal probability density function for a single variable is 1.]

5-5 LINEAR FUNCTIONS OF RANDOM VARIABLES

A random variable is sometimes defined as a function of one or more random variables. In this section, results for linear functions are highlighted because of their importance in the remainder of the book. For example, if the random variables X_1 and X_2 denote the length and width, respectively, of a manufactured part, $Y = 2X_1 + 2X_2$ is a random variable that represents the perimeter of the part. As another example, recall that the negative binomial random variable was represented as the sum of several geometric random variables.

In this section, we develop results for random variables that are linear combinations of random variables.

Linear Combination

Given random variables $X_1, X_2, \ldots, X_p$ and constants $c_1, c_2, \ldots, c_p$,

$$Y = c_1 X_1 + c_2 X_2 + \cdots + c_p X_p \qquad (5\text{-}34)$$

is a **linear combination** of $X_1, X_2, \ldots, X_p$.

Now, $E(Y)$ can be found from the joint probability distribution of $X_1, X_2, \ldots, X_p$ as follows. Assume $X_1, X_2, \ldots, X_p$ are continuous random variables. An analogous calculation can be used for discrete random variables.

$$E(Y) = \int_{-\infty}^{\infty}\int_{-\infty}^{\infty} \cdots \int_{-\infty}^{\infty} (c_1 x_1 + c_2 x_2 + \cdots + c_p x_p) f_{X_1 X_2 \ldots X_p}(x_1, x_2, \ldots, x_p)\, dx_1\, dx_2 \ldots dx_p$$

$$= c_1 \int_{-\infty}^{\infty}\int_{-\infty}^{\infty} \cdots \int_{-\infty}^{\infty} x_1 f_{X_1 X_2 \ldots X_p}(x_1, x_2, \ldots, x_p)\, dx_1\, dx_2 \ldots dx_p$$

$$+ c_2 \int_{-\infty}^{\infty}\int_{-\infty}^{\infty} \cdots \int_{-\infty}^{\infty} x_2 f_{X_1 X_2 \ldots X_p}(x_1, x_2, \ldots, x_p)\, dx_1\, dx_2 \ldots dx_p + \ldots ,$$

$$+ c_p \int_{-\infty}^{\infty}\int_{-\infty}^{\infty} \cdots \int_{-\infty}^{\infty} x_p f_{X_1 X_2 \ldots X_p}(x_1, x_2, \ldots, x_p)\, dx_1\, dx_2 \ldots dx_p$$

By using Equation 5-22 for each of the terms in this expression, we obtain the following.

Mean of a Linear Function

If $Y = c_1 X_1 + c_2 X_2 + \cdots + c_p X_p$,

$$E(Y) = c_1 E(X_1) + c_2 E(X_2) + \cdots + c_p E(X_p) \qquad (5\text{-}35)$$

5-5 LINEAR FUNCTIONS OF RANDOM VARIABLES

Furthermore, it is left as an exercise to show the following.

Variance of a Linear Function

If $X_1, X_2, \ldots, X_p$ are random variables, and $Y = c_1X_1 + c_2X_2 + \cdots + c_pX_p$, then in general

$$V(Y) = c_1^2 V(X_1) + c_2^2 V(X_2) + \cdots + c_p^2 V(X_p) + 2 \sum_{i<j} \sum c_i c_j \operatorname{cov}(X_i, X_j) \quad (5\text{-}36)$$

If $X_1, X_2, \ldots, X_p$ are **independent**,

$$V(Y) = c_1^2 V(X_1) + c_2^2 V(X_2) + \cdots + c_p^2 V(X_p) \quad (5\text{-}37)$$

Note that the result for the variance in Equation 5-37 requires the random variables to be independent. To see why the independence is important, consider the following simple example. Let X_1 denote any random variable and define $X_2 = -X_1$. Clearly, X_1 and X_2 are not independent. In fact, $\rho_{XY} = -1$. Now, $Y = X_1 + X_2$ is 0 with probability 1. Therefore, $V(Y) = 0$, regardless of the variances of X_1 and X_2.

EXAMPLE 5-32
Negative Binomial Distribution

In Chapter 3, we found that if Y is a negative binomial random variable with parameters p and r, $Y = X_1 + X_2 + \cdots + X_r$, where each X_i is a geometric random variable with parameter p and they are independent. Therefore, $E(X_i) = 1/p$ and $V(X_i) = (1-p)/p^2$. From Equation 5-35, $E(Y) = r/p$ and from Equation 5-37, $V(Y) = r(1-p)/p^2$.

An approach similar to the one applied in the above example can be used to verify the formulas for the mean and variance of an Erlang random variable in Chapter 4. An important use of equation 5-37 is in **error propagation** and this is presented in the following example.

EXAMPLE 5-33
Error Propagation

A semiconductor product consists of three layers. If the variances in thickness of the first, second, and third layers are 25, 40, and 30 nanometers squared, what is the variance of the thickness of the final product?

Let X_1, X_2, X_3, and X be random variables that denote the thickness of the respective layers, and the final product. Then

$$X = X_1 + X_2 + X_3$$

The variance of X is obtained from equaion 5-37

$$V(X) = V(X_1) + V(X_2) + V(X_3) = 25 + 40 + 30 = 95 \text{ nm}^2$$

Consequently, the standard deviation of thickness of the final product is $95^{1/2} = 9.75$ nm and this shows how the variation in each layer is propagated to the final product.

The particular linear function that represents the average of p random variables, with identical means and variances, is used quite often in subsequent chapters. We highlight the results for this special case.

Mean and Variance of an Average

If $\bar{X} = (X_1 + X_2 + \cdots + X_p)/p$ with $E(X_i) = \mu$ for $i = 1, 2, \ldots, p$

$$E(\bar{X}) = \mu \tag{5-38a}$$

if $X_1, X_2, \ldots, X_p$ are also independent with $V(X_i) = \sigma^2$ for $i = 1, 2, \ldots, p$,

$$V(\bar{X}) = \frac{\sigma^2}{p} \tag{5-38b}$$

The conclusion for $V(\bar{X})$ is obtained as follows. Using Equation 5-37, with $c_i = 1/p$ and $V(X_i) = \sigma^2$, yields

$$V(\bar{X}) = \underbrace{(1/p)^2\sigma^2 + \cdots + (1/p)^2\sigma^2}_{p \text{ terms}} = \sigma^2/p$$

Another useful result concerning linear functions of random variables is a reproductive property that holds for independent, normal random variables.

Reproductive Property of the Normal Distribution

If $X_1, X_2, \ldots, X_p$ are independent, normal random variables with $E(X_i) = \mu_i$ and $V(X_i) = \sigma_i^2$, for $i = 1, 2, \ldots, p$,

$$Y = c_1 X_1 + c_2 X_2 + \cdots + c_p X_p$$

is a normal random variable with

$$E(Y) = c_1 \mu_1 + c_2 \mu_2 + \cdots + c_p \mu_p$$

and

$$V(Y) = c_1^2 \sigma_1^2 + c_2^2 \sigma_2^2 + \cdots + c_p^2 \sigma_p^2 \tag{5-39}$$

The mean and variance of Y follow from Equations 5-35 and 5-37. The fact that Y has a normal distribution can be obtained from supplemental material on moment-generating functions on the Web site for the book.

EXAMPLE 5-34 Linear Function of Independent Normal Random Variables

Let the random variables X_1 and X_2 denote the length and width, respectively, of a manufactured part. Assume that X_1 is normal with $E(X_1) = 2$ centimeters and standard deviation 0.1 centimeter and that X_2 is normal with $E(X_2) = 5$ centimeters and standard deviation 0.2 centimeter. Also, assume that X_1 and X_2 are independent. Determine the probability that the perimeter exceeds 14.5 centimeters.

Then, $Y = 2X_1 + 2X_2$ is a normal random variable that represents the perimeter of the part. We obtain, $E(Y) = 14$ centimeters and the variance of Y is

$$V(Y) = 4 \times 0.1^2 + 4 \times 0.2^2 = 0.2$$

Now,

$$P(Y > 14.5) = P[(Y - \mu_Y)/\sigma_Y > (14.5 - 14)/\sqrt{0.2}]$$
$$= P(Z > 1.12) = 0.13$$

EXAMPLE 5-35
Beverage Volume

Soft-drink cans are filled by an automated filling machine. The mean fill volume is 12.1 fluid ounces, and the standard deviation is 0.1 fluid ounce. Assume that the fill volumes of the cans are independent, normal random variables. What is the probability that the average volume of 10 cans selected from this process is less than 12 fluid ounces?

Let $X_1, X_2, \ldots, X_{10}$ denote the fill volumes of the 10 cans. The average fill volume (denoted as $\overline{X}$) is a normal random variable with

$$E(\overline{X}) = 12.1 \quad \text{and} \quad V(\overline{X}) = \frac{0.1^2}{10} = 0.001$$

Consequently,

$$P(\overline{X} < 12) = P\left[\frac{\overline{X} - \mu_{\overline{X}}}{\sigma_{\overline{X}}} < \frac{12 - 12.1}{\sqrt{0.001}}\right]$$
$$= P(Z < -3.16) = 0.00079$$

EXERCISES FOR SECTION 5-5

5-54. X and Y are independent, normal random variables with $E(X) = 0$, $V(X) = 4$, $E(Y) = 10$, and $V(Y) = 9$.
Determine the following:
(a) $E(2X + 3Y)$ (b) $V(2X + 3Y)$
(c) $P(2X + 3Y < 30)$ (d) $P(2X + 3Y < 40)$

5-55. X and Y are independent, normal random variables with $E(X) = 2$, $V(X) = 5$, $E(Y) = 6$, and $V(Y) = 8$.
Determine the following:
(a) $E(3X + 2Y)$ (b) $V(3X + 2Y)$
(c) $P(3X + 2Y < 18)$ (d) $P(3X + 2Y < 28)$

5-56. Suppose that the random variable X represents the length of a punched part in centimeters. Let Y be the length of the part in millimeters. If $E(X) = 5$ and $V(X) = 0.25$, what are the mean and variance of Y?

5-57. A plastic casing for a magnetic disk is composed of two halves. The thickness of each half is normally distributed with a mean of 2 millimeters and a standard deviation of 0.1 millimeter and the halves are independent.
(a) Determine the mean and standard deviation of the total thickness of the two halves.
(b) What is the probability that the total thickness exceeds 4.3 millimeters?

5-58. Making handcrafted pottery generally takes two major steps: wheel throwing and firing. The time of wheel throwing and the time of firing are normally distributed random variables with means of 40 min and 60 min and standard deviations of 2 min and 3 min, respectively.

(a) What is the probability that a piece of pottery will be finished within 95 min?
(b) What is the probability that it will take longer than 110 min?

5-59. In the manufacture of electroluminescent lamps, several different layers of ink are deposited onto a plastic substrate. The thickness of these layers is critical if specifications regarding the final color and intensity of light are to be met. Let X and Y denote the thickness of two different layers of ink. It is known that X is normally distributed with a mean of 0.1 millimeter and a standard deviation of 0.00031 millimeter and Y is also normally distributed with a mean of 0.23 millimeter and a standard deviation of 0.00017 millimeter. Assume that these variables are independent.
(a) If a particular lamp is made up of these two inks only, what is the probability that the total ink thickness is less than 0.2337 millimeter?
(b) A lamp with a total ink thickness exceeding 0.2405 millimeters lacks the uniformity of color demanded by the customer. Find the probability that a randomly selected lamp fails to meet customer specifications.

5-60. The width of a casing for a door is normally distributed with a mean of 24 inches and a standard deviation of 1/8 inch. The width of a door is normally distributed with a mean of 23 and 7/8 inches and a standard deviation of 1/16 inch. Assume independence.
(a) Determine the mean and standard deviation of the difference between the width of the casing and the width of the door.

(b) What is the probability that the width of the casing minus the width of the door exceeds 1/4 inch?
(c) What is the probability that the door does not fit in the casing?

5-61. An article in *Knee Surgery Sports Traumatology, Arthroscopy*, "Effect of provider volume on resource utilization for surgical procedures," (2005, Vol. 13, pp. 273–279) showed a mean time of 129 minutes and a standard deviation of 14 minutes for ACL reconstruction surgery for high-volume hospitals (with more than 300 such surgeries per year). If a high-volume hospital needs to schedule 10 surgeries, what is the mean and variance of the total time to complete these surgeries? Assume the times of the surgeries are independent and normally distributed.

5-62. Soft-drink cans are filled by an automated filling machine and the standard deviation is 0.5 fluid ounce. Assume that the fill volumes of the cans are independent, normal random variables.
(a) What is the standard deviation of the average fill volume of 100 cans?
(b) If the mean fill volume is 12.1 ounces, what is the probability that the average fill volume of the 100 cans is below 12 fluid ounces?
(c) What should the mean fill volume equal so that the probability that the average of 100 cans is below 12 fluid ounces is 0.005?
(d) If the mean fill volume is 12.1 fluid ounces, what should the standard deviation of fill volume equal so that the probability that the average of 100 cans is below 12 fluid ounces is 0.005?
(e) Determine the number of cans that need to be measured such that the probability that the average fill volume is less than 12 fluid ounces is 0.01.

5-63. The photoresist thickness in semiconductor manufacturing has a mean of 10 micrometers and a standard deviation of 1 micrometer. Assume that the thickness is normally distributed and that the thicknesses of different wafers are independent.
(a) Determine the probability that the average thickness of 10 wafers is either greater than 11 or less than 9 micrometers.
(b) Determine the number of wafers that needs to be measured such that the probability that the average thickness exceeds 11 micrometers is 0.01.
(c) If the mean thickness is 10 micrometers, what should the standard deviation of thickness equal so that the probability that the average of 10 wafers is either greater than 11 or less than 9 micrometers is 0.001?

5-64. Assume that the weights of individuals are independent and normally distributed with a mean of 160 pounds and a standard deviation of 30 pounds. Suppose that 25 people squeeze into an elevator that is designed to hold 4300 pounds.
(a) What is the probability that the load (total weight) exceeds the design limit?
(b) What design limit is exceeded by 25 occupants with probability 0.0001?

5-65. Weights of parts are normally distributed with variance σ^2. Measurement error is normally distributed with mean zero and variance $0.5\sigma^2$, independent of the part weights, and adds to the part weight. Upper and lower specifications are centered at 3σ about the process mean.
(a) Without measurement error, what is the probability that a part exceeds the specifications?
(b) With measurement error, what is the probability that a part is measured as beyond specifications? Does this imply it is truly beyond specifications?
(c) What is the probability that a part is measured beyond specifications if the true weight of the part is one σ below the upper specification limit?

5-66. A U-shaped component is to be formed from the three parts A, B, and C. The picture is shown in Fig. 5-20. The length of A is normally distributed with a mean of 10 millimeters and a standard deviation of 0.1 millimeter. The thickness of parts B and C is normally distributed with a mean of 2 millimeters and a standard deviation of 0.05 millimeter. Assume all dimensions are independent.

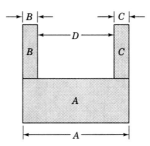

Figure 5-20 Figure for the U-shaped component.

(a) Determine the mean and standard deviation of the length of the gap D.
(b) What is the probability that the gap D is less than 5.9 millimeters?

5-6 GENERAL FUNCTIONS OF RANDOM VARIABLES

In many situations in statistics, it is necessary to derive the probability distribution of a function of one or more random variables. In this section, we present some results that are helpful in solving this problem.

6 Random Sampling and Data Description

CHAPTER OUTLINE

6-1 NUMERICAL SUMMARIES
6-2 STEM-AND-LEAF DIAGRAMS
6-3 FREQUENCY DISTRIBUTIONS AND HISTOGRAMS
6-4 BOX PLOTS
6-5 TIME SEQUENCE PLOTS
6-6 PROBABILITY PLOTS

LEARNING OBJECTIVES

After careful study of this chapter you should be able to do the following:
1. Compute and interpret the sample mean, sample variance, sample standard deviation, sample median, and sample range
2. Explain the concepts of sample mean, sample variance, population mean, and population variance
3. Construct and interpret visual data displays, including the stem-and-leaf display, the histogram, and the box plot
4. Explain the concept of random sampling
5. Construct and interpret normal probability plots
6. Explain how to use box plots and other data displays to visually compare two or more samples of data
7. Know how to use simple time series plots to visually display the important features of time-oriented data

6-1 NUMERICAL SUMMARIES

Well-constructed data summaries and displays are essential to good statistical thinking, because they can focus the engineer on important features of the data or provide insight about the type of model that should be used in solving the problem. The computer has become an important tool in the presentation and analysis of data. While many statistical techniques require only a hand-held calculator, much time and effort may be required by this approach, and a computer will perform the tasks much more efficiently.

Most statistical analysis is done using a prewritten library of statistical programs. The user enters the data and then selects the types of analysis and output displays that are of interest. Statistical software packages are available for both mainframe machines and personal computers. We will present examples of output from Minitab (one of the most widely-used PC packages), throughout the book. We will not discuss the hands-on use of Minitab for entering and editing data or using commands. This information is found in the software documentation.

We often find it useful to describe data features **numerically.** For example, we can characterize the location or central tendency in the data by the ordinary arithmetic average or mean. Because we almost always think of our data as a sample, we will refer to the arithmetic mean as the sample mean.

Sample Mean

> If the n observations in a sample are denoted by $x_1, x_2, \ldots, x_n$, the **sample mean** is
>
> $$\bar{x} = \frac{x_1 + x_2 + \cdots + x_n}{n} = \frac{\sum_{i=1}^{n} x_i}{n} \qquad (6\text{-}1)$$

EXAMPLE 6-1
Sample Mean

Let's consider the eight observations collected from the prototype engine connectors from Chapter 1. The eight observations are $x_1 = 12.6$, $x_2 = 12.9$, $x_3 = 13.4$, $x_4 = 12.3$, $x_5 = 13.6$, $x_6 = 13.5$, $x_7 = 12.6$, and $x_8 = 13.1$. The sample mean is

$$\bar{x} = \frac{x_1 + x_2 + \cdots + x_n}{n} = \frac{\sum_{i=1}^{8} x_i}{8} = \frac{12.6 + 12.9 + \cdots + 13.1}{8}$$

$$= \frac{104}{8} = 13.0 \text{ pounds}$$

A physical interpretation of the sample mean as a measure of location is shown in the dot diagram of the pull-off force data. See Figure 6-1. Notice that the sample mean $\bar{x} = 13.0$ can be thought of as a "balance point." That is, if each observation represents 1 pound of mass placed at the point on the x-axis, a fulcrum located at $\bar{x}$ would exactly balance this system of weights.

The sample mean is the average value of all the observations in the data set. Usually, these data are a sample of observations that have been selected from some larger **population** of observations. Here the population might consist of all the connectors that will be manufactured and sold to customers. Recall that this type of population is called a **conceptual** or **hypothetical population,** because it does not physically exist. Sometimes there is an actual physical population, such as a lot of silicon wafers produced in a semiconductor factory.

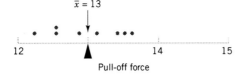

Figure 6-1 The sample mean as a balance point for a system of weights.

In previous chapters we have introduced the mean of a probability distribution, denoted μ. If we think of a probability distribution as a model for the population, one way to think of the mean is as the average of all the measurements in the population. For a finite population with N equally-likely values, the probability mass function is $f(x_i) = 1/N$ and the mean is

$$\mu = \sum_{i=1}^{N} x_i f(x_i) = \frac{\sum_{i=1}^{N} x_i}{N} \tag{6-2}$$

The sample mean, $\bar{x}$, is a reasonable estimate of the population mean, μ. Therefore, the engineer designing the connector using a 3/32-inch wall thickness would conclude, on the basis of the data, that an estimate of the mean pull-off force is 13.0 pounds.

Although the sample mean is useful, it does not convey all of the information about a sample of data. The variability or scatter in the data may be described by the **sample variance** or the **sample standard deviation**.

Sample Variance

If $x_1, x_2, \ldots, x_n$ is a sample of n observations, the sample variance is

$$s^2 = \frac{\sum_{i=1}^{n}(x_i - \bar{x})^2}{n-1} \tag{6-3}$$

The sample standard deviation, s, is the positive square root of the sample variance.

The units of measurements for the sample variance are the square of the original units of the variable. Thus, if x is measured in pounds, the units for the sample variance are (pounds)^2. The standard deviation has the desirable property of measuring variability in the original units of the variable of interest, x.

How Does the Sample Variance Measure Variability?

To see how the sample variance measures dispersion or variability, refer to Fig. 6-2, which shows the deviations $x_i - \bar{x}$ for the connector pull-off force data. The greater the amount of variability in the pull-off force data, the larger in absolute magnitude some of the deviations $x_i - \bar{x}$ will be. Since the deviations $x_i - \bar{x}$ always sum to zero, we must use a measure of variability that changes the negative deviations to nonnegative quantities. Squaring the deviations is the approach used in the sample variance. Consequently, if s^2 is small, there is relatively little variability in the data, but if s^2 is large, the variability is relatively large.

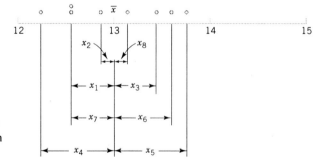

Figure 6-2 How the sample variance measures variability through the deviations $x_i - \bar{x}$.

Table 6-1 Calculation of Terms for the Sample Variance and Sample Standard Deviation

i	x_i	$x_i - \bar{x}$	$(x_i - \bar{x})^2$
1	12.6	−0.4	0.16
2	12.9	−0.1	0.01
3	13.4	0.4	0.16
4	12.3	−0.7	0.49
5	13.6	0.6	0.36
6	13.5	0.5	0.25
7	12.6	−0.4	0.16
8	13.1	0.1	0.01
	104.0	0.0	1.60

EXAMPLE 6-2
Sample Variance

Table 6-1 displays the quantities needed for calculating the sample variance and sample standard deviation for the pull-off force data. These data are plotted in Fig. 6-2. The numerator of s^2 is

$$\sum_{i=1}^{8} (x_i - \bar{x})^2 = 1.60$$

so the sample variance is

$$s^2 = \frac{1.60}{8 - 1} = \frac{1.60}{7} = 0.2286 \text{ (pounds)}^2$$

and the sample standard deviation is

$$s = \sqrt{0.2286} = 0.48 \text{ pounds}$$

Computation of s^2

The computation of s^2 requires calculation of $\bar{x}$, n subtractions, and n squaring and adding operations. If the original observations or the deviations $x_i - \bar{x}$ are not integers, the deviations $x_i - \bar{x}$ may be tedious to work with, and several decimals may have to be carried to ensure numerical accuracy. A more efficient computational formula for the sample variance is obtained as follows:

$$s^2 = \frac{\sum_{i=1}^{n}(x_i - \bar{x})^2}{n-1} = \frac{\sum_{i=1}^{n}(x_i^2 + \bar{x}^2 - 2\bar{x}x_i)}{n-1} = \frac{\sum_{i=1}^{n}x_i^2 + n\bar{x}^2 - 2\bar{x}\sum_{i=1}^{n}x_i}{n-1}$$

and since $\bar{x} = (1/n)\sum_{i=1}^{n} x_i$, this last equation reduces to

$$s^2 = \frac{\sum_{i=1}^{n} x_i^2 - \frac{\left(\sum_{i=1}^{n} x_i\right)^2}{n}}{n-1} \tag{6-4}$$

Note that Equation 6-4 requires squaring each individual x_i, then squaring the sum of the x_i, subtracting $(\sum x_i)^2/n$ from $\sum x_i^2$, and finally dividing by $n - 1$. Sometimes this is called the shortcut method for calculating s^2 (or s).

EXAMPLE 6-3

We will calculate the sample variance and standard deviation using the shortcut method, Equation 6-4. The formula gives

$$s^2 = \frac{\sum_{i=1}^{n} x_i^2 - \frac{\left(\sum_{i=1}^{n} x_i\right)^2}{n}}{n-1} = \frac{1353.6 - \frac{(104)^2}{8}}{7} = \frac{1.60}{7} = 0.2286 \text{ (pounds)}^2$$

and

$$s = \sqrt{0.2286} = 0.48 \text{ pounds}$$

These results agree exactly with those obtained previously.

Analogous to the sample variance s^2, the variability in the population is defined by the **population variance** (σ^2). As in earlier chapters, the positive square root of σ^2, or σ, will denote the **population standard deviation**. When the population is finite and consists of N equally-likely values, we may define the population variance as

$$\sigma^2 = \frac{\sum_{i=1}^{N} (x_i - \mu)^2}{N} \qquad (6\text{-}5)$$

We observed previously that the sample mean could be used as an estimate of the population mean. Similarly, the sample variance is an estimate of the population variance. In Chapter 7, we will discuss **estimation of parameters** more formally.

Note that the divisor for the sample variance is the sample size minus one ($n - 1$), while for the population variance it is the population size N. If we knew the true value of the population mean μ, we could find the *sample* variance as the average squared deviation of the sample observations about μ. In practice, the value of μ is almost never known, and so the sum of the squared deviations about the sample average $\bar{x}$ must be used instead. However, the observations x_i tend to be closer to their average, $\bar{x}$, than to the population mean, μ. Therefore, to compensate for this we use $n - 1$ as the divisor rather than n. If we used n as the divisor in the sample variance, we would obtain a measure of variability that is, on the average, consistently smaller than the true population variance σ^2.

Another way to think about this is to consider the sample variance s^2 as being based on $n - 1$ **degrees of freedom**. The term *degrees of freedom* results from the fact that the n deviations $x_1 - \bar{x}, x_2 - \bar{x}, \ldots, x_n - \bar{x}$ always sum to zero, and so specifying the values of any $n - 1$ of these quantities automatically determines the remaining one. This was illustrated in Table 6-1. Thus, only $n - 1$ of the n deviations, $x_i - \bar{x}$, are freely determined.

In addition to the sample variance and sample standard deviation, the **sample range**, or the difference between the largest and smallest observations, is a useful measure of variability. The sample range is defined as follows.

Sample Range

If the n observations in a sample are denoted by $x_1, x_2, \ldots, x_n$, the **sample range** is

$$r = \max(x_i) - \min(x_i) \qquad (6\text{-}6)$$

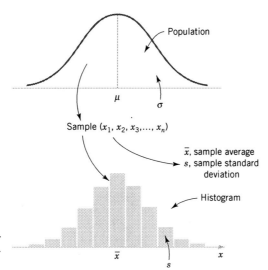

Figure 6-3 Relationship between a population and a sample.

For the pull-off force data, the sample range is $r = 13.6 - 12.3 = 1.3$. Generally, as the variability in sample data increases, the sample range increases.

The sample range is easy to calculate, but it ignores all of the information in the sample data between the largest and smallest values. For example, the two samples 1, 3, 5, 8, and 9 and 1, 5, 5, 5, and 9, both have the same range ($r = 8$). However, the standard deviation of the first sample is $s_1 = 3.35$, while the standard deviation of the second sample is $s_2 = 2.83$. The variability is actually less in the second sample.

Sometimes, when the sample size is small, say $n < 8$ or 10, the information loss associated with the range is not too serious. For example, the range is used widely in statistical quality control where sample sizes of 4 or 5 are fairly common. We will discuss some of these applications in Chapter 16.

In most statistics problems, we work with a sample of observations selected from the population that we are interested in studying. Figure 6-3 illustrates the relationship between the population and the sample.

EXERCISES FOR SECTIONS 6-1

6-1. Eight measurements were made on the inside diameter of forged piston rings used in an automobile engine. The data (in millimeters) are 74.001, 74.003, 74.015, 74.000, 74.005, 74.002, 74.005, and 74.004. Calculate the sample mean and sample standard deviation, construct a dot diagram, and comment on the data.

6-2. In *Applied Life Data Analysis* (Wiley, 1982), Wayne Nelson presents the breakdown time of an insulating fluid between electrodes at 34 kV. The times, in minutes, are as follows: 0.19, 0.78, 0.96, 1.31, 2.78, 3.16, 4.15, 4.67, 4.85, 6.50, 7.35, 8.01, 8.27, 12.06, 31.75, 32.52, 33.91, 36.71, and 72.89. Calculate the sample mean and sample standard deviation.

6-3. The January 1990 issue of *Arizona Trend* contains a supplement describing the 12 "best" golf courses in the state. The yardages (lengths) of these courses are as follows: 6981, 7099, 6930, 6992, 7518, 7100, 6935, 7518, 7013, 6800, 7041, and 6890. Calculate the sample mean and sample standard deviation. Construct a dot diagram of the data.

6-4. An article in the *Journal of Structural Engineering* (Vol. 115, 1989) describes an experiment to test the yield strength of circular tubes with caps welded to the ends. The first yields (in kN) are 96, 96, 102, 102, 102, 104, 104, 108, 126, 126, 128, 128, 140, 156, 160, 160, 164, and 170. Calculate the sample mean and sample standard deviation. Construct a dot diagram of the data.

6-5. An article in *Human Factors* (June 1989) presented data on visual accommodation (a function of eye movement) when recognizing a speckle pattern on a high-resolution CRT screen. The data are as follows: 36.45, 67.90, 38.77, 42.18, 26.72, 50.77, 39.30, and 49.71. Calculate the sample mean and sample standard deviation. Construct a dot diagram of the data.

6-6. The following data are direct solar intensity measurements (watts/m^2) on different days at a location in southern

Spain: 562, 869, 708, 775, 775, 704, 809, 856, 655, 806, 878, 909, 918, 558, 768, 870, 918, 940, 946, 661, 820, 898, 935, 952, 957, 693, 835, 905, 939, 955, 960, 498, 653, 730, and 753. Calculate the sample mean and sample standard deviation. Prepare a dot diagram of this data. Indicate where the sample mean falls on this diagram. Give a practical interpretation of the sample mean.

6-7. The April 22, 1991 issue of *Aviation Week and Space Technology* reports that during Operation Desert Storm, U.S. Air Force F-117A pilots flew 1270 combat sorties for a total of 6905 hours. What is the mean duration of an F-117A mission during this operation? Why is the parameter you have calculated a population mean?

6-8. Preventing fatigue crack propagation in aircraft structures is an important element of aircraft safety. An engineering study to investigate fatigue crack in $n = 9$ cyclically loaded wing boxes reported the following crack lengths (in mm): 2.13, 2.96, 3.02, 1.82, 1.15, 1.37, 2.04, 2.47, 2.60. Calculate the sample mean and sample standard deviation. Prepare a dot diagram of the data.

6-9. An article in the *Journal of Physiology* (2002, Vol. 545, pp. 27–41) "Response of rat muscle to acute resistance exercise defined by transcriptional and translational profiling" studied gene expression as a function of resistance exercise. Expression data (measures of gene activity) from one gene are shown in the following table. One group of rats was exercised for six hours while the other received no exercise. Compute the sample mean and standard deviation of the exercise and no-exercise groups separately. Construct a dot diagram for the exercise and no-exercise groups separately. Comment on any differences between the groups.

6 Hours of Exercise	6 Hours of Exercise	No Exercise	No Exercise
425.313	208.475	485.396	406.921
223.306	286.484	159.471	335.209
388.793	244.242	478.314	
139.262	408.099	245.782	
212.565	157.743	236.212	
324.024	436.37	252.773	

6-10. Exercise 6-5 describes data from an article in *Human Factors* on visual accommodation from an experiment involving a high-resolution CRT screen.

Data from a second experiment using a low-resolution screen were also reported in the article. They are 8.85, 35.80, 26.53, 64.63, 9.00, 15.38, 8.14, and 8.24. Prepare a dot diagram for this second sample and compare it to the one for the first sample. What can you conclude about CRT resolution in this situation?

6-11. The pH of a solution is measured eight times by one operator using the same instrument. She obtains the following data: 7.15, 7.20, 7.18, 7.19, 7.21, 7.20, 7.16, and 7.18. Calculate the sample mean and sample standard deviation. Comment on potential major sources of variability in this experiment.

6-12. An article in the *Journal of Aircraft* (1988) describes the computation of drag coefficients for the NASA 0012 airfoil. Different computational algorithms were used at $M_\infty = 0.7$ with the following results (drag coefficients are in units of drag counts; that is, one count is equivalent to a drag coefficient of 0.0001): 79, 100, 74, 83, 81, 85, 82, 80, and 84. Compute the sample mean, sample variance, and sample standard deviation, and construct a dot diagram.

6-13. The following data are the joint temperatures of the O-rings (°F) for each test firing or actual launch of the space shuttle rocket motor (from *Presidential Commission on the Space Shuttle Challenger Accident*, Vol. 1, pp. 129–131): 84, 49, 61, 40, 83, 67, 45, 66, 70, 69, 80, 58, 68, 60, 67, 72, 73, 70, 57, 63, 70, 78, 52, 67, 53, 67, 75, 61, 70, 81, 76, 79, 75, 76, 58, 31.

(a) Compute the sample mean and sample standard deviation and construct a dot diagram of the temperature data.

(b) Set aside the smallest observation (31°F) and recompute the quantities in part (a). Comment on your findings. How "different" are the other temperatures from this last value?

6-2 STEM-AND-LEAF DIAGRAMS

The dot diagram is a useful data display for small samples, up to (say) about 20 observations. However, when the number of observations is moderately large, other graphical displays may be more useful.

For example, consider the data in Table 6-2. These data are the compressive strengths in pounds per square inch (psi) of 80 specimens of a new aluminum-lithium alloy undergoing evaluation as a possible material for aircraft structural elements. The data were recorded in the order of testing, and in this format they do not convey much information about compressive strength.

Table 6-2 Compressive Strength (in psi) of 80 Aluminum-Lithium Alloy Specimens

105	221	183	186	121	181	180	143
97	154	153	174	120	168	167	141
245	228	174	199	181	158	176	110
163	131	154	115	160	208	158	133
207	180	190	193	194	133	156	123
134	178	76	167	184	135	229	146
218	157	101	171	165	172	158	169
199	151	142	163	145	171	148	158
160	175	149	87	160	237	150	135
196	201	200	176	150	170	118	149

Questions such as "What percent of the specimens fail below 120 psi?" are not easy to answer. Because there are many observations, constructing a dot diagram of these data would be relatively inefficient; more effective displays are available for large data sets.

A stem-and-leaf diagram is a good way to obtain an informative visual display of a data set $x_1, x_2, \ldots, x_n$, where each number x_i consists of at least two digits. To construct a stem-and-leaf diagram, use the following steps.

Steps to Construct a Stem-and-Leaf Diagram

(1) Divide each number x_i into two parts: a **stem**, consisting of one or more of the leading digits and a **leaf**, consisting of the remaining digit.
(2) List the stem values in a vertical column.
(3) Record the leaf for each observation beside its stem.
(4) Write the units for stems and leaves on the display.

To illustrate, if the data consist of percent defective information between 0 and 100 on lots of semiconductor wafers, we can divide the value 76 into the stem 7 and the leaf 6. In general, we should choose relatively few stems in comparison with the number of observations. It is usually best to choose between 5 and 20 stems.

EXAMPLE 6-4
Alloy Strength

To illustrate the construction of a stem-and-leaf diagram, consider the alloy compressive strength data in Table 6-2. We will select as stem values the numbers 7, 8, 9, ..., 24. The resulting stem-and-leaf diagram is presented in Fig. 6-4. The last column in the diagram is a frequency count of the number of leaves associated with each stem. Inspection of this display immediately reveals that most of the compressive strengths lie between 110 and 200 psi and that a central value is somewhere between 150 and 160 psi. Furthermore, the strengths are distributed approximately symmetrically about the central value. The stem-and-leaf diagram enables us to determine quickly some important features of the data that were not immediately obvious in the original display in Table 6-2.

In some data sets, it may be desirable to provide more classes or stems. One way to do this would be to modify the original stems as follows: Divide the stem 5 (say) into two new stems, 5L and 5U. The stem 5L has leaves 0, 1, 2, 3, and 4, and stem 5U has leaves 5, 6, 7, 8, and 9. This will double the number of original stems. We could increase the number of original stems by four by defining five new stems: 5z with leaves 0 and 1, 5t (for twos and three) with leaves 2 and 3, 5f (for fours and fives) with leaves 4 and 5, 5s (for six and seven) with leaves 6 and 7, and 5e with leaves 8 and 9.

Stem	Leaf	Frequency
7	6	1
8	7	1
9	7	1
10	5 1	2
11	5 8 0	3
12	1 0 3	3
13	4 1 3 5 3 5	6
14	2 9 5 8 3 1 6 9	8
15	4 7 1 3 4 0 8 8 6 8 0 8	12
16	3 0 7 3 0 5 0 8 7 9	10
17	8 5 4 4 1 6 2 1 0 6	10
18	0 3 6 1 4 1 0	7
19	9 6 0 9 3 4	6
20	7 1 0 8	4
21	8	1
22	1 8 9	3
23	7	1
24	5	1

Stem : Tens and hundreds digits (psi); Leaf: Ones digits (psi)

Figure 6-4 Stem-and-leaf diagram for the compressive strength data in Table 6-2.

EXAMPLE 6-5 Chemical Yield

Figure 6-5 illustrates the stem-and-leaf diagram for 25 observations on batch yields from a chemical process. In Fig. 6-5(a) we have used 6, 7, 8, and 9 as the stems. This results in too few stems, and the stem-and-leaf diagram does not provide much information about the data. In Fig. 6-5(b) we have divided each stem into two parts, resulting in a display that more adequately displays the data. Figure 6-5(c) illustrates a stem-and-leaf display with each stem divided into five parts. There are too many stems in this plot, resulting in a display that does not tell us much about the shape of the data.

Stem	Leaf
6	1 3 4 5 5 6
7	0 1 1 3 5 7 8 8 9
8	1 3 4 4 7 8 8
9	2 3 5

(a)

Stem	Leaf
6L	1 3 4
6U	5 5 6
7L	0 1 1 3
7U	5 7 8 8 9
8L	1 3 4 4
8U	7 8 8
9L	2 3
9U	5

(b)

Stem	Leaf
6z	1
6t	3
6f	4 5 5
6s	6
6e	
7z	0 1 1
7t	3
7f	5
7s	7
7e	8 8 9
8z	1
8t	3
8f	4 4
8s	7
8e	8 8
9z	
9t	2 3
9f	5
9s	
9e	

(c)

Figure 6-5 Stem-and-leaf displays for Example 6-5. Stem: Tens digits. Leaf: Ones digits.

Character Stem-and-Leaf Display

Stem-and-leaf of Strength
N = 80   Leaf Unit = 1.0

```
  1    7   6
  2    8   7
  3    9   7
  5   10   1 5
  8   11   0 5 8
 11   12   0 1 3
 17   13   1 3 3 4 5 5
 25   14   1 2 3 5 6 8 9 9
 37   15   0 0 1 3 4 4 6 7 8 8 8
(10)   16   0 0 0 3 3 5 7 7 8 9
 33   17   0 1 1 2 4 4 5 6 6 8
 23   18   0 0 1 1 3 4 6
 16   19   0 3 4 6 9 9
 10   20   0 1 7 8
  6   21   8
  5   22   1 8 9
  2   23   7
  1   24   5
```

Figure 6-6 A stem-and-leaf diagram from Minitab.

Figure 6-6 shows a stem-and-leaf display of the compressive strength data in Table 6-2 produced by Minitab. The software uses the same stems as in Fig. 6-4. Note also that the computer orders the leaves from smallest to largest on each stem. This form of the plot is usually called an **ordered stem-and-leaf diagram.** This is not usually done when the plot is constructed manually because it can be time consuming. The computer adds a column to the left of the stems that provides a count of the observations at and above each stem in the upper half of the display and a count of the observations at and below each stem in the lower half of the display. At the middle stem of 16, the column indicates the number of observations at this stem.

The ordered stem-and-leaf display makes it relatively easy to find data features such as percentiles, quartiles, and the median. The sample **median** is a measure of central tendency that divides the data into two equal parts, half below the median and half above. If the number of observations is even, the median is halfway between the two central values. From Fig. 6-6 we find the 40th and 41st values of strength as 160 and 163, so the median is $(160 + 163)/2 = 161.5$. If the number of observations is odd, the median is the central value. The sample mode is the most frequently occurring data value. Figure 6-6 indicates that the mode is 158; this value occurs four times, and no other value occurs as frequently in the sample. If there were more than one value that occurred four times, the data would have multiple modes.

We can also divide data into more than two parts. When an ordered set of data is divided into four equal parts, the division points are called quartiles. The *first* or *lower quartile*, q_1, is a value that has approximately 25% of the observations below it and approximately 75% of the observations above. The *second quartile*, q_2, has approximately 50% of the observations below its value. The second quartile is exactly equal to the median. The *third* or *upper quartile*, q_3, has approximately 75% of the observations below its value. As in the case of the median, the quartiles may not be unique. The compressive strength data in Fig. 6-6 contains $n = 80$ observations. Minitab software calculates the first and third quartiles as the $(n + 1)/4$ and $3(n + 1)/4$ ordered observations and interpolates as needed. For example, $(80 + 1)/4 = 20.25$ and $3(80 + 1)/4 = 60.75$. Therefore, Minitab interpolates between the 20th and 21st ordered observation to obtain $q_1 = 143.50$ and between the 60th and 61st observation to obtain $q_3 = 181.00$. In general, the 100kth percentile is a data value such that

Table 6-3 Summary Statistics for the Compressive Strength Data from Minitab

Variable	N	Mean	Median	StDev	SE Mean
	80	162.66	161.50	33.77	3.78
	Min	Max	Q1	Q3	
	76.00	245.00	143.50	181.00	

approximately $100k\%$ of the observations are at or below this value and approximately $100(1 - k)\%$ of them are above it. Finally, we may use the interquartile range, defined as IQR = $q_3 - q_1$, as a measure of variability. The interquartile range is less sensitive to the extreme values in the sample than is the ordinary sample range.

Many statistics software packages provide data summaries that include these quantities. The output obtained for the compressive strength data in Table 6-2 from Minitab is shown in Table 6-3.

EXERCISES FOR SECTION 6-2

6-14. An article in *Technometrics* (Vol. 19, 1977, p. 425) presents the following data on the motor fuel octane ratings of several blends of gasoline:

88.5	98.8	89.6	92.2	92.7	88.4	87.5	90.9
94.7	88.3	90.4	83.4	87.9	92.6	87.8	89.9
84.3	90.4	91.6	91.0	93.0	93.7	88.3	91.8
90.1	91.2	90.7	88.2	94.4	96.5	89.2	89.7
89.0	90.6	88.6	88.5	90.4	84.3	92.3	92.2
89.8	92.2	88.3	93.3	91.2	93.2	88.9	
91.6	87.7	94.2	87.4	86.7	88.6	89.8	
90.3	91.1	85.3	91.1	94.2	88.7	92.7	
90.0	86.7	90.1	90.5	90.8	92.7	93.3	
91.5	93.4	89.3	100.3	90.1	89.3	86.7	
89.9	96.1	91.1	87.6	91.8	91.0	91.0	

Construct a stem-and-leaf display for these data. Calculate the median and quartiles of these data.

6-15. The following data are the numbers of cycles to failure of aluminum test coupons subjected to repeated alternating stress at 21,000 psi, 18 cycles per second.

1115	865	1015	885	1594	1000	1416	1501
1310	2130	845	1223	2023	1820	1560	1238
1540	1421	1674	375	1315	1940	1055	990
1502	1109	1016	2265	1269	1120	1764	1468
1258	1481	1102	1910	1260	910	1330	1512
1315	1567	1605	1018	1888	1730	1608	1750
1085	1883	706	1452	1782	1102	1535	1642
798	1203	2215	1890	1522	1578	1781	
1020	1270	785	2100	1792	758	1750	

Construct a stem-and-leaf display for these data. Calculate the median and quartiles of these data. Does it appear likely that a coupon will "survive" beyond 2000 cycles? Justify your answer.

6-16. The percentage of cotton in material used to manufacture men's shirts follows. Construct a stem-and-leaf display for the data. Calculate the median and quartiles of these data.

34.2	37.8	33.6	32.6	33.8	35.8	34.7	34.6
33.1	36.6	34.7	33.1	34.2	37.6	33.6	33.6
34.5	35.4	35.0	34.6	33.4	37.3	32.5	34.1
35.6	34.6	35.4	35.9	34.7	34.6	34.1	34.7
36.3	33.8	36.2	34.7	34.6	35.5	35.1	35.7
35.1	37.1	36.8	33.6	35.2	32.8	36.8	36.8
34.7	34.0	35.1	32.9	35.0	32.1	37.9	34.3
33.6	34.1	35.3	33.5	34.9	34.5	36.4	32.7

6-17. The following data represent the yield on 90 consecutive batches of ceramic substrate to which a metal coating has been applied by a vapor-deposition process. Construct a stem-and-leaf display for these data. Calculate the median and quartiles of these data.

94.1	86.1	95.3	84.9	88.8	84.6	94.4	84.1
93.2	90.4	94.1	78.3	86.4	83.6	96.1	83.7
90.6	89.1	97.8	89.6	85.1	85.4	98.0	82.9
91.4	87.3	93.1	90.3	84.0	89.7	85.4	87.3
88.2	84.1	86.4	93.1	93.7	87.6	86.6	86.4
86.1	90.1	87.6	94.6	87.7	85.1	91.7	84.5
95.1	95.2	94.1	96.3	90.6	89.6	87.5	
90.0	86.1	92.1	94.7	89.4	90.0	84.2	
92.4	94.3	96.4	91.1	88.6	90.1	85.1	
87.3	93.2	88.2	92.4	84.1	94.3	90.5	
86.6	86.7	86.4	90.6	82.6	97.3	95.6	
91.2	83.0	85.0	89.1	83.1	96.8	88.3	

6-18. Calculate the sample median, mode, and mean of the data in Exercise 6-14. Explain how these three measures of location describe different features of the data.

6-19. Calculate the sample median, mode, and mean of the data in Exercise 6-15. Explain how these three measures of location describe different features in the data.

 6-20. Calculate the sample median, mode, and mean for the data in Exercise 6-16. Explain how these three measures of location describe different features of the data.

6-21. The net energy consumption (in billion of kilowatt hours) for countries in Asia in 2003 is as follows (source: U.S. Department of Energy Web site http://www.eia.doe.gov/emeu). Construct a stem-and-leaf diagram for these data and comment on any important features that you notice. Compute the sample mean, sample standard deviation, and sample median.

	Billion of Kilowatt Hours
Afghanistan	1.04
Australia	200.66
Bangladesh	16.20
Burma	6.88
China	1,671.23
Hong Kong	38.43
India	519.04
Indonesia	101.80
Japan	946.27
Korea, North	17.43
Korea, South	303.33
Laos	3.30
Malaysia	73.63
Mongolia	2.91
Nepal	2.30
New Zealand	37.03
Pakistan	71.54
Philippines	44.48
Singapore	30.89
Sri Lanka	6.80
Taiwan	154.34
Thailand	107.34
Vietnam	36.92
Asia & Oceania	**4,403.92**

6-22. The female students in an undergraduate engineering core course at ASU self-reported their heights to the nearest inch. The data are below. Construct a stem-and-leaf diagram for the height data and comment on any important features that you notice. Calculate the sample mean, sample standard deviation, and the sample median of height.

```
62  64  66  67  65  68  61  65  67  65  64  63  67
68  64  66  68  69  65  67  62  66  68  67  66  65
69  65  70  65  67  68  65  63  64  67  67
```

6-23. The shear strengths of 100 spot welds in a titanium alloy follow. Construct a stem-and-leaf diagram for the weld strength data and comment on any important features that you notice. What is the 95th percentile of strength?

```
5408 5431 5475 5442 5376 5388 5459 5422 5416 5435
5420 5429 5401 5446 5487 5416 5382 5357 5388 5457
5407 5469 5416 5377 5454 5375 5409 5459 5445 5429
5463 5408 5481 5453 5422 5354 5421 5406 5444 5466
5399 5391 5477 5447 5329 5473 5423 5441 5412 5384
5445 5436 5454 5453 5428 5418 5465 5427 5421 5396
5381 5425 5388 5388 5378 5481 5387 5440 5482 5406
5401 5411 5399 5431 5440 5413 5406 5342 5452 5420
5458 5485 5431 5416 5431 5390 5399 5435 5387 5462
5383 5401 5407 5385 5440 5422 5448 5366 5430 5418
```

6-24. An important quality characteristic of water is the concentration of suspended solid material. Following are 60 measurements on suspended solids from a certain lake. Construct a stem-and-leaf diagram for this data and comment on any important features that you notice. Compute the sample mean, sample standard deviation, and the sample median. What is the 90th percentile of concentration?

```
42.4  65.7  29.8  58.7  52.1  55.8  57.0  68.7  67.3  67.3
54.3  54.0  73.1  81.3  59.9  56.9  62.2  69.9  66.9  59.0
56.3  43.3  57.4  45.3  80.1  49.7  42.8  42.4  59.6  65.8
61.4  64.0  64.2  72.6  72.5  46.1  53.1  56.1  67.2  70.7
42.6  77.4  54.7  57.1  77.3  39.3  76.4  59.3  51.1  73.8
61.4  73.1  77.3  48.5  89.8  50.7  52.0  59.6  66.1  31.6
```

6-25. The United States Golf Association tests golf balls to ensure that they conform to the rules of golf. Balls are tested for weight, diameter, roundness, and overall distance. The overall distance test is conducted by hitting balls with a driver swung by a mechanical device nicknamed "Iron Byron" after the legendary great Byron Nelson, whose swing the machine is said to emulate. Following are 100 distances (in yards) achieved by a particular brand of golf ball in the overall distance test. Construct a stem-and-leaf diagram for this data and comment on any important features that you notice. Compute the sample mean, sample standard deviation, and the sample median. What is the 90th percentile of distances?

```
261.3  259.4  265.7  270.6  274.2  261.4  254.5  283.7
258.1  270.5  255.1  268.9  267.4  253.6  234.3  263.2
254.2  270.7  233.7  263.5  244.5  251.8  259.5  257.5
257.7  272.6  253.7  262.2  252.0  280.3  274.9  233.7
237.9  274.0  264.5  244.8  264.0  268.3  272.1  260.2
255.8  260.7  245.5  279.6  237.8  278.5  273.3  263.7
241.4  260.6  280.3  272.7  261.0  260.0  279.3  252.1
244.3  272.2  248.3  278.7  236.0  271.2  279.8  245.6
241.2  251.1  267.0  273.4  247.7  254.8  272.8  270.5
254.4  232.1  271.5  242.9  273.6  256.1  251.6
256.8  273.0  240.8  276.6  264.5  264.5  226.8
255.3  266.6  250.2  255.8  285.3  255.4  240.5
255.0  273.2  251.4  276.1  277.8  266.8  268.5
```

6-26. A semiconductor manufacturer produces devices used as central processing units in personal computers. The speed of the device (in megahertz) is important because it determines the price that the manufacturer can charge for the devices. The following table contains measurements on 120 devices. Construct a stem-and-leaf diagram for this data and comment on any important features that you notice. Compute the sample mean, sample standard deviation, and the sample median. What percentage of the devices has a speed exceeding 700 megahertz?

680 669 719 699 670 710 722 663 658 634 720 690
677 669 700 718 690 681 702 696 692 690 694 660
649 675 701 721 683 735 688 763 672 698 659 704
681 679 691 683 705 746 706 649 668 672 690 724
652 720 660 695 701 724 668 698 668 660 680 739
717 727 653 637 660 693 679 682 724 642 704 695
704 652 664 702 661 720 695 670 656 718 660 648
683 723 710 680 684 705 681 748 697 703 660 722
662 644 683 695 678 674 656 667 683 691 680 685
681 715 665 676 665 675 655 659 720 675 697 663

6-27. A group of wine enthusiasts taste-tested a pinot noir wine from Oregon. The evaluation was to grade the wine on a 0 to 100 point scale. The results follow. Construct a stem-and-leaf diagram for this data and comment on any important features that you notice. Compute the sample mean, sample standard deviation, and the sample median. A wine rated above 90 is considered truly exceptional. What proportion of the taste-tasters considered this particular pinot noir truly exceptional?

94 90 92 91 91 86 89 91 91 90
90 93 87 90 91 92 89 86 89 90
88 95 91 88 89 92 87 89 95 92
85 91 85 89 88 84 85 90 90 83

6-28. In their book *Introduction to Linear Regression Analysis* (3rd edition, Wiley, 2001) Montgomery, Peck, and Vining present measurements on $NbOCl_3$ concentration from a tube-flow reactor experiment. The data, in gram−mole per liter $\times$ 10^{-3}, are as follows. Construct a stem-and-leaf diagram for this data and comment on any important features you notice. Compute the sample mean, sample standard deviation, and the sample median.

450 450 473 507 457 452 453 1215 1256
1145 1085 1066 1111 1364 1254 1396 1575 1617
1733 2753 3186 3227 3469 1911 2588 2635 2725

6-29. In Exercise 6-22, we presented height data that was self-reported by female undergraduate engineering students in a core course at ASU. In the same class, the male students self-reported their heights as follows. Construct a comparative stem-and-leaf diagram by listing the stems in the center of the display and then placing the female leaves on the left and the male leaves on the right. Comment on any important features that you notice in this display.

69 67 69 70 65 68 69 70 71 69 66 67 69 75 68 67 68
69 70 71 72 68 69 69 70 71 68 72 69 69 68 69 73 70
73 68 69 71 67 68 65 68 68 69 70 74 71 69 70 69

6-3 FREQUENCY DISTRIBUTIONS AND HISTOGRAMS

A **frequency distribution** is a more compact summary of data than a stem-and-leaf diagram. To construct a frequency distribution, we must divide the range of the data into intervals, which are usually called **class intervals, cells,** or **bins.** If possible, the bins should be of equal width in order to enhance the visual information in the frequency distribution. Some judgment must be used in selecting the number of bins so that a reasonable display can be developed. The number of bins depends on the number of observations and the amount of scatter or dispersion in the data. A frequency distribution that uses either too few or too many bins will not be informative. We usually find that between 5 and 20 bins is satisfactory in most cases and that the number of bins should increase with n. Choosing the **number of bins** approximately equal to the square root of the number of observations often works well in practice.

A frequency distribution for the comprehensive strength data in Table 6-2 is shown in Table 6-4. Since the data set contains 80 observations, and since $\sqrt{80} \approx 9$, we suspect that about eight to nine bins will provide a satisfactory frequency distribution. The largest and smallest data values are 245 and 76, respectively, so the bins must cover a range of at least $245 - 76 = 169$ units on the psi scale. If we want the lower limit for the first bin to begin

Table 6-4 Frequency Distribution for the Compressive Strength Data in Table 6-2

Class	$70 \le x < 90$	$90 \le x < 110$	$110 \le x < 130$	$130 \le x < 150$	$150 \le x < 170$	$170 \le x < 190$	$190 \le x < 210$	$210 \le x < 230$	$230 \le x < 250$
Frequency	2	3	6	14	22	17	10	4	2
Relative frequency	0.0250	0.0375	0.0750	0.1750	0.2750	0.2125	0.1250	0.0500	0.0250
Cumulative relative frequency	0.0250	0.0625	0.1375	0.3125	0.5875	0.8000	0.9250	0.9750	1.0000

slightly below the smallest data value and the upper limit for the last bin to be slightly above the largest data value, we might start the frequency distribution at 70 and end it at 250. This is an interval or range of 180 psi units. Nine bins, each of width 20 psi, give a reasonable frequency distribution, so the frequency distribution in Table 6-4 is based on nine bins.

The second row of Table 6-4 contains a **relative frequency distribution.** The relative frequencies are found by dividing the observed frequency in each bin by the total number of observations. The last row of Table 6-4 expresses the relative frequencies on a cumulative basis. Frequency distributions are often easier to interpret than tables of data. For example, from Table 6-4 it is very easy to see that most of the specimens have compressive strengths between 130 and 190 psi and that 97.5 percent of the specimens fail below 230 psi.

The **histogram** is a visual display of the frequency distribution. The stages for constructing a histogram follow.

Constructing a Histogram (Equal Bin Widths)

(1) Label the bin (class interval) boundaries on a horizontal scale.
(2) Mark and label the vertical scale with the frequencies or the relative frequencies.
(3) Above each bin, draw a rectangle where height is equal to the frequency (or relative frequency) corresponding to that bin.

Figure 6-7 is the histogram for the compression strength data. The histogram, like the stem-and-leaf diagram, provides a visual impression of the shape of the distribution of the measurements and information about the central tendency and scatter or dispersion in the data. Notice the symmetric, bell-shaped distribution of the strength measurements in Fig. 6-7. This display often gives insight about possible choices of probability distributions to use as a model

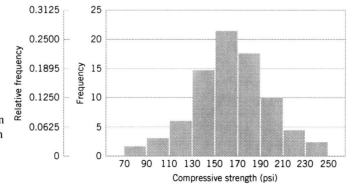

Figure 6-7 Histogram of compressive strength for 80 aluminum-lithium alloy specimens.

for the population. For example, here we would likely conclude that the normal distribution is a reasonable model for the population of compression strength measurements.

Sometimes a histogram with **unequal bin widths** will be employed. For example, if the data have several extreme observations or outliers, using a few equal-width bins will result in nearly all observations falling in just a few of the bins. Using many equal-width bins will result in many bins with zero frequency. A better choice is to use shorter intervals in the region where most of the data falls and a few wide intervals near the extreme observations. When the bins are of unequal width, the rectangle's **area** (not its height) should be proportional to the bin frequency. This implies that the rectangle height should be

$$\text{Rectangle height} = \frac{\text{bin frequency}}{\text{bin width}}$$

In passing from either the original data or stem-and-leaf diagram to a frequency distribution or histogram, we have lost some information because we no longer have the individual observations. However, this information loss is often small compared with the conciseness and ease of interpretation gained in using the frequency distribution and histogram.

Figure 6-8 shows a histogram of the compressive strength data from Minitab with 17 bins. We have noted that histograms may be relatively sensitive to the number of bins and their width. For small data sets, histograms may change dramatically in appearance if the number and/or width of the bins changes. Histograms are more stable for larger data sets, preferably of size 75 to 100 or more. Figure 6-9 shows the Minitab histogram for the compressive strength data with nine bins. This used Minitab's default setting and the result is similar to the original histogram shown in Fig. 6-7. Since the number of observations is moderately large ($n = 80$), the choice of the number of bins is not especially important, and both Figs. 6-8 and 6-9 convey similar information.

Figure 6-10 shows a variation of the histogram available in Minitab, the **cumulative frequency plot.** In this plot, the height of each bar is the total number of observations that are less than or equal to the upper limit of the bin. Cumulative distributions are also useful in data interpretation; for example, we can read directly from Fig. 6-10 that there are approximately 70 observations less than or equal to 200 psi.

When the sample size is large, the histogram can provide a reasonably reliable indicator of the general **shape** of the distribution or population of measurements from which the sample

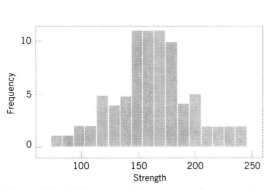

Figure 6-8 A histogram of the compressive strength data from Minitab with 17 bins.

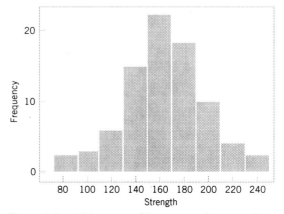

Figure 6-9 A histogram of the compressive strength data from Minitab with nine bins.

Figure 6-10 A cumulative distribution plot of the compressive strength data from Minitab.

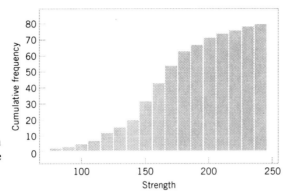

was drawn. Figure 6-11 presents three cases. The median is denoted as $\tilde{x}$. Generally, if the data are symmetric, as in Fig. 6-11(b), the mean and median coincide. If, in addition, the data have only one mode (we say the data are *unimodal*), the mean, median, and mode all coincide. If the data are *skewed* (asymmetric, with a long tail to one side), as in Fig. 6-11(a) and (c), the mean, median, and mode do not coincide. Usually, we find that mode < median < mean if the distribution is skewed to the right, whereas mode > median > mean if the distribution is skewed to the left.

Frequency distributions and histograms can also be used with qualitative or categorical data. In some applications there will be a natural ordering of the categories (such as freshman, sophomore, junior, and senior), whereas in others the order of the categories will be arbitrary (such as male and female). When using categorical data, the bins should have equal width.

EXAMPLE 6-6   Figure 6-12 on the next page presents the production of transport aircraft by the Boeing Company in 1985. Notice that the 737 was the most popular model, followed by the 757, 747, 767, and 707.

A chart of occurrences by category (in which the categories are ordered by the number of occurrences) is sometimes referred to as a **Pareto chart**. An exercise asks you to construct such a chart.

In this section we have concentrated on descriptive methods for the situation in which each observation in a data set is a single number or belongs to one category. In many cases, we work with data in which each observation consists of several measurements. For example, in a gasoline mileage study, each observation might consist of a measurement of miles per gallon, the size of the engine in the vehicle, engine horsepower, vehicle weight, and vehicle length. This is an example of **multivariate data.** In later chapters, we will discuss analyzing this type of data.

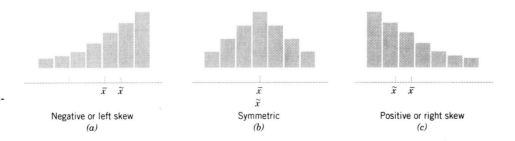

Figure 6-11 Histograms for symmetric and skewed distributions.

Negative or left skew
(a)

Symmetric
(b)

Positive or right skew
(c)

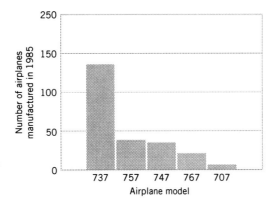

Figure 6-12 Airplane production in 1985. (*Source:* Boeing Company.)

EXERCISES FOR SECTION 6-3

6-30. Construct a frequency distribution and histogram for the motor fuel octane data from Exercise 6-14. Use eight bins.

6-31. Construct a frequency distribution and histogram using the failure data from Exercise 6-15.

6-32. Construct a frequency distribution and histogram for the cotton content data in Exercise 6-16.

6-33. Construct a frequency distribution and histogram for the yield data in Exercise 6-17.

6-34. Construct frequency distributions and histograms with 8 bins and 16 bins for the motor fuel octane data in Exercise 6-14. Compare the histograms. Do both histograms display similar information?

6-35. Construct histograms with 8 and 16 bins for the data in Exercise 6-15. Compare the histograms. Do both histograms display similar information?

6-36. Construct histograms with 8 and 16 bins for the data in Exercise 6-16. Compare the histograms. Do both histograms display similar information?

6-37. Construct a histogram for the energy consumption data in Exercise 6-21.

6-38. Construct a histogram for the female student height data in Exercise 6-22.

6-39. Construct a histogram for the spot weld shear strength data in Exercise 6-23. Comment on the shape of the histogram. Does it convey the same information as the stem-and-leaf display?

6-40. Construct a histogram for the water quality data in Exercise 6-24. Comment on the shape of the histogram. Does it convey the same information as the stem-and-leaf display?

6-41. Construct a histogram for the overall golf distance data in Exercise 6-25. Comment on the shape of the histogram. Does it convey the same information as the stem-and-leaf display?

6-42. Construct a histogram for the semiconductor speed data in Exercise 6-26. Comment on the shape of the histogram. Does it convey the same information as the stem-and-leaf display?

6-43. Construct a histogram for the pinot noir wine rating data in Exercise 6-27. Comment on the shape of the histogram. Does it convey the same information as the stem-and-leaf display?

6-44. **The Pareto Chart.** An important variation of a histogram for categorical data is the Pareto chart. This chart is widely used in quality improvement efforts, and the categories usually represent different types of defects, failure modes, or product/process problems. The categories are ordered so that the category with the largest frequency is on the left, followed by the category with the second largest frequency and so forth. These charts are named after the Italian economist V. Pareto, and they usually exhibit "Pareto's law"; that is, most of the defects can be accounted for by only a few categories. Suppose that the following information on structural defects in automobile doors is obtained: dents, 4; pits, 4; parts assembled out of sequence, 6; parts undertrimmed, 21; missing holes/slots, 8; parts not lubricated, 5; parts out of contour, 30; and parts not deburred, 3. Construct and interpret a Pareto chart.

6-4 BOX PLOTS

The stem-and-leaf display and the histogram provide general visual impressions about a data set, while numerical quantities such as $\bar{x}$ or s provide information about only one feature of the data. The **box plot** is a graphical display that simultaneously describes several important

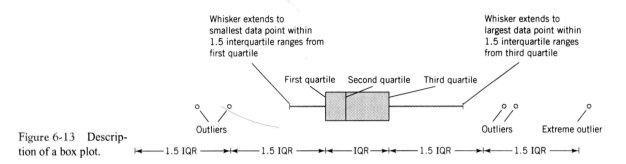

Figure 6-13 Description of a box plot.

features of a data set, such as center, spread, departure from symmetry, and identification of unusual observations or outliers.

A box plot displays the three quartiles, the minimum, and the maximum of the data on a rectangular box, aligned either horizontally or vertically. The box encloses the interquartile range with the left (or lower) edge at the first quartile, q_1, and the right (or upper) edge at the third quartile, q_3. A line is drawn through the box at the second quartile (which is the 50th percentile or the median), $q_2 = \bar{x}$. A line, or **whisker,** extends from each end of the box. The lower whisker is a line from the first quartile to the smallest data point within 1.5 interquartile ranges from the first quartile. The upper whisker is a line from the third quartile to the largest data point within 1.5 interquartile ranges from the third quartile. Data farther from the box than the whiskers are plotted as individual points. A point beyond a whisker, but less than 3 interquartile ranges from the box edge, is called an outlier. A point more than 3 interquartile ranges from the box edge is called an **extreme outlier.** See Fig. 6-13. Occasionally, different symbols, such as open and filled circles, are used to identify the two types of outliers. Sometimes box plots are called *box-and-whisker plots*.

Figure 6-14 presents the box plot from Minitab for the alloy compressive strength data shown in Table 6-2. This box plot indicates that the distribution of compressive strengths is fairly symmetric around the central value, because the left and right whiskers and the lengths of the left and right boxes around the median are about the same. There are also two mild outliers at lower strength and one at higher strength. The upper whisker extends to observation 237 because it is the greatest observation below the limit for upper outliers. This limit is $q_3 + 1.5\text{IQR} = 181 + 1.5(181 - 143.5) = 237.25$. The lower whisker extends to observation 97 because it is the smallest observation above the limit for lower outliers. This limit is $q_1 - 1.5\text{IQR} = 143.5 - 1.5(181 - 143.5) = 87.25$.

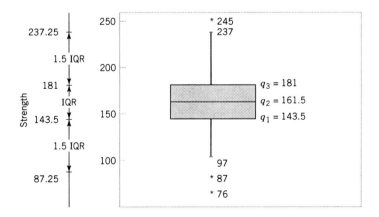

Figure 6-14 Box plot for compressive strength data in Table 6-2.

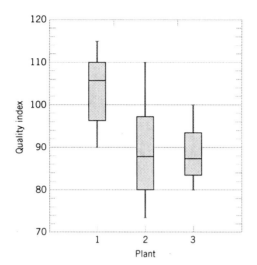

Figure 6-15 Comparative box plots of a quality index at three plants.

Box plots are very useful in graphical comparisons among data sets, because they have high visual impact and are easy to understand. For example, Fig. 6-15 shows the comparative box plots for a manufacturing quality index on semiconductor devices at three manufacturing plants. Inspection of this display reveals that there is too much variability at plant 2 and that plants 2 and 3 need to raise their quality index performance.

EXERCISES FOR SECTION 6-4

6-45. The "cold start ignition time" of an automobile engine is being investigated by a gasoline manufacturer. The following times (in seconds) were obtained for a test vehicle: 1.75, 1.92, 2.62, 2.35, 3.09, 3.15, 2.53, 1.91.
(a) Calculate the sample mean, sample variance, and sample standard deviation.
(b) Construct a box plot of the data.

6-46. An article in the *Transactions of the Institution of Chemical Engineers* (Vol. 34, 1956, pp. 280–293) reported data from an experiment investigating the effect of several process variables on the vapor phase oxidation of naphthalene. A sample of the percentage mole conversion of naphthalene to maleic anhydride follows: 4.2, 4.7, 4.7, 5.0, 3.8, 3.6, 3.0, 5.1, 3.1, 3.8, 4.8, 4.0, 5.2, 4.3, 2.8, 2.0, 2.8, 3.3, 4.8, 5.0.
(a) Calculate the sample mean, sample variance, and sample standard deviation.
(b) Construct a box plot of the data.

6-47. The nine measurements that follow are furnace temperatures recorded on successive batches in a semiconductor manufacturing process (units are °F): 953, 950, 948, 955, 951, 949, 957, 954, 955.
(a) Calculate the sample mean, sample variance, and standard deviation.
(b) Find the median. How much could the largest temperature measurement increase without changing the median value?
(c) Construct a box plot of the data.

6-48. Exercise 6-12 presents drag coefficients for the NASA 0012 airfoil. You were asked to calculate the sample mean, sample variance, and sample standard deviation of those coefficients.
(a) Find the median and the upper and lower quartiles of the drag coefficients.
(b) Construct a box plot of the data.
(c) Set aside the largest observation (100) and rework parts a and b. Comment on your findings.

6-49. Exercise 6-13 presented the joint temperatures of the O-rings (°F) for each test firing or actual launch of the space shuttle rocket motor. In that exercise you were asked to find the sample mean and sample standard deviation of temperature.
(a) Find the median and the upper and lower quartiles of temperature.
(b) Set aside the smallest observation (31°F) and recompute the quantities in part (a). Comment on your findings. How "different" are the other temperatures from this smallest value?
(c) Construct a box plot of the data and comment on the possible presence of outliers.

6-50. Reconsider the motor fuel octane rating data in Exercise 6-14. Construct a box plot of the data and write an interpretation of the plot. How does the box plot compare in interpretive value to the original stem-and-leaf diagram.

6-51. Reconsider the energy consumption data in Exercise 6-21. Construct a box plot of the data and write an interpretation of the plot. How does the box plot compare in interpretive value to the original stem-and-leaf diagram?

6-52. Reconsider the water quality data in Exercise 6-24. Construct a box plot of the concentrations and write an interpretation of the plot. How does the box plot compare in interpretive value to the original stem-and-leaf diagram?

6-53. Reconsider the weld strength data in Exercise 6-23. Construct a box plot of the data and write an interpretation of the plot. How does the box plot compare in interpretive value to the original stem-and-leaf diagram?

6-54. Reconsider the semiconductor speed data in Exercise 6-26. Construct a box plot of the data and write an interpretation of the plot. How does the box plot compare in interpretive value to the original stem-and-leaf diagram?

6-55. Use the data on heights of female and male engineering students from Exercises 6-22 and 6-29 to construct comparative box plots. Write an interpretation of the information that you see in these plots.

6-56. In Exercise 6-45, data was presented on the cold start ignition time of a particular gasoline used in a test vehicle. A second formulation of the gasoline was tested in the same vehicle, with the following times (in seconds): 1.83, 1.99, 3.13, 3.29, 2.65, 2.87, 3.40, 2.46, 1.89, and 3.35. Use this new data along with the cold start times reported in Exercise 6-45 to construct comparative box plots. Write an interpretation of the information that you see in these plots.

6-57. An article in *Nature Genetics* (2003, Vol. 34(1), pp. 85–90) "Treatment-specific changes in gene expression discriminate in vivo drug response in human leukemia cells" studied gene expression as a function of treatments for leukemia. One group received a high dose of the drug while the control group received no treatment. Expression data (measures of gene activity) from one gene are shown in the following table. Construct a box plot for each group of patients. Write an interpretation to compare the information in these plots.

Gene Expression

High Dose	Control	Control	Control
16.1	297.1	820.1	166.5
134.9	491.8	82.5	2258.4
52.7	1332.9	713.9	497.5
14.4	1172	785.6	263.4
124.3	1482.7	114	252.3
99	335.4	31.9	351.4
24.3	528.9	86.3	678.9
16.3	24.1	646.6	3010.2
15.2	545.2	169.9	67.1
47.7	92.9	20.2	318.2
12.9	337.1	280.2	2476.4
72.7	102.3	194.2	181.4
126.7	255.1	408.4	2081.5
46.4	100.5	155.5	424.3
60.3	159.9	864.6	188.1
23.5	168	355.4	563
43.6	95.2	634	149.1
79.4	132.5	2029.9	2122.9
38	442.6	362.1	1295.9
58.2	15.8		
26.5	175.6		
25.1	131.1		

6-5 TIME SEQUENCE PLOTS

The graphical displays that we have considered thus far such as histograms, stem-and-leaf plots, and box plots are very useful visual methods for showing the variability in data. However, we noted in Chapter 1 that time is an important factor that contributes to variability in data, and those graphical methods do not take this into account. A **time series** or **time sequence** is a data set in which the observations are recorded in the order in which they occur. A **time series plot** is a graph in which the vertical axis denotes the observed value of the variable (say x) and the horizontal axis denotes the time (which could be minutes, days, years, etc.) When measurements are plotted as a time series, we often see trends, cycles, or other broad features of the data that could not be seen otherwise.

For example, consider Fig. 6-16(a), which presents a time series plot of the annual sales of a company for the last 10 years. The general impression from this display is that sales show an upward **trend**. There is some variability about this trend, with some years' sales increasing over those of the last year and some years' sales decreasing. Figure 6-16(b) shows the last three years of sales reported by quarter. This plot clearly shows that the annual sales in this

218   CHAPTER 6 RANDOM SAMPLING AND DATA DESCRIPTION

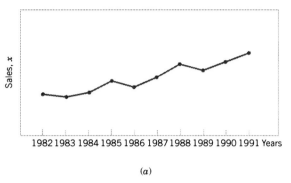
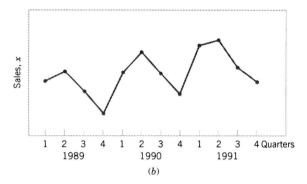

Figure 6-16   Company sales by year (*a*) and by quarter (*b*).

business exhibit a **cyclic** variability by quarter, with the first- and second-quarter sales being generally greater than sales during the third and fourth quarters.

Sometimes it can be very helpful to combine a time series plot with some of the other graphical displays that we have considered previously. J. Stuart Hunter (*The American Statistician*, Vol. 42, 1988, p. 54) has suggested combining the stem-and-leaf plot with a time series plot to form a **digidot plot.**

Figure 6-17 shows a digidot plot for the observations on compressive strength from Table 6-2, assuming that these observations are recorded in the order in which they occurred. This plot effectively displays the overall variability in the compressive strength data and simultaneously shows the variability in these measurements over time. The general impression is that compressive strength varies around the mean value of 162.66, and there is no strong obvious pattern in this variability over time.

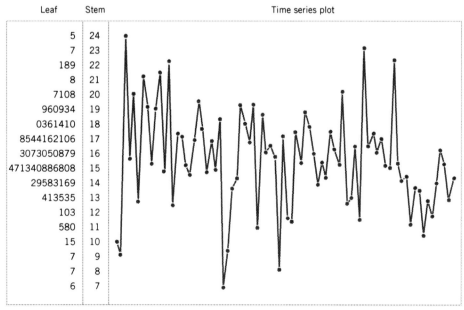

Figure 6-17   A digidot plot of the compressive strength data in Table 6-2.

Figure 6-18 A digidot plot of chemical process concentration readings, observed hourly.

Leaf	Stem
8	9e
6	9s
45	9f
2333	9t
0010000	9z
99998	8e
66677	8s
45	8f
23	8t
1	8z

The digidot plot in Fig. 6-18 tells a different story. This plot summarizes 30 observations on concentration of the output product from a chemical process, where the observations are recorded at one-hour time intervals. This plot indicates that during the first 20 hours of operation this process produced concentrations generally above 85 grams per liter, but that following sample 20, something may have occurred in the process that results in lower concentrations. If this variability in output product concentration can be reduced, operation of this process can be improved.

EXERCISES FOR SECTION 6-5

6-58. The following data are the viscosity measurements for a chemical product observed hourly (read down, then left to right). Construct and interpret either a digidot plot or a separate stem-and-leaf and time series plot of these data. Specifications on product viscosity are at 48 ± 2. What conclusions can you make about process performance?

47.9	48.6	48.0	48.1	43.0	43.2
47.9	48.8	47.5	48.0	42.9	43.6
48.6	48.1	48.6	48.3	43.6	43.2
48.0	48.3	48.0	43.2	43.3	43.5
48.4	47.2	47.9	43.0	43.0	43.0
48.1	48.9	48.3	43.5	42.8	
48.0	48.6	48.5	43.1	43.1	

6-59. The pull-off force for a connector is measured in a laboratory test. Data for 40 test specimens follow (read down, then left to right). Construct and interpret either a digidot plot or a separate stem-and-leaf and time series plot of the data.

241	203	201	251	236	190
258	195	195	238	245	175
237	249	255	210	209	178
210	220	245	198	212	175
194	194	235	199	185	190
225	245	220	183	187	
248	209	249	213	218	

6-60. In their book *Time Series Analysis, Forecasting, and Control* (Prentice Hall, 1994), G. E. P. Box, G. M. Jenkins, and G. C. Reinsel present chemical process concentration readings made every two hours. Some of these data follow (read down, then left to right).

17.0	16.7	17.1	17.5	17.6
16.6	17.4	17.4	18.1	17.5
16.3	17.2	17.4	17.5	16.5
16.1	17.4	17.5	17.4	17.8
17.1	17.4	17.4	17.4	17.3
16.9	17.0	17.6	17.1	17.3
16.8	17.3	17.4	17.6	17.1
17.4	17.2	17.3	17.7	17.4
17.1	17.4	17.0	17.4	16.9
17.0	16.8	17.8	17.8	17.3

Construct and interpret either a digidot plot or a stem-and-leaf plot of these data.

6-61. The 100 annual Wolfer sunspot numbers from 1770 to 1869 follow. (For an interesting analysis and interpretation of these numbers, see the book by Box, Jenkins, and Reinsel referenced in Exercise 6-60. Their analysis requires some advanced knowledge of statistics and statistical model building.) Read down, then left to right. The 1869 result is 74. Construct and interpret either a digidot plot or a stem-and-leaf and time series plot of these data.

101	31	154	38	83	90	1916	25	1942	27	1968	30	1994	13
82	7	125	23	132	67	1917	21	1943	41	1969	27	1995	20
66	20	85	10	131	60	1918	21	1944	31	1970	29	1996	15
35	92	68	24	118	47	1919	14	1945	27	1971	23	1997	16
41	10	16	8	62	94	1920	8	1946	35	1972	20	1998	12
21	8	7	13	98	96	1921	11	1947	26	1973	16	1999	18
16	2	4	57	124	77	1922	14	1948	28	1974	21	2000	15
6	0	2	122	96	59	1923	23	1949	36	1975	21	2001	16
4	1	8	138	66	44	1924	18	1950	39	1976	25	2002	13
7	5	17	103	64	47	1925	17	1951	21	1977	16	2003	15
14	12	36	86	54	30	1926	19	1952	17	1978	18	2004	15
34	14	50	63	39	16	1927	20	1953	22	1979	15		
45	35	62	37	21	7	1928	22	1954	17	1980	18		
43	46	67	24	7	37	1929	19	1955	19	1981	14		
48	41	71	11	4	74	1930	13	1956	15	1982	10		
42	30	48	15	23		1931	26	1957	34	1983	15		
28	24	28	40	55		1932	13	1958	10	1984	8		
						1933	14	1959	15	1985	15		

6-62. In their book *Forecasting and Time Series Analysis*, 2nd edition (McGraw-Hill, 1990), D. C. Montgomery, L. A. Johnson, and J. S. Gardiner analyze the data in Table 6-5, which are the monthly total passenger airline miles flown in the United Kingdom from 1964–1970 (in millions of miles). Comment on any features of the data that are apparent. Construct and interpret either a digidot plot or a separate stem-and-leaf and time series plot of these data.

6-63. The following table shows the number of earthquakes per year of magnitude 7.0 and greater since 1900 (source: Earthquake Data Base System of the U.S. Geological Survey, National Earthquake Information Center, Golden CO).

1900	13	1904	16	1908	18	1912	22
1901	14	1905	26	1909	32	1913	23
1902	8	1906	32	1910	36	1914	22
1903	10	1907	27	1911	24	1915	18

1934	22	1960	22	1986	6
1935	24	1961	18	1987	11
1936	21	1962	15	1988	8
1937	22	1963	20	1989	7
1938	26	1964	15	1990	18
1939	21	1965	22	1991	16
1940	23	1966	19	1992	13
1941	24	1967	16	1993	12

Construct and interpret either a digidot plot or a separate stem-and-leaf and time series plot of these data.

6-64. The following table shows U.S. petroleum imports, imports as a percentage of total, and Persian Gulf imports as a percentage of all imports by year since 1973 (source: U.S. Department of Energy Web site http://www.eia.doe.gov/). Construct and interpret either a digidot plot or a separate stem-and-leaf and time series plot for each column of data.

Table 6-5 United Kingdom Passenger Airline Miles Flown

Month	1964	1965	1966	1967	1968	1969	1970
Jan.	7.269	8.350	8.186	8.334	8.639	9.491	10.840
Feb.	6.775	7.829	7.444	7.899	8.772	8.919	10.436
Mar.	7.819	8.829	8.484	9.994	10.894	11.607	13.589
Apr.	8.371	9.948	9.864	10.078	10.455	8.852	13.402
May	9.069	10.638	10.252	10.801	11.179	12.537	13.103
June	10.248	11.253	12.282	12.953	10.588	14.759	14.933
July	11.030	11.424	11.637	12.222	10.794	13.667	14.147
Aug.	10.882	11.391	11.577	12.246	12.770	13.731	14.057
Sept.	10.333	10.665	12.417	13.281	13.812	15.110	16.234
Oct.	9.109	9.396	9.637	10.366	10.857	12.185	12.389
Nov.	7.685	7.775	8.094	8.730	9.290	10.645	11.594
Dec.	7.682	7.933	9.280	9.614	10.925	12.161	12.772

Year	Petroleum Imports (thousand barrels per day)	Total Petroleum Imports as Percent of Petroleum Products Supplied (tenth percent)	Petroleum Imports from Persian Gulf as Percent of Total Petroleum Imports (tenth percent)
1973	6256.145	361	135
1974	6112.184	367	170
1975	6055.712	371	192
1976	7312.598	418	251
1977	8807.249	477	278
1978	8363.411	443	265
1979	8456.129	456	244
1980	6909.025	405	219
1981	5995.673	373	203
1982	5113.311	334	136
1983	5051.353	331	87
1984	5436.982	345	93
1985	5067.144	322	61
1986	6223.512	382	146
1987	6677.696	400	161
1988	7402.021	428	208
1989	8060.545	465	230
1990	8017.521	471	245
1991	7626.748	456	241
1992	7887.697	463	225
1993	8620.422	500	206
1994	8996.222	507	192
1995	8834.94	498	178
1996	9478.492	517	169
1997	10161.562	545	172
1998	10708.071	566	199
1999	10852.258	555	227
2000	11459.251	581	217
2001	11871.337	604	232
2002	11530.241	583	196
2003	12264.386	612	203
2004	13145.093	634	189

6-6 PROBABILITY PLOTS

How do we know if a particular probability distribution is a reasonable model for data? Sometimes, this is an important question because many of the statistical techniques presented in subsequent chapters are based on an assumption that the population distribution is of a specific type. Thus, we can think of determining whether data come from a specific probability distribution as **verifying assumptions.** In other cases, the form of the distribution can give insight into the underlying physical mechanism generating the data. For example, in reliability engineering, verifying that time-to-failure data come from an exponential distribution identifies the **failure mechanism** in the sense that the failure rate is constant with respect to time.

Some of the visual displays we have used earlier, such as the histogram, can provide insight about the form of the underlying distribution. However, histograms are usually not really reliable indicators of the distribution form unless the sample size is very large. A **probability plot** is a graphical method for determining whether sample data conform to a hypothesized distribution based on a subjective visual examination of the data. The general procedure is very simple and can be performed quickly. It is also more reliable than the histogram for small to moderate size samples. Probability plotting typically uses special axes that have been scaled for the hypothesized distribution. Software is widely available for the normal, lognormal, Weibull, and various chi-square and gamma distributions. We focus primarily on normal probability plots because many statistical techniques are appropriate only when the population is (at least approximately) normal.

To construct a probability plot, the observations in the sample are first ranked from smallest to largest. That is, the sample $x_1, x_2, \ldots, x_n$ is arranged as $x_{(1)}, x_{(2)}, \ldots, x_{(n)}$, where $x_{(1)}$ is the smallest observation, $x_{(2)}$ is the second smallest observation, and so forth, with $x_{(n)}$ the largest. The ordered observations $x_{(j)}$ are then plotted against their observed cumulative frequency $(j - 0.5)/n$ on the appropriate probability paper. If the hypothesized distribution adequately describes the data, the plotted points will fall approximately along a straight line; if the plotted points deviate significantly from a straight line, the hypothesized model is not appropriate. Usually, the determination of whether or not the data plot as a straight line is subjective. The procedure is illustrated in the following example.

EXAMPLE 6-7
Battery Life

Ten observations on the effective service life in minutes of batteries used in a portable personal computer are as follows: 176, 191, 214, 220, 205, 192, 201, 190, 183, 185. We hypothesize that battery life is adequately modeled by a normal distribution. To use probability plotting to investigate this hypothesis, first arrange the observations in ascending order and calculate their cumulative frequencies $(j - 0.5)/10$ as shown in Table 6-6.

The pairs of values $x_{(j)}$ and $(j - 0.5)/10$ are now plotted on normal probability axes. This plot is shown in Fig. 6-19. Most normal probability plots $100(j - 0.5)/n$ on the left vertical scale and $100[1 - (j - 0.5)/n]$ on the right vertical scale, with the variable value plotted on the horizontal scale. A straight line, chosen subjectively, has been drawn through the plotted points. In drawing the straight line, you should be influenced more by the points near the middle of the plot than by the extreme points. A good rule of thumb is to draw the line approximately between the 25th and 75th percentile points. This is how the line in Fig. 6-19 was determined. In assessing the "closeness" of the points to the straight line, imagine a "fat pencil" lying along the line. If all the points are covered by this imaginary pencil, a

Table 6-6 Calculation for Constructing a Normal Probability Plot

j	$x_{(j)}$	$(j - 0.5)/10$	z_j
1	176	0.05	−1.64
2	183	0.15	−1.04
3	185	0.25	−0.67
4	190	0.35	−0.39
5	191	0.45	−0.13
6	192	0.55	0.13
7	201	0.65	0.39
8	205	0.75	0.67
9	214	0.85	1.04
10	220	0.95	1.64

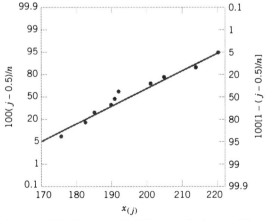

Figure 6-19 Normal probability plot for battery life.

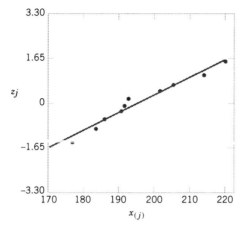

Figure 6-20 Normal probability plot obtained from standardized normal scores.

normal distribution adequately describes the data. Since the points in Fig. 6-19 would pass the "fat pencil" test, we conclude that the normal distribution is an appropriate model.

A normal probability plot can also be constructed on ordinary axes by plotting the standardized normal scores z_j against $x_{(j)}$, where the standardized normal scores satisfy

$$\frac{j - 0.5}{n} = P(Z \le z_j) = \Phi(z_j)$$

For example, if $(j - 0.5)/n = 0.05$, $\Phi(z_j) = 0.05$ implies that $z_j = -1.64$. To illustrate, consider the data from Example 6-4. In the last column of Table 6-6 we show the standardized normal scores. Figure 6-20 presents the plot of z_j versus $x_{(j)}$. This normal probability plot is equivalent to the one in Fig. 6-19.

We have constructed our probability plots with the probability scale (or the z-scale) on the vertical axis. Some computer packages "flip" the axis and put the probability scale on the horizontal axis.

The normal probability plot can be useful in identifying distributions that are symmetric but that have tails that are "heavier" or "lighter" than the normal. They can also be useful in identifying skewed distributions. When a sample is selected from a light-tailed distribution (such as the uniform distribution), the smallest and largest observations will not be as extreme as would be expected in a sample from a normal distribution. Thus if we consider the straight line drawn through the observations at the center of the normal probability plot, observations on the left side will tend to fall below the line, whereas observations on the right side will tend to fall above the line. This will produce an S-shaped normal probability plot such as shown in Fig. 6-21(a). A heavy-tailed distribution will result in data that also produces an S-shaped normal probability plot, but now the observations on the left will be above the straight line and the observations on the right will lie below the line. See Fig. 6-19(b). A positively skewed distribution will tend to produce a pattern such as shown in Fig. 6-19(c), where points on both ends of the plot tend to fall below the line, giving a curved shape to the plot. This occurs because both the smallest and the largest observations from this type of distribution are larger than expected in a sample from a normal distribution.

Even when the underlying population is exactly normal, the sample data will not plot exactly on a straight line. Some judgment and experience are required to evaluate the plot. Generally, if the sample size is $n < 30$, there can be a lot of deviation from linearity in normal

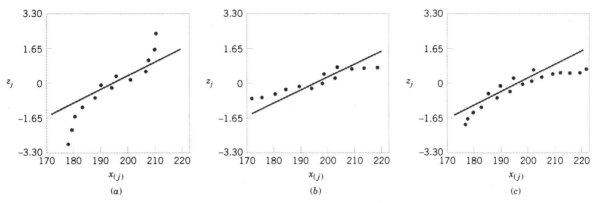

Figure 6-21 Normal probability plots indicating a nonnormal distribution. (a) Light-tailed distribution. (b) Heavy-tailed distribution. (c) A distribution with positive (or right) skew.

plots, so in these cases only a very severe departure from linearity should be interpreted as a strong indication of nonnormality. As n increases, the linear pattern will tend to become stronger, and the normal probability plot will be easier to interpret and more reliable as an indicator of the form of the distribution.

EXERCISES FOR SECTION 6-6

6-65. Construct a normal probability plot of the piston ring diameter data in Exercise 6-1. Does it seem reasonable to assume that piston ring diameter is normally distributed?

6-66. Construct a normal probability plot of the insulating fluid breakdown time data in Exercise 6-2. Does it seem reasonable to assume that breakdown time is normally distributed?

6-67. Construct a normal probability plot of the visual accommodation data in Exercise 6-5. Does it seem reasonable to assume that visual accommodation is normally distributed?

6-68. Construct a normal probability plot of the solar intensity data in Exercise 6-6. Does it seem reasonable to assume that solar intensity is normally distributed?

6-69. Construct a normal probability plot of the O-ring joint temperature data in Exercise 6-13. Does it seem reasonable to assume that O-ring joint temperature is normally distributed? Discuss any interesting features that you see on the plot.

6-70. Construct a normal probability plot of the octane rating data in Exercise 6-14. Does it seem reasonable to assume that octane rating is normally distributed?

6-71. Construct a normal probability plot of the cycles to failure data in Exercise 6-15. Does it seem reasonable to assume that cycles to failure is normally distributed?

6-72. Construct a normal probability plot of the suspended solids concentration data in Exercise 6-24. Does it seem reasonable to assume that the concentration of suspended solids in water from this particular lake is normally distributed?

6-73. Construct two normal probability plots for the height data in Exercises 6-22 and 6-29. Plot the data for female and male students on the same axes. Does height seem to be normally distributed for either group of students? If both populations have the same variance, the two normal probability plots should have identical slopes. What conclusions would you draw about the heights of the two groups of students from visual examination of the normal probability plots?

6-74. It is possible to obtain a "quick and dirty" estimate of the mean of a normal distribution from the fiftieth percentile value on a normal probability plot. Provide an argument why this is so. It is also possible to obtain an estimate of the standard deviation of a normal distribution by subtracting the eighty-fourth percentile value from the fiftieth percentile value. Provide an argument why this is so.

Supplemental Exercises

6-75. The National Oceanic and Atmospheric Administration provided the monthly absolute estimates of global (land and ocean combined) temperature index (degrees C) from 2000–2004. Read January to December from left to right. (source: http://www.ncdc.noaa.gov/oa/climate/research/anomalies/anomalies.html) Construct and interpret either a digidot plot or a separate stem-and-leaf and time series plot of these data.

Year	Global Monthly Temperature											
2000	12.28	12.63	13.22	14.21	15.13	15.82	16.05	16.02	15.29	14.29	13.16	12.47
2001	12.44	12.55	13.35	14.22	15.28	15.99	16.23	16.17	15.44	14.52	13.52	12.61
2002	12.71	12.88	13.49	14.30	15.26	16.02	16.28	16.06	15.44	14.45	13.43	12.56
2003	12.60	12.62	13.30	14.21	15.28	16.02	16.28	16.19	15.54	14.62	13.49	12.77
2004	12.52	12.77	13.41	14.28	15.15	15.92	16.17	16.08	15.48	14.59	13.58	12.71

6-76. The concentration of a solution is measured six times by one operator using the same instrument. She obtains the following data: 63.2, 67.1, 65.8, 64.0, 65.1, and 65.3 (grams per liter).
(a) Calculate the sample mean. Suppose that the desirable value for this solution has been specified to be 65.0 grams per liter. Do you think that the sample mean value computed here is close enough to the target value to accept the solution as conforming to target? Explain your reasoning.
(b) Calculate the sample variance and sample standard deviation.
(c) Suppose that in measuring the concentration, the operator must set up an apparatus and use a reagent material. What do you think the major sources of variability are in this experiment? Why is it desirable to have a small variance of these measurements?

6-77. The table below shows unemployment data for the U.S. that is seasonally adjusted. Construct a time series plot of this data and comment on any features (source: U.S. Bureau of Labor Web site http://data.bls.gov).

6-78. A sample of six resistors yielded the following resistances (ohms): $x_1 = 45$, $x_2 = 38$, $x_3 = 47$, $x_4 = 41$, $x_5 = 35$, and $x_6 = 43$.

(a) Compute the sample variance and sample standard deviation.
(b) Subtract 35 from each of the original resistance measurements and compute s^2 and s. Compare your results with those obtained in part (a) and explain your findings.
(c) If the resistances were 450, 380, 470, 410, 350, and 430 ohms, could you use the results of previous parts of this problem to find s^2 and s?

6-79. Consider the following two samples:

Sample 1: 10, 9, 8, 7, 8, 6, 10, 6

Sample 2: 10, 6, 10, 6, 8, 10, 8, 6

(a) Calculate the sample range for both samples. Would you conclude that both samples exhibit the same variability? Explain.
(b) Calculate the sample standard deviations for both samples. Do these quantities indicate that both samples have the same variability? Explain.
(c) Write a short statement contrasting the sample range versus the sample standard deviation as a measure of variability.

6-80. An article in *Quality Engineering* (Vol. 4, 1992, pp. 487–495) presents viscosity data from a batch chemical process. A sample of these data follows on p. 226:

					Unemployment Percentage							
Year	Jan	Feb	Mar	Apr	May	Jun	Jul	Aug	Sep	Oct	Nov	Dec
1995	5.6	5.4	5.4	5.8	5.6	5.6	5.7	5.7	5.6	5.5	5.6	5.6
1996	5.6	5.5	5.5	5.6	5.6	5.3	5.5	5.1	5.2	5.2	5.4	5.4
1997	5.3	5.2	5.2	5.1	4.9	5.0	4.9	4.8	4.9	4.7	4.6	4.7
1998	4.6	4.6	4.7	4.3	4.4	4.5	4.5	4.5	4.6	4.5	4.4	4.4
1999	4.3	4.4	4.2	4.3	4.2	4.3	4.3	4.2	4.2	4.1	4.1	4.0
2000	4.0	4.1	4.0	3.8	4.0	4.0	4.0	4.1	3.9	3.9	3.9	3.9
2001	4.2	4.2	4.3	4.4	4.3	4.5	4.6	4.9	5.0	5.3	5.6	5.7
2002	5.7	5.7	5.7	5.9	5.8	5.8	5.8	5.7	5.7	5.7	5.9	6.0
2003	5.8	5.9	5.8	6.0	6.1	6.3	6.2	6.1	6.1	6.0	5.9	5.7
2004	5.7	5.6	5.7	5.5	5.6	5.6	5.5	5.4	5.4	5.5	5.4	5.4
2005	5.2	5.4	5.2	5.2	5.1	5.0	5.0					

13.3	14.3	14.9	15.2	15.8	14.2	16.0	14.0
14.5	16.1	13.7	15.2	13.7	16.9	14.9	14.4
15.3	13.1	15.2	15.9	15.1	14.9	13.6	13.7
15.3	15.5	14.5	16.5	13.4	15.2	15.3	13.8
14.3	12.6	15.3	14.8	14.1	14.4	14.3	15.6
14.8	14.6	15.6	15.1	14.8	15.2	15.6	14.5
15.2	14.3	15.8	17.0	14.3	14.6	16.1	12.8
14.5	15.4	13.3	14.9	14.3	16.4	13.9	16.1
14.6	15.2	14.1	14.8	16.4	14.2	15.2	16.6
14.1	16.8	15.4	14.0	16.9	15.7	14.4	15.6

(a) Reading down and left to right, draw a time series plot of all the data and comment on any features of the data that are revealed by this plot.
(b) Consider the notion that the first 40 observations were generated from a specific process, whereas the last 40 observations were generated from a different process. Does the plot indicate that the two processes generate similar results?
(c) Compute the sample mean and sample variance of the first 40 observations; then compute these values for the second 40 observations. Do these quantities indicate that both processes yield the same mean level? The same variability? Explain.

6-81. The total net electricity consumption of the U.S. by year from 1980–2002 (in billion kilowatt hours) follows. Net consumption excludes the energy consumed by the generating units. Read left to right.

2094.45, 2147.10, 2086.44, 2150.96, 2285.80, 2323.97, 2368.75, 2457.27, 2578.06, 2754.95, 2826.59, 2988.97, 3068.67, 3157.35, 3246.98, 3294.04, 3425.13, 3494.60, 3604.68, 3553.79, 3659.99

(source: U.S. Department of Energy Web site http://www/eia.doe.gov/emeu/international/electric.html#Consumption Total)

Construct a time series plot of these data. Construct and interpret a stem-and-leaf display of these data.

6-82. Reconsider the data from Exercise 6-80. Prepare comparative box plots for two groups of observations: the first 40 and the last 40. Comment on the information in the box plots.

6-83. The data shown in Table 6-7 are monthly champagne sales in France (1962-1969) in thousands of bottles.
(a) Construct a time series plot of the data and comment on any features of the data that are revealed by this plot.
(b) Speculate on how you would use a graphical procedure to forecast monthly champagne sales for the year 1970.

6-84. The following data are the temperatures of effluent at discharge from a sewage treatment facility on consecutive days:

43	47	51	48	52	50	46	49
45	52	46	51	44	49	46	51
49	45	44	50	48	50	49	50

(a) Calculate the sample mean, sample median, sample variance and sample standard deviation.
(b) Construct a box plot of the data and comment on the information in this display.

6-85. A manufacturer of coil springs is interested in implementing a quality control system to monitor his production process. As part of this quality system, it is decided to record the number of nonconforming coil springs in each production batch of size 50. During 40 days of production, 40 batches of data were collected as follows:

Read data across.

9	12	6	9	7	14	12	4	6	7
8	5	9	7	8	11	3	6	7	7
11	4	4	8	7	5	6	4	5	8
19	19	18	12	11	17	15	17	13	13

(a) Construct a stem-and-leaf plot of the data.
(b) Find the sample average and standard deviation.

Table 6-7 Champagne Sales in France

Month	1962	1963	1964	1965	1966	1967	1968	1969
Jan.	2.851	2.541	3.113	5.375	3.633	4.016	2.639	3.934
Feb.	2.672	2.475	3.006	3.088	4.292	3.957	2.899	3.162
Mar.	2.755	3.031	4.047	3.718	4.154	4.510	3.370	4.286
Apr.	2.721	3.266	3.523	4.514	4.121	4.276	3.740	4.676
May	2.946	3.776	3.937	4.520	4.647	4.968	2.927	5.010
June	3.036	3.230	3.986	4.539	4.753	4.677	3.986	4.874
July	2.282	3.028	3.260	3.663	3.965	3.523	4.217	4.633
Aug.	2.212	1.759	1.573	1.643	1.723	1.821	1.738	1.659
Sept.	2.922	3.595	3.528	4.739	5.048	5.222	5.221	5.591
Oct.	4.301	4.474	5.211	5.428	6.922	6.873	6.424	6.981
Nov.	5.764	6.838	7.614	8.314	9.858	10.803	9.842	9.851
Dec.	7.132	8.357	9.254	10.651	11.331	13.916	13.076	12.670

(c) Construct a time series plot of the data. Is there evidence that there was an increase or decrease in the average number of nonconforming springs made during the 40 days? Explain.

6-86. A communication channel is being monitored by recording the number of errors in a string of 1000 bits. Data for 20 of these strings follow:

Read data across.

3 1 0 1 3 2 4 1 3 1
1 1 2 3 3 2 0 2 0 1

(a) Construct a stem-and-leaf plot of the data.
(b) Find the sample average and standard deviation.
(c) Construct a time series plot of the data. Is there evidence that there was an increase or decrease in the number of errors in a string? Explain.

6-87. Reconsider the golf course yardage data in Exercise 6-3. Construct a box plot of the yardages and write an interpretation of the plot.

6-88. Reconsider the data in Exercise 6-80. Construct normal probability plots for two groups of the data: the first 40 and the last 40 observations. Construct both plots on the same axes. What tentative conclusions can you draw?

6-89. Construct a normal probability plot of the effluent discharge temperature data from Exercise 6-84. Based on the plot, what tentative conclusions can you draw?

6-90. Construct normal probability plots of the cold start ignition time data presented in Exercises 6-45 and 6-56. Construct a separate plot for each gasoline formulation, but arrange the plots on the same axes. What tentative conclusions can you draw?

6-91. Reconsider the golf ball overall distance data in Exercise 6-25. Construct a box plot of the yardage distance and write an interpretation of the plot. How does the box plot compare in interpretive value to the original stem-and-leaf diagram.

6-92. **Transformations.** In some data sets, a transformation by some mathematical function applied to the original data, such as $\sqrt{y}$ or log y, can result in data that are simpler to work with statistically than the original data. To illustrate the effect of a transformation, consider the following data, which represent cycles to failure for a yarn product: 675, 3650, 175, 1150, 290, 2000, 100, 375.

(a) Construct a normal probability plot and comment on the shape of the data distribution.
(b) Transform the data using logarithms; that is, let y^* (new value) = log y (old value). Construct a normal probability plot of the transformed data and comment on the effect of the transformation.

6-93. In 1879, A. A. Michelson made 100 determinations of the velocity of light in air using a modification of a method proposed by the French physicist Foucault. He made the measurements in five trials of 20 measurements each. The observations (in kilometers per second) follow. Each value has 299,000 subtracted from it.

Trial 1

850	900	930	950	980
1000	930	760	1000	960
740	1070	850	980	880
980	650	810	1000	960

Trial 2

960	960	880	850	900
830	810	880	800	760
940	940	800	880	840
790	880	830	790	800

Trial 3

880	880	720	620	970
880	850	840	850	840
880	860	720	860	950
910	870	840	840	840

Trial 4

890	810	800	760	750
910	890	880	840	850
810	820	770	740	760
920	860	720	850	780

Trial 5

890	780	760	790	820
870	810	810	950	810
840	810	810	810	850
870	740	940	800	870

The currently accepted true velocity of light in a vacuum is 299,792.5 kilometers per second. Stigler (1977, *The Annals of Statistics*) reports that the "true" value for comparison to these measurements is 734.5. Construct comparative box plots of these measurements. Does it seem that all five trials are consistent with respect to the variability of the measurements? Are all five trials centered on the same value? How does each group of trials compare to the true value? Could there have been "startup" effects in the experiment that Michelson performed? Could there have been bias in the measuring instrument?

6-94. In 1789, Henry Cavendish estimated the density of the earth by using a torsion balance. His 29 measurements follow, expressed as a multiple of the density of water.

5.50	5.30	5.47	5.10	5.29	5.65
5.55	5.61	5.75	5.63	5.27	5.44
5.57	5.36	4.88	5.86	5.34	5.39
5.34	5.53	5.29	4.07	5.85	5.46
5.42	5.79	5.62	5.58	5.26	

(a) Calculate the sample mean, sample standard deviation, and median of the Cavendish density data.

(b) Construct a normal probability plot of the data. Comment on the plot. Does there seem to be a "low" outlier in the data?

(c) Would the sample median be a better estimate of the density of the earth than the sample mean? Why?

MIND-EXPANDING EXERCISES

6-95. Consider the airfoil data in Exercise 6-12. Subtract 30 from each value and then multiply the resulting quantities by 10. Now compute s^2 for the new data. How is this quantity related to s^2 for the *original* data? Explain why.

6-96. Consider the quantity $\sum_{i=1}^{n}(x_i - a)^2$. For what value of a is this quantity minimized?

6-97. Using the results of Exercise 6-96, which of the two quantities $\sum_{i=1}^{n}(x_i - \bar{x})^2$ and $\sum_{i=1}^{n}(x_i - \mu)^2$ will be smaller, provided that $\bar{x} \neq \mu$?

6-98. **Coding the Data.** Let $y_i = a + bx_i$, $i = 1, 2, \ldots, n$, where a and b are nonzero constants. Find the relationship between $\bar{x}$ and $\bar{y}$, and between s_x and s_y.

6-99. A sample of temperature measurements in a furnace yielded a sample average (°F) of 835.00 and a sample standard deviation of 10.5. Using the results from Exercise 6-98, what are the sample average and sample standard deviations expressed in °C?

6-100. Consider the sample $x_1, x_2, \ldots, x_n$ with sample mean $\bar{x}$ and sample standard deviation s. Let $z_i = (x_i - \bar{x})/s$, $i = 1, 2, \ldots, n$. What are the values of the sample mean and sample standard deviation of the z_i?

6-101. An experiment to investigate the survival time in hours of an electronic component consists of placing the parts in a test cell and running them for 100 hours under elevated temperature conditions. (This is called an "accelerated" life test.) Eight components were tested with the following resulting failure times:

75, 63, 100$^+$, 36, 51, 45, 80, 90

The observation 100$^+$ indicates that the unit still functioned at 100 hours. Is there any meaningful measure of location that can be calculated for these data? What is its numerical value?

6-102. Suppose that we have a sample $x_1, x_2, \ldots, x_n$ and we have calculated $\bar{x}_n$ and s_n^2 for the sample. Now an $(n + 1)$st observation becomes available. Let $\bar{x}_{n+1}$ and s_{n+1}^2 be the sample mean and sample variance for the sample using all $n + 1$ observations.

(a) Show how $\bar{x}_{n+1}$ can be computed using $\bar{x}_n$ and x_{n+1}.

(b) Show that $ns_{n+1}^2 = (n - 1)s_n^2 + \dfrac{n(x_{n+1} - \bar{x}_n)^2}{n + 1}$

(c) Use the results of parts (a) and (b) to calculate the new sample average and standard deviation for the data of Exercise 6-22, when the new observation is $x_{38} = 64$.

6-103. **Trimmed Mean.** Suppose that the data are arranged in increasing order, $T\%$ of the observations are removed from each end and the sample mean of the remaining numbers is calculated. The resulting quantity is called a *trimmed mean*. The trimmed mean generally lies between the sample mean $\bar{x}$ and the sample median $\tilde{x}$. Why?

(a) Calculate the 10% trimmed mean for the yield data in Exercise 6-17.

(b) Calculate the 20% trimmed mean for the yield data in Exercise 6-17 and compare it with the quantity found in part (a).

(c) Compare the values calculated in parts (a) and (b) with the sample mean and median for the yield data. Is there much difference in these quantities? Why?

6-104. **Trimmed Mean.** Suppose that the sample size n is such that the quantity $nT/100$ is not an integer. Develop a procedure for obtaining a trimmed mean in this case.

IMPORTANT TERMS AND CONCEPTS

Box plot
Frequency distribution and histogram
Median, quartiles and percentiles
Multivariable data

Normal probability plot
Pareto chart
Population mean
Population standard deviation
Population variance

Probability plot
Relative frequency distribution
Sample mean
Sample standard deviation

Sample variance
Stem-and-leaf diagram
Time series plots

Sampling Distributions and Point Estimation of Parameters

CHAPTER OUTLINE

7-1 INTRODUCTION

7-2 SAMPLING DISTRIBUTIONS AND THE CENTRAL LIMIT THEOREM

7-3 GENERAL CONCEPTS OF POINT ESTIMATION

    7-3.1 Unbiased Estimators

    7-3.2 Variance of a Point Estimator

    7-3.3 Standard Error: Reporting a Point Estimate

    7-3.4 Mean Squared Error of an Estimator

7-4 METHODS OF POINT ESTIMATION

    7-4.1 Method of Moments

    7-4.2 Method of Maximum Likelihood

    7-4.3 Bayesian Estimation of Parameters

LEARNING OBJECTIVES

After careful study of this chapter you should be able to do the following:

1. Explain the general concepts of estimating the parameters of a population or a probability distribution
2. Explain the important role of the normal distribution as a sampling distribution
3. Understand the central limit theorem
4. Explain important properties of point estimators, including bias, variance, and mean square error
5. Know how to construct point estimators using the method of moments and the method of maximum likelihood
6. Know how to compute and explain the precision with which a parameter is estimated
7. Know how to construct a point estimator using the Bayesian approach

7-1 INTRODUCTION

The field of statistical inference consists of those methods used to make decisions or to draw conclusions about a **population**. These methods utilize the information contained in a **sample** from the population in drawing conclusions. This chapter begins our study of the statistical methods used for inference and decision making.

Statistical inference may be divided into two major areas: **parameter estimation** and **hypothesis testing**. As an example of a parameter estimation problem, suppose that a structural engineer is analyzing the tensile strength of a component used in an automobile chassis. Since variability in tensile strength is naturally present between the individual components because of differences in raw material batches, manufacturing processes, and measurement procedures (for example), the engineer is interested in estimating the mean tensile strength of the components. In practice, the engineer will use sample data to compute a number that is in some sense a reasonable value (or guess) of the true mean. This number is called a **point estimate**. We will see that it is possible to establish the precision of the estimate.

Now consider a situation in which two different reaction temperatures can be used in a chemical process, say t_1 and t_2. The engineer conjectures that t_1 results in higher yields than does t_2. Statistical hypothesis testing is a framework for solving problems of this type. In this case, the hypothesis would be that the mean yield using temperature t_1 is greater than the mean yield using temperature t_2. Notice that there is no emphasis on estimating yields; instead, the focus is on drawing conclusions about a stated hypothesis.

Suppose that we want to obtain a point estimate of a population parameter. We know that before the data is collected, the observations are considered to be random variables, say $X_1, X_2, \ldots, X_n$. Therefore, any function of the observation, or any **statistic**, is also a random variable. For example, the sample mean $\overline{X}$ and the sample variance S^2 are statistics and they are also random variables.

Since a statistic is a random variable, it has a probability distribution. We call the probability distribution of a statistic a **sampling distribution**. The notion of a sampling distribution is very important and will be discussed and illustrated later in the chapter.

When discussing inference problems, it is convenient to have a general symbol to represent the parameter of interest. We will use the Greek symbol θ (theta) to represent the parameter. The objective of point estimation is to select a single number, based on sample data, that is the most plausible value for θ. A numerical value of a sample statistic will be used as the point estimate.

In general, if X is a random variable with probability distribution $f(x)$, characterized by the unknown parameter θ, and if $X_1, X_2, \ldots, X_n$ is a random sample of size n from X, the statistic $\hat{\Theta} = h(X_1, X_2, \ldots, X_n)$ is called a **point estimator** of θ. Note that $\hat{\Theta}$ is a random variable because it is a function of random variables. After the sample has been selected, $\hat{\Theta}$ takes on a particular numerical value $\hat{\theta}$ called the **point estimate** of θ.

Point Estimator

> A point estimate of some population parameter θ is a single numerical value $\hat{\theta}$ of a statistic $\hat{\Theta}$. The statistic $\hat{\Theta}$ is called the **point estimator**.

As an example, suppose that the random variable X is normally distributed with an unknown mean μ. The sample mean is a point estimator of the unknown population mean μ.

That is, $\hat{\mu} = \overline{X}$. After the sample has been selected, the numerical value $\bar{x}$ is the point estimate of μ. Thus, if $x_1 = 25$, $x_2 = 30$, $x_3 = 29$, and $x_4 = 31$, the point estimate of μ is

$$\bar{x} = \frac{25 + 30 + 29 + 31}{4} = 28.75$$

Similarly, if the population variance σ^2 is also unknown, a point estimator for σ^2 is the sample variance S^2, and the numerical value $s^2 = 6.9$ calculated from the sample data is called the point estimate of σ^2.

Estimation problems occur frequently in engineering. We often need to estimate

- The mean μ of a single population
- The variance σ^2 (or standard deviation σ) of a single population
- The proportion p of items in a population that belong to a class of interest
- The difference in means of two populations, $\mu_1 - \mu_2$
- The difference in two population proportions, $p_1 - p_2$

Reasonable point estimates of these parameters are as follows:

- For μ, the estimate is $\hat{\mu} = \bar{x}$, the sample mean.
- For σ^2, the estimate is $\hat{\sigma}^2 = s^2$, the sample variance.
- For p, the estimate is $\hat{p} = x/n$, the sample proportion, where x is the number of items in a random sample of size n that belong to the class of interest.
- For $\mu_1 - \mu_2$, the estimate is $\hat{\mu}_1 - \hat{\mu}_2 = \bar{x}_1 - \bar{x}_2$, the difference between the sample means of two independent random samples.
- For $p_1 - p_2$, the estimate is $\hat{p}_1 - \hat{p}_2$, the difference between two sample proportions computed from two independent random samples.

We may have several different choices for the point estimator of a parameter. For example, if we wish to estimate the mean of a population, we might consider the sample mean, the sample median, or perhaps the average of the smallest and largest observations in the sample as point estimators. In order to decide which point estimator of a particular parameter is the best one to use, we need to examine their statistical properties and develop some criteria for comparing estimators.

7-2 SAMPLING DISTRIBUTIONS AND THE CENTRAL LIMIT THEOREM

Statistical inference is concerned with making **decisions** about a population based on the information contained in a random sample from that population. For instance, we may be interested in the mean fill volume of a can of soft drink. The mean fill volume in the population is required to be 300 milliliters. An engineer takes a random sample of 25 cans and computes the sample average fill volume to be $\bar{x} = 298$ milliliters. The engineer will probably decide that the population mean is $\mu = 300$ milliliters, even though the sample mean was 298 milliliters because he or she knows that the sample mean is a reasonable estimate of μ and that a sample mean of 298 milliliters is very likely to occur, even if the true population mean is $\mu = 300$ milliliters. In fact, if the true mean is 300 milliliters, tests of

25 cans made repeatedly, perhaps every five minutes, would produce values of $\bar{x}$ that vary both above and below $\mu = 300$ milliliters.

The link between the probability models in the earlier chapters and the data is made as follows. Each numerical value in the data is the observed value of a random variable. Furthermore, the random variables are usually assumed to be independent and identically distributed. These random variables are known as a random sample.

Random Sample

> The random variables $X_1, X_2, \ldots, X_n$ are a **random sample** of size n if (a) the X_i's are independent random variables, and (b) every X_i has the same probability distribution.

The observed data is also referred to as a random sample, but the use of the same phrase should not cause any confusion.

The primary purpose in taking a random sample is to obtain information about the unknown population parameters. Suppose, for example, that we wish to reach a conclusion about the proportion of people in the United States who prefer a particular brand of soft drink. Let p represent the unknown value of this proportion. It is impractical to question every individual in the population to determine the true value of p. In order to make an inference regarding the true proportion p, a more reasonable procedure would be to select a random sample (of an appropriate size) and use the observed proportion $\hat{p}$ of people in this sample favoring the brand of soft drink.

The sample proportion, $\hat{p}$ is computed by dividing the number of individuals in the sample who prefer the brand of soft drink by the total sample size n. Thus, $\hat{p}$ is a function of the observed values in the random sample. Since many random samples are possible from a population, the value of $\hat{p}$ will vary from sample to sample. That is, $\hat{p}$ is a random variable. Such a random variable is called a **statistic**.

Statistic

> A **statistic** is any function of the observations in a random sample.

We have encountered statistics before. For example, if $X, X_2, \ldots, X_n$ is a random sample of size n, the sample mean $\bar{X}$, the sample variance S^2, and the sample standard deviation S are statistics. Since a statistic is a random variable, it has a probability distribution.

Sampling Distribution

> The probability distribution of a statistic is called a **sampling distribution**.

For example, the probability distribution of $\bar{X}$ is called the **sampling distribution of the mean**. The sampling distribution of a statistic depends on the distribution of the population, the size of the sample, and the method of sample selection. We now present perhaps the most important sampling distribution. Other sampling distributions and their applications will be illustrated extensively in the following two chapters.

Consider determining the sampling distribution of the sample mean $\bar{X}$. Suppose that a random sample of size n is taken from a normal population with mean μ and variance σ^2. Now each observation in this sample, say, $X_1, X_2, \ldots, X_n$, is a normally and independently distributed random variable with mean μ and variance σ^2. Then because linear functions of independent normally distributed random variables are also normally distributed (Chapter 5), we conclude that the sample mean

$$\overline{X} = \frac{X_1 + X_2 + \cdots + X_n}{n}$$

has a normal distribution with mean

$$\mu_{\overline{X}} = \frac{\mu + \mu + \cdots + \mu}{n} = \mu$$

and variance

$$\sigma^2_{\overline{X}} = \frac{\sigma^2 + \sigma^2 + \cdots + \sigma^2}{n^2} = \frac{\sigma^2}{n}$$

If we are sampling from a population that has an unknown probability distribution, the sampling distribution of the sample mean will still be approximately normal with mean μ and variance σ^2/n, if the sample size n is large. This is one of the most useful theorems in statistics, called the central limit theorem. The statement is as follows:

Central Limit Theorem

If $X_1, X_2, \ldots, X_n$ is a random sample of size n taken from a population (either finite or infinite) with mean μ and finite variance σ^2, and if $\overline{X}$ is the sample mean, the limiting form of the distribution of

$$Z = \frac{\overline{X} - \mu}{\sigma/\sqrt{n}} \tag{7-1}$$

as $n \to \infty$, is the standard normal distribution.

The normal approximation for $\overline{X}$ depends on the sample size n. Figure 7-1(a) shows the distribution obtained for throws of a single, six-sided true die. The probabilities are equal (1/6) for all the values obtained, 1, 2, 3, 4, 5, or 6. Figure 7-1(b) shows the distribution of the average score obtained when tossing two dice, and Fig. 7-1(c), 7-1(d), and 7-1(e) show the distributions of average scores obtained when tossing three, five, and ten dice, respectively. Notice that, while the population (one die) is relatively far from normal, the distribution of averages is approximated reasonably well by the normal distribution for sample sizes as small as five. (The dice throw distributions are discrete, however, while the normal is continuous). Although the central limit theorem will work well for small samples ($n = 4, 5$) in most cases, particularly where the population is continuous, unimodal, and symmetric, larger samples will be required in other situations, depending on the shape of the population. In many cases of practical interest, if $n \geq 30$, the normal approximation will be satisfactory regardless of the shape of the population. If $n < 30$, the central limit theorem will work if the distribution of the population is not severely nonnormal.

234 CHAPTER 7 SAMPLING DISTRIBUTIONS AND POINT ESTIMATION OF PARAMETERS

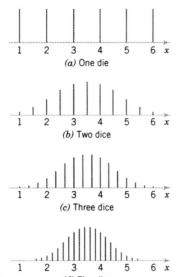

Figure 7-1 Distributions of average scores from throwing dice. [Adapted with permission from Box, Hunter, and Hunter (1978).]

EXAMPLE 7-1
Resistors

An electronics company manufactures resistors that have a mean resistance of 100 ohms and a standard deviation of 10 ohms. The distribution of resistance is normal. Find the probability that a random sample of $n = 25$ resistors will have an average resistance less than 95 ohms.

Note that the sampling distribution of $\overline{X}$ is normal, with mean $\mu_{\overline{X}} = 100$ ohms and a standard deviation of

$$\sigma_{\overline{X}} = \frac{\sigma}{\sqrt{n}} = \frac{10}{\sqrt{25}} = 2$$

Therefore, the desired probability corresponds to the shaded area in Fig. 7-1. Standardizing the point $\overline{X} = 95$ in Fig. 7-2, we find that

$$z = \frac{95 - 100}{2} = -2.5$$

and therefore,

$$P(\overline{X} < 95) = P(Z < -2.5)$$
$$= 0.0062$$

The following example makes use of the central limit theorem.

EXAMPLE 7-2
Central Limit Theorem

Suppose that a random variable X has a continuous uniform distribution

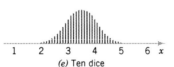

Find the distribution of the sample mean of a random sample of size $n = 40$.

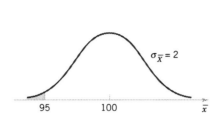
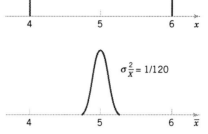

Figure 7-2 Probability for Example 7-1.

Figure 7-3 The distributions of X and $\overline{X}$ for Example 7-2.

The mean and variance of X are $\mu = 5$ and $\sigma^2 = (6-4)^2/12 = 1/3$. The central limit theorem indicates that the distribution of $\overline{X}$ is approximately normal with mean $\mu_{\overline{X}} = 5$ and variance $\sigma_{\overline{X}}^2 = \sigma^2/n = 1/[3(40)] = 1/120$. The distributions of X and $\overline{X}$ are shown in Fig. 7-3.

Now consider the case in which we have two independent populations. Let the first population have mean μ_1 and variance σ_1^2 and the second population have mean μ_2 and variance σ_2^2. Suppose that both populations are normally distributed. Then, using the fact that linear combinations of independent normal random variables follow a normal distribution (see Chapter 5), we can say that the sampling distribution of $\overline{X}_1 - \overline{X}_2$ is normal with mean

$$\mu_{\overline{X}_1 - \overline{X}_2} = \mu_{\overline{X}_1} - \mu_{\overline{X}_2} = \mu_1 - \mu_2 \tag{7-2}$$

and variance

$$\sigma_{\overline{X}_1 - \overline{X}_2}^2 = \sigma_{\overline{X}_1}^2 + \sigma_{\overline{X}_2}^2 = \frac{\sigma_1^2}{n_1} + \frac{\sigma_2^2}{n_2} \tag{7-3}$$

If the two populations are not normally distributed and if both sample sizes n_1 and n_2 are greater than 30, we may use the central limit theorem and assume that $\overline{X}_1$ and $\overline{X}_2$ follow approximately independent normal distributions. Therefore, the sampling distribution of $\overline{X}_1 - \overline{X}_2$ is approximately normal with mean and variance given by Equations 7-2 and 7-3, respectively. If either n_1 or n_2 is less than 30, the sampling distribution of $\overline{X}_1 - \overline{X}_2$ will still be approximately normal with mean and variance given by Equations 7-2 and 7-3, provided that the population from which the small sample is taken is not dramatically different from the normal. We may summarize this with the following definition.

Approximate Sampling Distribution of a Difference in Sample Means

If we have two independent populations with means μ_1 and μ_2 and variances σ_1^2 and σ_2^2 and if $\overline{X}_1$ and $\overline{X}_2$ are the sample means of two independent random samples of sizes n_1 and n_2 from these populations, then the sampling distribution of

$$Z = \frac{\overline{X}_1 - \overline{X}_2 - (\mu_1 - \mu_2)}{\sqrt{\sigma_1^2/n_1 + \sigma_2^2/n_2}} \tag{7-4}$$

is approximately standard normal, if the conditions of the central limit theorem apply. If the two populations are normal, the sampling distribution of Z is exactly standard normal.

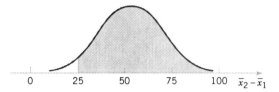

Figure 7-4 The sampling distribution of $\bar{X}_2 - \bar{X}_1$ in Example 7-3.

EXAMPLE 7-3
Aircraft Engine Life

The effective life of a component used in a jet-turbine aircraft engine is a random variable with mean 5000 hours and standard deviation 40 hours. The distribution of effective life is fairly close to a normal distribution. The engine manufacturer introduces an improvement into the manufacturing process for this component that increases the mean life to 5050 hours and decreases the standard deviation to 30 hours. Suppose that a random sample of $n_1 = 16$ components is selected from the "old" process and a random sample of $n_2 = 25$ components is selected from the "improved" process. What is the probability that the difference in the two sample means $\bar{X}_2 - \bar{X}_1$ is at least 25 hours? Assume that the old and improved processes can be regarded as independent populations.

To solve this problem, we first note that the distribution of $\bar{X}_1$ is normal with mean $\mu_1 = 5000$ hours and standard deviation $\sigma_1/\sqrt{n_1} = 40/\sqrt{16} = 10$ hours, and the distribution of $\bar{X}_2$ is normal with mean $\mu_2 = 5050$ hours and standard deviation $\sigma_2/\sqrt{n_2} = 30/\sqrt{25} = 6$ hours. Now the distribution of $\bar{X}_2 - \bar{X}_1$ is normal with mean $\mu_2 - \mu_1 = 5050 - 5000 = 50$ hours and variance $\sigma_2^2/n_2 + \sigma_1^2/n_1 = (6)^2 + (10)^2 = 136$ hours^2. This sampling distribution is shown in Fig. 7-4. The probability that $\bar{X}_2 - \bar{X}_1 \geq 25$ is the shaded portion of the normal distribution in this figure.

Corresponding to the value $\bar{x}_2 - \bar{x}_1 = 25$ in Fig. 7-4, we find that

$$z = \frac{25 - 50}{\sqrt{136}} = -2.14$$

and we find that

$$P(\bar{X}_2 - \bar{X}_1 \geq 25) = P(Z \geq -2.14)$$
$$= 0.9838$$

EXERCISES FOR SECTION 7-2

7-1. PVC pipe is manufactured with a mean diameter of 1.01 inch and a standard deviation of 0.003 inch. Find the probability that a random sample of $n = 9$ sections of pipe will have a sample mean diameter greater than 1.009 inch and less than 1.012 inch.

7-2. Suppose that samples of size $n = 25$ are selected at random from a normal population with mean 100 and standard deviation 10. What is the probability that the sample mean falls in the interval from $\mu_{\bar{X}} - 1.8\sigma_{\bar{X}}$ to $\mu_{\bar{X}} + 1.0\sigma_{\bar{X}}$?

7-3. A synthetic fiber used in manufacturing carpet has tensile strength that is normally distributed with mean 75.5 psi and standard deviation 3.5 psi. Find the probability that a random sample of $n = 6$ fiber specimens will have sample mean tensile strength that exceeds 75.75 psi.

7-4. Consider the synthetic fiber in the previous exercise. How is the standard deviation of the sample mean changed when the sample size is increased from $n = 6$ to $n = 49$?

7-5. The compressive strength of concrete is normally distributed with $\mu = 2500$ psi and $\sigma = 50$ psi. Find the probability that a random sample of $n = 5$ specimens will have a sample mean diameter that falls in the interval from 2499 psi to 2510 psi.

7-6. Consider the concrete specimens in the previous exercise. What is the standard error of the sample mean?

7-7. A normal population has mean 100 and variance 25. How large must the random sample be if we want the standard error of the sample average to be 1.5?

7-8. Suppose that the random variable X has the continuous uniform distribution

$$f(x) = \begin{cases} 1, & 0 \leq x \leq 1 \\ 0, & \text{otherwise} \end{cases}$$

Suppose that a random sample of $n = 12$ observations is selected from this distribution. What is the approximate probability distribution of $\bar{X} - 6$? Find the mean and variance of this quantity.

7-9. Suppose that X has a discrete uniform distribution

$$f(x) = \begin{cases} 1/3, & x = 1, 2, 3 \\ 0, & \text{otherwise} \end{cases}$$

A random sample of $n = 36$ is selected from this population. Find the probability that the sample mean is greater than 2.1 but less than 2.5, assuming that the sample mean would be measured to the nearest tenth.

7-10. The amount of time that a customer spends waiting at an airport check-in counter is a random variable with mean 8.2 minutes and standard deviation 1.5 minutes. Suppose that a random sample of $n = 49$ customers is observed. Find the probability that the average time waiting in line for these customers is
(a) Less than 10 minutes
(b) Between 5 and 10 minutes
(c) Less than 6 minutes

7-11. A random sample of size $n_1 = 16$ is selected from a normal population with a mean of 75 and a standard deviation of 8. A second random sample of size $n_2 = 9$ is taken from another normal population with mean 70 and standard deviation 12. Let $\bar{X}_1$ and $\bar{X}_2$ be the two sample means. Find
(a) The probability that $\bar{X}_1 - \bar{X}_2$ exceeds 4
(b) The probability that $3.5 \le \bar{X}_1 - \bar{X}_2 \le 5.5$

7-12. A consumer electronics company is comparing the brightness of two different types of picture tubes for use in its television sets. Tube type A has mean brightness of 100 and standard deviation of 16, while tube type B has unknown mean brightness, but the standard deviation is assumed to be identical to that for type A. A random sample of $n = 25$ tubes of each type is selected, and $\bar{X}_B - \bar{X}_A$ is computed. If μ_B equals or exceeds μ_A, the manufacturer would like to adopt type B for use. The observed difference is $\bar{x}_B - \bar{x}_A = 3.5$. What decision would you make, and why?

7-13. The elasticity of a polymer is affected by the concentration of a reactant. When low concentration is used, the true mean elasticity is 55, and when high concentration is used the mean elasticity is 60. The standard deviation of elasticity is 4, regardless of concentration. If two random samples of size 16 are taken, find the probability that $\bar{X}_{high} - \bar{X}_{low} \ge 2$.

7-3 GENERAL CONCEPTS OF POINT ESTIMATION

7-3.1 Unbiased Estimators

An estimator should be "close" in some sense to the true value of the unknown parameter. Formally, we say that $\hat{\Theta}$ is an unbiased estimator of θ if the expected value of $\hat{\Theta}$ is equal to θ. This is equivalent to saying that the mean of the probability distribution of $\hat{\Theta}$ (or the mean of the sampling distribution of $\hat{\Theta}$) is equal to θ.

Bias of an Estimator

The point estimator $\hat{\Theta}$ is an unbiased estimator for the parameter θ if

$$E(\hat{\Theta}) = \theta \tag{7-5}$$

If the estimator is not unbiased, then the difference

$$E(\hat{\Theta}) - \theta \tag{7-6}$$

is called the **bias** of the estimator $\hat{\Theta}$.

When an estimator is unbiased, the bias is zero; that is, $E(\hat{\Theta}) - \theta = 0$.

EXAMPLE 7-4
Sample Mean and Variance Are Unbiased

Suppose that X is a random variable with mean μ and variance σ^2. Let $X_1, X_2, \ldots, X_n$ be a random sample of size n from the population represented by X. Show that the sample mean $\bar{X}$ and sample variance S^2 are unbiased estimators of μ and σ^2, respectively.

First consider the sample mean. In Section 5.5 in Chapter 5, we showed that $E(\bar{X}) = \mu$. Therefore, the sample mean $\bar{X}$ is an unbiased estimator of the population mean μ.

Now consider the sample variance. We have

$$E(S^2) = E\left[\frac{\sum_{i=1}^{n}(X_i - \overline{X})^2}{n-1}\right] = \frac{1}{n-1} E \sum_{i=1}^{n}(X_i - \overline{X})^2$$

$$= \frac{1}{n-1} E \sum_{i=1}^{n}(X_i^2 + \overline{X}^2 - 2\overline{X}X_i) = \frac{1}{n-1} E \left(\sum_{i=1}^{n} X_i^2 - n\overline{X}^2\right)$$

$$= \frac{1}{n-1}\left[\sum_{i=1}^{n} E(X_i^2) - nE(\overline{X}^2)\right]$$

The last equality follows the equation for the mean of a linear function in Chapter 5. However, since $E(X_i^2) = \mu^2 + \sigma^2$ and $E(\overline{X}^2) = \mu^2 + \sigma^2/n$, we have

$$E(S^2) = \frac{1}{n-1}\left[\sum_{i=1}^{n}(\mu^2 + \sigma^2) - n(\mu^2 + \sigma^2/n)\right]$$

$$= \frac{1}{n-1}(n\mu^2 + n\sigma^2 - n\mu^2 - \sigma^2) = \sigma^2$$

Therefore, the sample variance S^2 is an unbiased estimator of the population variance σ^2.

Although S^2 is unbiased for σ^2, S is a biased estimator of σ. For large samples, the bias is very small. However, there are good reasons for using S as an estimator of σ in samples from normal distributions, as we will see in the next three chapters when we discuss confidence intervals and hypothesis testing.

Sometimes there are several unbiased estimators of the sample population parameter. For example, suppose we take a random sample of size $n = 10$ from a normal population and obtain the data $x_1 = 12.8$, $x_2 = 9.4$, $x_3 = 8.7$, $x_4 = 11.6$, $x_5 = 13.1$, $x_6 = 9.8$, $x_7 = 14.1$, $x_8 = 8.5$, $x_9 = 12.1$, $x_{10} = 10.3$. Now the sample mean is

$$\overline{x} = \frac{12.8 + 9.4 + 8.7 + 11.6 + 13.1 + 9.8 + 14.1 + 8.5 + 12.1 + 10.3}{10} = 11.04$$

the sample median is

$$\tilde{x} = \frac{10.3 + 11.6}{2} = 10.95$$

and a 10% trimmed mean (obtained by discarding the smallest and largest 10% of the sample before averaging) is

$$\overline{x}_{tr(10)} = \frac{8.7 + 9.4 + 9.8 + 10.3 + 11.6 + 12.1 + 12.8 + 13.1}{8} = 10.98$$

We can show that all of these are unbiased estimates of μ. Since there is not a unique unbiased estimator, we cannot rely on the property of unbiasedness alone to select our estimator. We need a method to select among unbiased estimators. We suggest a method in the following section.

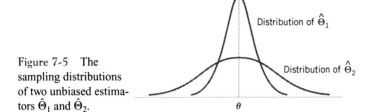

Figure 7-5 The sampling distributions of two unbiased estimators $\hat{\Theta}_1$ and $\hat{\Theta}_2$.

7-3.2 Variance of a Point Estimator

Suppose that $\hat{\Theta}_1$ and $\hat{\Theta}_2$ are unbiased estimators of θ. This indicates that the distribution of each estimator is centered at the true value of θ. However, the variance of these distributions may be different. Figure 7-5 illustrates the situation. Since $\hat{\Theta}_1$ has a smaller variance than $\hat{\Theta}_2$, the estimator $\hat{\Theta}_1$ is more likely to produce an estimate close to the true value θ. A logical principle of estimation, when selecting among several estimators, is to choose the estimator that has minimum variance.

Minimum Variance Unbiased Estimator

If we consider all unbiased estimators of θ, the one with the smallest variance is called the **minimum variance unbiased estimator** (MVUE).

In a sense, the MVUE is most likely among all unbiased estimators to produce an estimate $\hat{\theta}$ that is close to the true value of θ. It has been possible to develop methodology to identify the MVUE in many practical situations. While this methodology is beyond the scope of this book, we give one very important result concerning the normal distribution.

If $X_1, X_2, \ldots, X_n$ is a random sample of size n from a normal distribution with mean μ and variance σ^2, the sample mean $\overline{X}$ is the MVUE for μ.

In situations in which we do not know whether an MVUE exists, we could still use a minimum variance principle to choose among competing estimators. Suppose, for example, we wish to estimate the mean of a population (not necessarily a *normal* population). We have a random sample of n observations $X_1, X_2, \ldots, X_n$ and we wish to compare two possible estimators for μ: the sample mean $\overline{X}$ and a single observation from the sample, say, X_i. Note that both $\overline{X}$ and X_i are unbiased estimators of μ; for the sample mean, we have $V(\overline{X}) = \sigma^2/n$ from Chapter 5 and the variance of any observation is $V(X_i) = \sigma^2$. Since $V(\overline{X}) < V(X_i)$ for sample sizes $n \geq 2$, we would conclude that the sample mean is a better estimator of μ than a single observation X_i.

7-3.3 Standard Error: Reporting a Point Estimate

When the numerical value or point estimate of a parameter is reported, it is usually desirable to give some idea of the precision of estimation. The measure of precision usually employed is the standard error of the estimator that has been used.

Standard Error of an Estimator

The standard error of an estimator $\hat{\Theta}$ is its standard deviation, given by $\sigma_{\hat{\Theta}} = \sqrt{V(\hat{\Theta})}$. If the standard error involves unknown parameters that can be estimated, substitution of those values into $\sigma_{\hat{\Theta}}$ produces an **estimated standard error**, denoted by $\hat{\sigma}_{\hat{\Theta}}$.

Sometimes the estimated standard error is denoted by $s_{\hat{\Theta}}$ or $se(\hat{\Theta})$.

Suppose we are sampling from a normal distribution with mean μ and variance σ^2. Now the distribution of $\overline{X}$ is normal with mean μ and variance σ^2/n, so the standard error of $\overline{X}$ is

$$\sigma_{\overline{X}} = \frac{\sigma}{\sqrt{n}}$$

If we did not know σ but substituted the sample standard deviation S into the above equation, the estimated standard error of $\overline{X}$ would be

$$\hat{\sigma}_{\overline{X}} = \frac{S}{\sqrt{n}}$$

When the estimator follows a normal distribution, as in the above situation, we can be reasonably confident that the true value of the parameter lies within two standard errors of the estimate. Since many point estimators are normally distributed (or approximately so) for large n, this is a very useful result. Even in cases in which the point estimator is not normally distributed, we can state that so long as the estimator is unbiased, the estimate of the parameter will deviate from the true value by as much as four standard errors at most 6 percent of the time. Thus a very conservative statement is that the true value of the parameter differs from the point estimate by at most four standard errors. See Chebyshev's inequality in the supplemental material on the Web site.

EXAMPLE 7-5 Thermal Conductivity

An article in the *Journal of Heat Transfer* (Trans. ASME, Sec. C, 96, 1974, p. 59) described a new method of measuring the thermal conductivity of Armco iron. Using a temperature of 100°F and a power input of 550 watts, the following 10 measurements of thermal conductivity (in Btu/hr-ft-°F) were obtained:

$$41.60, 41.48, 42.34, 41.95, 41.86,$$
$$42.18, 41.72, 42.26, 41.81, 42.04$$

A point estimate of the mean thermal conductivity at 100°F and 550 watts is the sample mean or

$$\overline{x} = 41.924 \text{ Btu/hr-ft-°F}$$

The standard error of the sample mean is $\sigma_{\overline{X}} = \sigma/\sqrt{n}$, and since σ is unknown, we may replace it by the sample standard deviation $s = 0.284$ to obtain the estimated standard error of $\overline{X}$ as

$$\hat{\sigma}_{\overline{X}} = \frac{s}{\sqrt{n}} = \frac{0.284}{\sqrt{10}} = 0.0898$$

Notice that the standard error is about 0.2 percent of the sample mean, implying that we have obtained a relatively precise point estimate of thermal conductivity. If we can assume that thermal conductivity is normally distributed, 2 times the standard error is $2\hat{\sigma}_{\bar{X}} = 2(0.0898) = 0.1796$, and we are highly confident that the true mean thermal conductivity is within the interval 41.924 ± 0.1796, or between 41.744 and 42.104.

7-3.4 Mean Squared Error of an Estimator

Sometimes it is necessary to use a biased estimator. In such cases, the mean squared error of the estimator can be important. The mean squared error of an estimator $\hat{\Theta}$ is the expected squared difference between $\hat{\Theta}$ and θ.

Mean Squared Error of an Estimator

The **mean squared error** of an estimator $\hat{\Theta}$ of the parameter θ is defined as

$$\text{MSE}(\hat{\Theta}) = E(\hat{\Theta} - \theta)^2 \qquad (7\text{-}7)$$

The mean squared error can be rewritten as follows:

$$\text{MSE}(\hat{\Theta}) = E[\hat{\Theta} - E(\hat{\Theta})]^2 + [\theta - E(\hat{\Theta})]^2$$
$$= V(\hat{\Theta}) + (\text{bias})^2$$

That is, the mean squared error of $\hat{\Theta}$ is equal to the variance of the estimator plus the squared bias. If $\hat{\Theta}$ is an unbiased estimator of θ, the mean squared error of $\hat{\Theta}$ is equal to the variance of $\hat{\Theta}$.

The mean squared error is an important criterion for comparing two estimators. Let $\hat{\Theta}_1$ and $\hat{\Theta}_2$ be two estimators of the parameter θ, and let MSE $(\hat{\Theta}_1)$ and MSE $(\hat{\Theta}_2)$ be the mean squared errors of $\hat{\Theta}_1$ and $\hat{\Theta}_2$. Then the **relative efficiency** of $\hat{\Theta}_2$ to $\hat{\Theta}_1$ is defined as

$$\frac{\text{MSE}(\hat{\Theta}_1)}{\text{MSE}(\hat{\Theta}_2)} \qquad (7\text{-}8)$$

If this relative efficiency is less than 1, we would conclude that $\hat{\Theta}_1$ is a more efficient estimator of θ than $\hat{\Theta}_2$, in the sense that it has a smaller mean square error.

Sometimes we find that biased estimators are preferable to unbiased estimators because they have smaller mean squared error. That is, we may be able to reduce the variance of the estimator considerably by introducing a relatively small amount of bias. As long as the reduction in variance is greater than the squared bias, an improved estimator from a mean squared error viewpoint will result. For example, Fig. 7-6 shows the probability distribution of a biased estimator $\hat{\Theta}_1$ that has a smaller variance than the unbiased estimator $\hat{\Theta}_2$. An estimate based on $\hat{\Theta}_1$ would more likely be close to the true value of θ than would an estimate based on $\hat{\Theta}_2$. Linear regression analysis (Chapters 11 and 12) is an area in which biased estimators are occasionally used.

An estimator $\hat{\Theta}$ that has a mean squared error that is less than or equal to the mean squared error of any other estimator, for all values of the parameter θ, is called an **optimal** estimator of θ. Optimal estimators rarely exist.

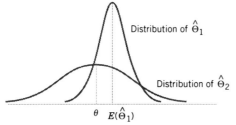

Figure 7-6 A biased estimator $\hat{\Theta}_1$ that has smaller variance than the unbiased estimator $\hat{\Theta}_2$.

EXERCISES FOR SECTION 7-3

7-14. Suppose we have a random sample of size $2n$ from a population denoted by X, and $E(X) = \mu$ and $V(X) = \sigma^2$. Let

$$\overline{X}_1 = \frac{1}{2n} \sum_{i=1}^{2n} X_i \quad \text{and} \quad \overline{X}_2 = \frac{1}{n} \sum_{i=1}^{n} X_i$$

be two estimators of μ. Which is the better estimator of μ? Explain your choice.

7-15. Let $X_1, X_2, \ldots, X_7$ denote a random sample from a population having mean μ and variance σ^2. Consider the following estimators of μ:

$$\hat{\Theta}_1 = \frac{X_1 + X_2 + \cdots + X_7}{7}$$

$$\hat{\Theta}_2 = \frac{2X_1 - X_6 + X_4}{2}$$

(a) Is either estimator unbiased?
(b) Which estimator is best? In what sense is it best? Calculate the relative efficiency of the two estimators.

7-16. Suppose that $\hat{\Theta}_1$ and $\hat{\Theta}_2$ are unbiased estimators of the parameter θ. We know that $V(\hat{\Theta}_1) = 10$ and $V(\hat{\Theta}_2) = 4$. Which estimator is best and in what sense is it best? Calculate the relative efficiency of the two estimators.

7-17. Suppose that $\hat{\Theta}_1$ and $\hat{\Theta}_2$ are estimators of the parameter θ. We know that $E(\hat{\Theta}_1) = \theta$, $E(\hat{\Theta}_2) = \theta/2$, $V(\hat{\Theta}_1) = 10$, $V(\hat{\Theta}_2) = 4$. Which estimator is best? In what sense is it best?

7-18. Suppose that $\hat{\Theta}_1$, $\hat{\Theta}_2$, and $\hat{\Theta}_3$ are estimators of θ. We know that $E(\hat{\Theta}_1) = E(\hat{\Theta}_2) = \theta$, $E(\hat{\Theta}_3) \neq \theta$, $V(\hat{\Theta}_1) = 12$, $V(\hat{\Theta}_2) = 10$, and $E(\hat{\Theta}_3 - \theta)^2 = 6$. Compare these three estimators. Which do you prefer? Why?

7-19. Let three random samples of sizes $n_1 = 20$, $n_2 = 10$, and $n_3 = 8$ be taken from a population with mean μ and variance σ^2. Let S_1^2, S_2^2, and S_3^2 be the sample variances. Show that $S^2 = (20S_1^2 + 10S_2^2 + 8S_3^2)/38$ is an unbiased estimator of σ^2.

7-20. (a) Show that $\sum_{i=1}^{n} (X_i - \overline{X})^2/n$ is a biased estimator of σ^2.
(b) Find the amount of bias in the estimator.
(c) What happens to the bias as the sample size n increases?

7-21. Let $X_1, X_2, \ldots, X_n$ be a random sample of size n from a population with mean μ and variance σ^2.
(a) Show that $\overline{X}^2$ is a biased estimator for μ^2.
(b) Find the amount of bias in this estimator.
(c) What happens to the bias as the sample size n increases?

7-22. Data on pull-off force (pounds) for connectors used in an automobile engine application are as follows: 79.3, 75.1, 78.2, 74.1, 73.9, 75.0, 77.6, 77.3, 73.8, 74.6, 75.5, 74.0, 74.7, 75.9, 72.9, 73.8, 74.2, 78.1, 75.4, 76.3, 75.3, 76.2, 74.9, 78.0, 75.1, 76.8.
(a) Calculate a point estimate of the mean pull-off force of all connectors in the population. State which estimator you used and why.
(b) Calculate a point estimate of the pull-off force value that separates the weakest 50% of the connectors in the population from the strongest 50%.
(c) Calculate point estimates of the population variance and the population standard deviation.
(d) Calculate the standard error of the point estimate found in part (a). Provide an interpretation of the standard error.
(e) Calculate a point estimate of the proportion of all connectors in the population whose pull-off force is less than 73 pounds.

7-23. Data on oxide thickness of semiconductors are as follows: 425, 431, 416, 419, 421, 436, 418, 410, 431, 433, 423, 426, 410, 435, 436, 428, 411, 426, 409, 437, 422, 428, 413, 416.
(a) Calculate a point estimate of the mean oxide thickness for all wafers in the population.
(b) Calculate a point estimate of the standard deviation of oxide thickness for all wafers in the population.
(c) Calculate the standard error of the point estimate from part (a).
(d) Calculate a point estimate of the median oxide thickness for all wafers in the population.
(e) Calculate a point estimate of the proportion of wafers in the population that have oxide thickness greater than 430 angstrom.

7-24. Suppose that X is the number of observed "successes" in a sample of n observations where p is the probability of success on each observation.
(a) Show that $\hat{P} = X/n$ is an unbiased estimator of p.
(b) Show that the standard error of $\hat{P}$ is $\sqrt{p(1-p)/n}$. How would you estimate the standard error?

7-25. $\overline{X}_1$ and S_1^2 are the sample mean and sample variance from a population with mean μ_1 and variance σ_1^2. Similarly, $\overline{X}_2$ and S_2^2 are the sample mean and sample variance from a second independent population with mean μ_2 and variance σ_2^2. The sample sizes are n_1 and n_2, respectively.

(a) Show that $\overline{X}_1 - \overline{X}_2$ is an unbiased estimator of $\mu_1 - \mu_2$.

(b) Find the standard error of $\overline{X}_1 - \overline{X}_2$. How could you estimate the standard error?

(c) Suppose that both populations have the same variance; that is, $\sigma_1^2 = \sigma_2^2 = \sigma^2$. Show that

$$S_p^2 = \frac{(n_1 - 1)S_1^2 + (n_2 - 1)S_2^2}{n_1 + n_2 - 2}$$

is an unbiased estimator of σ^2.

7-26. Two different plasma etchers in a semiconductor factory have the same mean etch rate μ. However, machine 1 is newer than machine 2 and consequently has smaller variability in etch rate. We know that the variance of etch rate for machine 1 is σ_1^2 and for machine 2 it is $\sigma_2^2 = a\sigma_1^2$. Suppose that we have n_1 independent observations on etch rate from machine 1 and n_2 independent observations on etch rate from machine 2.

(a) Show that $\hat{\mu} = \alpha \overline{X}_1 + (1 - \alpha) \overline{X}_2$ is an unbiased estimator of μ for any value of α between 0 and 1.

(b) Find the standard error of the point estimate of μ in part (a).

(c) What value of α would minimize the standard error of the point estimate of μ?

(d) Suppose that $a = 4$ and $n_1 = 2n_2$. What value of α would you select to minimize the standard error of the point estimate of μ. How "bad" would it be to arbitrarily choose $\alpha = 0.5$ in this case?

7-27. Of n_1 randomly selected engineering students at ASU, X_1 owned an HP calculator, and of n_2 randomly selected engineering students at Virginia Tech X_2 owned an HP calculator. Let p_1 and p_2 be the probability that randomly selected ASU and Va. Tech engineering students, respectively, own HP calculators.

(a) Show that an unbiased estimate for $p_1 - p_2$ is $(X_1/n_1) - (X_2/n_2)$.

(b) What is the standard error of the point estimate in part (a)?

(c) How would you compute an estimate of the standard error found in part (b)?

(d) Suppose that $n_1 = 200$, $X_1 = 150$, $n_2 = 250$, and $X_2 = 185$. Use the results of part (a) to compute an estimate of $p_1 - p_2$.

(e) Use the results in parts (b) through (d) to compute an estimate of the standard error of the estimate.

7-4 METHODS OF POINT ESTIMATION

The definitions of unbiasedness and other properties of estimators do not provide any guidance about how good estimators can be obtained. In this section, we discuss methods for obtaining point estimators: the method of moments and the method of maximum likelihood. We also discuss Bayesian estimation of parameters. Maximum likelihood estimates are generally preferable to moment estimators because they have better efficiency properties. However, moment estimators are sometimes easier to compute. Both methods can produce unbiased point estimators.

7-4.1 Method of Moments

The general idea behind the method of moments is to equate population moments, which are defined in terms of expected values, to the corresponding sample moments. The population moments will be functions of the unknown parameters. Then these equations are solved to yield estimators of the unknown parameters.

Moments

Let $X_1, X_2, \ldots, X_n$ be a random sample from the probability distribution $f(x)$, where $f(x)$ can be a discrete probability mass function or a continuous probability density function. The kth **population moment** (or **distribution moment**) is $E(X^k)$, $k = 1, 2, \ldots$. The corresponding kth **sample moment** is $(1/n)\sum_{i=1}^{n} X_i^k$, $k = 1, 2, \ldots$.

Statistical Intervals for a Single Sample

CHAPTER OUTLINE

8-1 INTRODUCTION

8-2 CONFIDENCE INTERVAL ON THE MEAN OF A NORMAL DISTRIBUTION, VARIANCE KNOWN

    8-2.1 Development of the Confidence Interval and its Basic Properties

    8-2.2 Choice of Sample Size

    8-2.3 One-Sided Confidence Bounds

    8-2.4 General Method to Derive a Confidence Interval

    8-2.5 Large-Sample Confidence Interval for μ

8-3 CONFIDENCE INTERVAL ON THE MEAN OF A NORMAL DISTRIBUTION, VARIANCE UNKNOWN

    8-3.1 t Distribution

    8-3.2 t Confidence Interval on μ

8-4 CONFIDENCE INTERVAL ON THE VARIANCE AND STANDARD DEVIATION OF A NORMAL DISTRIBUTION

8-5 LARGE-SAMPLE CONFIDENCE INTERVAL FOR A POPULATION PROPORTION

8-6 GUIDELINES FOR CONSTRUCTING CONFIDENCE INTERVALS

8-7 TOLERANCE AND PREDICTION INTERVALS

    8-7.1 Prediction Interval for a Future Observation

    8-7.2 Tolerance Interval for a Normal Distribution

LEARNING OBJECTIVES

After careful study of this chapter, you should be able to do the following:

1. Construct confidence intervals on the mean of a normal distribution, using either the normal distribution or the t distribution method
2. Construct confidence intervals on the variance and standard deviation of a normal distribution
3. Construct confidence intervals on a population proportion
4. Use a general method for constructing an approximate confidence interval on a parameter
5. Construct prediction intervals for a future observation

6. Construct a tolerance interval for a normal population
7. Explain the three types of interval estimates: confidence intervals, prediction intervals, and tolerance intervals

8-1 INTRODUCTION

In the previous chapter we illustrated how a parameter can be estimated from sample data. However, it is important to understand how good is the estimate obtained. For example, suppose that we estimate the mean viscosity of a chemical product to be $\hat{\mu} = \bar{x} = 1000$. Now because of sampling variability, it is almost never the case that $\mu = \bar{x}$. The point estimate says nothing about how close $\hat{\mu}$ is to μ. Is the process mean likely to be between 900 and 1100? Or is it likely to be between 990 and 1010? The answer to these questions affects our decisions regarding this process. Bounds that represent an interval of plausible values for a parameter are an example of an interval estimate. Surprisingly, it is easy to determine such intervals in many cases, and the same data that provided the point estimate are typically used.

An interval estimate for a population parameter is called a **confidence interval**. We cannot be certain that the interval contains the true, unknown population parameter—we only use a sample from the full population to compute the point estimate and the interval. However, the confidence interval is constructed so that we have high confidence that it does contain the unknown population parameter. Confidence intervals are widely used in engineering and the sciences.

A **tolerance interval** is another important type of interval estimate. For example, the chemical product viscosity data might be assumed to be normally distributed. We might like to calculate limits that bound 95% of the viscosity values. For a normal distribution, we know that 95% of the distribution is in the interval

$$\mu - 1.96\sigma, \mu + 1.96\sigma \tag{8-1}$$

However, this is not a useful tolerance interval because the parameters μ and σ are unknown. Point estimates such as $\bar{x}$ and s can be used in Equation 8-1 for μ and σ. However, we need to account for the potential error in each point estimate to form a tolerance interval for the distribution. The result is an interval of the form

$$\bar{x} - ks, \bar{x} + ks \tag{8-2}$$

where k is an appropriate constant (that is larger than 1.96 to account for the estimation error). As for a confidence interval, it is not certain that Equation 8-2 bounds 95% of the distribution, but the interval is constructed so that we have high confidence that it does. Tolerance intervals are widely used and, as we will subsequently see, they are easy to calculate for normal distributions.

Confidence and tolerance intervals bound unknown elements of a distribution. In this chapter you will learn to appreciate the value of these intervals. A **prediction interval** provides bounds on one (or more) future observations from the population. For example, a prediction interval could be used to bound a single, new measurement of viscosity—another useful interval. With a large sample size, the prediction interval for normally distributed data tends to the tolerance interval in Equation 8-1, but for more modest sample sizes the prediction and tolerance intervals are different.

Keep the purpose of the three types of interval estimates clear:

- A confidence interval bounds population or distribution parameters (such as the mean viscosity).
- A tolerance interval bounds a selected proportion of a distribution.
- A prediction interval bounds future observations from the population or distribution.

8-2 CONFIDENCE INTERVAL ON THE MEAN OF A NORMAL DISTRIBUTION, VARIANCE KNOWN

The basic ideas of a confidence interval (CI) are most easily understood by initially considering a simple situation. Suppose that we have a normal population with unknown mean μ and known variance σ^2. This is a somewhat unrealistic scenario because typically both the mean and variance are unknown. However, in subsequent sections we will present confidence intervals for more general situations.

8-2.1 Development of the Confidence Interval and Its Basic Properties

Suppose that $X_1, X_2, \ldots, X_n$ is a random sample from a normal distribution with unknown mean μ and known variance σ^2. From the results of Chapter 5 we know that the sample mean $\overline{X}$ is normally distributed with mean μ and variance σ^2/n. We may **standardize** $\overline{X}$ by subtracting the mean and dividing by the standard deviation, which results in the variable

$$Z = \frac{\overline{X} - \mu}{\sigma/\sqrt{n}} \tag{8-3}$$

The random variable Z has a standard normal distribution.

A **confidence interval** estimate for μ is an interval of the form $l \leq \mu \leq u$, where the end-points l and u are computed from the sample data. Because different samples will produce different values of l and u, these end-points are values of random variables L and U, respectively. Suppose that we can determine values of L and U such that the following probability statement is true:

$$P\{L \leq \mu \leq U\} = 1 - \alpha \tag{8-4}$$

where $0 \leq \alpha \leq 1$. There is a probability of $1 - \alpha$ of selecting a sample for which the CI will contain the true value of μ. Once we have selected the sample, so that $X_1 = x_1, X_2 = x_2, \ldots, X_n = x_n$, and computed l and u, the resulting confidence interval for μ is

$$l \leq \mu \leq u \tag{8-5}$$

The end-points or bounds l and u are called the **lower-** and **upper-confidence limits**, respectively, and $1 - \alpha$ is called the **confidence coefficient**.

In our problem situation, because $Z = (\overline{X} - \mu)/(\sigma/\sqrt{n})$ has a standard normal distribution, we may write

$$P\left\{-z_{\alpha/2} \leq \frac{\overline{X} - \mu}{\sigma/\sqrt{n}} \leq z_{\alpha/2}\right\} = 1 - \alpha$$

Now manipulate the quantities inside the brackets by (1) multiplying through by $\sigma/\sqrt{n}$, (2) subtracting $\overline{X}$ from each term, and (3) multiplying through by -1. This results in

$$P\left\{\overline{X} - z_{\alpha/2}\frac{\sigma}{\sqrt{n}} \leq \mu \leq \overline{X} + z_{\alpha/2}\frac{\sigma}{\sqrt{n}}\right\} = 1 - \alpha \qquad (8\text{-}6)$$

From consideration of Equation 8-4, the lower and upper limits of the inequalities in Equation 8-6 are the lower- and upper-confidence limits L and U, respectively. This leads to the following definition.

Confidence Interval on the Mean, Variance Known

> If $\bar{x}$ is the sample mean of a random sample of size n from a normal population with known variance σ^2, a $100(1 - \alpha)\%$ CI on μ is given by
>
> $$\bar{x} - z_{\alpha/2}\sigma/\sqrt{n} \leq \mu \leq \bar{x} + z_{\alpha/2}\sigma/\sqrt{n} \qquad (8\text{-}7)$$
>
> where $z_{\alpha/2}$ is the upper $100\alpha/2$ percentage point of the standard normal distribution.

EXAMPLE 8-1
Metallic Material Transition

ASTM Standard E23 defines standard test methods for notched bar impact testing of metallic materials. The Charpy V-notch (CVN) technique measures impact energy and is often used to determine whether or not a material experiences a ductile-to-brittle transition with decreasing temperature. Ten measurements of impact energy (J) on specimens of A238 steel cut at 60°C are as follows: 64.1, 64.7, 64.5, 64.6, 64.5, 64.3, 64.6, 64.8, 64.2, and 64.3. Assume that impact energy is normally distributed with $\sigma = 1J$. We want to find a 95% CI for μ, the mean impact energy. The required quantities are $z_{\alpha/2} = z_{0.025} = 1.96$, $n = 10$, $\sigma = 1$, and $\bar{x} = 64.46$. The resulting 95% CI is found from Equation 8-7 as follows:

$$\bar{x} - z_{\alpha/2}\frac{\sigma}{\sqrt{n}} \leq \mu \leq \bar{x} + z_{\alpha/2}\frac{\sigma}{\sqrt{n}}$$

$$64.46 - 1.96\frac{1}{\sqrt{10}} \leq \mu \leq 64.46 + 1.96\frac{1}{\sqrt{10}}$$

$$63.84 \leq \mu \leq 65.08$$

That is, based on the sample data, a range of highly plausible values for mean impact energy for A238 steel at 60°C is $63.84J \leq \mu \leq 65.08J$.

Interpreting a Confidence Interval
How does one interpret a confidence interval? In the impact energy estimation problem in Example 8-1 the 95% CI is $63.84 \leq \mu \leq 65.08$, so it is tempting to conclude that μ is within this interval with probability 0.95. However, with a little reflection, it's easy to see that this cannot be correct; the true value of μ is unknown and the statement $63.84 \leq \mu \leq 65.08$ is either correct (true with probability 1) or incorrect (false with probability 1). The correct interpretation lies in the realization that a CI is a *random interval* because in the probability statement defining the end-points of the interval (Equation 8-4), L and U are random variables. Consequently, the correct interpretation of a $100(1 - \alpha)\%$ CI depends on the relative frequency view of probability. Specifically, if an infinite number of random samples are collected and a $100(1 - \alpha)\%$ confidence interval for μ is computed from each sample, $100(1 - \alpha)\%$ of these intervals will contain the true value of μ.

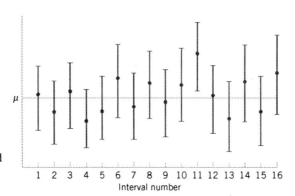

Figure 8-1 Repeated construction of a confidence interval for μ.

The situation is illustrated in Fig. 8-1, which shows several $100(1 - \alpha)\%$ confidence intervals for the mean μ of a normal distribution. The dots at the center of the intervals indicate the point estimate of μ (that is, $\bar{x}$). Notice that one of the intervals fails to contain the true value of μ. If this were a 95% confidence interval, in the long run only 5% of the intervals would fail to contain μ.

Now in practice, we obtain only one random sample and calculate one confidence interval. Since this interval either will or will not contain the true value of μ, it is not reasonable to attach a probability level to this specific event. The appropriate statement is the observed interval $[l, u]$ brackets the true value of μ with **confidence** $100(1 - \alpha)$. This statement has a frequency interpretation; that is, we don't know if the statement is true for this specific sample, but the *method* used to obtain the interval $[l, u]$ yields correct statements $100(1 - \alpha)\%$ of the time.

Confidence Level and Precision of Estimation

Notice in Example 8-1 that our choice of the 95% level of confidence was essentially arbitrary. What would have happened if we had chosen a higher level of confidence, say, 99%? In fact, doesn't it seem reasonable that we would want the higher level of confidence? At $\alpha = 0.01$, we find $z_{\alpha/2} = z_{0.01/2} = z_{0.005} = 2.58$, while for $\alpha = 0.05$, $z_{0.025} = 1.96$. Thus, the **length** of the 95% confidence interval is

$$2(1.96\sigma/\sqrt{n}) = 3.92\sigma/\sqrt{n}$$

whereas the length of the 99% CI is

$$2(2.58\sigma/\sqrt{n}) = 5.16\sigma/\sqrt{n}$$

Thus, the 99% CI is longer than the 95% CI. This is why we have a higher level of confidence in the 99% confidence interval. Generally, for a fixed sample size n and standard deviation σ, the higher the confidence level, the longer the resulting CI.

The length of a confidence interval is a measure of the **precision** of estimation. From the preceeding discussion, we see that precision is inversely related to the confidence level. It is desirable to obtain a confidence interval that is short enough for decision-making purposes and that also has adequate confidence. One way to achieve this is by choosing the sample size n to be large enough to give a CI of specified length or precision with prescribed confidence.

Figure 8-2 Error in estimating μ with $\bar{x}$.

8-2.2 Choice of Sample Size

The precision of the confidence interval in Equation 8-7 is $2z_{\alpha/2}\sigma/\sqrt{n}$. This means that in using $\bar{x}$ to estimate μ, the error $E = |\bar{x} - \mu|$ is less than or equal to $z_{\alpha/2}\sigma/\sqrt{n}$ with confidence $100(1 - \alpha)$. This is shown graphically in Fig. 8-2. In situations where the sample size can be controlled, we can choose n so that we are $100(1 - \alpha)$ percent confident that the error in estimating μ is less than a specified bound on the error E. The appropriate sample size is found by choosing n such that $z_{\alpha/2}\sigma/\sqrt{n} = E$. Solving this equation gives the following formula for n.

Sample Size for Specified Error on the Mean, Variance Known

If $\bar{x}$ is used as an estimate of μ, we can be $100(1 - \alpha)\%$ confident that the error $|\bar{x} - \mu|$ will not exceed a specified amount E when the sample size is

$$n = \left(\frac{z_{\alpha/2}\sigma}{E}\right)^2 \qquad (8\text{-}8)$$

If the right-hand side of Equation 8-8 is not an integer, it must be rounded up. This will ensure that the level of confidence does not fall below $100(1 - \alpha)\%$. Notice that $2E$ is the length of the resulting confidence interval.

EXAMPLE 8-2
Metallic Material Transition

To illustrate the use of this procedure, consider the CVN test described in Example 8-1, and suppose that we wanted to determine how many specimens must be tested to ensure that the 95% CI on μ for A238 steel cut at 60°C has a length of at most $1.0J$. Since the bound on error in estimation E is one-half of the length of the CI, to determine n we use Equation 8-8 with $E = 0.5$, $\sigma = 1$, and $z_{\alpha/2} = 1.96$. The required sample size is 16

$$n = \left(\frac{z_{\alpha/2}\sigma}{E}\right)^2 = \left[\frac{(1.96)1}{0.5}\right]^2 = 15.37$$

and because n must be an integer, the required sample size is $n = 16$.

Notice the general relationship between sample size, desired length of the confidence interval $2E$, confidence level $100(1 - \alpha)$, and standard deviation σ:

- As the desired length of the interval $2E$ decreases, the required sample size n increases for a fixed value of σ and specified confidence.
- As σ increases, the required sample size n increases for a fixed desired length $2E$ and specified confidence.
- As the level of confidence increases, the required sample size n increases for fixed desired length $2E$ and standard deviation σ.

8-2.3 One-Sided Confidence Bounds

The confidence interval in Equation 8-7 gives both a lower confidence bound and an upper confidence bound for μ. Thus it provides a two-sided CI. It is also possible to obtain one-sided confidence bounds for μ by setting either $l = -\infty$ or $u = \infty$ and replacing $z_{\alpha/2}$ by z_α.

One-Sided Confidence Bounds on the Mean, Variance Known

A $100(1 - \alpha)\%$ **upper-confidence bound** for μ is

$$\mu \leq u = \bar{x} + z_\alpha \sigma/\sqrt{n} \qquad (8\text{-}9)$$

and a $100(1 - \alpha)\%$ **lower-confidence bound** for μ is

$$\bar{x} - z_\alpha \sigma/\sqrt{n} = l \leq \mu \qquad (8\text{-}10)$$

EXAMPLE 8-3
One-Sided Confidence Bound

The same data for impact testing from Example 8-1 is used to construct a lower, one-sided 95% confidence interval for the mean impact energy. Recall that $\bar{x} = 64.46$, $\sigma = 1J$, and $n = 10$. The interval is

$$\bar{x} - z_\alpha \frac{\sigma}{\sqrt{n}} \leq \mu$$

$$64.46 - 1.64 \frac{1}{\sqrt{10}} \leq \mu$$

$$63.94 \leq \mu$$

The lower limit for the two-sided interval in Example 8-1 was 63.84. Because $z_\alpha < z_{\alpha/2}$, the lower limit of a one-sided interval is always greater than the lower limit of a two-sided interval of equal confidence. The one-sided interval does not bound μ from above so that it still achieves 95% confidence with a slightly greater lower limit. If our interest is only in the lower limit for μ, then the one-sided interval is preferred because it provides equal confidence with a greater lower limit. Similarly, a one-sided upper limit is always less than a two-sided upper limit of equal confidence.

8-2.4 General Method to Derive a Confidence Interval

It is easy to give a general method for finding a confidence interval for an unknown parameter θ. Let $X_1, X_2, \ldots, X_n$ be a random sample of n observations. Suppose we can find a statistic $g(X_1, X_2, \ldots, X_n; \theta)$ with the following properties:

1. $g(X_1, X_2, \ldots, X_n; \theta)$ depends on both the sample and θ.
2. The probability distribution of $g(X_1, X_2, \ldots, X_n; \theta)$ does not depend on θ or any other unknown parameter.

In the case considered in this section, the parameter $\theta = \mu$. The random variable $g(X_1, X_2, \ldots, X_n; \mu) = (\bar{X} - \mu)/(\sigma/\sqrt{n})$ and satisfies both conditions above; it depends on the sample and on μ, and it has a standard normal distribution since σ is known. Now one must find constants C_L and C_U so that

$$P[C_L \le g(X_1, X_2, \ldots, X_n; \theta) \le C_U] = 1 - \alpha \qquad (8\text{-}11)$$

Because of property 2, C_L and C_U do not depend on θ. In our example, $C_L = -z_{\alpha/2}$ and $C_U = z_{\alpha/2}$. Finally, you must manipulate the inequalities in the probability statement so that

$$P[L(X_1, X_2, \ldots, X_n) \le \theta \le U(X_1, X_2, \ldots, X_n)] = 1 - \alpha \qquad (8\text{-}12)$$

This gives $L(X_1, X_2, \ldots, X_n)$ and $U(X_1, X_2, \ldots, X_n)$ as the lower and upper confidence limits defining the $100(1 - \alpha)\%$ confidence interval for θ. The quantity $g(X_1, X_2, \ldots, X_n; \theta)$ is often called a "pivotal quantity" because we pivot on this quantity in Equation 8-11 to produce Equation 8-12. In our example, we manipulated the pivotal quantity $(\overline{X} - \mu)/(\sigma/\sqrt{n})$ to obtain $L(X_1, X_2, \ldots, X_n) = \overline{X} - z_{\alpha/2}\sigma/\sqrt{n}$ and $U(X_1, X_2, \ldots, X_n) = \overline{X} + z_{\alpha/2}\sigma/\sqrt{n}$.

8-2.5 Large-Sample Confidence Interval for μ

We have assumed that the population distribution is normal with unknown mean and known standard deviation σ. We now present a **large-sample CI** for μ that does not require these assumptions. Let $X_1, X_2, \ldots, X_n$ be a random sample from a population with unknown mean μ and variance σ^2. Now if the sample size n is large, the central limit theorem implies that $\overline{X}$ has approximately a normal distribution with mean μ and variance σ^2/n. Therefore $Z = (\overline{X} - \mu)/(\sigma/\sqrt{n})$ has approximately a standard normal distribution. This ratio could be used as a pivotal quantity and manipulated as in Section 8-2.1 to produce an approximate CI for μ. However, the standard deviation σ is unknown. It turns out that when n is large, replacing σ by the sample standard deviation S has little effect on the distribution of Z. This leads to the following useful result.

Large-Sample Confidence Interval on the Mean

When n is large, the quantity

$$\frac{\overline{X} - \mu}{S/\sqrt{n}}$$

has an approximate standard normal distribution. Consequently,

$$\overline{x} - z_{\alpha/2}\frac{s}{\sqrt{n}} \le \mu \le \overline{x} + z_{\alpha/2}\frac{s}{\sqrt{n}} \qquad (8\text{-}13)$$

is a **large sample confidence interval** for μ, with confidence level of approximately $100(1 - \alpha)\%$.

Equation 8-13 holds regardless of the shape of the population distribution. Generally n should be at least 40 to use this result reliably. The central limit theorem generally holds for $n \ge 30$, but the larger sample size is recommended here because replacing σ by S in Z results in additional variability.

EXAMPLE 8-4
Mercury Contamination

An article in the 1993 volume of the *Transactions of the American Fisheries Society* reports the results of a study to investigate the mercury contamination in largemouth bass. A sample of fish was selected from 53 Florida lakes and mercury concentration in the muscle tissue was measured (ppm). The mercury concentration values are

1.230	0.490	0.490	1.080	0.590	0.280	0.180	0.100	0.940
1.330	0.190	1.160	0.980	0.340	0.340	0.190	0.210	0.400
0.040	0.830	0.050	0.630	0.340	0.750	0.040	0.860	0.430
0.044	0.810	0.150	0.560	0.840	0.870	0.490	0.520	0.250
1.200	0.710	0.190	0.410	0.500	0.560	1.100	0.650	0.270
0.270	0.500	0.770	0.730	0.340	0.170	0.160	0.270	

The summary statistics from Minitab are displayed below:

Descriptive Statistics: Concentration

Variable	N	Mean	Median	TrMean	StDev	SE Mean
Concentration	53	0.5250	0.4900	0.5094	0.3486	0.0479

Variable	Minimum	Maximum	Q1	Q3
Concentration	0.0400	1.3300	0.2300	0.7900

Figure 8-3(a) and (b) presents the histogram and normal probability plot of the mercury concentration data. Both plots indicate that the distribution of mercury concentration is not normal and is positively skewed. We want to find an approximate 95% CI on μ. Because $n > 40$, the assumption of normality is not necessary to use Equation 8-13. The required quantities are $n = 53$, $\bar{x} = 0.5250$, $s = 0.3486$, and $z_{0.025} = 1.96$. The approximate 95% CI on μ is

$$\bar{x} - z_{0.025} \frac{s}{\sqrt{n}} \leq \mu \leq \bar{x} + z_{0.025} \frac{s}{\sqrt{n}}$$

$$0.5250 - 1.96 \frac{0.3486}{\sqrt{53}} \leq \mu \leq 0.5250 + 1.96 \frac{0.3486}{\sqrt{53}}$$

$$0.4311 \leq \mu \leq 0.6189$$

This interval is fairly wide because there is a lot of variability in the mercury concentration measurements.

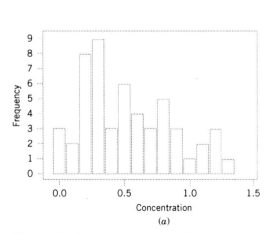
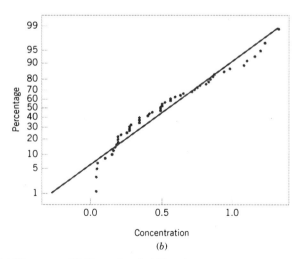

Figure 8-3 Mercury concentration in largemouth bass (a) Histogram. (b) Normal probability plot.

Large-Sample Confidence Interval for a Parameter

The large-sample confidence interval for μ in Equation 8-13 is a special case of a more general result. Suppose that θ is a parameter of a probability distribution and let $\hat{\Theta}$ be an estimator of θ. If $\hat{\Theta}$ (1) has an approximate normal distribution, (2) is approximately unbiased for θ, and (3) has standard deviation $\sigma_{\hat{\Theta}}$ that can be estimated from the sample data, then the quantity $(\hat{\Theta} - \theta)/\sigma_{\hat{\Theta}}$ has an approximate standard normal distribution. Then a large-sample approximate CI for θ is given by

Large-Sample Approximate Confidence Interval

$$\hat{\theta} - z_{\alpha/2}\sigma_{\hat{\Theta}} \leq \theta \leq \hat{\theta} + z_{\alpha/2}\sigma_{\hat{\Theta}} \tag{8-14}$$

Maximum likelihood estimators usually satisfy the three conditions listed above, so Equation 8-14 is often used when $\hat{\Theta}$ is the maximum likelihood estimator of θ. Finally, note that Equation 8-14 can be used even when $\sigma_{\hat{\Theta}}$ is a function of other unknown parameters (or of θ). Essentially, all one does is to use the sample data to compute estimates of the unknown parameters and substitute those estimates into the expression for $\sigma_{\hat{\Theta}}$.

EXERCISES FOR SECTION 8-2

8-1. For a normal population with known variance σ^2, answer the following questions:
(a) What is the confidence level for the interval $\bar{x} - 2.14\sigma/\sqrt{n} \leq \mu \leq \bar{x} + 2.14\sigma/\sqrt{n}$?
(b) What is the confidence level for the interval $\bar{x} - 2.49\sigma/\sqrt{n} \leq \mu \leq \bar{x} + 2.49\sigma/\sqrt{n}$?
(c) What is the confidence level for the interval $\bar{x} - 1.85\sigma/\sqrt{n} \leq \mu \leq \bar{x} + 1.85\sigma/\sqrt{n}$?

8-2. For a normal population with known variance σ^2:
(a) What value of $z_{\alpha/2}$ in Equation 8-7 gives 98% confidence?
(b) What value of $z_{\alpha/2}$ in Equation 8-7 gives 80% confidence?
(c) What value of $z_{\alpha/2}$ in Equation 8-7 gives 75% confidence?

8-3. Consider the one-sided confidence interval expressions for a mean of a normal population.
(a) What value of z_α would result in a 90% CI?
(b) What value of z_α would result in a 95% CI?
(c) What value of z_α would result in a 99% CI?

8-4. A confidence interval estimate is desired for the gain in a circuit on a semiconductor device. Assume that gain is normally distributed with standard deviation $\sigma = 20$.
(a) Find a 95% CI for μ when $n = 10$ and $\bar{x} = 1000$.
(b) Find a 95% CI for μ when $n = 25$ and $\bar{x} = 1000$.
(c) Find a 99% CI for μ when $n = 10$ and $\bar{x} = 1000$.
(d) Find a 99% CI for μ when $n = 25$ and $\bar{x} = 1000$.
(e) How does the length of the CIs computed above change with the changes in sample size and confidence level?

8-5. Consider the gain estimation problem in Exercise 8-4.
(a) How large must n be if the length of the 95% CI is to be 40?
(b) How large must n be if the length of the 99% CI is to be 40?

8-6. Following are two confidence interval estimates of the mean μ of the cycles to failure of an automotive door latch mechanism (the test was conducted at an elevated stress level to accelerate the failure).

$$3124.9 \leq \mu \leq 3215.7 \qquad 3110.5 \leq \mu \leq 3230.1$$

(a) What is the value of the sample mean cycles to failure?
(b) The confidence level for one of these CIs is 95% and the confidence level for the other is 99%. Both CIs are calculated from the same sample data. Which is the 95% CI? Explain why.

8-7. Suppose that $n = 100$ random samples of water from a fresh water lake were taken and the calcium concentration (milligrams per liter) measured. A 95% CI on the mean calcium concentration is $0.49 \leq \mu \leq 0.82$.
(a) Would a 99% CI calculated from the same sample data be longer or shorter?
(b) Consider the following statement: There is a 95% chance that μ is between 0.49 and 0.82. Is this statement correct? Explain your answer.
(c) Consider the following statement: If $n = 100$ random samples of water from the lake were taken and the 95% CI on μ computed, and this process was repeated 1000 times, 950 of the CIs will contain the true value of μ. Is this statement correct? Explain your answer.

268 CHAPTER 8 STATISTICAL INTERVALS FOR A SINGLE SAMPLE

8-8. Past experience has indicated that the breaking strength of yarn used in manufacturing drapery material is normally distributed and that $\sigma = 2$ psi. A random sample of nine specimens is tested, and the average breaking strength is found to be 98 psi. Find a 95% two-sided confidence interval on the true mean breaking strength.

8-9. The yield of a chemical process is being studied. From previous experience yield is known to be normally distributed and $\sigma = 3$. The past five days of plant operation have resulted in the following percent yields: 91.6, 88.75, 90.8, 89.95, and 91.3. Find a 95% two-sided confidence interval on the true mean yield.

8-10. The diameter of holes for a cable harness is known to have a normal distribution with $\sigma = 0.01$ inch. A random sample of size 10 yields an average diameter of 1.5045 inch. Find a 99% two-sided confidence interval on the mean hole diameter.

8-11. A manufacturer produces piston rings for an automobile engine. It is known that ring diameter is normally distributed with $\sigma = 0.001$ millimeters. A random sample of 15 rings has a mean diameter of $\bar{x} = 74.036$ millimeters.
(a) Construct a 99% two-sided confidence interval on the mean piston ring diameter.
(b) Construct a 99% lower-confidence bound on the mean piston ring diameter. Compare the lower bound of this confidence interval with the one in part (a).

8-12. The life in hours of a 75-watt light bulb is known to be normally distributed with $\sigma = 25$ hours. A random sample of 20 bulbs has a mean life of $\bar{x} = 1014$ hours.
(a) Construct a 95% two-sided confidence interval on the mean life.
(b) Construct a 95% lower-confidence bound on the mean life. Compare the lower bound of this confidence interval with the one in part (a).

8-13. A civil engineer is analyzing the compressive strength of concrete. Compressive strength is normally distributed with $\sigma^2 = 1000(\text{psi})^2$. A random sample of 12 specimens has a mean compressive strength of $\bar{x} = 3250$ psi.
(a) Construct a 95% two-sided confidence interval on mean compressive strength.
(b) Construct a 99% two-sided confidence interval on mean compressive strength. Compare the width of this confidence interval with the width of the one found in part (a).

8-14. Suppose that in Exercise 8-12 we wanted the error in estimating the mean life from the two-sided confidence interval to be five hours at 95% confidence. What sample size should be used?

8-15. Suppose that in Exercise 8-12 we wanted the total width of the two-sided confidence interval on mean life to be six hours at 95% confidence. What sample size should be used?

8-16. Suppose that in Exercise 8-13 it is desired to estimate the compressive strength with an error that is less than 15 psi at 99% confidence. What sample size is required?

8-17. By how much must the sample size n be increased if the length of the CI on μ in Equation 8-7 is to be halved?

8-18. If the sample size n is doubled, by how much is the length of the CI on μ in Equation 8-7 reduced? What happens to the length of the interval if the sample size is increased by a factor of four?

8-19. An article in the *Journal of Agricultural Science,* "The use of residual maximum likelihood to model grain quality characteristics of wheat with variety, climatic and nitrogen fertilizer effects" (1997, Vol. 128, pp. 135–142) investigated means of wheat grain crude protein content (CP) and Hagberg falling number (HFN) surveyed in the UK. The analysis used a variety of nitrogen fertilizer application (kg N/ha), temperature (°C), and total monthly rainfall (mm). The data shown below describe temperatures for wheat grown at Harper Adams Agricultural College between 1982 and 1993. The temperatures measured in June were obtained as followed:

15.2	14.2	14.0	12.2	14.4	12.5
14.3	14.2	13.5	11.8	15.2	

Assume that the standard deviation is known to be $\sigma = 0.5$.
(a) Construct a 99% two-sided confidence interval on the mean temperature.
(b) Construct a 95% lower-confidence bound on the mean temperature.
(c) Suppose that we wanted to be 95% confident that the error in estimating the mean temperature is less than 2 degrees Celsius. What sample size should be used?
(d) Suppose that we wanted the total width of the two-sided confidence interval on mean temperature to be 1.5 degrees Celsius at 95% confidence. What sample size should be used?

8-3 CONFIDENCE INTERVAL ON THE MEAN OF A NORMAL DISTRIBUTION, VARIANCE UNKNOWN

When we are constructing confidence intervals on the mean μ of a normal population when σ^2 is known, we can use the procedure in Section 8-2.1. This CI is also approximately valid (because of the central limit theorem) regardless of whether or not the underlying population is normal, so long as n is reasonably large ($n \geq 40$, say). As noted in Section 8-2.5, we can even handle the case of unknown variance for the large-sample-size situation. However, when the sample is small and σ^2 is unknown, we must make an assumption about the form of the underlying distribution to obtain a valid CI procedure. A reasonable assumption in many cases is that the underlying distribution is normal.

Many populations encountered in practice are well approximated by the normal distribution, so this assumption will lead to confidence interval procedures of wide applicability. In fact, moderate departure from normality will have little effect on validity. When the assumption is unreasonable, an alternate is to use the nonparametric procedures in Chapter 15 that are valid for any underlying distribution.

Suppose that the population of interest has a normal distribution with unknown mean μ and unknown variance σ^2. Assume that a random sample of size n, say $X_1, X_2, \ldots, X_n$, is available, and let $\overline{X}$ and S^2 be the sample mean and variance, respectively.

We wish to construct a two-sided CI on μ. If the variance σ^2 is known, we know that $Z = (\overline{X} - \mu)/(\sigma/\sqrt{n})$ has a standard normal distribution. When σ^2 is unknown, a logical procedure is to replace σ with the sample standard deviation S. The random variable Z now becomes $T = (\overline{X} - \mu)/(S/\sqrt{n})$. A logical question is what effect does replacing σ by S have on the distribution of the random variable T? If n is large, the answer to this question is "very little," and we can proceed to use the confidence interval based on the normal distribution from Section 8-2.5. However, n is usually small in most engineering problems, and in this situation a different distribution must be employed to construct the CI.

8-3.1 t Distribution

t Distribution

Let $X_1, X_2, \ldots, X_n$ be a random sample from a normal distribution with unknown mean μ and unknown variance σ^2. The random variable

$$T = \frac{\overline{X} - \mu}{S/\sqrt{n}} \tag{8-15}$$

has a t distribution with $n - 1$ degrees of freedom.

The t probability density function is

$$f(x) = \frac{\Gamma[(k+1)/2]}{\sqrt{\pi k}\,\Gamma(k/2)} \cdot \frac{1}{[(x^2/k) + 1]^{(k+1)/2}} \qquad -\infty < x < \infty \tag{8-16}$$

where k is the number of degrees of freedom. The mean and variance of the t distribution are zero and $k/(k-2)$ (for $k > 2$), respectively.

Several t distributions are shown in Fig. 8-4. The general appearance of the t distribution is similar to the standard normal distribution in that both distributions are symmetric and unimodal, and the maximum ordinate value is reached when the mean $\mu = 0$. However, the t distribution has heavier tails than the normal; that is, it has more probability in the tails than the normal distribution. As the number of degrees of freedom $k \to \infty$, the limiting form of the t distribution is the standard normal distribution. Generally, the number of degrees of freedom for t are the number of degrees of freedom associated with the estimated standard deviation.

Appendix Table V provides percentage points of the t distribution. We will let $t_{\alpha,k}$ be the value of the random variable T with k degrees of freedom above which we find an area

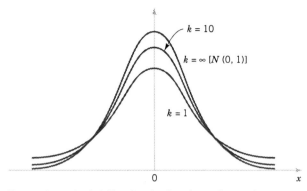

Figure 8-4 Probability density functions of several t distributions.

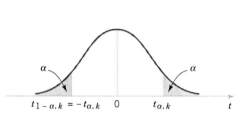

Figure 8-5 Percentage points of the t distribution.

(or probability) α. Thus, $t_{\alpha,k}$ is an upper-tail 100α percentage point of the t distribution with k degrees of freedom. This percentage point is shown in Fig. 8-5. In the Appendix Table V the α values are the column headings, and the degrees of freedom are listed in the left column. To illustrate the use of the table, note that the t-value with 10 degrees of freedom having an area of 0.05 to the right is $t_{0.05,10} = 1.812$. That is,

$$P(T_{10} > t_{0.05,10}) = P(T_{10} > 1.812) = 0.05$$

Since the t distribution is symmetric about zero, we have $t_{1-\alpha,n} = -t_{\alpha,n}$; that is, the t-value having an area of $1 - \alpha$ to the right (and therefore an area of α to the left) is equal to the negative of the t-value that has area α in the right tail of the distribution. Therefore, $t_{0.95,10} = -t_{0.05,10} = -1.812$. Finally, because $t_{\alpha,\infty}$ is the standard normal distribution, the familiar z_α values appear in the last row of Appendix Table V.

8-3.2 t Confidence Interval on μ

It is easy to find a $100(1 - \alpha)$ percent confidence interval on the mean of a normal distribution with unknown variance by proceeding essentially as we did in Section 8-2.1. We know that the distribution of $T = (\overline{X} - \mu)/(S/\sqrt{n})$ is t with $n - 1$ degrees of freedom. Letting $t_{\alpha/2,n-1}$ be the upper $100\alpha/2$ percentage point of the t distribution with $n - 1$ degrees of freedom, we may write:

$$P(-t_{\alpha/2,n-1} \leq T \leq t_{\alpha/2,n-1}) = 1 - \alpha$$

or

$$P\left(-t_{\alpha/2,n-1} \leq \frac{\overline{X} - \mu}{S/\sqrt{n}} \leq t_{\alpha/2,n-1}\right) = 1 - \alpha$$

Rearranging this last equation yields

$$P(\overline{X} - t_{\alpha/2,n-1}S/\sqrt{n} \leq \mu \leq \overline{X} + t_{\alpha/2,n-1}S/\sqrt{n}) = 1 - \alpha \qquad (8\text{-}17)$$

This leads to the following definition of the $100(1 - \alpha)$ percent two-sided confidence interval on μ.

8-3 CONFIDENCE INTERVAL ON THE MEAN OF A NORMAL DISTRIBUTION, VARIANCE UNKNOWN

Confidence Interval on the Mean, Variance Unknown

If $\bar{x}$ and s are the mean and standard deviation of a random sample from a normal distribution with unknown variance σ^2, a **100(1 − α) percent confidence interval on μ** is given by

$$\bar{x} - t_{\alpha/2,n-1}s/\sqrt{n} \leq \mu \leq \bar{x} + t_{\alpha/2,n-1}s/\sqrt{n} \quad (8\text{-}18)$$

where $t_{\alpha/2,n-1}$ is the upper $100\alpha/2$ percentage point of the t distribution with $n - 1$ degrees of freedom.

One-sided confidence bounds on the mean of a normal distribution are also of interest and are easy to find. Simply use only the appropriate lower or upper confidence limit from Equation 8-18 and replace $t_{\alpha/2,n-1}$ by $t_{\alpha,n-1}$.

EXAMPLE 8-5 Alloy Adhesion

An article in the journal *Materials Engineering* (1989, Vol. II, No. 4, pp. 275–281) describes the results of tensile adhesion tests on 22 U-700 alloy specimens. The load at specimen failure is as follows (in megapascals):

19.8	10.1	14.9	7.5	15.4	15.4
15.4	18.5	7.9	12.7	11.9	11.4
11.4	14.1	17.6	16.7	15.8	
19.5	8.8	13.6	11.9	11.4	

The sample mean is $\bar{x} = 13.71$, and the sample standard deviation is $s = 3.55$. Figures 8-6 and 8-7 show a box plot and a normal probability plot of the tensile adhesion test data, respectively. These displays provide good support for the assumption that the population is normally distributed. We want to find a 95% CI on μ. Since $n = 22$, we have $n - 1 = 21$ degrees of freedom for t, so $t_{0.025,21} = 2.080$. The resulting CI is

$$\bar{x} - t_{\alpha/2,n-1}s/\sqrt{n} \leq \mu \leq \bar{x} + t_{\alpha/2,n-1}s/\sqrt{n}$$
$$13.71 - 2.080(3.55)/\sqrt{22} \leq \mu \leq 13.71 + 2.080(3.55)/\sqrt{22}$$
$$13.71 - 1.57 \leq \mu \leq 13.71 + 1.57$$
$$12.14 \leq \mu \leq 15.28$$

The CI is fairly wide because there is a lot of variability in the tensile adhesion test measurements.

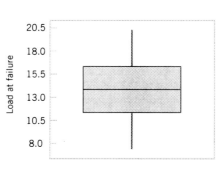

Figure 8-6 Box and whisker plot for the load at failure data in Example 8-5.

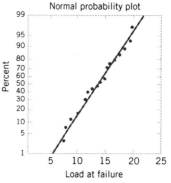

Figure 8-7 Normal probability plot of the load at failure data from Example 8-5.

EXERCISES FOR SECTION 8-3

8-20. Find the values of the following percentiles: $t_{0.025,15}$, $t_{0.05,10}$, $t_{0.10,20}$, $t_{0.005,25}$, and $t_{0.001,30}$.

8-21. Determine the t-percentile that is required to construct each of the following two-sided confidence intervals:
(a) Confidence level = 95%, degrees of freedom = 12
(b) Confidence level = 95%, degrees of freedom = 24
(c) Confidence level = 99%, degrees of freedom = 13
(d) Confidence level = 99.9%, degrees of freedom = 15

8-22. Determine the t-percentile that is required to construct each of the following one-sided confidence intervals:
(a) Confidence level = 95%, degrees of freedom = 14
(b) Confidence level = 99%, degrees of freedom = 19
(c) Confidence level = 99.9%, degrees of freedom = 24

8-23. A research engineer for a tire manufacturer is investigating tire life for a new rubber compound and has built 16 tires and tested them to end-of-life in a road test. The sample mean and standard deviation are 60,139.7 and 3645.94 kilometers. Find a 95% confidence interval on mean tire life.

8-24. An Izod impact test was performed on 20 specimens of PVC pipe. The sample mean is $\bar{x} = 1.25$ and the sample standard deviation is $s = 0.25$. Find a 99% lower confidence bound on Izod impact strength.

8-25. A postmix beverage machine is adjusted to release a certain amount of syrup into a chamber where it is mixed with carbonated water. A random sample of 25 beverages was found to have a mean syrup content of $\bar{x} = 1.10$ fluid ounces and a standard deviation of $s = 0.015$ fluid ounces. Find a 95% CI on the mean volume of syrup dispensed.

8-26. An article in *Medicine and Science in Sports and Exercise* "Maximal Leg-Strength Training Improves Cycling Economy in Previously Untrained Men," (2005, Vol. 37 pp. 131–1236) studied cycling performance before and after eight weeks of leg-strength training. Seven previously untrained males performed leg-strength training three days per week for eight weeks (with four sets of five replications at 85% of one repetition maximum). Peak power during incremental cycling increased to a mean of 315 watts with a standard deviation of 16 watts. Construct a 95% confidence interval for the mean peak power after training.

8-27. An article in *Obesity Research* "Impaired pressure natriuresis in obese youths," (2003, Vol. 11, pp. 745–751) described a study in which all meals were provided for 14 lean boys for three days followed by one stress (with a video-game task). The average systolic blood pressure (SBP) during the test was 118.3 mm HG with a standard deviation of 9.9 mm HG. Construct a 99% one-sided upper confidence interval for mean SBP.

8-28. An article in the *Journal of Composite Materials* (December 1989, Vol. 23, p. 1200) describes the effect of delamination on the natural frequency of beams made from composite laminates. Five such delaminated beams were subjected to loads, and the resulting frequencies were as follows (in hertz):

230.66, 233.05, 232.58, 229.48, 232.58

Check the assumption of normality in the population. Calculate a 90% two-sided confidence interval on mean natural frequency.

8-29. The Bureau of Meteorology of the Australian Government provided the mean annual rainfall (in millimeters) in Australia 1983–2002 as follows (http://www.bom.gov.au/climate/change/rain03.txt)

499.2, 555.2, 398.8, 391.9, 453.4, 459.8, 483.7, 417.6, 469.2, 452.4, 499.3, 340.6, 522.8, 469.9, 527.2, 565.5, 584.1, 727.3, 558.6, 338.6

Check the assumption of normality in the population. Construct a 95% confidence interval for the mean annual rainfall.

8-30. The solar energy consumed (in trillion BTU) in the U.S. by year from 1989–2004 (source: U.S. Department of Energy Web site http://www.eia.doe.gov/emeu) is shown in the table below. Read down, then right for year.

55.291	66.458	70.237	65.454
59.718	68.548	69.787	64.391
62.688	69.857	68.793	63.62
63.886	70.833	66.388	63.287

Check the assumption of normality in the population. Construct a 95% confidence interval for the mean solar energy consumed.

8-31. The brightness of a television picture tube can be evaluated by measuring the amount of current required to achieve a particular brightness level. A sample of 10 tubes results in $\bar{x} = 317.2$ and $s = 15.7$. Find (in microamps) a 99% confidence interval on mean current required. State any necessary assumptions about the underlying distribution of the data.

8-32. A particular brand of diet margarine was analyzed to determine the level of polyunsaturated fatty acid (in percentages). A sample of six packages resulted in the following data: 16.8, 17.2, 17.4, 16.9, 16.5, 17.1.
(a) Check the assumption that the level of polyunsaturated fatty acid is normally distributed.

(b) Calculate a 99% confidence interval on the mean μ. Provide a practical interpretation of this interval.

(c) Calculate a 99% lower confidence bound on the mean. Compare this bound with the lower bound of the two-sided confidence interval and discuss why they are different.

8-33. The compressive strength of concrete is being tested by a civil engineer. He tests 12 specimens and obtains the following data.

2216	2237	2249	2204
2225	2301	2281	2263
2318	2255	2275	2295

(a) Check the assumption that compressive strength is normally distributed. Include a graphical display in your answer.

(b) Construct a 95% two-sided confidence interval on the mean strength.

(c) Construct a 95% lower-confidence bound on the mean strength. Compare this bound with the lower bound of the two-sided confidence interval and discuss why they are different.

8-34. A machine produces metal rods used in an automobile suspension system. A random sample of 15 rods is selected, and the diameter is measured. The resulting data (in millimeters) are as follows:

8.24	8.25	8.20	8.23	8.24
8.21	8.26	8.26	8.20	8.25
8.23	8.23	8.19	8.28	8.24

(a) Check the assumption of normality for rod diameter.

(b) Calculate a 95% two-sided confidence interval on mean rod diameter.

(c) Calculate a 95% upper confidence bound on the mean. Compare this bound with the upper bound of the two-sided confidence interval and discuss why they are different.

8-35. An article in *Computers & Electrical Engineering*, "Parallel simulation of cellular neural networks" (1996, Vol. 22, pp. 61–84) considered the speed-up of cellular neural networks (CNN) for a parallel general-purpose computing architecture based on six transputers in different areas. The data follow:

3.775302	3.350679	4.217981	4.030324	4.639692
4.139665	4.395575	4.824257	4.268119	4.584193
4.930027	4.315973	4.600101		

(a) Is there evidence to support the assumption that speed-up of CNN is normally distributed? Include a graphical display in your answer.

(b) Construct a 95% two-sided confidence interval on the mean speed-up.

(c) Construct a 95% lower confidence bound on the mean speed-up.

8-36. The wall thickness of 25 glass 2-liter bottles was measured by a quality-control engineer. The sample mean was $\bar{x} = 4.05$ millimeters, and the sample standard deviation was $s = 0.08$ millimeter. Find a 95% lower confidence bound for mean wall thickness. Interpret the interval you have obtained.

8-37. An article in *Nuclear Engineering International* (February 1988, p. 33) describes several characteristics of fuel rods used in a reactor owned by an electric utility in Norway. Measurements on the percentage of enrichment of 12 rods were reported as follows:

| 2.94 | 3.00 | 2.90 | 2.75 | 3.00 | 2.95 |
| 2.90 | 2.75 | 2.95 | 2.82 | 2.81 | 3.05 |

(a) Use a normal probability plot to check the normality assumption.

(b) Find a 99% two-sided confidence interval on the mean percentage of enrichment. Are you comfortable with the statement that the mean percentage of enrichment is 2.95 percent? Why?

8-4 CONFIDENCE INTERVAL ON THE VARIANCE AND STANDARD DEVIATION OF A NORMAL DISTRIBUTION

Sometimes confidence intervals on the population variance or standard deviation are needed. When the population is modeled by a normal distribution, the tests and intervals described in this section are applicable. The following result provides the basis of constructing these confidence intervals.

χ^2 Distribution

Let $X_1, X_2, \ldots, X_n$ be a random sample from a normal distribution with mean μ and variance σ^2, and let S^2 be the sample variance. Then the random variable

$$X^2 = \frac{(n-1)S^2}{\sigma^2} \qquad (8\text{-}19)$$

has a chi-square (χ^2) distribution with $n - 1$ degrees of freedom.

274 CHAPTER 8 STATISTICAL INTERVALS FOR A SINGLE SAMPLE

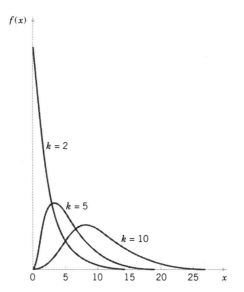

Figure 8-8 Probability density functions of several χ^2 distributions.

The probability density function of a χ^2 random variable is

$$f(x) = \frac{1}{2^{k/2}\Gamma(k/2)} x^{(k/2)-1} e^{-x/2} \qquad x > 0 \tag{8-20}$$

where k is the number of degrees of freedom. The mean and variance of the χ^2 distribution are k and $2k$, respectively. Several chi-square distributions are shown in Fig. 8-8. Note that the chi-square random variable is nonnegative and that the probability distribution is skewed to the right. However, as k increases, the distribution becomes more symmetric. As $k \to \infty$, the limiting form of the chi-square distribution is the normal distribution.

The percentage points of the χ^2 distribution are given in Table IV of the Appendix. Define $\chi^2_{\alpha,k}$ as the percentage point or value of the chi-square random variable with k degrees of freedom such that the probability that X^2 exceeds this value is α. That is,

$$P(X^2 > \chi^2_{\alpha,k}) = \int_{\chi^2_{\alpha,k}}^{\infty} f(u)\, du = \alpha$$

This probability is shown as the shaded area in Fig. 8-9(a). To illustrate the use of Table IV, note that the areas α are the column headings and the degrees of freedom k are given in the left column. Therefore, the value with 10 degrees of freedom having an area (probability) of 0.05

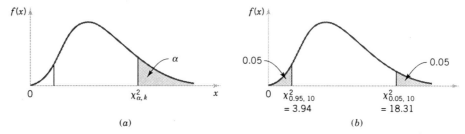

Figure 8-9 Percentage point of the χ^2 distribution. (a) The percentage point $\chi^2_{\alpha,k}$. (b) The upper percentage point $\chi^2_{0.05,10} = 18.31$ and the lower percentage point $\chi^2_{0.95,10} = 3.94$.

to the right is $\chi^2_{0.05,10} = 18.31$. This value is often called an **upper** 5% point of chi-square with 10 degrees of freedom. We may write this as a probability statement as follows:

$$P(X^2 > \chi^2_{0.05,10}) = P(X^2 > 18.31) = 0.05$$

Conversely, a **lower** 5% point of chi-square with 10 degrees of freedom would be $\chi^2_{0.95,10} = 3.94$ (from Appendix A). Both of these percentage points are shown in Figure 8-9(b).

The construction of the $100(1 - \alpha)\%$ CI for σ^2 is straightforward. Because

$$X^2 = \frac{(n-1)S^2}{\sigma^2}$$

is chi-square with $n - 1$ degrees of freedom, we may write

$$P(\chi^2_{1-\alpha/2,n-1} \leq X^2 \leq \chi^2_{\alpha/2,n-1}) = 1 - \alpha$$

so that

$$P\left(\chi^2_{1-\alpha/2,n-1} \leq \frac{(n-1)S^2}{\sigma^2} \leq \chi^2_{\alpha/2,n-1}\right) = 1 - \alpha$$

This last equation can be rearranged as

$$P\left(\frac{(n-1)S^2}{\chi^2_{\alpha/2,n-1}} \leq \sigma^2 \leq \frac{(n-1)S^2}{\chi^2_{1-\alpha/2,n-1}}\right) = 1 - \alpha$$

This leads to the following definition of the confidence interval for σ^2.

Confidence Interval on the Variance

If s^2 is the sample variance from a random sample of n observations from a normal distribution with unknown variance σ^2, then a **$100(1 - \alpha)\%$ confidence interval on σ^2 is**

$$\frac{(n-1)s^2}{\chi^2_{\alpha/2,n-1}} \leq \sigma^2 \leq \frac{(n-1)s^2}{\chi^2_{1-\alpha/2,n-1}} \quad (8\text{-}21)$$

where $\chi^2_{\alpha/2,n-1}$ and $\chi^2_{1-\alpha/2,n-1}$ are the upper and lower $100\alpha/2$ percentage points of the chi-square distribution with $n - 1$ degrees of freedom, respectively. A **confidence interval for σ** has lower and upper limits that are the square roots of the corresponding limits in Equation 8-21.

It is also possible to find a $100(1 - \alpha)\%$ lower confidence bound or upper confidence bound on σ^2.

One-Sided Confidence Bounds on the Variance

The $100(1 - \alpha)\%$ lower and upper confidence bounds on σ^2 are

$$\frac{(n-1)s^2}{\chi^2_{\alpha,n-1}} \leq \sigma^2 \quad \text{and} \quad \sigma^2 \leq \frac{(n-1)s^2}{\chi^2_{1-\alpha,n-1}} \quad (8\text{-}22)$$

respectively.

EXAMPLE 8-6
Detergent Filling

An automatic filling machine is used to fill bottles with liquid detergent. A random sample of 20 bottles results in a sample variance of fill volume of $s^2 = 0.0153$ (fluid ounces)^2. If the variance of fill volume is too large, an unacceptable proportion of bottles will be under- or overfilled. We will assume that the fill volume is approximately normally distributed. A 95% upper-confidence interval is found from Equation 8-22 as follows:

$$\sigma^2 \leq \frac{(n-1)s^2}{\chi^2_{0.95,19}}$$

or

$$\sigma^2 \leq \frac{(19)0.0153}{10.117} = 0.0287 \text{ (fluid ounce)}^2$$

This last expression may be converted into a confidence interval on the standard deviation σ by taking the square root of both sides, resulting in

$$\sigma \leq 0.17$$

Therefore, at the 95% level of confidence, the data indicate that the process standard deviation could be as large as 0.17 fluid ounce.

EXERCISES FOR SECTION 8-4

8-38. Determine the values of the following percentiles: $\chi^2_{0.05,10}$, $\chi^2_{0.025,15}$, $\chi^2_{0.01,12}$, $\chi^2_{0.95,20}$, $\chi^2_{0.99,18}$, $\chi^2_{0.995,16}$, and $\chi^2_{0.005,25}$.

8-34. Determine the χ^2 percentile that is required to construct each of the following CIs:
(a) Confidence level = 95%, degrees of freedom = 24, one-sided (upper)
(b) Confidence level = 99%, degrees of freedom = 9, one-sided (lower)
(c) Confidence level = 90%, degrees of freedom = 19, two-sided.

8-39. A rivet is to be inserted into a hole. A random sample of $n = 15$ parts is selected, and the hole diameter is measured. The sample standard deviation of the hole diameter measurements is $s = 0.008$ millimeters. Construct a 99% lower confidence bound for σ^2.

8-40. The sugar content of the syrup in canned peaches is normally distributed. A random sample of $n = 10$ cans yields a sample standard deviation of $s = 4.8$ milligrams. Calculate a 95% two-sided confidence interval for σ.

8-41. The percentage of titanium in an alloy used in aerospace castings is measured in 51 randomly selected parts. The sample standard deviation is $s = 0.37$. Construct a 95% two-sided confidence interval for σ.

8-42. An article in *Medicine and Science in Sports and Exercise* "Electrostimulation Training Effects on the Physical Performance of Ice Hockey Players," (2005, Vol. 37, pp. 455–460) considered the use of electromyostimulation (EMS) as a method to train healthy skeletal muscle. EMS sessions consisted of 30 contractions (four-second duration, 85 Hz) and were carried out three times per week for three weeks on 17 ice hockey players. The ten-meter skating performance test showed a standard deviation of 0.09 seconds. Construct a 95% confidence interval of the standard deviation of the skating performance test.

8-43. An article in *Urban Ecosystems*, "Urbanization and warming of Phoenix (Arizona, USA): Impacts, feedbacks and mitigation," (2002, Vol. 6, pp. 183–203) mentions that Phoenix is ideal to study the effects of an urban heat island because it has grown from a population of 300,000 to approximately 3 million over the last 50 years and this is a period with a continuous, detailed climate record. The 50 year averages of the mean annual temperatures at eight sites in Phoenix are shown below. Check the assumption of normality in the population with a probability plot. Construct a 95% confidence interval for the standard deviation over the sites of the mean annual temperatures.

Site	Average Mean Temperature (°C)
Sky Harbor Airport	23.3
Phoenix Greenway	21.7
Phoenix Encanto	21.6
Waddell	21.7
Litchfield	21.3
Laveen	20.7
Maricopa	20.9
Harlquahala	20.1

8-44. An article in *Cancer Research* "Analyses of litter-matched time-to-response data, with modifications for recovery of interlitter information," (1977, Vol. 37, pp. 3863–3868) tested the tumorigenesis of a drug. Rats were randomly selected from litters and given the drug. The times of tumor appearance were recorded as follows:

101, 104, 104, 77, 89, 88, 104, 96, 82, 70, 89, 91, 39, 103, 93, 85, 104, 104, 81, 67, 104, 104, 104, 87, 104, 89, 78, 104, 86, 76, 103, 102, 80, 45, 94, 104, 104, 76, 80, 72, 73

Calculate a 95% confidence interval on the standard deviation of time until a tumor appearance. Check the assumption of normality of the population and comment on the assumptions for the confidence interval.

8-45. An article in *Technometrics* (1999, Vol. 41, pp. 202–211) studied the capability of a gauge by measuring the weight of paper. The data for repeated measurements of one sheet of paper are in the following table. Construct a 95% one-sided, upper confidence interval for the standard deviation of these measurements. Check the assumption of normality of the data and comment on the assumptions for the confidence interval.

Observations

3.481	3.448	3.485	3.475	3.472
3.477	3.472	3.464	3.472	3.470
3.470	3.470	3.477	3.473	3.474

8-46. An article in the *Australian Journal of Agricultural Research*, "Non-Starch Polysaccharides and Broiler Performance on Diets Containing Soyabean Meal as the Sole Protein Concentrate" (1993, Vol. 44, No. 8, pp. 1483–1499) determined the essential amino acid (Lysine) composition level of soybean meals are as shown below (g/kg):

| 22.2 | 24.7 | 20.9 | 26.0 | 27.0 |
| 24.8 | 26.5 | 23.8 | 25.6 | 23.9 |

(a) Construct a 99% two-sided confidence interval for σ^2.
(b) Calculate a 99% lower confidence bound for σ^2.
(c) Calculate a 90% lower confidence bound for σ.
(d) Compare the intervals that you have computed.

8-5 LARGE-SAMPLE CONFIDENCE INTERVAL FOR A POPULATION PROPORTION

It is often necessary to construct confidence intervals on a population proportion. For example, suppose that a random sample of size n has been taken from a large (possibly infinite) population and that $X(\leq n)$ observations in this sample belong to a class of interest. Then $\hat{P} = X/n$ is a point estimator of the proportion of the population p that belongs to this class. Note that n and p are the parameters of a binomial distribution. Furthermore, from Chapter 4 we know that the sampling distribution of $\hat{P}$ is approximately normal with mean p and variance $p(1 - p)/n$, if p is not too close to either 0 or 1 and if n is relatively large. Typically, to apply this approximation we require that np and $n(1 - p)$ be greater than or equal to 5. We will make use of the normal approximation in this section.

Normal Approximation for a Binomial Proportion

If n is large, the distribution of

$$Z = \frac{X - np}{\sqrt{np(1 - p)}} = \frac{\hat{P} - p}{\sqrt{\frac{p(1 - p)}{n}}}$$

is approximately standard normal.

To construct the confidence interval on p, note that

$$P(-z_{\alpha/2} \leq Z \leq z_{\alpha/2}) \simeq 1 - \alpha$$

so

$$P\left(-z_{\alpha/2} \leq \frac{\hat{P} - p}{\sqrt{\frac{p(1-p)}{n}}} \leq z_{\alpha/2}\right) \simeq 1 - \alpha$$

This may be rearranged as

$$P\left(\hat{P} - z_{\alpha/2}\sqrt{\frac{p(1-p)}{n}} \leq p \leq \hat{P} + z_{\alpha/2}\sqrt{\frac{p(1-p)}{n}}\right) \simeq 1 - \alpha \quad (8\text{-}23)$$

The quantity $\sqrt{p(1-p)/n}$ in Equation 8-23 is called the *standard error of the point estimator* $\hat{P}$. This was discussed in Chapter 7. Unfortunately, the upper and lower limits of the confidence interval obtained from Equation 8-23 contain the unknown parameter p. However, as suggested at the end of Section 8-2.5, a satisfactory solution is to replace p by $\hat{P}$ in the standard error, which results in

$$P\left(\hat{P} - z_{\alpha/2}\sqrt{\frac{\hat{P}(1-\hat{P})}{n}} \leq p \leq \hat{P} + z_{\alpha/2}\sqrt{\frac{\hat{P}(1-\hat{P})}{n}}\right) \simeq 1 - \alpha \quad (8\text{-}24)$$

This leads to the approximate $100(1 - \alpha)\%$ confidence interval on p.

Approximate Confidence Interval on a Binomial Proportion

> If $\hat{p}$ is the proportion of observations in a random sample of size n that belongs to a class of interest, an approximate $100(1 - \alpha)\%$ confidence interval on the proportion p of the population that belongs to this class is
>
> $$\hat{p} - z_{\alpha/2}\sqrt{\frac{\hat{p}(1-\hat{p})}{n}} \leq p \leq \hat{p} + z_{\alpha/2}\sqrt{\frac{\hat{p}(1-\hat{p})}{n}} \quad (8\text{-}25)$$
>
> where $z_{\alpha/2}$ is the upper $\alpha/2$ percentage point of the standard normal distribution.

This procedure depends on the adequacy of the normal approximation to the binomial. To be reasonably conservative, this requires that np and $n(1 - p)$ be greater than or equal to 5. In situations where this approximation is inappropriate, particularly in cases where n is small, other methods must be used. Tables of the binomial distribution could be used to obtain a confidence interval for p. However, we could also use numerical methods based on the binomial probability mass function that are implemented in computer programs.

EXAMPLE 8-7
Crankshaft Bearings

In a random sample of 85 automobile engine crankshaft bearings, 10 have a surface finish that is rougher than the specifications allow. Therefore, a point estimate of the proportion of bearings in the population that exceeds the roughness specification is $\hat{p} = x/n = 10/85 = 0.12$. A 95% two-sided confidence interval for p is computed from Equation 8-25 as

$$\hat{p} - z_{0.025}\sqrt{\frac{\hat{p}(1-\hat{p})}{n}} \leq p \leq \hat{p} + z_{0.025}\sqrt{\frac{\hat{p}(1-\hat{p})}{n}}$$

or

$$0.12 - 1.96\sqrt{\frac{0.12(0.88)}{85}} \leq p \leq 0.12 + 1.96\sqrt{\frac{0.12(0.88)}{85}}$$

which simplifies to

$$0.05 \leq p \leq 0.19$$

Choice of Sample Size

Since $\hat{P}$ is the point estimator of p, we can define the error in estimating p by $\hat{P}$ as $E = |p - \hat{P}|$. Note that we are approximately $100(1 - \alpha)\%$ confident that this error is less than $z_{\alpha/2}\sqrt{p(1-p)/n}$. For instance, in Example 8-7, we are 95% confident that the sample proportion $\hat{p} = 0.12$ differs from the true proportion p by an amount not exceeding 0.07.

In situations where the sample size can be selected, we may choose n to be $100(1 - \alpha)\%$ confident that the error is less than some specified value E. If we set $E = z_{\alpha/2}\sqrt{p(1-p)/n}$ and solve for n, the appropriate sample size is

Sample Size for a Specified Error on a Binomial Proportion

$$n = \left(\frac{z_{\alpha/2}}{E}\right)^2 p(1-p) \tag{8-26}$$

An estimate of p is required to use Equation 8-26. If an estimate $\hat{p}$ from a previous sample is available, it can be substituted for p in Equation 8-26, or perhaps a subjective estimate can be made. If these alternatives are unsatisfactory, a preliminary sample can be taken, $\hat{p}$ computed, and then Equation 8-26 used to determine how many additional observations are required to estimate p with the desired accuracy. Another approach to choosing n uses the fact that the sample size from Equation 8-26 will always be a maximum for $p = 0.5$ [that is, $p(1 - p) \leq 0.25$ with equality for $p = 0.5$], and this can be used to obtain an upper bound on n. In other words, we are at least $100(1 - \alpha)\%$ confident that the error in estimating p by $\hat{p}$ is less than E if the sample size is

$$n = \left(\frac{z_{\alpha/2}}{E}\right)^2 (0.25) \tag{8-27}$$

EXAMPLE 8-8
Crankshaft Bearings

Consider the situation in Example 8-7. How large a sample is required if we want to be 95% confident that the error in using $\hat{p}$ to estimate p is less than 0.05? Using $\hat{p} = 0.12$ as an initial estimate of p, we find from Equation 8-26 that the required sample size is

$$n = \left(\frac{z_{0.025}}{E}\right)^2 \hat{p}(1 - \hat{p}) = \left(\frac{1.96}{0.05}\right)^2 0.12(0.88) \cong 163$$

If we wanted to be *at least* 95% confident that our estimate $\hat{p}$ of the true proportion p was within 0.05 regardless of the value of p, we would use Equation 8-27 to find the sample size

$$n = \left(\frac{z_{0.025}}{E}\right)^2 (0.25) = \left(\frac{1.96}{0.05}\right)^2 (0.25) \cong 385$$

One-Sided Confidence Bounds

We may find approximate one-sided confidence bounds on p by a simple modification of Equation 8-25.

Approximate One-Sided Confidence Bounds on a Binomial Proportion

The approximate $100(1 - \alpha)\%$ lower and upper confidence bounds are

$$\hat{p} - z_\alpha \sqrt{\frac{\hat{p}(1 - \hat{p})}{n}} \leq p \quad \text{and} \quad p \leq \hat{p} + z_\alpha \sqrt{\frac{\hat{p}(1 - \hat{p})}{n}} \quad (8\text{-}28)$$

respectively.

EXERCISES FOR SECTION 8-5

8-47. The fraction of defective integrated circuits produced in a photolithography process is being studied. A random sample of 300 circuits is tested, revealing 13 defectives.
(a) Calculate a 95% two-sided CI on the fraction of defective circuits produced by this particular tool.
(b) Calculate a 95% upper confidence bound on the fraction of defective circuits.

8-48. An article in *Knee Surgery, Sports Traumatology, Arthroscopy* (2005, Vol. 13, pp. 273–279) "Arthroscopic meniscal repair with an absorbable screw: results and surgical technique" showed that only 25 out of 37 tears (67.6%) located between 3 and 6 mm from the meniscus rim were healed.
(a) Calculate a two-sided 95% confidence interval on the proportion of such tears that will heal.
(b) Calculate a 95% lower confidence bound on the proportion of such tears that will heal.

8-49. The 2004 presidential election exit polls from the critical state of Ohio provided the following results. There were 2020 respondents in the exit polls and 768 were college graduates. Of the college graduates, 412 voted for George Bush.
(a) Calculate a 95% confidence interval for the proportion of college graduates in Ohio that voted for George Bush.
(b) Calculate a 95% lower confidence bound for the proportion of college graduates in Ohio that voted for George Bush.

8-50. Of 1000 randomly selected cases of lung cancer, 823 resulted in death within 10 years.
(a) Calculate a 95% two-sided confidence interval on the death rate from lung cancer.
(b) Using the point estimate of p obtained from the preliminary sample, what sample size is needed to be 95% confident that the error in estimating the true value of p is less than 0.03?
(c) How large must the sample be if we wish to be at least 95% confident that the error in estimating p is less than 0.03, regardless of the true value of p?

8-51. An article in *Journal of the American Statistical Association* (1990, Vol. 85, pp. 972–985) measured weight of 30 rats under experiment controls. Suppose that there are 12 underweight rats.
(a) Calculate a 95% two-sided confidence interval on the true proportion of rats that would show underweight from the experiment.
(b) Using the point estimate of p obtained from the preliminary sample, what sample size is needed to be 95% confident that the error in estimating the true value of p is less than 0.02?
(c) How large must the sample be if we wish to be at least 95% confident that the error in estimating p is less than 0.02, regardless of the true value of p?

8-52. A random sample of 50 suspension helmets used by motorcycle riders and automobile race-car drivers was subjected to an impact test, and on 18 of these helmets some damage was observed.
(a) Find a 95% two-sided confidence interval on the true proportion of helmets of this type that would show damage from this test.
(b) Using the point estimate of p obtained from the preliminary sample of 50 helmets, how many helmets must be tested to be 95% confident that the error in estimating the true value of p is less than 0.02?
(c) How large must the sample be if we wish to be at least 95% confident that the error in estimating p is less than 0.02, regardless of the true value of p?

8-53. The Arizona Department of Transportation wishes to survey state residents to determine what proportion of the population would like to increase statewide highway speed limits to 75 mph from 65 mph. How many residents do they need to survey if they want to be at least 99% confident that the sample proportion is within 0.05 of the true proportion?

8-54. A study is to be conducted of the percentage of homeowners who own at least two television sets. How large a sample is required if we wish to be 99% confident that the error in estimating this quantity is less than 0.017?

8-6 GUIDELINES FOR CONSTRUCTING CONFIDENCE INTERVALS

The most difficult step in constructing a confidence interval is often the match of the appropriate calculation to the objective of the study. Common cases are listed in Table 8-1 along with the reference to the section that covers the appropriate calculation for a confidence interval test. Table 8-1 provides a simple road map to help select the appropriate analysis. Two primary comments can help identify the analysis:

1. Determine the parameter (and the distribution of the data) that will be bounded by the confidence interval or tested by the hypothesis.
2. Check if other parameters are known or need to be estimated.

In Chapter 9, we will study a procedure closely related to confidence intervals called hypothesis testing. Table 8-1 can be used for those procedures also. This road map will be extended to more cases in Chapter 10.

8-7 TOLERANCE AND PREDICTION INTERVALS

8-7.1 Prediction Interval for a Future Observation

In some problem situations, we may be interested in predicting a future observation of a variable. This is a different problem than estimating the mean of that variable, so a confidence interval is not appropriate. In this section we show how to obtain a $100(1 - \alpha)\%$ prediction interval on a future value of a normal random variable.

Table 8-1 The Roadmap for Constructing Confidence Intervals and Performing Hypothesis Tests, One-Sample Case

Parameter to Be Bounded by the Confidence Interval or Tested with a Hypothesis?	Symbol	Other Parameters?	Confidence Interval Section	Hypothesis Test Section	Comments
Mean of normal distribution	μ	Standard deviation σ known	8-2	9-2	
Mean of arbitrary distribution with large sample size	μ	Sample size large enough that central limit theorem applies and σ is essentially known	8-2.5	9-2.5	Large sample size is often taken to be $n \geq 40$
Mean of normal distribution	μ	Standard deviation σ unknown and estimated	8-3	9-3	
Variance (or standard deviation) of normal distribution	σ^2	Mean μ unknown and estimated	8-4	9-4	
Population Proportion	p	None	8-5	9-5	

Suppose that $X_1, X_2, \ldots, X_n$ is a random sample from a normal population. We wish to predict the value X_{n+1}, a single **future** observation. A point prediction of X_{n+1} is $\overline{X}$, the sample mean. The prediction error is $X_{n+1} - \overline{X}$. The expected value of the prediction error is

$$E(X_{n+1} - \overline{X}) = \mu - \mu = 0$$

and the variance of the prediction error is

$$V(X_{n+1} - \overline{X}) = \sigma^2 + \frac{\sigma^2}{n} = \sigma^2\left(1 + \frac{1}{n}\right)$$

because the future observation, X_{n+1} is independent of the mean of the current sample $\overline{X}$. The prediction error $X_{n+1} - \overline{X}$ is normally distributed. Therefore

$$Z = \frac{X_{n+1} - \overline{X}}{\sigma\sqrt{1 + \frac{1}{n}}}$$

has a standard normal distribution. Replacing σ with S results in

$$T = \frac{X_{n+1} - \overline{X}}{S\sqrt{1 + \frac{1}{n}}}$$

which has a t distribution with $n - 1$ degrees of freedom. Manipulating T as we have done previously in the development of a CI leads to a prediction interval on the future observation X_{n+1}.

Prediction Interval

A $100(1 - \alpha)\%$ prediction interval on a single future observation from a normal distribution is given by

$$\bar{x} - t_{\alpha/2, n-1}s\sqrt{1 + \frac{1}{n}} \leq X_{n+1} \leq \bar{x} + t_{\alpha/2, n-1}s\sqrt{1 + \frac{1}{n}} \quad (8\text{-}29)$$

The prediction interval for X_{n+1} will always be longer than the confidence interval for μ because there is more variability associated with the prediction error than with the error of estimation. This is easy to see because the prediction error is the difference between two random variables $(X_{n+1} - \overline{X})$, and the estimation error in the CI is the difference between one random variable and a constant $(\overline{X} - \mu)$. As n gets larger $(n \to \infty)$, the length of the CI decreases to zero, essentially becoming the single value μ, but the length of the prediction interval approaches $2z_{\alpha/2}\sigma$. So as n increases, the uncertainty in estimating μ goes to zero, although there will always be uncertainty about the future value X_{n+1} even when there is no need to estimate any of the distribution parameters.

EXAMPLE 8-9
Alloy Adhesion

Reconsider the tensile adhesion tests on specimens of U-700 alloy described in Example 8-5. The load at failure for $n = 22$ specimens was observed, and we found that $\bar{x} = 13.71$ and $s = 3.55$. The 95% confidence interval on μ was $12.14 \leq \mu \leq 15.28$. We plan to test a twenty-third specimen.

A 95% prediction interval on the load at failure for this specimen is

$$\bar{x} - t_{\alpha/2,n-1}s\sqrt{1 + \frac{1}{n}} \le X_{n+1} \le \bar{x} + t_{\alpha/2,n-1}s\sqrt{1 + \frac{1}{n}}$$

$$13.71 - (2.080)3.55\sqrt{1 + \frac{1}{22}} \le X_{23} \le 13.71 + (2.080)3.55\sqrt{1 + \frac{1}{22}}$$

$$6.16 \le X_{23} \le 21.26$$

Notice that the prediction interval is considerably longer than the CI.

8-7.2 Tolerance Interval for a Normal Distribution

Consider a population of semiconductor processors. Suppose that the speed of these processors has a normal distribution with mean $\mu = 600$ megahertz and standard deviation $\sigma = 30$ megahertz. Then the interval from $600 - 1.96(30) = 541.2$ to $600 + 1.96(30) = 658.8$ megahertz captures the speed of 95% of the processors in this population because the interval from -1.96 to 1.96 captures 95% of the area under the standard normal curve. The interval from $\mu - z_{\alpha/2}\sigma$ to $\mu + z_{\alpha/2}\sigma$ is called a **tolerance interval**.

If μ and σ are unknown, we can use the data from a random sample of size n to compute $\bar{x}$ and s, and then form the interval $(\bar{x} - 1.96s, \bar{x} + 1.96s)$. However, because of sampling variability in $\bar{x}$ and s, it is likely that this interval will contain less than 95% of the values in the population. The solution to this problem is to replace 1.96 by some value that will make the proportion of the distribution contained in the interval 95% with some level of confidence. Fortunately, it is easy to do this.

Tolerance Interval

> A **tolerance interval** for capturing at least $\gamma\%$ of the values in a normal distribution with confidence level $100(1 - \alpha)\%$ is
>
> $$\bar{x} - ks, \quad \bar{x} + ks$$
>
> where k is a tolerance interval factor found in Appendix Table XII. Values are given for $\gamma = 90\%$, 95%, and 99% and for 90%, 95%, and 99% confidence.

One-sided tolerance bounds can also be computed. The tolerance factors for these bounds are also given in Appendix Table XII.

EXAMPLE 8-10
Alloy Adhesion

Let's reconsider the tensile adhesion tests originally described in Example 8-5. The load at failure for $n = 22$ specimens was observed, and we found that $\bar{x} = 13.71$ and $s = 3.55$. We want to find a tolerance interval for the load at failure that includes 90% of the values in the population with 95% confidence. From Appendix Table XII the tolerance factor k for $n = 22$, $\gamma = 0.90$, and 95% confidence is $k = 2.264$. The desired tolerance interval is

$$(\bar{x} - ks, \bar{x} + ks) \quad \text{or} \quad [13.71 - (2.264)3.55, \, 13.71 + (2.264)3.55]$$

which reduces to (5.67, 21.74). We can be 95% confident that at least 90% of the values of load at failure for this particular alloy lie between 5.67 and 21.74 megapascals.

284 CHAPTER 8 STATISTICAL INTERVALS FOR A SINGLE SAMPLE

From Appendix Table XII, we note that as $n \to \infty$, the value of k goes to the z-value associated with the desired level of containment for the normal distribution. For example, if we want 90% of the population to fall in the two-sided tolerance interval, k approaches $z_{0.05} = 1.645$ as $n \to \infty$. Note that as $n \to \infty$, a $100(1 - \alpha)\%$ prediction interval on a future value approaches a tolerance interval that contains $100(1 - \alpha)\%$ of the distribution.

EXERCISES FOR SECTION 8-7

8-55. Consider the tire-testing data described in Exercise 8-23. Compute a 95% prediction interval on the life of the next tire of this type tested under conditions that are similar to those employed in the original test. Compare the length of the prediction interval with the length of the 95% CI on the population mean.

8-56. Consider the Izod impact test described in Exercise 8-24. Compute a 99% prediction interval on the impact strength of the next specimen of PVC pipe tested. Compare the length of the prediction interval with the length of the 99% CI on the population mean.

8-57. Consider the syrup dispensing measurements described in Exercise 8-25. Compute a 95% prediction interval on the syrup volume in the next beverage dispensed. Compare the length of the prediction interval with the length of the 95% CI on the population mean.

8-58. Consider the natural frequency of beams described in Exercise 8-28. Compute a 90% prediction interval on the diameter of the natural frequency of the next beam of this type that will be tested. Compare the length of the prediction interval with the length of the 90% CI on the population mean.

8-59. Consider the rainfall in Exercise 8-29. Compute a 95% prediction interval on the rainfall for the next year. Compare the length of the prediction interval with the length of the 95% CI on the population mean.

8-60. Consider the margarine test described in Exercise 8-32. Compute a 99% prediction interval on the polyunsaturated fatty acid in the next package of margarine that is tested. Compare the length of the prediction interval with the length of the 99% CI on the population mean.

8-61. Consider the television tube brightness test described in Exercise 8-31. Compute a 99% prediction interval on the brightness of the next tube tested. Compare the length of the prediction interval with the length of the 99% CI on the population mean.

8-62. Consider the suspension rod diameter measurements described in Exercise 8-34. Compute a 95% prediction interval on the diameter of the next rod tested. Compare the length of the prediction interval with the length of the 95% CI on the population mean.

8-63. Consider the test on the compressive strength of concrete described in Exercise 8-33. Compute a 90% prediction interval on the next specimen of concrete tested.

8-64. Consider the bottle wall thickness measurements described in Exercise 8-36. Compute a 90% prediction interval on the wall thickness of the next bottle tested.

8-65. Consider the fuel rod enrichment data described in Exercise 8-37. Compute a 90% prediction interval on the enrichment of the next rod tested. Compare the length of the prediction interval with the length of the 99% CI on the population mean.

8-66. How would you obtain a one-sided prediction bound on a future observation? Apply this procedure to obtain a 95% one-sided prediction bound on the wall thickness of the next bottle for the situation described in Exercise 8-36.

8-67. Consider the tire-testing data in Exercise 8-23. Compute a 95% tolerance interval on the life of the tires that has confidence level 95%. Compare the length of the tolerance interval with the length of the 95% CI on the population mean. Which interval is shorter? Discuss the difference in interpretation of these two intervals.

8-68. Consider the Izod impact test described in Exercise 8-24. Compute a 99% tolerance interval on the impact strength of PVC pipe that has confidence level 90%. Compare the length of the tolerance interval with the length of the 99% CI on the population mean. Which interval is shorter? Discuss the difference in interpretation of these two intervals.

8-69. Consider the syrup volume data in Exercise 8-25. Compute a 95% tolerance interval on the syrup volume that has confidence level 90%. Compare the length of the tolerance interval with the length of the 95% CI on the population mean.

8-70. Consider the margarine test described in Exercise 8-32. Compute a 99% tolerance interval on the polyunsaturated fatty acid in this particular type of margarine that has confidence level 95%. Compare the length of the tolerance interval with the length of the 99% CI on the population mean. Which interval is shorter? Discuss the difference in interpretation of these two intervals.

8-71. Consider the rainfall data in Exercise 8-29. Compute a 95% tolerance interval that has confidence level 95%. Compare the length of the tolerance interval with the length of the 95% CI on the population mean. Discuss the difference in interpretation of these two intervals.

8-72. Consider the diameter data in Exercise 8-34. Compute a 95% tolerance interval on the diameter of the rods described that has 90% confidence. Compare the length of the tolerance interval with the length of the 95% CI on the population mean. Which interval is shorter? Discuss the difference in interpretation of these two intervals.

8-73. Consider the television tube brightness data in Exercise 8-31. Compute a 99% tolerance interval on the brightness of the television tubes that has confidence level 95%. Compare the length of the tolerance interval with the length of the 99% CI on the population mean. Which interval is shorter? Discuss the difference in interpretation of these two intervals.

8-74. Consider the strength of concrete data in Exercise 8-33. Compute a 90% tolerance interval on the compressive strength of the concrete that has 90% confidence.

8-75. Consider the fuel rod enrichment data described in Exercise 8-37. Compute a 99% tolerance interval on rod enrichment that has confidence level 95%. Compare the length of the tolerance interval with the length of the 95% CI on the population mean.

8-76. Consider the bottle wall thickness measurements described in Exercise 8-36.
(a) Compute a 90% tolerance interval on bottle wall thickness that has confidence level 90%.
(b) Compute a 90% lower tolerance bound on bottle wall thickness that has confidence level 90%. Why would a lower tolerance bound likely be of interest here?

Supplemental Exercises

8-77. Consider the confidence interval for μ with known standard deviation σ:

$$\bar{x} - z_{\alpha_1}\sigma/\sqrt{n} \leq \mu \leq \bar{x} + z_{\alpha_2}\sigma/\sqrt{n}$$

where $\alpha_1 + \alpha_2 = \alpha$. Let $\alpha = 0.05$ and find the interval for $\alpha_1 = \alpha_2 = \alpha/2 = 0.025$. Now find the interval for the case $\alpha_1 = 0.01$ and $\alpha_2 = 0.04$. Which interval is shorter? Is there any advantage to a "symmetric" confidence interval?

8-78. A normal population has a known mean 50 and unknown variance.
(a) A random sample of $n = 16$ is selected from this population, and the sample results are $\bar{x} = 52$ and $s = 8$. How unusual are these results? That is, what is the probability of observing a sample average as large as 52 (or larger) if the known, underlying mean is actually 50?
(b) A random sample of $n = 30$ is selected from this population, and the sample results are $\bar{x} = 52$ and $s = 8$. How unusual are these results?
(c) A random sample of $n = 100$ is selected from this population, and the sample results are $\bar{x} = 52$ and $s = 8$. How unusual are these results?
(d) Compare your answers to parts (a)–(c) and explain why they are the same or differ.

8-79. A normal population has known mean $\mu = 50$ and variance $\sigma^2 = 5$. What is the approximate probability that the sample variance is greater than or equal to 7.44? less than or equal to 2.56?
(a) For a random sample of $n = 16$.
(b) For a random sample of $n = 30$.
(c) For a random sample of $n = 71$.
(d) Compare your answers to parts (a)–(c) for the approximate probability that the sample variance is greater than or equal to 7.44. Explain why this tail probability is increasing or decreasing with increased sample size.
(e) Compare your answers to parts (a)–(c) for the approximate probability that the sample variance is less than or equal to 2.56. Explain why this tail probability is increasing or decreasing with increased sample size.

8-80. An article in the *Journal of Sports Science* (1987, Vol. 5, pp. 261–271) presents the results of an investigation of the hemoglobin level of Canadian Olympic ice hockey players. The data reported are as follows (in g/dl):

15.3	16.0	14.4	16.2	16.2
14.9	15.7	15.3	14.6	15.7
16.0	15.0	15.7	16.2	14.7
14.8	14.6	15.6	14.5	15.2

(a) Given the following probability plot of the data, what is a logical assumption about the underlying distribution of the data?

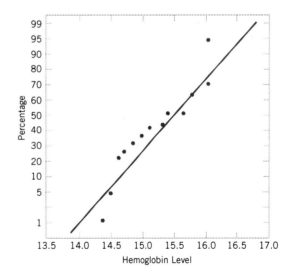

(b) Explain why this check of the distribution underlying the sample data is important if we want to construct a confidence interval on the mean.
(c) Based on this sample data, a 95% confidence interval for the mean is (15.04, 15.62). Is it reasonable to infer that the true mean could be 14.5? Explain your answer.

(d) Explain why this check of the distribution underlying the sample data is important if we want to construct a confidence interval on the variance.
(e) Based on this sample data, a 95% confidence interval for the variance is (0.22, 0.82). Is it reasonable to infer that the true variance could be 0.35? Explain your answer.
(f) Is it reasonable to use these confidence intervals to draw an inference about the mean and variance of hemoglobin levels
  (i) of Canadian doctors? Explain your answer.
  (ii) of Canadian children ages 6–12? Explain your answer.

8-81. The article "Mix Design for Optimal Strength Development of Fly Ash Concrete" (*Cement and Concrete Research*, 1989, Vol. 19, No. 4, pp. 634–640) investigates the compressive strength of concrete when mixed with fly ash (a mixture of silica, alumina, iron, magnesium oxide, and other ingredients). The compressive strength for nine samples in dry conditions on the twenty-eighth day are as follows (in megapascals):

40.2   30.4   28.9   30.5   22.4
25.8   18.4   14.2   15.3

(a) Given the following probability plot of the data, what is a logical assumption about the underlying distribution of the data?

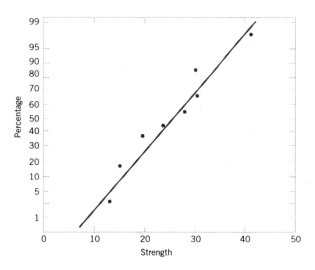

(b) Find a 99% lower one-sided confidence interval on mean compressive strength. Provide a practical interpretation of this interval.
(c) Find a 98% two-sided confidence interval on mean compressive strength. Provide a practical interpretation of this interval and explain why the lower end-point of the interval is or is not the same as in part (b).
(d) Find a 99% upper one-sided confidence interval on the variance of compressive strength. Provide a practical interpretation of this interval.
(e) Find a 98% two-sided confidence interval on the variance of compression strength. Provide a practical interpretation of this interval and explain why the upper end-point of the interval is or is not the same as in part (d).
(f) Suppose that it was discovered that the largest observation 40.2 was misrecorded and should actually be 20.4. Now the sample mean $\bar{x} = 23$ and the sample variance $s^2 = 39.8$. Use these new values and repeat parts (c) and (e). Compare the original computed intervals and the newly computed intervals with the corrected observation value. How does this mistake affect the values of the sample mean, sample variance, and the width of the two-sided confidence intervals?
(g) Suppose, instead, that it was discovered that the largest observation 40.2 is correct, but that the observation 25.8 is incorrect and should actually be 24.8. Now the sample mean $\bar{x} = 25$ and the standard deviation $s = 8.41$. Use these new values and repeat parts (c) and (e). Compare the original computed intervals and the newly computed intervals with the corrected observation value. How does this mistake affect the values of the sample mean, sample variance, and the width of the two-sided confidence intervals?
(h) Use the results from parts (f) and (g) to explain the effect of mistakenly recorded values on sample estimates. Comment on the effect when the mistaken values are near the sample mean and when they are not.

8-82. An operating system for a personal computer has been studied extensively, and it is known that the standard deviation of the response time following a particular command is $\sigma = 8$ milliseconds. A new version of the operating system is installed, and we wish to estimate the mean response time for the new system to ensure that a 95% confidence interval for μ has length at most 5 milliseconds.
(a) If we can assume that response time is normally distributed and that $\sigma = 8$ for the new system, what sample size would you recommend?
(b) Suppose that we are told by the vendor that the standard deviation of the response time of the new system is smaller, say $\sigma = 6$; give the sample size that you recommend and comment on the effect the smaller standard deviation has on this calculation.

8-83. Consider the hemoglobin data in Exercise 8-80. Find the following:
(a) An interval that contains 95% of the hemoglobin values with 90% confidence.
(b) An interval that contains 99% of the hemoglobin values with 90% confidence.

8-84. Consider the compressive strength of concrete data from Exercise 8-81. Find a 95% prediction interval on the next sample that will be tested.

8-85. The maker of a shampoo knows that customers like this product to have a lot of foam. Ten sample bottles of the product are selected at random and the foam heights observed

are as follows (in millimeters): 210, 215, 194, 195, 211, 201, 198, 204, 208, and 196.
(a) Is there evidence to support the assumption that foam height is normally distributed?
(b) Find a 95% CI on the mean foam height.
(c) Find a 95% prediction interval on the next bottle of shampoo that will be tested.
(d) Find an interval that contains 95% of the shampoo foam heights with 99% confidence.
(e) Explain the difference in the intervals computed in parts (b), (c), and (d).

8-86. During the 1999 and 2000 baseball seasons, there was much speculation that the unusually large number of home runs that were hit was due at least in part to a livelier ball. One way to test the "liveliness" of a baseball is to launch the ball at a vertical surface with a known velocity V_L and measure the ratio of the outgoing velocity V_O of the ball to V_L. The ratio $R = V_O/V_L$ is called the coefficient of restitution. Following are measurements of the coefficient of restitution for 40 randomly selected baseballs. The balls were thrown from a pitching machine at an oak surface.

0.6248	0.6237	0.6118	0.6159	0.6298	0.6192
0.6520	0.6368	0.6220	0.6151	0.6121	0.6548
0.6226	0.6280	0.6096	0.6300	0.6107	0.6392
0.6230	0.6131	0.6223	0.6297	0.6435	0.5978
0.6351	0.6275	0.6261	0.6262	0.6262	0.6314
0.6128	0.6403	0.6521	0.6049	0.6170	
0.6134	0.6310	0.6065	0.6214	0.6141	

(a) Is there evidence to support the assumption that the coefficient of restitution is normally distributed?
(b) Find a 99% CI on the mean coefficient of restitution.
(c) Find a 99% prediction interval on the coefficient of restitution for the next baseball that will be tested.
(d) Find an interval that will contain 99% of the values of the coefficient of restitution with 95% confidence.
(e) Explain the difference in the three intervals computed in parts (b), (c), and (d).

8-87. Consider the baseball coefficient of restitution data in Exercise 8-86. Suppose that any baseball that has a coefficient of restitution that exceeds 0.635 is considered too lively. Based on the available data, what proportion of the baseballs in the sampled population are too lively? Find a 95% lower confidence bound on this proportion.

8-88. An article in the *ASCE Journal of Energy Engineering* ("Overview of Reservoir Release Improvements at 20 TVA Dams," Vol. 125, April 1999, pp. 1–17) presents data on dissolved oxygen concentrations in streams below 20 dams in the Tennessee Valley Authority system. The observations are (in milligrams per liter): 5.0, 3.4, 3.9, 1.3, 0.2, 0.9, 2.7, 3.7, 3.8, 4.1, 1.0, 1.0, 0.8, 0.4, 3.8, 4.5, 5.3, 6.1, 6.9, and 6.5.
(a) Is there evidence to support the assumption that the dissolved oxygen concentration is normally distributed?
(b) Find a 95% CI on the mean dissolved oxygen concentration.
(c) Find a 95% prediction interval on the dissolved oxygen concentration for the next stream in the system that will be tested.
(d) Find an interval that will contain 95% of the values of the dissolved oxygen concentration with 99% confidence.
(e) Explain the difference in the three intervals computed in parts (b), (c), and (d).

8-89. The tar content in 30 samples of cigar tobacco follows:

1.542	1.585	1.532	1.466	1.499	1.611
1.622	1.466	1.546	1.494	1.548	1.626
1.440	1.608	1.520	1.478	1.542	1.511
1.459	1.533	1.532	1.523	1.397	1.487
1.598	1.498	1.600	1.504	1.545	1.558

(a) Is there evidence to support the assumption that the tar content is normally distributed?
(b) Find a 99% CI on the mean tar content.
(c) Find a 99% prediction interval on the tar content for the next observation that will be taken on this particular type of tobacco.
(d) Find an interval that will contain 99% of the values of the tar content with 95% confidence.
(e) Explain the difference in the three intervals computed in parts (b), (c), and (d).

8-90. A manufacturer of electronic calculators takes a random sample of 1200 calculators and finds that there are eight defective units.
(a) Construct a 95% confidence interval on the population proportion.
(b) Is there evidence to support a claim that the fraction of defective units produced is 1% or less?

8-91. An article in *The Engineer* ("Redesign for Suspect Wiring," June 1990) reported the results of an investigation into wiring errors on commercial transport aircraft that may produce faulty information to the flight crew. Such a wiring error may have been responsible for the crash of a British Midland Airways aircraft in January 1989 by causing the pilot to shut down the wrong engine. Of 1600 randomly selected aircraft, eight were found to have wiring errors that could display incorrect information to the flight crew.
(a) Find a 99% confidence interval on the proportion of aircraft that have such wiring errors.
(b) Suppose we use the information in this example to provide a preliminary estimate of p. How large a sample would be required to produce an estimate of p that we are 99% confident differs from the true value by at most 0.008?
(c) Suppose we did not have a preliminary estimate of p. How large a sample would be required if we wanted to be at least 99% confident that the sample proportion differs

from the true proportion by at most 0.008 regardless of the true value of p?

(d) Comment on the usefulness of preliminary information in computing the needed sample size.

8-92. An article in *Engineering Horizons* (Spring 1990, p. 26) reported that 117 of 484 new engineering graduates were planning to continue studying for an advanced degree. Consider this as a random sample of the 1990 graduating class.

(a) Find a 90% confidence interval on the proportion of such graduates planning to continue their education.

(b) Find a 95% confidence interval on the proportion of such graduates planning to continue their education.

(c) Compare your answers to parts (a) and (b) and explain why they are the same or different.

(d) Could you use either of these confidence intervals to determine whether the proportion is actually 0.25? Explain your answer. *Hint:* Use the normal approximation to the binomial.

8-93. An article in the *Journal of Applied Physiology*, "Humidity does not affect central nervous system oxygen toxicity" (2001, Vol. 91, pp. 1327–1333) reported that central nervous system (CNS) oxygen toxicity can appear in humans on exposure to oxygen pressures >180 kPa. CNS oxygen toxicity can occur as convulsions (similar to epileptic seizures, grand mal) and loss of consciousness without any warning symptoms. CNS oxygen toxicity is a risk encountered in several fields of human activity, such as combat diving with closed-circuit breathing apparatus and diving with mixtures of nitrogen and oxygen (nitrox) or nitrogen, oxygen, and helium (trimix) in sport and professional diving to depths >30 m. The risk of oxygen toxicity is always considered when deep diving is planned. The data shown below demonstrates shortened latencies in a dry atmosphere (<10% humidity) in 11 rats at O_2 of 507 kPa. The data collected is as follows:

22 26 19 27 37 27
14 19 23 18 18

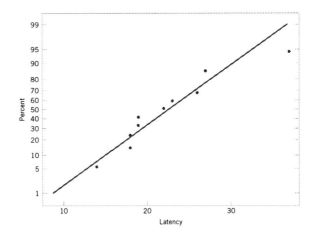

(a) Given the above probability plot of the data, what is a logical assumption about the underlying distribution of the data?

(b) Explain why this check of the distribution underlying the sample data is important if we want to construct a confidence interval on the mean.

(c) Find the 95% confidence interval for the mean.

(d) Explain why this check of the distribution underlying the sample data is important if we want to construct a confidence interval on the variance.

(e) Find the 95% confidence interval for the variance.

8-94. An article in the *Journal of Human Nutrition and Dietetics*, "The Validation of Energy and Protein Intakes by Doubly Labeled Water and 24-Hour Urinary Nitrogen Excretion in Post-Obese Subjects" (1995, Vol. 8, pp. 51–64) showed the energy intake expressed as a basal metabolic rate, BMR (MJ).

5.40 5.67 5.79 6.85 6.92
5.70 6.08 5.48 5.44 5.51

(a) Use a normal probability plot to check the normality assumption.

(b) Find a 99% two-sided confidence interval on the mean BMR.

MIND-EXPANDING EXERCISES

8-95. An electrical component has a time-to-failure (or lifetime) distribution that is exponential with parameter λ, so the mean lifetime is $\mu = 1/\lambda$. Suppose that a sample of n of these components is put on test, and let X_i be the observed lifetime of component i. The test continues only until the rth unit fails, where $r < n$. This results in a **censored** life test. Let X_1 denote the time at which the first failure occurred, X_2 denote the time at which the second failure occurred, and so on. Then the total lifetime that has been accumulated at test termination is

$$T_r = \sum_{i=1}^{r} X_i + (n - r)X_r$$

We have previously shown in Exercise 7-66 that T_r/r is an unbiased estimator for μ.

(a) It can be shown that $2\lambda T_r$ has a chi-square distribution with $2r$ degrees of freedom. Use this fact to develop a $100(1 - \alpha)\%$ confidence interval for mean lifetime $\mu = 1/\lambda$.

(b) Suppose 20 units were put on test, and the test terminated after 10 failures occurred. The failure times (in hours) are 15, 18, 19, 20, 21, 21, 22, 27, 28, 29. Find a 95% confidence interval on mean lifetime.

8-96. Consider a two-sided confidence interval for the mean μ when σ is known:

$$\bar{x} - z_{\alpha_1}\sigma/\sqrt{n} \leq \mu \leq \bar{x} + z_{\alpha_2}\sigma/\sqrt{n}$$

where $\alpha_1 + \alpha_2 = \alpha$. If $\alpha_1 = \alpha_2 = \alpha/2$, we have the usual $100(1 - \alpha)\%$ confidence interval for μ. In the above, when $\alpha_1 \neq \alpha_2$, the interval is not symmetric about μ. The length of the interval is $L = \sigma(z_{\alpha_1} + z_{\alpha_2})/\sqrt{n}$. Prove that the length of the interval L is minimized when $\alpha_1 = \alpha_2 = \alpha/2$. Hint: Remember that $\Phi(z_\alpha) = 1 - \alpha$, so $\Phi^{-1}(1 - \alpha) = z_\alpha$, and the relationship between the derivative of a function $y = f(x)$ and the inverse $x = f^{-1}(y)$ is $(d/dy)f^{-1}(y) = 1/[(d/dx)f(x)]$.

8-97. It is possible to construct a **nonparametric tolerance interval** that is based on the extreme values in a random sample of size n from any continuous population. If p is the minimum proportion of the population contained between the smallest and largest sample observations with confidence $1 - \alpha$, it can be shown that

$$np^{n-1} - (n - 1)p^n = \alpha$$

and n is approximately

$$n = \frac{1}{2} + \left(\frac{1 + p}{1 - p}\right)\left(\frac{\chi^2_{\alpha,4}}{4}\right)$$

(a) In order to be 95% confident that at least 90% of the population will be included between the extreme values of the sample, what sample size will be required?

(b) A random sample of 10 transistors gave the following measurements on saturation current (in milliamps): 10.25, 10.41, 10.30, 10.26, 10.19, 10.37, 10.29, 10.34, 10.23, 10.38. Find the limits that contain a proportion p of the saturation current measurements at 95% confidence. What is the proportion p contained by these limits?

8-98. Suppose that $X_1, X_2, \ldots, X_n$ is a random sample from a continuous probability distribution with median $\tilde{\mu}$.

(a) Show that

$$P\{\min(X_i) < \tilde{\mu} < \max(X_i)\}$$
$$= 1 - \left(\frac{1}{2}\right)^{n-1}$$

Hint: The complement of the event $[\min(X_i) < \tilde{\mu} < \max(X_i)]$ is $[\max(X_i) \leq \tilde{\mu}] \cup [\min(X_i) \leq \tilde{\mu}]$, but $\max(X_i) \leq \tilde{\mu}$ if and only if $X_i \leq \tilde{\mu}$ for all i.]

(b) Write down a $100(1 - \alpha)\%$ confidence interval for the median $\tilde{\mu}$, where

$$\alpha = \left(\frac{1}{2}\right)^{n-1}.$$

8-99. Students in the industrial statistics lab at ASU calculate a lot of confidence intervals on μ. Suppose all these CIs are independent of each other. Consider the next one thousand 95% confidence intervals that will be calculated. How many of these CIs do you expect to capture the true value of μ? What is the probability that between 930 and 970 of these intervals contain the true value of μ?

IMPORTANT TERMS AND CONCEPTS

Chi-squared distribution
Confidence coefficient
Confidence interval
Confidence interval for a population proportion
Confidence interval for the variance of a normal distribution
Confidence intervals for the mean of a normal distribution
Confidence level
Error in estimation
Large sample confidence interval
One-sided confidence bounds
Precision of parameter estimation
Prediction interval
Tolerance interval
Two-sided confidence interval
t distribution

Tests of Hypotheses for a Single Sample

CHAPTER OUTLINE

9-1 HYPOTHESIS TESTING
    9-1.1 Statistical Hypotheses
    9-1.2 Tests of Statistical Hypotheses
    9-1.3 One-Sided and Two-Sided Hypotheses
    9-1.4 P-Values in Hypothesis Tests
    9-1.5 Connection between Hypothesis Tests and Confidence Intervals
    9-1.6 General Procedure for Hypothesis Tests

9-2 TESTS ON THE MEAN OF A NORMAL DISTRIBUTION, VARIANCE KNOWN
    9-2.1 Hypothesis Tests on the Mean
    9-2.2 Type II Error and Choice of Sample Size
    9-2.3 Large-Sample Test

9-3 TESTS ON THE MEAN OF A NORMAL DISTRIBUTION, VARIANCE UNKNOWN
    9-3.1 Hypothesis Tests on the Mean
    9-3.2 P-Value for a t-Test
    9-3.3 Type II Error and Choice of Sample Size

9-4 TESTS ON THE VARIANCE AND STANDARD DEVIATION OF A NORMAL DISTRIBUTION
    9-4.1 Hypothesis Tests on the Variance
    9-4.2 Type II Error and Choice of Sample Size

9-5 TESTS ON A POPULATION PROPORTION
    9-5.1 Large-Sample Tests on a Proportion
    9-5.2 Type II Error and Choice of Sample Size

9-6 SUMMARY TABLE OF INFERENCE PROCEDURES FOR A SINGLE SAMPLE

9-7 TESTING FOR GOODNESS OF FIT

9-8 CONTINGENCY TABLE TESTS

LEARNING OBJECTIVES

After careful study of this chapter, you should be able to do the following:
1. Structure engineering decision-making problems as hypothesis tests
2. Test hypotheses on the mean of a normal distribution using either a Z-test or a t-test procedure
3. Test hypotheses on the variance or standard deviation of a normal distribution

4. Test hypotheses on a population proportion
5. Use the P-value approach for making decisions in hypotheses tests
6. Compute power, type II error probability, and make sample size selection decisions for tests on means, variances, and proportions
7. Explain and use the relationship between confidence intervals and hypothesis tests
8. Use the chi-square goodness of fit test to check distributional assumptions
9. Use contingency table tests

9-1 HYPOTHESIS TESTING

9-1.1 Statistical Hypotheses

In the previous chapter we illustrated how to construct a confidence interval estimate of a parameter from sample data. However, many problems in engineering require that we decide whether to accept or reject a statement about some parameter. The statement is called a **hypothesis**, and the decision-making procedure about the hypothesis is called **hypothesis testing**. This is one of the most useful aspects of statistical inference, since many types of decision-making problems, tests, or experiments in the engineering world can be formulated as hypothesis-testing problems. Furthermore, as we will see, there is a very close connection between hypothesis testing and confidence intervals.

Statistical hypothesis testing and confidence interval estimation of parameters are the fundamental methods used at the data analysis stage of a **comparative experiment,** in which the engineer is interested, for example, in comparing the mean of a population to a specified value. These simple comparative experiments are frequently encountered in practice and provide a good foundation for the more complex experimental design problems that we will discuss in Chapters 13 and 14. In this chapter we discuss comparative experiments involving a single population, and our focus is on testing hypotheses concerning the parameters of the population.

We now give a formal definition of a statistical hypothesis.

Statistical Hypothesis

> A statistical hypothesis is a statement about the parameters of one or more populations.

Since we use probability distributions to represent populations, a statistical hypothesis may also be thought of as a statement about the probability distribution of a random variable. The hypothesis will usually involve one or more parameters of this distribution.

For example, suppose that we are interested in the burning rate of a solid propellant used to power aircrew escape systems. Now burning rate is a random variable that can be described by a probability distribution. Suppose that our interest focuses on the mean burning rate (a parameter of this distribution). Specifically, we are interested in deciding whether or not the mean burning rate is 50 centimeters per second. We may express this formally as

$$H_0: \mu = 50 \text{ centimeters per second}$$
$$H_1: \mu \neq 50 \text{ centimeters per second} \tag{9-1}$$

The statement H_0: $\mu = 50$ centimeters per second in Equation 9-1 is called the **null hypothesis**, and the statement H_1: $\mu \neq 50$ centimeters per second is called the **alternative hypothesis**. Since the alternative hypothesis specifies values of μ that could be either greater or less than 50 centimeters per second, it is called a **two-sided alternative hypothesis**. In some situations, we may wish to formulate a **one-sided alternative hypothesis**, as in

$$H_0: \mu = 50 \text{ centimeters per second} \qquad H_0: \mu = 50 \text{ centimeters per second}$$
$$\text{or} \qquad (9\text{-}2)$$
$$H_1: \mu < 50 \text{ centimeters per second} \qquad H_1: \mu > 50 \text{ centimeters per second}$$

It is important to remember that hypotheses are always statements about the population or distribution under study, not statements about the sample. The value of the population parameter specified in the null hypothesis (50 centimeters per second in the above example) is usually determined in one of three ways. First, it may result from past experience or knowledge of the process, or even from previous tests or experiments. The objective of hypothesis testing then is usually to determine whether the parameter value has changed. Second, this value may be determined from some theory or model regarding the process under study. Here the objective of hypothesis testing is to verify the theory or model. A third situation arises when the value of the population parameter results from external considerations, such as design or engineering specifications, or from contractual obligations. In this situation, the usual objective of hypothesis testing is conformance testing.

A procedure leading to a decision about a particular hypothesis is called a **test of a hypothesis**. Hypothesis-testing procedures rely on using the information in a random sample from the population of interest. If this information is consistent with the hypothesis, we will not reject the hypothesis; however, if this information is inconsistent with the hypothesis, we will conclude that the hypothesis is false. We emphasize that the truth or falsity of a particular hypothesis can never be known with certainty, unless we can examine the entire population. This is usually impossible in most practical situations. Therefore, a hypothesis-testing procedure should be developed with the probability of reaching a wrong conclusion in mind. In our treatment of hypothesis testing, the null hypothesis will always be stated so that it specifies an exact value of the parameter (as in the statement H_0: $\mu = 50$ centimeters per second in Equation 9-1). Testing the hypothesis involves taking a random sample, computing a **test statistic** from the sample data, and then using the test statistic to make a decision about the null hypothesis.

9-1.2 Tests of Statistical Hypotheses

To illustrate the general concepts, consider the propellant burning rate problem introduced earlier. The null hypothesis is that the mean burning rate is 50 centimeters per second, and the alternate is that it is not equal to 50 centimeters per second. That is, we wish to test

$$H_0: \mu = 50 \text{ centimeters per second}$$
$$H_1: \mu \neq 50 \text{ centimeters per second}$$

Suppose that a sample of $n = 10$ specimens is tested and that the sample mean burning rate $\bar{x}$ is observed. The sample mean is an estimate of the true population mean μ. A value of the sample mean $\bar{x}$ that falls close to the hypothesized value of $\mu = 50$ centimeters per second does not conflict with the null hypothesis that the true mean μ is really 50 centimeters per second. On the other hand, a sample mean that is considerably different from 50 centimeters per second is evidence in support of the alternative hypothesis H_1. Thus, the sample mean is the test statistic in this case.

The sample mean can take on many different values. Suppose that if $48.5 \leq \bar{x} \leq 51.5$, we will not reject the null hypothesis H_0: $\mu = 50$, and if either $\bar{x} < 48.5$ or $\bar{x} > 51.5$, we will reject the null hypothesis in favor of the alternative hypothesis H_1: $\mu \neq 50$. This is illustrated in Fig. 9-1. The values of $\bar{x}$ that are less than 48.5 and greater than 51.5 constitute the **critical region** for the test, while all values that are in the interval $48.5 \leq \bar{x} \leq 51.5$ form a region for which we will fail to reject the null hypothesis. By convention, this is usually called the **acceptance region**. The boundaries between the critical regions and the acceptance region are called the **critical values**. In our example the critical values are 48.5 and 51.5. It is customary to state conclusions relative to the null hypothesis H_0. Therefore, we reject H_0 in favor of H_1 if the test statistic falls in the critical region and fail to reject H_0 otherwise.

This decision procedure can lead to either of two wrong conclusions. For example, the true mean burning rate of the propellant could be equal to 50 centimeters per second. However, for the randomly selected propellant specimens that are tested, we could observe a value of the test statistic $\bar{x}$ that falls into the critical region. We would then reject the null hypothesis H_0 in favor of the alternate H_1 when, in fact, H_0 is really true. This type of wrong conclusion is called a **type I error**.

Type I Error

> Rejecting the null hypothesis H_0 when it is true is defined as a **type I error**.

Now suppose that the true mean burning rate is different from 50 centimeters per second, yet the sample mean $\bar{x}$ falls in the acceptance region. In this case we would fail to reject H_0 when it is false. This type of wrong conclusion is called a **type II error**.

Type II Error

> Failing to reject the null hypothesis when it is false is defined as a **type II error**.

Thus, in testing any statistical hypothesis, four different situations determine whether the final decision is correct or in error. These situations are presented in Table 9-1.

Because our decision is based on random variables, probabilities can be associated with the type I and type II errors in Table 9-1. The probability of making a type I error is denoted by the Greek letter α.

Probability of Type I Error

> $\alpha = P(\text{type I error}) = P(\text{reject } H_0 \text{ when } H_0 \text{ is true})$ (9-3)

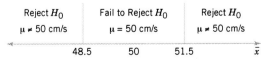

Figure 9-1 Decision criteria for testing H_0: $\mu = 50$ centimeters per second versus H_1: $\mu \neq 50$ centimeters per second.

Table 9-1 Decisions in Hypothesis Testing

Decision	H_0 Is True	H_0 Is False
Fail to reject H_0	no error	type II error
Reject H_0	type I error	no error

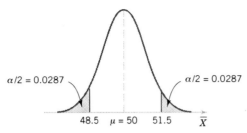

Figure 9-2 The critical region for $H_0: \mu = 50$ versus $H_1: \mu \neq 50$ and $n = 10$.

Sometimes the type I error probability is called the **significance level**, or the **α-error**, or the **size of the test**. In the propellant burning rate example, a type I error will occur when either $\bar{x} > 51.5$ or $\bar{x} < 48.5$ when the true mean burning rate is $\mu = 50$ centimeters per second. Suppose that the standard deviation of burning rate is $\sigma = 2.5$ centimeters per second and that the burning rate has a distribution for which the conditions of the central limit theorem apply, so the distribution of the sample mean is approximately normal with mean $\mu = 50$ and standard deviation $\sigma/\sqrt{n} = 2.5/\sqrt{10} = 0.79$. The probability of making a type I error (or the significance level of our test) is equal to the sum of the areas that have been shaded in the tails of the normal distribution in Fig. 9-2. We may find this probability as

$$\alpha = P(\bar{X} < 48.5 \text{ when } \mu = 50) + P(\bar{X} > 51.5 \text{ when } \mu = 50)$$

The z-values that correspond to the critical values 48.5 and 51.5 are

$$z_1 = \frac{48.5 - 50}{0.79} = -1.90 \quad \text{and} \quad z_2 = \frac{51.5 - 50}{0.79} = 1.90$$

Therefore

$$\alpha = P(Z < -1.90) + P(Z > 1.90) = 0.0287 + 0.0287 = 0.0574$$

This implies that 5.74% of all random samples would lead to rejection of the hypothesis $H_0: \mu = 50$ centimeters per second when the true mean burning rate is really 50 centimeters per second.

From inspection of Fig. 9-2, notice that we can reduce α by widening the acceptance region. For example, if we make the critical values 48 and 52, the value of α is

$$\alpha = P\left(Z < \frac{48 - 50}{0.79}\right) + P\left(Z > \frac{52 - 50}{0.79}\right) = P(Z < -2.53) + P(Z > 2.53)$$
$$= 0.0057 + 0.0057 = 0.0114$$

We could also reduce α by increasing the sample size. If $n = 16$, $\sigma/\sqrt{n} = 2.5/\sqrt{16} = 0.625$, and using the original critical region from Fig. 9-1, we find

$$z_1 = \frac{48.5 - 50}{0.625} = -2.40 \quad \text{and} \quad z_2 = \frac{51.5 - 50}{0.625} = 2.40$$

Therefore

$$\alpha = P(Z < -2.40) + P(Z > 2.40) = 0.0082 + 0.0082 = 0.0164$$

In evaluating a hypothesis-testing procedure, it is also important to examine the probability of a type II error, which we will denote by β. That is,

Probability of Type II Error

$$\beta = P(\text{type II error}) = P(\text{fail to reject } H_0 \text{ when } H_0 \text{ is false}) \quad (9\text{-}4)$$

To calculate β (sometimes called the β-error), we must have a specific alternative hypothesis; that is, we must have a particular value of μ. For example, suppose that it is important to reject the null hypothesis H_0: $\mu = 50$ whenever the mean burning rate μ is greater than 52 centimeters per second or less than 48 centimeters per second. We could calculate the probability of a type II error β for the values $\mu = 52$ and $\mu = 48$ and use this result to tell us something about how the test procedure would perform. Specifically, how will the test procedure work if we wish to detect, that is, reject H_0, for a mean value of $\mu = 52$ or $\mu = 48$? Because of symmetry, it is necessary only to evaluate one of the two cases—say, find the probability of accepting the null hypothesis H_0: $\mu = 50$ centimeters per second when the true mean is $\mu = 52$ centimeters per second.

Figure 9-3 will help us calculate the probability of type II error β. The normal distribution on the left in Fig. 9-3 is the distribution of the test statistic $\overline{X}$ when the null hypothesis H_0: $\mu = 50$ is true (this is what is meant by the expression "under H_0: $\mu = 50$"), and the normal distribution on the right is the distribution of $\overline{X}$ when the alternative hypothesis is true and the value of the mean is 52 (or "under H_1: $\mu = 52$"). Now a type II error will be committed if the sample mean $\overline{X}$ falls between 48.5 and 51.5 (the critical region boundaries) when $\mu = 52$. As seen in Fig. 9-3, this is just the probability that $48.5 \leq \overline{X} \leq 51.5$ when the true mean is $\mu = 52$, or the shaded area under the normal distribution centered at $\mu = 52$. Therefore, referring to Fig. 9-3, we find that

$$\beta = P(48.5 \leq \overline{X} \leq 51.5 \text{ when } \mu = 52)$$

The z-values corresponding to 48.5 and 51.5 when $\mu = 52$ are

$$z_1 = \frac{48.5 - 52}{0.79} = -4.43 \quad \text{and} \quad z_2 = \frac{51.5 - 52}{0.79} = -0.63$$

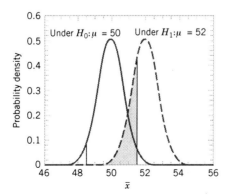

Figure 9-3 The probability of type II error when $\mu = 52$ and $n = 10$.

Therefore

$$\beta = P(-4.43 \leq Z \leq -0.63) = P(Z \leq -0.63) - P(Z \leq -4.43)$$
$$= 0.2643 - 0.0000 = 0.2643$$

Thus, if we are testing H_0: $\mu = 50$ against H_1: $\mu \neq 50$ with $n = 10$, and the true value of the mean is $\mu = 52$, the probability that we will fail to reject the false null hypothesis is 0.2643. By symmetry, if the true value of the mean is $\mu = 48$, the value of β will also be 0.2643.

The probability of making a type II error β increases rapidly as the true value of μ approaches the hypothesized value. For example, see Fig. 9-4, where the true value of the mean is $\mu = 50.5$ and the hypothesized value is H_0: $\mu = 50$. The true value of μ is very close to 50, and the value for β is

$$\beta = P(48.5 \leq \overline{X} \leq 51.5 \text{ when } \mu = 50.5)$$

As shown in Fig. 9-4, the z-values corresponding to 48.5 and 51.5 when $\mu = 50.5$ are

$$z_1 = \frac{48.5 - 50.5}{0.79} = -2.53 \quad \text{and} \quad z_2 = \frac{51.5 - 50.5}{0.79} = 1.27$$

Therefore

$$\beta = P(-2.53 \leq Z \leq 1.27) = P(Z \leq 1.27) - P(Z \leq -2.53)$$
$$= 0.8980 - 0.0057 = 0.8923$$

Thus, the type II error probability is much higher for the case where the true mean is 50.5 centimeters per second than for the case where the mean is 52 centimeters per second. Of course, in many practical situations we would not be as concerned with making a type II error if the mean were "close" to the hypothesized value. We would be much more interested in detecting large differences between the true mean and the value specified in the null hypothesis.

The type II error probability also depends on the sample size n. Suppose that the null hypothesis is H_0: $\mu = 50$ centimeters per second and that the true value of the mean is $\mu = 52$. If the sample size is increased from $n = 10$ to $n = 16$, the situation of Fig. 9-5 results.

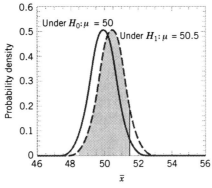

Figure 9-4 The probability of type II error when $\mu = 50.5$ and $n = 10$.

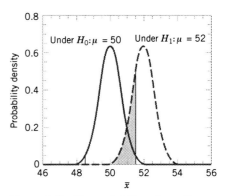

Figure 9-5 The probability of type II error when $\mu = 52$ and $n = 16$.

The normal distribution on the left is the distribution of $\overline{X}$ when the mean $\mu = 50$, and the normal distribution on the right is the distribution of $\overline{X}$ when $\mu = 52$. As shown in Fig. 9-5, the type II error probability is

$$\beta = P(48.5 \leq \overline{X} \leq 51.5 \text{ when } \mu = 52)$$

When $n = 16$, the standard deviation of $\overline{X}$ is $\sigma/\sqrt{n} = 2.5/\sqrt{16} = 0.625$, and the z-values corresponding to 48.5 and 51.5 when $\mu = 52$ are

$$z_1 = \frac{48.5 - 52}{0.625} = -5.60 \quad \text{and} \quad z_2 = \frac{51.5 - 52}{0.625} = -0.80$$

Therefore

$$\beta = P(-5.60 \leq Z \leq -0.80) = P(Z \leq -0.80) - P(Z \leq -5.60)$$
$$= 0.2119 - 0.0000 = 0.2119$$

Recall that when $n = 10$ and $\mu = 52$, we found that $\beta = 0.2643$; therefore, increasing the sample size results in a decrease in the probability of type II error.

The results from this section and a few other similar calculations are summarized in the following table. The critical values are adjusted to maintain equal α for $n = 10$ and $n = 16$. This type of calculation is discussed later in the chapter.

Acceptance Region	Sample Size	α	β at $\mu = 52$	β at $\mu = 50.5$
$48.5 < \bar{x} < 51.5$	10	0.0576	0.2643	0.8923
$48 < \bar{x} < 52$	10	0.0114	0.5000	0.9705
$48.81 < \bar{x} < 51.19$	16	0.0576	0.0966	0.8606
$48.42 < \bar{x} < 51.58$	16	0.0114	0.2515	0.9578

The results in boxes were not calculated in the text but can easily be verified by the reader. This display and the discussion above reveal four important points:

1. The size of the critical region, and consequently the probability of a type I error α, can always be reduced by appropriate selection of the critical values.

2. Type I and type II errors are related. A decrease in the probability of one type of error always results in an increase in the probability of the other, provided that the sample size n does not change.

3. An increase in sample size reduces β, provided that α is held constant.

4. When the null hypothesis is false, β increases as the true value of the parameter approaches the value hypothesized in the null hypothesis. The value of β decreases as the difference between the true mean and the hypothesized value increases.

Generally, the analyst controls the type I error probability α when he or she selects the critical values. Thus, it is usually easy for the analyst to set the type I error probability at (or near) any desired value. Since the analyst can directly control the probability of wrongly rejecting H_0, we always think of rejection of the null hypothesis H_0 as a **strong conclusion.**

On the other hand, the probability of type II error β is not a constant, but depends on the true value of the parameter. It also depends on the sample size that we have selected. Because the type II error probability β is a function of both the sample size and the extent to which the null hypothesis H_0 is false, it is customary to think of the decision to accept H_0 as a **weak conclusion**, unless we know that β is acceptably small. Therefore, rather than saying we "accept H_0", we prefer the terminology "fail to reject H_0". Failing to reject H_0 implies that we have not found sufficient evidence to reject H_0, that is, to make a strong statement. Failing to reject H_0 does not necessarily mean that there is a high probability that H_0 is true. It may simply mean that more data are required to reach a strong conclusion. This can have important implications for the formulation of hypotheses.

An important concept that we will make use of is the power of a statistical test.

Power

> The **power** of a statistical test is the probability of rejecting the null hypothesis H_0 when the alternative hypothesis is true.

The power is computed as $1 - \beta$, and power can be interpreted as **the probability of correctly rejecting a false null hypothesis**. We often compare statistical tests by comparing their power properties. For example, consider the propellant burning rate problem when we are testing H_0: $\mu = 50$ centimeters per second against H_1: $\mu \neq 50$ centimeters per second. Suppose that the true value of the mean is $\mu = 52$. When $n = 10$, we found that $\beta = 0.2643$, so the power of this test is $1 - \beta = 1 - 0.2643 = 0.7357$ when $\mu = 52$.

Power is a very descriptive and concise measure of the **sensitivity** of a statistical test, where by sensitivity we mean the ability of the test to detect differences. In this case, the sensitivity of the test for detecting the difference between a mean burning rate of 50 centimeters per second and 52 centimeters per second is 0.7357. That is, if the true mean is really 52 centimeters per second, this test will correctly reject H_0: $\mu = 50$ and "detect" this difference 73.57% of the time. If this value of power is judged to be too low, the analyst can increase either α or the sample size n.

9-1.3 One-Sided and Two-Sided Hypotheses

In constructing hypotheses, we will always state the null hypothesis as an equality so that the probability of type I error α can be controlled at a specific value. The alternative hypothesis might be either one-sided or two-sided, depending on the conclusion to be drawn if H_0 is rejected. If the objective is to make a claim involving statements such as greater than, less than, superior to, exceeds, at least, and so forth, a one-sided alternative is appropriate. If no direction is implied by the claim, or if the claim not equal to is to be made, a two-sided alternative should be used.

EXAMPLE 9-1
Propellant
Burning Rate

Consider the propellant burning rate problem. Suppose that if the burning rate is less than 50 centimeters per second, we wish to show this with a strong conclusion. The hypotheses should be stated as

H_0: $\mu = 50$ centimeters per second
H_1: $\mu < 50$ centimeters per second

Here the critical region lies in the lower tail of the distribution of $\overline{X}$. Since the rejection of H_0 is always a strong conclusion, this statement of the hypotheses will produce the desired outcome if H_0 is rejected. Notice that, although the null hypothesis is stated with an equal sign, it is understood to include any value of μ not specified by the alternative hypothesis. Therefore, failing to reject H_0 does not mean that $\mu = 50$ centimeters per second exactly, but only that we do not have strong evidence in support of H_1.

In some real-world problems where one-sided test procedures are indicated, it is occasionally difficult to choose an appropriate formulation of the alternative hypothesis. For example, suppose that a soft-drink beverage bottler purchases 10-ounce bottles from a glass company. The bottler wants to be sure that the bottles meet the specification on mean internal pressure or bursting strength, which for 10-ounce bottles is a minimum strength of 200 psi. The bottler has decided to formulate the decision procedure for a specific lot of bottles as a hypothesis testing problem. There are two possible formulations for this problem, either

$$H_0: \mu = 200 \text{ psi}$$
$$H_1: \mu > 200 \text{ psi} \tag{9-5}$$

or

$$H_0: \mu = 200 \text{ psi}$$
$$H_1: \mu < 200 \text{ psi} \tag{9-6}$$

Consider the formulation in Equation 9-5. If the null hypothesis is rejected, the bottles will be judged satisfactory; if H_0 is not rejected, the implication is that the bottles do not conform to specifications and should not be used. Because rejecting H_0 is a strong conclusion, this formulation forces the bottle manufacturer to "demonstrate" that the mean bursting strength of the bottles exceeds the specification. Now consider the formulation in Equation 9-6. In this situation, the bottles will be judged satisfactory unless H_0 is rejected. That is, we conclude that the bottles are satisfactory unless there is strong evidence to the contrary.

Which formulation is correct, the one of Equation 9-5 or Equation 9-6? The answer is it depends on the objective of the analysis. For Equation 9-5, there is some probability that H_0 will not be rejected (i.e., we would decide that the bottles are not satisfactory), even though the true mean is slightly greater than 200 psi. This formulation implies that we want the bottle manufacturer to demonstrate that the product meets or exceeds our specifications. Such a formulation could be appropriate if the manufacturer has experienced difficulty in meeting specifications in the past or if product safety considerations force us to hold tightly to the 200 psi specification. On the other hand, for the formulation of Equation 9-6 there is some probability that H_0 will be accepted and the bottles judged satisfactory, even though the true mean is slightly less than 200 psi. We would conclude that the bottles are unsatisfactory only when there is strong evidence that the mean does not exceed 200 psi, that is, when $H_0: \mu = 200$ psi is rejected. This formulation assumes that we are relatively happy with the bottle manufacturer's past performance and that small deviations from the specification of $\mu \geq 200$ psi are not harmful.

In formulating one-sided alternative hypotheses, we should remember that rejecting H_0 is always a strong conclusion. Consequently, we should put the statement about which it is important to make a strong conclusion in the alternative hypothesis. In real-world problems, this will often depend on our point of view and experience with the situation.

9-1.4 P-Values in Hypothesis Tests

One way to report the results of a hypothesis test is to state that the null hypothesis was or was not rejected at a specified α-value or level of significance. For example, in the propellant problem above, we can say that H_0: μ = 50 was rejected at the 0.05 level of significance. This statement of conclusions is often inadequate because it gives the decision maker no idea about whether the computed value of the test statistic was just barely in the rejection region or whether it was very far into this region. Furthermore, stating the results this way imposes the predefined level of significance on other users of the information. This approach may be unsatisfactory because some decision makers might be uncomfortable with the risks implied by $\alpha = 0.05$.

To avoid these difficulties the *P-value* approach has been adopted widely in practice. The P-value is the probability that the test statistic will take on a value that is at least as extreme as the observed value of the statistic when the null hypothesis H_0 is true. Thus, a P-value conveys much information about the weight of evidence against H_0, and so a decision maker can draw a conclusion at *any* specified level of significance. We now give a formal definition of a P-value.

P-Value

> The *P*-value is the smallest level of significance that would lead to rejection of the null hypothesis H_0 with the given data.

It is customary to call the test statistic (and the data) significant when the null hypothesis H_0 is rejected; therefore, we may think of the P-value as the smallest level α at which the data are significant. Once the P-value is known, the decision maker can determine how significant the data are without the data analyst formally imposing a preselected level of significance.

Consider the two-sided hypothesis test for burning rate

$$H_0 : \mu = 50 \quad H_1 : \mu \neq 50$$

with $n = 16$ and $\sigma = 2.5$. Suppose that the observed sample mean is $\bar{x} = 51.3$ centimeters per second. Figure 9-6 shows a critical region for this test with critical values at 51.3 and the symmetric value 48.7. The P-value of the test is the α associated with this critical region. Any smaller value for α decreases the critical region and the test fails to reject the

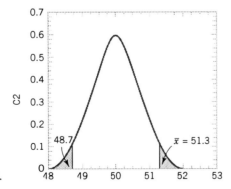

Figure 9-6 *P*-value is area of shaded region when $\bar{x} = 51.3$.

null hypothesis when $\bar{x} = 51.3$. The *P*-value is easy to compute after the test statistic is observed. In this example

$$\begin{aligned}
P\text{-value} &= 1 - P(48.7 < \bar{X} < 51.3) \\
&= 1 - P\left(\frac{48.7 - 50}{2.5/\sqrt{16}} < Z < \frac{51.3 - 50}{2.5/\sqrt{16}}\right) \\
&= 1 - P(-2.08 < Z < 2.08) \\
&= 1 - 0.962 = 0.038
\end{aligned}$$

This null hypothesis would be rejected for any $0.038 < \alpha$, but not rejected for a smaller significance level such as 0.01. *P*-values are often calculated in this manner. The observed value of the test statistic is used to define a critical region, and the α associated with this critical region is computed.

It is not always easy to compute the exact *P*-value for a test. However, most modern computer programs for statistical analysis report *P*-values. We will also show how to approximate the *P*-value.

9-1.5 Connection between Hypothesis Tests and Confidence Intervals

There is a close relationship between the test of a hypothesis about any parameter, say θ, and the confidence interval for θ. If $[l, u]$ is a $100(1 - \alpha)\%$ confidence interval for the parameter θ, the test of size α of the hypothesis

$$H_0: \theta = \theta_0$$
$$H_1: \theta \neq \theta_0$$

will lead to rejection of H_0 if and only if θ_0 is **not** in the $100(1 - \alpha)\%$ CI $[l, u]$. As an illustration, consider the escape system propellant problem with $\bar{x} = 51.3$, $\sigma = 2.5$, and $n = 16$. The null hypothesis $H_0: \mu = 50$ was rejected, using $\alpha = 0.05$. The 95% two-sided CI on μ can be calculated using Equation 8-7. This CI is $51.3 \pm 1.96(2.5/\sqrt{16})$ and this is $50.075 \leq \mu \leq 52.525$. Because the value $\mu_0 = 50$ is not included in this interval, the null hypothesis $H_0: \mu = 50$ is rejected.

Although hypothesis tests and CIs are equivalent procedures insofar as decision making or **inference** about μ is concerned, each provides somewhat different insights. For instance, the confidence interval provides a range of likely values for μ at a stated confidence level, whereas hypothesis testing is an easy framework for displaying the **risk levels** such as the *P*-value associated with a specific decision. We will continue to illustrate the connection between the two procedures throughout the text.

9-1.6 General Procedure for Hypothesis Tests

This chapter develops hypothesis-testing procedures for many practical problems. Use of the following sequence of steps in applying hypothesis-testing methodology is recommended.

1. From the problem context, identify the parameter of interest.
2. State the null hypothesis, H_0.
3. Specify an appropriate alternative hypothesis, H_1.

4. Choose a significance level α.
5. Determine an appropriate test statistic.
6. State the rejection region for the statistic. This step is not necessary if the P-value approach is used.
7. Compute any necessary sample quantities, substitute these into the equation for the test statistic, and compute that value.
8. Decide whether or not H_0 should be rejected and report that in the problem context.

Steps 1–4 should be completed prior to examination of the sample data. This sequence of steps will be illustrated in subsequent sections.

In practice, such a formal and (seemingly) rigid procedure is not always necessary. Generally, once the experimenter (or decision maker) has decided on the question of interest and has determined the *design of the experiment* (that is, how the data are to be collected, how the measurements are to be made, and how many observations are required), only three steps are really required:

1. Specify the test statistic to be used (such as Z_0).
2. Specify the location of the critical region (two-tailed, upper-tailed, or lower-tailed).
3. Specify the criteria for rejection (typically, the value of α, or the P-value at which rejection should occur).

These steps are often completed almost simultaneously in solving real-world problems, although we emphasize that it is important to think carefully about each step. That is why we present and use the eight-step process: it seems to reinforce the essentials of the correct approach. While you may not use it every time in solving real problems, it is a helpful framework when you are first learning about hypothesis testing.

Statistical versus Practical Significance
We noted previously that reporting the results of a hypothesis test in terms of a P-value is very useful because it conveys more information than just the simple statement "reject H_0" or "fail to reject H_0". That is, rejection of H_0 at the 0.05 level of significance is much more meaningful if the value of the test statistic is well into the critical region, greatly exceeding the 5% critical value, than if it barely exceeds that value.

Even a very small P-value can be difficult to interpret from a practical viewpoint when we are making decisions because, while a small P-value indicates statistical significance in the sense that H_0 should be rejected in favor of H_1, the actual departure from H_0 that has been detected may have little (if any) practical significance (engineers like to say "engineering significance"). This is particularly true when the sample size n is large.

For example, consider the propellant burning rate problem of Example 9-1 where we are testing H_0: $\mu = 50$ centimeters per second versus H_1: $\mu \neq 50$ centimeters per second with $\sigma = 2.5$. If we suppose that the mean rate is really 50.5 centimeters per second, this is not a serious departure from H_0: $\mu = 50$ centimeters per second in the sense that if the mean really is 50.5 centimeters per second there is no practical observable effect on the performance of the aircrew escape system. In other words, concluding that $\mu = 50$ centimeters per second when it is really 50.5 centimeters per second is an inexpensive error and has no practical significance. For a reasonably large sample size, a true value of $\mu = 50.5$ will lead to a sample $\bar{x}$ that is close to 50.5 centimeters per second, and we would not want this value of $\bar{x}$ from the sample to result in rejection of H_0. The following display shows the P-value for testing H_0: $\mu = 50$ when we observe $\bar{x} = 50.5$ centimeters per second and the power of the test at $\alpha = 0.05$ when the true mean is 50.5 for various sample sizes n:

Sample Size n	P-value When $\bar{x} = 50.5$	Power (at $\alpha = 0.05$) When True $\mu = 50.5$
10	0.527	0.097
25	0.317	0.170
50	0.157	0.293
100	0.046	0.516
400	6.3×10^{-5}	0.979
1000	2.5×10^{-10}	1.000

The P-value column in this display indicates that for large sample sizes the observed sample value of $\bar{x} = 50.5$ would strongly suggest that $H_0: \mu = 50$ should be rejected, even though the observed sample results imply that from a practical viewpoint the true mean does not differ much at all from the hypothesized value $\mu_0 = 50$. The power column indicates that if we test a hypothesis at a fixed significance level α and even if there is little practical difference between the true mean and the hypothesized value, a large sample size will almost always lead to rejection of H_0. The moral of this demonstration is clear:

> Be careful when interpreting the results from hypothesis testing when the sample size is large, because any small departure from the hypothesized value μ_0 will probably be detected, even when the difference is of little or no practical significance.

EXERCISES FOR SECTION 9-1

9-1. In each of the following situations, state whether it is a correctly stated hypothesis testing problem and why.
(a) $H_0: \mu = 25, H_1: \mu \neq 25$
(b) $H_0: \sigma > 10, H_1: \sigma = 10$
(c) $H_0: \bar{x} = 50, H_1: \bar{x} \neq 50$
(d) $H_0: p = 0.1, H_1: p = 0.5$
(e) $H_0: s = 30, H_1: s > 30$

9-2. A semiconductor manufacturer collects data from a new tool and conducts a hypothesis test with the null hypothesis that a critical dimension mean width equals 100 nm. The conclusion is to not reject the null hypothesis. Does this result provide strong evidence that the critical dimension mean equals 100 nm? Explain.

9-3. The standard deviation of critical dimension thickness in semiconductor manufacturing is $\sigma = 20$ nm.
(a) State the null and alternative hypotheses used to demonstrate that the standard deviation is reduced.
(b) Assume that the previous test does not reject the null hypothesis. Does this result provide strong evidence that the standard deviation has not been reduced? Explain.

9-4. The mean pull-off force of a connector depends on cure time.
(a) State the null and alternative hypotheses used to demonstrate that the pull-off force is below 25 newtons.
(b) Assume that the previous test does not reject the null hypothesis. Does this result provide strong evidence that the pull-off force is greater than or equal to 25 newtons? Explain.

9-5. A textile fiber manufacturer is investigating a new drapery yarn, which the company claims has a mean thread elongation of 12 kilograms with a standard deviation of 0.5 kilograms. The company wishes to test the hypothesis $H_0: \mu = 12$ against $H_1: \mu < 12$, using a random sample of four specimens.
(a) What is the type I error probability if the critical region is defined as $\bar{x} < 11.5$ kilograms?

(b) Find β for the case where the true mean elongation is 11.25 kilograms.
(c) Find β for the case where the true mean is 11.5 kilograms.

9-6. Repeat Exercise 9-5 using a sample size of $n = 16$ and the same critical region.

9-7. In Exercise 9-5, find the boundary of the critical region if the type I error probability is
(a) $\alpha = 0.01$ and $n = 4$ (c) $\alpha = 0.01$ and $n = 16$
(b) $\alpha = 0.05$ and $n = 4$ (d) $\alpha = 0.05$ and $n = 16$

9-8. In Exercise 9-5, calculate the probability of a type II error if the true mean elongation is 11.5 kilograms and
(a) $\alpha = 0.05$ and $n = 4$
(b) $\alpha = 0.05$ and $n = 16$
(c) Compare the values of β calculated in the previous parts. What conclusion can you draw?

9-9. In Exercise 9-5, calculate the P-value if the observed statistic is
(a) $\bar{x} = 11.25$ (b) $\bar{x} = 11.0$ (c) $\bar{x} = 11.75$

9-10. The heat evolved in calories per gram of a cement mixture is approximately normally distributed. The mean is thought to be 100 and the standard deviation is 2. We wish to test $H_0: \mu = 100$ versus $H_1: \mu \neq 100$ with a sample of $n = 9$ specimens.
(a) If the acceptance region is defined as $98.5 \leq \bar{x} \leq 101.5$, find the type I error probability α.
(b) Find β for the case where the true mean heat evolved is 103.
(c) Find β for the case where the true mean heat evolved is 105. This value of β is smaller than the one found in part (b) above. Why?

9-11. Repeat Exercise 9-10 using a sample size of $n = 5$ and the same acceptance region.

9-12. In Exercise 9-10, find the boundary of the critical region if the type I error probability is
(a) $\alpha = 0.01$ and $n = 9$ (c) $\alpha = 0.01$ and $n = 5$
(b) $\alpha = 0.05$ and $n = 9$ (d) $\alpha = 0.05$ and $n = 5$

9-13. In Exercise 9-10, calculate the probability of a type II error if the true mean heat evolved is 103 and
(a) $\alpha = 0.05$ and $n = 9$
(b) $\alpha = 0.05$ and $n = 5$
(c) Compare the values of β calculated in the previous parts. What conclusion can you draw?

9-14. In Exercise 9-10, calculate the P-value if the observed statistic is
(a) $\bar{x} = 98$ (b) $\bar{x} = 101$ (c) $\bar{x} = 102$

9-15. A consumer products company is formulating a new shampoo and is interested in foam height (in millimeters). Foam height is approximately normally distributed and has a standard deviation of 20 millimeters. The company wishes to test $H_0: \mu = 175$ millimeters versus $H_1: \mu > 175$ millimeters, using the results of $n = 10$ samples.
(a) Find the type I error probability α if the critical region is $\bar{x} > 185$.
(b) What is the probability of type II error if the true mean foam height is 185 millimeters?
(c) Find β for the true mean of 195 millimeters.

9-16. Repeat Exercise 9-15 assuming that the sample size is $n = 16$ and the boundary of the critical region is the same.

9-17. In Exercise 9-15, find the boundary of the critical region if the type I error probability is
(a) $\alpha = 0.01$ and $n = 10$ (c) $\alpha = 0.01$ and $n = 16$
(b) $\alpha = 0.05$ and $n = 10$ (d) $\alpha = 0.05$ and $n = 16$

9-18. In Exercise 9-15, calculate the probability of a type II error if the true mean foam height is 185 millimeters and
(a) $\alpha = 0.05$ and $n = 10$
(b) $\alpha = 0.05$ and $n = 16$
(c) Compare the values of β calculated in the previous parts. What conclusion can you draw?

9-19. In Exercise 9-15, calculate the P-value if the observed statistic is
(a) $\bar{x} = 180$ (b) $\bar{x} = 190$ (c) $\bar{x} = 170$

9-20. A manufacturer is interested in the output voltage of a power supply used in a PC. Output voltage is assumed to be normally distributed, with standard deviation 0.25 Volts, and the manufacturer wishes to test $H_0: \mu = 5$ Volts against $H_1: \mu \neq 5$ Volts, using $n = 8$ units.
(a) The acceptance region is $4.85 \leq \bar{x} \leq 5.15$. Find the value of α.
(b) Find the power of the test for detecting a true mean output voltage of 5.1 Volts.

9-21. Rework Exercise 9-20 when the sample size is 16 and the boundaries of the acceptance region do not change. What impact does the change in sample size have on the results of parts (a) and (b)?

9-22. In Exercise 9-20, find the boundary of the critical region if the type I error probability is
(a) $\alpha = 0.01$ and $n = 8$ (c) $\alpha = 0.01$ and $n = 16$
(b) $\alpha = 0.05$ and $n = 8$ (d) $\alpha = 0.05$ and $n = 16$

9-23. In Exercise 9-20, calculate the P-value if the observed statistic is
(a) $\bar{x} = 5.2$ (b) $\bar{x} = 4.7$ (c) $\bar{x} = 5.1$

9-24. In Exercise 9-20, calculate the probability of a type II error if the true mean output is 5.05 volts and
(a) $\alpha = 0.05$ and $n = 10$
(b) $\alpha = 0.05$ and $n = 16$
(c) Compare the values of β calculated in the previous parts. What conclusion can you draw?

9-25. The proportion of adults living in Tempe, Arizona, who are college graduates is estimated to be $p = 0.4$. To test this hypothesis, a random sample of 15 Tempe adults is selected. If the number of college graduates is between 4 and 8, the hypothesis will be accepted; otherwise, we will conclude that $p \neq 0.4$.
(a) Find the type I error probability for this procedure, assuming that $p = 0.4$.

(b) Find the probability of committing a type II error if the true proportion is really $p = 0.2$.

9-26. The proportion of residents in Phoenix favoring the building of toll roads to complete the freeway system is believed to be $p = 0.3$. If a random sample of 10 residents shows that 1 or fewer favor this proposal, we will conclude that $p < 0.3$.
(a) Find the probability of type I error if the true proportion is $p = 0.3$.
(b) Find the probability of committing a type II error with this procedure if $p = 0.2$.
(c) What is the power of this procedure if the true proportion is $p = 0.2$?

9-27. A random sample of 500 registered voters in Phoenix is asked if they favor the use of oxygenated fuels year-round to reduce air pollution. If more than 400 voters respond positively, we will conclude that at least 60% of the voters favor the use of these fuels.

(a) Find the probability of type I error if exactly 60% of the voters favor the use of these fuels.
(b) What is the type II error probability β if 75% of the voters favor this action?
Hint: use the normal approximation to the binomial.

9-28. If we plot the probability of accepting $H_0: \mu = \mu_0$ versus various values of μ and connect the points with a smooth curve, we obtain the **operating characteristic curve** (or the **OC curve**) of the test procedure. These curves are used extensively in industrial applications of hypothesis testing to display the sensitivity and relative performance of the test. When the true mean is really equal to μ_0, the probability of accepting H_0 is $1 - \alpha$.
(a) Construct an OC curve for Exercise 9-15, using values of the true mean μ of 178, 181, 184, 187, 190, 193, 196, and 199.
(b) Convert the OC curve into a plot of the **power function** of the test.

9-2 TESTS ON THE MEAN OF A NORMAL DISTRIBUTION, VARIANCE KNOWN

In this section, we consider hypothesis testing about the mean μ of a single normal population where the variance of the population σ^2 is known. We will assume that a random sample $X_1, X_2, \ldots, X_n$ has been taken from the population. Based on our previous discussion, the sample mean $\overline{X}$ is an **unbiased point estimator** of μ with variance σ^2/n.

9-2.1 Hypothesis Tests on the Mean

Suppose that we wish to test the hypotheses

$$H_0: \mu = \mu_0$$
$$H_1: \mu \neq \mu_0 \tag{9-7}$$

where μ_0 is a specified constant. We have a random sample $X_1, X_2, \ldots, X_n$ from a normal population. Since $\overline{X}$ has a normal distribution (i.e., the sampling distribution of $\overline{X}$ is normal) with mean μ_0 and standard deviation $\sigma/\sqrt{n}$ if the null hypothesis is true, we could construct a critical region based on the computed value of the sample mean $\overline{X}$, as in Section 9-1.2.

It is usually more convenient to *standardize* the sample mean and use a test statistic based on the standard normal distribution. That is, the test procedure for $H_0: \mu = \mu_0$ uses the test statistic

Test Statistic

$$Z_0 = \frac{\overline{X} - \mu_0}{\sigma/\sqrt{n}} \tag{9-8}$$

If the null hypothesis $H_0: \mu = \mu_0$ is true, $E(\overline{X}) = \mu_0$, and it follows that the distribution of Z_0 is the standard normal distribution [denoted $N(0, 1)$]. Consequently, if $H_0: \mu = \mu_0$ is true, the probability is $1 - \alpha$ that the test statistic Z_0 falls between $-z_{\alpha/2}$ and $z_{\alpha/2}$, where $z_{\alpha/2}$ is the

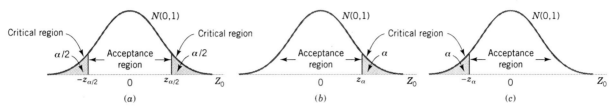

Figure 9-7 The distribution of Z_0 when $H_0: \mu = \mu_0$ is true, with critical region for (a) the two-sided alternative $H_1: \mu \neq \mu_0$, (b) the one-sided alternative $H_1: \mu > \mu_0$, and (c) the one-sided alternative $H_1: \mu < \mu_0$.

$100\alpha/2$ percentage point of the standard normal distribution. The regions associated with $z_{\alpha/2}$ and $-z_{\alpha/2}$ are illustrated in Fig. 9-7(a). Note that the probability is α that the test statistic Z_0 will fall in the region $Z_0 > z_{\alpha/2}$ or $Z_0 < -z_{\alpha/2}$ when $H_0: \mu = \mu_0$ is true. Clearly, a sample producing a value of the test statistic that falls in the tails of the distribution of Z_0 would be unusual if $H_0: \mu = \mu_0$ is true; therefore, it is an indication that H_0 is false. Thus, we should reject H_0 if the observed value of the test statistic z_0 is either

$$z_0 > z_{\alpha/2} \quad \text{or} \quad z_0 < -z_{\alpha/2} \tag{9-9}$$

and we should fail to reject H_0 if

$$-z_{\alpha/2} \leq z_0 \leq z_{\alpha/2} \tag{9-10}$$

The inequalities in Equation 9-10 define the acceptance region for H_0, and the two inequalities in Equation 9-9 define the critical region or rejection region. The type I error probability for this test procedure is α.

It is easier to understand the critical region and the test procedure, in general, when the test statistic is Z_0 rather than $\overline{X}$. However, the same critical region can always be written in terms of the computed value of the sample mean $\bar{x}$. A procedure identical to the above is as follows:

Reject $H_0: \mu = \mu_0$ if either $\bar{x} > a$ or $\bar{x} < b$

where

$$a = \mu_0 + z_{\alpha/2}\sigma/\sqrt{n} \quad \text{and} \quad b = \mu_0 - z_{\alpha/2}\sigma/\sqrt{n}$$

EXAMPLE 9-2
Propellant
Burning Rate

Aircrew escape systems are powered by a solid propellant. The burning rate of this propellant is an important product characteristic. Specifications require that the mean burning rate must be 50 centimeters per second. We know that the standard deviation of burning rate is $\sigma = 2$ centimeters per second. The experimenter decides to specify a type I error probability or significance level of $\alpha = 0.05$ and selects a random sample of $n = 25$ and obtains a sample average burning rate of $\bar{x} = 51.3$ centimeters per second. What conclusions should be drawn?

We may solve this problem by following the eight-step procedure outlined in Section 9-1.6. This results in

1. The parameter of interest is μ, the mean burning rate.
2. $H_0: \mu = 50$ centimeters per second
3. $H_1: \mu \neq 50$ centimeters per second
4. $\alpha = 0.05$

5. The test statistic is

$$z_0 = \frac{\bar{x} - \mu_0}{\sigma/\sqrt{n}}$$

6. Reject H_0 if $z_0 > 1.96$ or if $z_0 < -1.96$. Note that this results from step 4, where we specified $\alpha = 0.05$, and so the boundaries of the critical region are at $z_{0.025} = 1.96$ and $-z_{0.025} = -1.96$.

7. Computations: Since $\bar{x} = 51.3$ and $\sigma = 2$,

$$z_0 = \frac{51.3 - 50}{2/\sqrt{25}} = 3.25$$

8. Conclusion: Since $z_0 = 3.25 > 1.96$, we reject $H_0: \mu = 50$ at the 0.05 level of significance. Stated more completely, we conclude that the mean burning rate differs from 50 centimeters per second, based on a sample of 25 measurements. In fact, there is strong evidence that the mean burning rate exceeds 50 centimeters per second.

We may also develop procedures for testing hypotheses on the mean μ where the alternative hypothesis is one-sided. Suppose that we specify the hypotheses as

$$H_0: \mu = \mu_0$$
$$H_1: \mu > \mu_0 \tag{9-11}$$

In defining the critical region for this test, we observe that a negative value of the test statistic Z_0 would never lead us to conclude that $H_0: \mu = \mu_0$ is false. Therefore, we would place the critical region in the **upper tail** of the standard normal distribution and reject H_0 if the computed value of z_0 is too large. That is, we would reject H_0 if

$$z_0 > z_\alpha \tag{9-12}$$

as shown in Figure 9-7(b). Similarly, to test

$$H_0: \mu = \mu_0$$
$$H_1: \mu < \mu_0 \tag{9-13}$$

we would calculate the test statistic Z_0 and reject H_0 if the value of z_0 is too small. That is, the critical region is in the **lower tail** of the standard normal distribution as shown in Figure 9-7(c), and we reject H_0 if

$$z_0 < -z_\alpha \tag{9-14}$$

Tests on the Mean, Variance Known

Null hypothesis: $H_0: \mu = \mu_0$

Test statistic: $Z_0 = \dfrac{\bar{X} - \mu_0}{\sigma/\sqrt{n}}$

Alternative hypothesis	Rejection criteria
$H_1: \mu \neq \mu_0$	$z_0 > z_{\alpha/2, n-1}$ or $z_0 < -z_{\alpha/2, n-1}$
$H_1: \mu > \mu_0$	$z_0 > z_{\alpha, n-1}$
$H_1: \mu < \mu_0$	$z_0 < -z_{\alpha, n-1}$

P-Values

For the foregoing normal distribution tests it is relatively easy to compute the P-value. If z_0 is the computed value of the test statistic, the P-value is

$$P = \begin{cases} 2[1 - \Phi(|z_0|)] & \text{for a two-tailed test: } H_0: \mu = \mu_0 \quad H_1: \mu \neq \mu_0 \\ 1 - \Phi(z_0) & \text{for a upper-tailed test: } H_0: \mu = \mu_0 \quad H_1: \mu > \mu_0 \\ \Phi(z_0) & \text{for a lower-tailed test: } H_0: \mu = \mu_0 \quad H_1: \mu < \mu_0 \end{cases} \quad (9\text{-}15)$$

Here, $\Phi(z)$ is the standard normal cumulative distribution function defined in Chapter 4. Recall that $\Phi(z) = P(Z \leq z)$, where Z is $N(0, 1)$. To illustrate this, consider the propellant problem in Example 9-2. The computed value of the test statistic is $z_0 = 3.25$ and since the alternative hypothesis is two-tailed, the P-value is

$$P\text{-value} = 2[1 - \Phi(3.25)] = 0.0012$$

Thus, $H_0: \mu = 50$ would be rejected at any level of significance $\alpha \geq P\text{-value} = 0.0012$. For example, H_0 would be rejected if $\alpha = 0.01$, but it would not be rejected if $\alpha = 0.001$.

9-2.2 Type II Error and Choice of Sample Size

In testing hypotheses, the analyst directly selects the type I error probability. However, the probability of type II error β depends on the choice of sample size. In this section, we will show how to calculate the probability of type II error β. We will also show how to select the sample size to obtain a specified value of β.

Finding the Probability of Type II Error β
Consider the two-sided hypothesis

$$H_0: \mu = \mu_0$$
$$H_1: \mu \neq \mu_0$$

Suppose that the null hypothesis is false and that the true value of the mean is $\mu = \mu_0 + \delta$, say, where $\delta > 0$. The test statistic Z_0 is

$$Z_0 = \frac{\bar{X} - \mu_0}{\sigma/\sqrt{n}} = \frac{\bar{X} - (\mu_0 + \delta)}{\sigma/\sqrt{n}} + \frac{\delta\sqrt{n}}{\sigma}$$

Therefore, the distribution of Z_0 when H_1 is true is

$$Z_0 \sim N\left(\frac{\delta\sqrt{n}}{\sigma}, 1\right) \quad (9\text{-}16)$$

The distribution of the test statistic Z_0 under both the null hypothesis H_0 and the alternate hypothesis H_1 is shown in Fig. 9-8. From examining this figure, we note that if H_1 is true, a type II error will be made only if $-z_{\alpha/2} \leq Z_0 \leq z_{\alpha/2}$ where $Z_0 \sim N(\delta\sqrt{n}/\sigma, 1)$. That is, the probability of the type II error β is the probability that Z_0 falls between $-z_{\alpha/2}$ and $z_{\alpha/2}$ *given that H_1 is true*. This probability is shown as the shaded portion of Fig. 9-8. Expressed mathematically, this probability is

Figure 9-8 The distribution of Z_0 under H_0 and H_1.

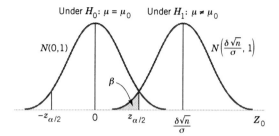

Probability of a Type II Error for a Two-Sided Test on the Mean, Variance Known

$$\beta = \Phi\left(z_{\alpha/2} - \frac{\delta\sqrt{n}}{\sigma}\right) - \Phi\left(-z_{\alpha/2} - \frac{\delta\sqrt{n}}{\sigma}\right) \quad (9\text{-}17)$$

where $\Phi(z)$ denotes the probability to the left of z in the standard normal distribution. Note that Equation 9-17 was obtained by evaluating the probability that Z_0 falls in the interval $[-z_{\alpha/2}, z_{\alpha/2}]$ when H_1 is true. Furthermore, note that Equation 9-17 also holds if $\delta < 0$, due to the symmetry of the normal distribution. It is also possible to derive an equation similar to Equation 9-17 for a one-sided alternative hypothesis.

Sample Size Formulas
One may easily obtain formulas that determine the appropriate sample size to obtain a particular value of β for a given δ and α. For the two-sided alternative hypothesis, we know from Equation 9-17 that

$$\beta = \Phi\left(z_{\alpha/2} - \frac{\delta\sqrt{n}}{\sigma}\right) - \Phi\left(-z_{\alpha/2} - \frac{\delta\sqrt{n}}{\sigma}\right)$$

or if $\delta > 0$,

$$\beta \simeq \Phi\left(z_{\alpha/2} - \frac{\delta\sqrt{n}}{\sigma}\right) \quad (9\text{-}18)$$

since $\Phi(-z_{\alpha/2} - \delta\sqrt{n}/\sigma) \simeq 0$ when δ is positive. Let z_β be the 100β upper percentile of the standard normal distribution. Then, $\beta = \Phi(-z_\beta)$. From Equation 9-18

$$-z_\beta \simeq z_{\alpha/2} - \frac{\delta\sqrt{n}}{\sigma}$$

or

Sample Size for a Two-Sided Test on the Mean, Variance Known

$$n \simeq \frac{(z_{\alpha/2} + z_\beta)^2 \sigma^2}{\delta^2} \quad \text{where} \quad \delta = \mu - \mu_0 \quad (9\text{-}19)$$

If n is not an integer, the convention is to round the sample size up to the next integer. This approximation is good when $\Phi(-z_{\alpha/2} - \delta\sqrt{n}/\sigma)$ is small compared to β. For either of the one-sided alternative hypotheses the sample size required to produce a specified type II error with probability β given δ and α is

Sample Size for a One-Sided Test on the Mean, Variance Known

$$n = \frac{(z_\alpha + z_\beta)^2 \sigma^2}{\delta^2} \quad \text{where} \quad \delta = \mu - \mu_0 \quad (9\text{-}20)$$

EXAMPLE 9-3
Propellant Burning Rate Type II Error

Consider the rocket propellant problem of Example 9-2. Suppose that the true burning rate is 49 centimeters per second. What is β for the two-sided test with $\alpha = 0.05$, $\sigma = 2$, and $n = 25$?

Here $\delta = 1$ and $z_{\alpha/2} = 1.96$. From Equation 9-17

$$\beta = \Phi\left(1.96 - \frac{\sqrt{25}}{\sigma}\right) - \Phi\left(-1.96 - \frac{\sqrt{25}}{\sigma}\right) = \Phi(-0.54) - \Phi(-4.46) = 0.295$$

The probability is about 0.3 that this difference from 50 centimeters per second will not be detected. That is, the probability is about 0.3 that the test will fail to reject the **null** hypothesis when the true burning rate is 49 centimeters per second.

Suppose that the analyst wishes to design the test so that **if the true mean burning rate differs from 50 centimeters per second by as much as 1 centimeter per second, the test will detect this** (i.e., reject $H_0: \mu = 50$) with a high probability, say 0.90. Now, we note that $\sigma = 2$, $\delta = 51 - 50 = 1$, $\alpha = 0.05$, and $\beta = 0.10$. Since $z_{\alpha/2} = z_{0.025} = 1.96$ and $z_\beta = z_{0.10} = 1.28$, the sample size required to detect this departure from $H_0: \mu = 50$ is found by Equation 9-19 as

$$n \simeq \frac{(z_{\alpha/2} + z_\beta)^2 \sigma^2}{\delta^2} = \frac{(1.96 + 1.28)^2 2^2}{(1)^2} \simeq 42$$

The approximation is good here, since $\Phi(-z_{\alpha/2} - \delta\sqrt{n}/\sigma) = \Phi(-1.96 - (1)\sqrt{42}/2) = \Phi(-5.20) \simeq 0$, which is small relative to β.

Using Operating Characteristic Curves
When performing sample size or type II error calculations, it is sometimes more convenient to use the **operating characteristic (OC) curves** in Appendix Charts VIa and VIb. These curves plot β as calculated from Equation 9-17 against a parameter d for various sample sizes n. Curves are provided for both $\alpha = 0.05$ and $\alpha = 0.01$. The parameter d is defined as

$$d = \frac{|\mu - \mu_0|}{\sigma} = \frac{|\delta|}{\sigma} \quad (9\text{-}21)$$

so one set of operating characteristic curves can be used for all problems regardless of the values of μ_0 and σ. From examining the operating characteristic curves or Equation 9-17 and Fig. 9-8, we note that

1. The further the true value of the mean μ is from μ_0, the smaller the probability of type II error β for a given n and α. That is, we see that for a specified sample size and α, large differences in the mean are easier to detect than small ones.

9-2 TESTS ON THE MEAN OF A NORMAL DISTRIBUTION, VARIANCE KNOWN   311

2. For a given δ and α, the probability of type II error β decreases as n increases. That is, to detect a specified difference δ in the mean, we may make the test more powerful by increasing the sample size.

EXAMPLE 9-4
Propellant
Burning Rate
Type II Error
from OC Curve

Consider the propellant problem in Example 9-2. Suppose that the analyst is concerned about the probability of type II error if the true mean burning rate is $\mu = 51$ centimeters per second. We may use the operating characteristic curves to find β. Note that $\delta = 51 - 50 = 1$, $n = 25$, $\sigma = 2$, and $\alpha = 0.05$. Then using Equation 9-21 gives

$$d = \frac{|\mu - \mu_0|}{\sigma} = \frac{|\delta|}{\sigma} = \frac{1}{2}$$

and from Appendix Chart VIIa, with $n = 25$, we find that $\beta = 0.30$. That is, if the true mean burning rate is $\mu = 51$ centimeters per second, there is approximately a 30% chance that this will not be detected by the test with $n = 25$.

EXAMPLE 9-5
Propellant
Burning Rate
Sample Size
from OC Curve

Once again, consider the propellant problem in Example 9-2. Suppose that the analyst would like to design the test so that if the true mean burning rate differs from 50 centimeters per second by as much as 1 centimeter per second, the test will detect this (i.e., reject H_0: $\mu = 50$) with a high probability, say, 0.90. This is exactly the same requirement as in Example 9-3, where we used Equation 9-19 to find the required sample size to be $n = 42$. The operating characteristic curves can also be used to find the sample size for this test. Since $d = |\mu - \mu_0|/\sigma = 1/2$, $\alpha = 0.05$, and $\beta = 0.10$, we find from Appendix Chart VIIa that the required sample size is approximately $n = 40$. This closely agrees with the sample size calculated from Equation 9-19.

In general, the operating characteristic curves involve three parameters: β, d, and n. Given any two of these parameters, the value of the third can be determined. There are two typical applications of these curves:

1. For a given n and d, find β (as illustrated in Example 9-4). This kind of problem is often encountered when the analyst is concerned about the sensitivity of an experiment already performed, or when sample size is restricted by economic or other factors.
2. For a given β and d, find n. This was illustrated in Example 9-5. This kind of problem is usually encountered when the analyst has the opportunity to select the sample size at the outset of the experiment.

Operating characteristic curves are given in Appendix Charts VIIc and VIId for the one-sided alternatives. If the alternative hypothesis is either H_1: $\mu > \mu_0$ or H_1: $\mu < \mu_0$, the abscissa scale on these charts is

$$d = \frac{|\mu - \mu_0|}{\sigma} \tag{9-22}$$

Using the Computer
Many statistics software packages will calculate sample sizes and type II error probabilities. To illustrate, here are some computations from Minitab for the propellant burning rate problem.

Minitab Computations

Power and Sample Size

1-Sample Z Test
Testing mean = null (versus not = null)
Calculating power for mean = null + difference
Alpha = 0.05 Sigma = 2

Difference	Sample Size	Target Power	Actual Power
1	43	0.9000	0.9064

Power and Sample Size

1-Sample Z Test
Testing mean = null (versus not = null)
Calculating power for mean = null + difference
Alpha = 0.05 Sigma = 2

Difference	Sample Size	Target Power	Actual Power
1	28	0.7500	0.7536

Power and Sample Size

1-Sample Z Test
Testing mean = null (versus not = null)
Calculating power for mean = null + difference
Alpha = 0.05 Sigma = 2

Difference	Sample Size	Power
1	25	0.7054

In the first part of the boxed display, we asked Minitab to work Example 9-3, that is, to find the sample size n that would allow detection of a difference from $\mu_0 = 50$ of 1 centimeter per second with power of 0.9 and $\alpha = 0.05$. The answer, $n = 43$, agrees closely with the calculated value from Equation 9-19 in Example 9-3, which was $n = 42$. The difference is due to Minitab using a value of z_β that has more than two decimal places. The second part of the computer output relaxes the power requirement to 0.75. Note that the effect is to reduce the required sample size to $n = 28$. The third part of the output is the solution to Example 9-4, where we wish to determine the type II error probability of (β) or the power = $1 - \beta$ for the sample size $n = 25$. Note that Minitab computes the power to be 0.7054, which agrees closely with the answer obtained from the OC curve in Example 9-4. Generally, however, the computer calculations will be more accurate than visually reading values from an OC curve.

9-2.3 Large-Sample Test

We have developed the test procedure for the null hypothesis $H_0: \mu = \mu_0$ assuming that the population is normally distributed and that σ^2 is known. In many if not most practical situations σ^2 will be unknown. Furthermore, we may not be certain that the population is well modeled by a normal distribution. In these situations if n is large (say $n > 40$) the sample standard deviation s can be substituted for σ in the test procedures with little effect. Thus, while we have given a test

EXERCISES FOR SECTION 9-2

9-29. State the null and alternative hypothesis in each case.
(a) A hypothesis test will be used to potentially provide evidence that the population mean is greater than 10.
(b) A hypothesis test will be used to potentially provide evidence that the population mean is not equal to 7.
(c) A hypothesis test will be used to potentially provide evidence that the population mean is less than 5.

9-30. A hypothesis will be used to test that a population mean equals 7 against the alternative that the population mean does not equal 7 with known variance σ. What are the critical values for the test statistic Z_0 for the following significance levels?
(a) 0.01 (b) 0.05 (c) 0.10

9-31. A hypothesis will be used to test that a population mean equals 10 against the alternative that the population mean is greater than 10 with known variance σ. What is the critical value for the test statistic Z_0 for the following significance levels?
(a) 0.01 (b) 0.05 (c) 0.10

9-32. A hypothesis will be used to test that a population mean equals 5 against the alternative that the population mean is less than 5 with known variance σ. What is the critical value for the test statistic Z_0 for the following significance levels?
(a) 0.01 (b) 0.05 (c) 0.10

9-33. For the hypothesis test $H_0: \mu = 7$ against $H_1: \mu \neq 7$ and variance known, calculate the P-value for each of the following test statistics.
(a) $z_0 = 2.05$ (b) $z_0 = -1.84$ (c) $z_0 = 0.4$

9-34. For the hypothesis test $H_0: \mu = 10$ against $H_1: \mu > 10$ and variance known, calculate the P-value for each of the following test statistics.
(a) $z_0 = 2.05$ (b) $z_0 = -1.84$ (c) $z_0 = 0.4$

9-35. For the hypothesis test $H_0: \mu = 5$ against $H_1: \mu < 5$ and variance known, calculate the P-value for each of the following test statistics.
(a) $z_0 = 2.05$ (b) $z_0 = -1.84$ (c) $z_0 = 0.4$

9-36. The mean water temperature downstream from a power plant cooling tower discharge pipe should be no more than 100°F. Past experience has indicated that the standard deviation of temperature is 2°F. The water temperature is measured on nine randomly chosen days, and the average temperature is found to be 98°F.
(a) Is there evidence that the water temperature is acceptable at $\alpha = 0.05$?
(b) What is the P-value for this test?
(c) What is the probability of accepting the null hypothesis at $\alpha = 0.05$ if the water has a true mean temperature of 104°F?

9-37. A manufacturer produces crankshafts for an automobile engine. The wear of the crankshaft after 100,000 miles (0.0001 inch) is of interest because it is likely to have an impact on warranty claims. A random sample of $n = 15$ shafts is tested and $\bar{x} = 2.78$. It is known that $\sigma = 0.9$ and that wear is normally distributed.
(a) Test $H_0: \mu = 3$ versus $H_1: \mu \neq 3$ using $\alpha = 0.05$.
(b) What is the power of this test if $\mu = 3.25$?
(c) What sample size would be required to detect a true mean of 3.75 if we wanted the power to be at least 0.9?

9-38. A melting point test of $n = 10$ samples of a binder used in manufacturing a rocket propellant resulted in $\bar{x} = 154.2°F$. Assume that melting point is normally distributed with $\sigma = 1.5°F$.
(a) Test $H_0: \mu = 155$ versus $H_1: \mu \neq 155$ using $\alpha = 0.01$.
(b) What is the P-value for this test?
(c) What is the β-error if the true mean is $\mu = 150$?
(d) What value of n would be required if we want $\beta < 0.1$ when $\mu = 150$? Assume that $\alpha = 0.01$.

9-39. The life in hours of a battery is known to be approximately normally distributed, with standard deviation $\sigma = 1.25$ hours. A random sample of 10 batteries has a mean life of $\bar{x} = 40.5$ hours.
(a) Is there evidence to support the claim that battery life exceeds 40 hours? Use $\alpha = 0.05$.
(b) What is the P-value for the test in part (a)?
(c) What is the β-error for the test in part (a) if the true mean life is 42 hours?
(d) What sample size would be required to ensure that β does not exceed 0.10 if the true mean life is 44 hours?
(e) Explain how you could answer the question in part (a) by calculating an appropriate confidence bound on life.

9-40. An engineer who is studying the tensile strength of a steel alloy intended for use in golf club shafts knows that tensile strength is approximately normally distributed with $\sigma = 60$ psi. A random sample of 12 specimens has a mean tensile strength of $\bar{x} = 3450$ psi.
(a) Test the hypothesis that mean strength is 3500 psi. Use $\alpha = 0.01$.

(b) What is the smallest level of significance at which you would be willing to reject the null hypothesis?
(c) What is the β-error for the test in part (a) if the true mean is 3470?
(d) Suppose that we wanted to reject the null hypothesis with probability at least 0.8 if mean strength $\mu = 3500$. What sample size should be used?
(e) Explain how you could answer the question in part (a) with a two-sided confidence interval on mean tensile strength.

9-41. Supercavitation is a propulsion technology for undersea vehicles that can greatly increase their speed. It occurs above approximately 50 meters per second, when pressure drops sufficiently to allow the water to dissociate into water vapor, forming a gas bubble behind the vehicle. When the gas bubble completely encloses the vehicle, supercavitation is said to occur. Eight tests were conducted on a scale model of an undersea vehicle in a towing basin with the average observed speed $\bar{x} = 102.2$ meters per second. Assume that speed is normally distributed with known standard deviation $\sigma = 4$ meters per second.
(a) Test the hypotheses H_0: $\mu = 100$ versus H_1: $\mu < 100$ using $\alpha = 0.05$.
(b) What is the P-value for the test in part (a)?
(c) Compute the power of the test if the true mean speed is as low as 95 meters per second.
(d) What sample size would be required to detect a true mean speed as low as 95 meters per second if we wanted the power of the test to be at least 0.85?
(e) Explain how the question in part (a) could be answered by constructing a one-sided confidence bound on the mean speed.

9-42. A bearing used in an automotive application is suppose to have a nominal inside diameter of 1.5 inches. A random sample of 25 bearings is selected and the average inside diameter of these bearings is 1.4975 inches. Bearing diameter is known to be normally distributed with standard deviation $\sigma = 0.01$ inch.
(a) Test the hypotheses H_0: $\mu = 1.5$ versus H_1: $\mu \neq 1.5$ using $\alpha = 0.01$.
(b) What is the P-value for the test in part (a)?
(c) Compute the power of the test if the true mean diameter is 1.495 inches.
(d) What sample size would be required to detect a true mean diameter as low as 1.495 inches if we wanted the power of the test to be at least 0.9?
(e) Explain how the question in part (a) could be answered by constructing a two-sided confidence interval on the mean diameter.

9-43. Medical researchers have developed a new artificial heart constructed primarily of titanium and plastic. The heart will last and operate almost indefinitely once it is implanted in the patient's body, but the battery pack needs to be recharged about every four hours. A random sample of 50 battery packs is selected and subjected to a life test. The average life of these batteries is 4.05 hours. Assume that battery life is normally distributed with standard deviation $\sigma = 0.2$ hour.
(a) Is there evidence to support the claim that mean battery life exceeds 4 hours? Use $\alpha = 0.05$.
(b) What is the P-value for the test in part (a)?
(c) Compute the power of the test if the true mean battery life is 4.5 hours.
(d) What sample size would be required to detect a true mean battery life of 4.5 hours if we wanted the power of the test to be at least 0.9?
(e) Explain how the question in part (a) could be answered by constructing a one-sided confidence bound on the mean life.

9-3 TESTS ON THE MEAN OF A NORMAL DISTRIBUTION, VARIANCE UNKNOWN

9-3.1 Hypothesis Tests on the Mean

We now consider the case of hypothesis testing on the mean of a population with **unknown variance** σ^2. The situation is analogous to Section 8-3, where we considered a confidence interval on the mean for the same situation. As in that section, the validity of the test procedure we will describe rests on the assumption that the population distribution is at least approximately normal. The important result upon which the test procedure relies is that if $X_1, X_2, \ldots, X_n$ is a random sample from a normal distribution with mean μ and variance σ^2, the random variable

$$T = \frac{\bar{X} - \mu}{S/\sqrt{n}}$$

has a t distribution with $n - 1$ degrees of freedom. Recall that we used this result in Section 8-3 to devise the t-confidence interval for μ. Now consider testing the hypotheses

$$H_0: \mu = \mu_0$$
$$H_1: \mu \neq \mu_0$$

We will use the **test statistic**

$$T_0 = \frac{\bar{X} - \mu_0}{S/\sqrt{n}} \tag{9-23}$$

If the null hypothesis is true, T_0 has a t distribution with $n - 1$ degrees of freedom. When we know the distribution of the test statistic when H_0 is true (this is often called the **reference distribution** or the **null distribution**), we can locate the critical region to control the type I error probability at the desired level. In this case we would use the t percentage points $-t_{\alpha/2,n-1}$ and $t_{\alpha/2,n-1}$ as the boundaries of the critical region so that we would reject $H_0: \mu = \mu_0$ if

$$t_0 > t_{\alpha/2,n-1} \quad \text{or if} \quad t_0 < -t_{\alpha/2,n-1}$$

where t_0 is the observed value of the test statistic T_0. The test procedure is very similar to the test on the mean with known variance described in Section 9-2, except that T_0 is used as the test statistic instead of Z_0 and the t_{n-1} distribution is used to define the critical region instead of the standard normal distribution. A summary of the test procedures for both two- and one-sided alternative hypotheses follows:

Tests on the Mean, Variance Unknown

Null hypothesis: $H_0: \mu = \mu_0$

Test statistic: $T_0 = \dfrac{\bar{X} - \mu_0}{S/\sqrt{n}}$

Alternative hypothesis	Rejection criteria
$H_1: \mu \neq \mu_0$	$t_0 > t_{\alpha/2,n-1}$ or $t_0 < -t_{\alpha/2,n-1}$
$H_1: \mu > \mu_0$	$t_0 > t_{\alpha,n-1}$
$H_1: \mu < \mu_0$	$t_0 < -t_{\alpha,n-1}$

Figure 9-9 shows the location of the critical region for these situations.

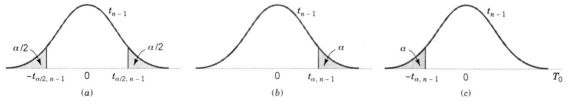

Figure 9-9 The reference distribution for $H_0: \mu = \mu_0$ with critical region for (a) $H_1: \mu \neq \mu_0$, (b) $H_1: \mu > \mu_0$, and (c) $H_1: \mu < \mu_0$.

EXAMPLE 9-6
Golf Club Design

The increased availability of light materials with high strength has revolutionized the design and manufacture of golf clubs, particularly drivers. Clubs with hollow heads and very thin faces can result in much longer tee shots, especially for players of modest skills. This is due partly to the "spring-like effect" that the thin face imparts to the ball. Firing a golf ball at the head of the club and measuring the ratio of the outgoing velocity of the ball to the incoming velocity can quantify this spring-like effect. The ratio of velocities is called the coefficient of restitution of the club. An experiment was performed in which 15 drivers produced by a particular club maker were selected at random and their coefficients of restitution measured. In the experiment the golf balls were fired from an air cannon so that the incoming velocity and spin rate of the ball could be precisely controlled. It is of interest to determine if there is evidence (with $\alpha = 0.05$) to support a claim that the mean coefficient of restitution exceeds 0.82. The observations follow:

0.8411	0.8191	0.8182	0.8125	0.8750
0.8580	0.8532	0.8483	0.8276	0.7983
0.8042	0.8730	0.8282	0.8359	0.8660

The sample mean and sample standard deviation are $\bar{x} = 0.83725$ and $s = 0.02456$. The normal probability plot of the data in Fig. 9-10 supports the assumption that the coefficient of restitution is normally distributed. Since the objective of the experimenter is to demonstrate that the mean coefficient of restitution exceeds 0.82, a one-sided alternative hypothesis is appropriate.

The solution using the eight-step procedure for hypothesis testing is as follows:

1. The parameter of interest is the mean coefficient of restitution, μ.
2. H_0: $\mu = 0.82$
3. H_1: $\mu > 0.82$. We want to reject H_0 if the mean coefficient of restitution exceeds 0.82.
4. $\alpha = 0.05$
5. The test statistic is

$$t_0 = \frac{\bar{x} - \mu_0}{s/\sqrt{n}}$$

6. Reject H_0 if $t_0 > t_{0.05,14} = 1.761$

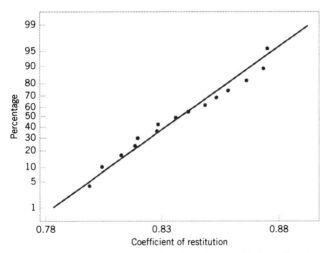

Figure 9-10. Normal probability plot of the coefficient of restitution data from Example 9-6.

7. Computations: Since $\bar{x} = 0.83725$, $s = 0.02456$, $\mu_0 = 0.82$, and $n = 15$, we have

$$t_0 = \frac{0.83725 - 0.82}{0.02456/\sqrt{15}} = 2.72$$

8. Conclusions: Since $t_0 = 2.72 > 1.761$, we reject H_0 and conclude at the 0.05 level of significance that the mean coefficient of restitution exceeds 0.82.

Minitab will conduct the one-sample t-test. The output from this software package is in the following display:

Minitab Computations

One-Sample T: COR

Test of mu = 0.82 vs mu > 0.82

Variable	N	Mean	StDev	SE Mean
COR	15	0.83725	0.02456	0.00634

Variable	95.0% Lower Bound	T	P
COR	0.82608	2.72	0.008

Notice that Minitab computes both the test statistic T_0 and a 95% lower confidence bound for the coefficient of restitution. Because the 95% lower confidence bound exceeds 0.82, we would reject the hypothesis that H_0: $\mu = 0.82$ and conclude that the alternative hypothesis H_1: $\mu > 0.82$ is true. Minitab also calculates a P-value for the test statistic T_0. In the next section we explain how this is done.

9-3.2 P-Value for a t-Test

The P-value for a t-test is just the smallest level of significance at which the null hypothesis would be rejected. That is, it is the tail area beyond the value of the test statistic t_0 for a one-sided test or twice this area for a two-sided test. Because the t-table in Appendix A contains only 10 critical values for each t distribution, computation of the exact P-value directly from the table is usually impossible. However, it is easy to find upper and lower bounds on the P-value from this table.

To illustrate, consider the t-test based on 14 degrees of freedom in Example 9-6. The relevant critical values from Appendix Table IV are as follows:

| Critical Value: | 0.258 | 0.692 | 1.345 | 1.761 | 2.145 | 2.624 | 2.977 | 3.326 | 3.787 | 4.140 |
| Tail Area: | 0.40 | 0.25 | 0.10 | 0.05 | 0.025 | 0.01 | 0.005 | 0.0025 | 0.001 | 0.0005 |

Notice that $t_0 = 2.72$ in Example 9-6, and that this is between two tabulated values, 2.624 and 2.977. Therefore, the P-value must be between 0.01 and 0.005. These are effectively the upper and lower bounds on the P-value.

Example 9-6 is an upper-tailed test. If the test is lower-tailed, just change the sign of t_0 and proceed as above. Remember that for a two-tailed test the level of significance associated with a particular critical value is twice the corresponding tail area in the column heading. This fact must be taken into account when we compute the bound on the P-value. For example, suppose that $t_0 = 2.72$ for a two-tailed alternate based on 14 degrees of freedom. The value $2.624 < t_0 < 2.977$, so the lower and upper bounds on the P-value would be $0.01 = 2(0.005) < P\text{-value} < 0.02 = 2(0.01)$ for this case.

Finally, most computer programs report *P*-values along with the computed value of the test statistic. Some hand-held calculators also have this capability. In Example 9-6, Minitab gave the *P*-value for the value $t_0 = 2.72$ in Example 9-6 as 0.008.

9-3.3 Type II Error and Choice of Sample Size

The type II error probability for tests on the mean of a normal distribution with unknown variance depends on the distribution of the test statistic in Equation 9-23 when the null hypothesis $H_0: \mu = \mu_0$ is false. When the true value of the mean is $\mu = \mu_0 + \delta$, the distribution for T_0 is called the **noncentral *t* distribution** with $n - 1$ degrees of freedom and noncentrality parameter $\delta\sqrt{n}/\sigma$. Note that if $\delta = 0$, the noncentral *t* distribution reduces to the usual **central *t* distribution**. Therefore, the type II error of the two-sided alternative (for example) would be

$$\beta = P\{-t_{\alpha/2, n-1} \leq T_0 \leq t_{\alpha/2, n-1} \mid \delta \neq 0\}$$
$$= P\{-t_{\alpha/2, n-1} \leq T_0' \leq t_{\alpha/2, n-1}\}$$

where T_0' denotes the noncentral *t* random variable. Finding the type II error probability β for the *t*-test involves finding the probability contained between two points of the noncentral *t* distribution. Because the noncentral *t*-random variable has a messy density function, this integration must be done numerically.

Fortunately, this ugly task has already been done, and the results are summarized in a series of O.C. curves in Appendix Charts VII*e*, VII*f*, VII*g*, and VII*h* that plot β for the *t*-test against a parameter *d* for various sample sizes *n*. Curves are provided for two-sided alternatives on Charts VII*e* and VII*f*. The abscissa scale factor *d* on these charts is defined as

$$d = \frac{|\mu - \mu_0|}{\sigma} = \frac{|\delta|}{\sigma} \tag{9-24}$$

For the one-sided alternative $\mu > \mu_0$ or $\mu < \mu_0$, we use charts VIG and VIH with

$$d = \frac{|\mu - \mu_0|}{\sigma} = \frac{|\delta|}{\sigma} \tag{9-25}$$

We note that *d* depends on the unknown parameter σ^2. We can avoid this difficulty in several ways. In some cases, we may use the results of a previous experiment or prior information to make a rough initial estimate of σ^2. If we are interested in evaluating test performance after the data have been collected, we could use the sample variance s^2 to estimate σ^2. If there is no previous experience on which to draw in estimating σ^2, we then define the difference in the mean *d* that we wish to detect relative to σ. For example, if we wish to detect a small difference in the mean, we might use a value of $d = |\delta|/\sigma \leq 1$ (for example), whereas if we are interested in detecting only moderately large differences in the mean, we might select $d = |\delta|/\sigma = 2$ (for example). That is, it is the value of the ratio $|\delta|/\sigma$ that is important in determining sample size, and if it is possible to specify the relative size of the difference in means that we are interested in detecting, then a proper value of *d* can usually be selected.

EXAMPLE 9-7
Golf Club Design
Sample Size

Consider the golf club testing problem from Example 9-6. If the mean coefficient of restitution exceeds 0.82 by as much as 0.02, is the sample size $n = 15$ adequate to ensure that $H_0: \mu = 0.82$ will be rejected with probability at least 0.8?

9-3 TESTS ON THE MEAN OF A NORMAL DISTRIBUTION, VARIANCE UNKNOWN

To solve this problem, we will use the sample standard deviation $s = 0.02456$ to estimate σ. Then $d = |\delta|/\sigma = 0.02/0.02456 = 0.81$. By referring to the operating characteristic curves in Appendix Chart VIIg (for $\alpha = 0.05$) with $d = 0.81$ and $n = 15$, we find that $\beta = 0.10$, approximately. Thus, the probability of rejecting H_0: $\mu = 0.82$ if the true mean exceeds this by 0.02 is approximately $1 - \beta = 1 - 0.10 = 0.90$, and we conclude that a sample size of $n = 15$ is adequate to provide the desired sensitivity.

Minitab will also perform power and sample size computations for the one-sample t-test. Below are several calculations based on the golf club testing problem:

Minitab Computations

Power and Sample Size

1-Sample t Test
Testing mean = null (versus > null)
Calculating power for mean = null + difference
Alpha = 0.05 Sigma = 0.02456

Difference	Sample Size	Power
0.02	15	0.9117

Power and Sample Size

1-Sample t Test
Testing mean = null (versus > null)
Calculating power for mean = null + difference
Alpha = 0.05 Sigma = 0.02456

Difference	Sample Size	Power
0.01	15	0.4425

Power and Sample Size

1-Sample t Test
Testing mean = null (versus > null)
Calculating power for mean = null + difference
Alpha = 0.05 Sigma = 0.02456

Difference	Sample Size	Target Power	Actual Power
0.01	39	0.8000	0.8029

In the first portion of the computer output, Minitab reproduces the solution to Example 9-7, verifying that a sample size of $n = 15$ is adequate to give power of at least 0.8 if the mean coefficient of restitution exceeds 0.82 by at least 0.02. In the middle section of the output, we used Minitab to compute the power to detect a difference between μ and $\mu_0 = 0.82$ of 0.01. Notice that with $n = 15$, the power drops considerably to 0.4425. The final portion of the output is the sample size required for a power of at least 0.8 if the difference between μ and μ_0 of interest is actually 0.01. A much larger n is required to detect this smaller difference.

EXERCISES FOR SECTION 9-3

9-44. A hypothesis will be used to test that a population mean equals 7 against the alternative that the population mean does not equal 7 with unknown variance σ. What are the critical values for the test statistic T_0 for the following significance levels and sample sizes?
(a) $\alpha = 0.01$ and $n = 20$
(b) $\alpha = 0.05$ and $n = 12$
(c) $\alpha = 0.10$ and $n = 15$

9-45. A hypothesis will be used to test that a population mean equals 10 against the alternative that the population mean is greater than 10 with known variance σ. What is the critical value for the test statistic Z_0 for the following significance levels?
(a) $\alpha = 0.01$ and $n = 20$
(b) $\alpha = 0.05$ and $n = 12$
(c) $\alpha = 0.10$ and $n = 15$

9-46. A hypothesis will be used to test that a population mean equals 5 against the alternative that the population mean is less than 5 with known variance σ. What is the critical value for the test statistic Z_0 for the following significance levels?
(a) $\alpha = 0.01$ and $n = 20$
(b) $\alpha = 0.05$ and $n = 12$
(c) $\alpha = 0.10$ and $n = 15$

9-47. For the hypothesis test $H_0: \mu = 7$ against $H_1: \mu \neq 7$ with variance unknown and $n = 20$, approximate the P-value for each of the following test statistics.
(a) $t_0 = 2.05$ (b) $t_0 = -1.84$ (c) $t_0 = 0.4$

9-48. For the hypothesis test $H_0: \mu = 10$ against $H_1: \mu > 10$ with variance unknown and $n = 15$, approximate the P-value for each of the following test statistics.
(a) $t_0 = 2.05$ (b) $t_0 = -1.84$ (c) $t_0 = 0.4$

9-49. For the hypothesis test $H_0: \mu = 5$ against $H_1: \mu < 5$ with variance unknown and $n = 12$, approximate the P-value for each of the following test statistics.
(a) $t_0 = 2.05$ (b) $t_0 = -1.84$ (c) $t_0 = 0.4$

9-50. An article in the *ASCE Journal of Energy Engineering* (1999, Vol. 125, pp. 59–75) describes a study of the thermal inertia properties of autoclaved aerated concrete used as a building material. Five samples of the material were tested in a structure, and the average interior temperature (°C) reported was as follows: 23.01, 22.22, 22.04, 22.62, and 22.59.
(a) Test the hypotheses $H_0: \mu = 22.5$ versus $H_1: \mu \neq 22.5$, using $\alpha = 0.05$. Find the P-value.
(b) Check the assumption that interior temperature is normally distributed.
(c) Compute the power of the test if the true mean interior temperature is as high as 22.75.
(d) What sample size would be required to detect a true mean interior temperature as high as 22.75 if we wanted the power of the test to be at least 0.9?
(e) Explain how the question in part (a) could be answered by constructing a two-sided confidence interval on the mean interior temperature.

9-51. A 1992 article in the *Journal of the American Medical Association* ("A Critical Appraisal of 98.6 Degrees F, the Upper Limit of the Normal Body Temperature, and Other Legacies of Carl Reinhold August Wundrlich") reported body temperature, gender, and heart rate for a number of subjects. The body temperatures for 25 female subjects follow: 97.8, 97.2, 97.4, 97.6, 97.8, 97.9, 98.0, 98.0, 98.0, 98.1, 98.2, 98.3, 98.3, 98.4, 98.4, 98.4, 98.5, 98.6, 98.6, 98.7, 98.8, 98.8, 98.9, 98.9, and 99.0.
(a) Test the hypotheses $H_0: \mu = 98.6$ versus $H_1: \mu \neq 98.6$, using $\alpha = 0.05$. Find the P-value.
(b) Check the assumption that female body temperature is normally distributed.
(c) Compute the power of the test if the true mean female body temperature is as low as 98.0.
(d) What sample size would be required to detect a true mean female body temperature as low as 98.2 if we wanted the power of the test to be at least 0.9?
(e) Explain how the question in part (a) could be answered by constructing a two-sided confidence interval on the mean female body temperature.

9-52. Cloud seeding has been studied for many decades as a weather modification procedure (for an interesting study of this subject, see the article in *Technometrics*, "A Bayesian Analysis of a Multiplicative Treatment Effect in Weather Modification", Vol. 17, pp. 161–166). The rainfall in acre-feet from 20 clouds that were selected at random and seeded with silver nitrate follows: 18.0, 30.7, 19.8, 27.1, 22.3, 18.8, 31.8, 23.4, 21.2, 27.9, 31.9, 27.1, 25.0, 24.7, 26.9, 21.8, 29.2, 34.8, 26.7, and 31.6.
(a) Can you support a claim that mean rainfall from seeded clouds exceeds 25 acre-feet? Use $\alpha = 0.01$. Find the P-value.
(b) Check that rainfall is normally distributed.
(c) Compute the power of the test if the true mean rainfall is 27 acre-feet.
(d) What sample size would be required to detect a true mean rainfall of 27.5 acre-feet if we wanted the power of the test to be at least 0.9?
(e) Explain how the question in part (a) could be answered by constructing a one-sided confidence bound on the mean diameter.

9-53. The sodium content of thirty 300-gram boxes of organic corn flakes was determined. The data (in milligrams) are as follows: 131.15, 130.69, 130.91, 129.54, 129.64, 128.77, 130.72, 128.33, 128.24, 129.65, 130.14, 129.29, 128.71, 129.00, 129.39, 130.42, 129.53, 130.12, 129.78, 130.92, 131.15, 130.69, 130.91, 129.54, 129.64, 128.77, 130.72, 128.33, 128.24, and 129.65.

(a) Can you support a claim that mean sodium content of this brand of cornflakes differs from 130 milligrams? Use $\alpha = 0.05$. Find the P-value.
(b) Check that sodium content is normally distributed.
(c) Compute the power of the test if the true mean sodium content is 130.5 milligrams.
(d) What sample size would be required to detect a true mean sodium content of 130.1 milligrams if we wanted the power of the test to be at least 0.75?
(e) Explain how the question in part (a) could be answered by constructing a two-sided confidence interval on the mean sodium content.

9-54. Consider the baseball coefficient of restitution data first presented in Exercise 8-86.
(a) Does the data support the claim that the mean coefficient of restitution of baseballs exceeds 0.635? Use $\alpha = 0.05$. Find the P-value.
(b) Check the normality assumption.
(c) Compute the power of the test if the true mean coefficient of restitution is as high as 0.64.
(d) What sample size would be required to detect a true mean coefficient of restitution as high as 0.64 if we wanted the power of the test to be at least 0.75?
(e) Explain how the question in part (a) could be answered with a confidence interval.

9-55. Consider the dissolved oxygen concentration at TVA dams first presented in Exercise 8-88.
(a) Test the hypotheses H_0: $\mu = 4$ versus H_1: $\mu \neq 4$. Use $\alpha = 0.01$. Find the P-value.
(b) Check the normality assumption.
(c) Compute the power of the test if the true mean dissolved oxygen concentration is as low as 3.
(d) What sample size would be required to detect a true mean dissolved oxygen concentration as low as 2.5 if we wanted the power of the test to be at least 0.9?
(e) Explain how the question in part (a) could be answered with a confidence interval.

9-56. Reconsider the data from *Medicine and Science in Sports and Exercise* described in Exercise 8-26. The sample size was seven and the sample mean and sample standard deviation were 315 watts and 16 watts, respectively.
(a) Is there evidence that leg strength exceeds 300 watts at significance level 0.05? Find the P-value.
(b) Compute the power of the test if the true strength is 305 watts.
(c) What sample size would be required to detect a true mean of 305 watts if the power of the test should be at least 0.90?
(d) Explain how the question in part (a) could be answered with a confidence interval.

9-57. Reconsider the tire testing experiment described in Exercise 8-23.
(a) The engineer would like to demonstrate that the mean life of this new tire is in excess of 60,000 kilometers. Formulate and test appropriate hypotheses, and draw conclusions using $\alpha = 0.05$.
(b) Suppose that if the mean life is as long as 61,000 kilometers, the engineer would like to detect this difference with probability at least 0.90. Was the sample size $n = 16$ used in part (a) adequate?

9-58. Reconsider the Izod impact test on PVC pipe described in Exercise 8-24. Suppose that you want to use the data from this experiment to support a claim that the mean impact strength exceeds the ASTM standard (one foot-pound per inch). Formulate and test the appropriate hypotheses using $\alpha = 0.05$.

9-59. Reconsider the television tube brightness experiment in Exercise 8-31. Suppose that the design engineer claims that this tube will require at least 300 microamps of current to produce the desired brightness level. Formulate and test an appropriate hypothesis to confirm this claim using $\alpha = 0.05$. Find the P-value for this test. State any necessary assumptions about the underlying distribution of the data.

9-60. Exercise 6-22 gave data on the heights of female engineering students at ASU.
(a) Can you support a claim that mean height of female engineering students at ASU is at least 65 inches? Use $\alpha = 0.05$. Find the P-value.
(b) Check the normality assumption.
(c) Compute the power of the test if the true mean height is 62 inches.
(d) What sample size would be required to detect a true mean height of 64 inches if we wanted the power of the test to be at least 0.8?

9-61. Exercise 6-25 describes testing golf balls for an overall distance standard.
(a) Can you support a claim that mean distance achieved by this particular golf ball exceeds 280 yards? Use $\alpha = 0.05$. Find the P-value.
(b) Check the normality assumption.
(c) Compute the power of the test if the true mean distance is 290 yards.
(d) What sample size would be required to detect a true mean distance of 290 yards if we wanted the power of the test to be at least 0.8?

9-62. Exercise 6-24 presented data on the concentration of suspended solids in lake water.
(a) Test the hypotheses H_0: $\mu = 55$ versus H_1: $\mu \neq 55$, use $\alpha = 0.05$. Find the P-value.
(b) Check the normality assumption.
(c) Compute the power of the test if the true mean concentration is as low as 50.
(d) What sample size would be required to detect a true mean concentration as low as 50 if we wanted the power of the test to be at least 0.9?

9-4 TESTS ON THE VARIANCE AND STANDARD DEVIATION OF A NORMAL DISTRIBUTION

Sometimes hypothesis tests on the population variance or standard deviation are needed. When the population is modeled by a normal distribution, the tests and intervals described in this section are applicable.

9-4.1 Hypothesis Tests on the Variance

Suppose that we wish to test the hypothesis that the variance of a normal population σ^2 equals a specified value, say σ_0^2, or equivalently, that the standard deviation σ is equal to σ_0. Let $X_1, X_2, \ldots, X_n$ be a random sample of n observations from this population. To test

$$H_0: \sigma^2 = \sigma_0^2$$
$$H_1: \sigma^2 \neq \sigma_0^2$$
(9-26)

we will use the test statistic:

Test Statistic

$$X_0^2 = \frac{(n-1)S^2}{\sigma_0^2}$$
(9-27)

If the null hypothesis $H_0: \sigma^2 = \sigma_0^2$ is true, the test statistic X_0^2 defined in Equation 9-27 follows the chi-square distribution with $n - 1$ degrees of freedom. This is the reference distribution for this test procedure. Therefore, we calculate χ_0^2, the value of the test statistic X_0^2, and the null hypothesis $H_0: \sigma^2 = \sigma_0^2$ would be rejected if

$$\chi_0^2 > \chi_{\alpha/2, n-1}^2 \quad \text{or if} \quad \chi_0^2 < \chi_{1-\alpha/2, n-1}^2$$

where $\chi_{\alpha/2, n-1}^2$ and $\chi_{1-\alpha/2, n-1}^2$ are the upper and lower $100\alpha/2$ percentage points of the chi-square distribution with $n - 1$ degrees of freedom, respectively. Figure 9-11(a) shows the critical region.

The same test statistic is used for one-sided alternative hypotheses. For the one-sided hypothesis

$$H_0: \sigma^2 = \sigma_0^2$$
$$H_1: \sigma^2 > \sigma_0^2$$
(9-28)

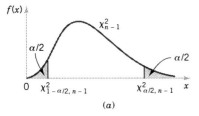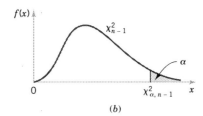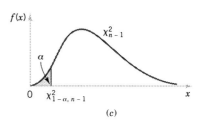

Figure 9-11 Reference distribution for the test of $H_0: \sigma^2 = \sigma_0^2$ with critical region values for (a) $H_1: \sigma^2 \neq \sigma_0^2$, (b) $H_1: \sigma^2 > \sigma_0^2$, and (c) $H_1: \sigma^2 < \sigma_0^2$.

we would reject H_0 if $\chi_0^2 > \chi_{\alpha,n-1}^2$, whereas for the other one-sided hypothesis

$$H_0: \sigma^2 = \sigma_0^2$$
$$H_1: \sigma^2 < \sigma_0^2 \qquad (9\text{-}29)$$

we would reject H_0 if $\chi_0^2 < \chi_{1-\alpha,n-1}^2$. The one-sided critical regions are shown in Figure 9-11(b) and (c).

Tests on the Variance of a Normal Distribution

Null hypothesis: $H_0: \sigma^2 = \sigma_0^2$

Test statistic: $\chi_0^2 = \dfrac{(n-1)S^2}{\sigma_0^2}$

Alternative hypothesis	Rejection criteria
$H_1: \sigma^2 \neq \sigma_0^2$	$\chi_0^2 > \chi_{\alpha/2,n-1}^2$ or $\chi_0^2 < -\chi_{\alpha/2,n-1}^2$
$H_1: \sigma^2 > \sigma_0^2$	$\chi_0^2 > \chi_{\alpha,n-1}^2$
$H_1: \sigma^2 < \sigma_0^2$	$\chi_0^2 < -\chi_{\alpha,n-1}^2$

EXAMPLE 9-8
Automated Filling

An automated filling machine is used to fill bottles with liquid detergent. A random sample of 20 bottles results in a sample variance of fill volume of $s^2 = 0.0153$ (fluid ounces)^2. If the variance of fill volume exceeds 0.01 (fluid ounces)^2, an unacceptable proportion of bottles will be underfilled or overfilled. Is there evidence in the sample data to suggest that the manufacturer has a problem with underfilled or overfilled bottles? Use $\alpha = 0.05$, and assume that fill volume has a normal distribution.

Using the eight-step procedure results in the following:

1. The parameter of interest is the population variance σ^2.
2. $H_0: \sigma^2 = 0.01$
3. $H_1: \sigma^2 > 0.01$
4. $\alpha = 0.05$
5. The test statistic is

$$\chi_0^2 = \frac{(n-1)s^2}{\sigma_0^2}$$

6. Reject H_0 if $\chi_0^2 > \chi_{0.05,19}^2 = 30.14$.
7. Computations:

$$\chi_0^2 = \frac{19(0.0153)}{0.01} = 29.07$$

8. Conclusions: Since $\chi_0^2 = 29.07 < \chi_{0.05,19}^2 = 30.14$, we conclude that there is no strong evidence that the variance of fill volume exceeds 0.01 (fluid ounces)^2.

Using Appendix Table III, it is easy to place bounds on the P-value of a chi-square test. From inspection of the table, we find that $\chi_{0.10,19}^2 = 27.20$ and $\chi_{0.05,19}^2 = 30.14$. Since $27.20 < 29.07 < 30.14$, we conclude that the P-value for the test in Example 9-8 is in the interval $0.05 < P\text{-value} < 0.10$. The actual P-value from computer software is $P\text{-value} = 0.0649$.

9-4.2 Type II Error and Choice of Sample Size

Operating characteristic curves for the chi-square tests in Section 9-4.1 are provided in Appendix Charts VIi through VIn for $\alpha = 0.05$ and $\alpha = 0.01$. For the two-sided alternative hypothesis of Equation 9-26, Charts VIIi and VIIj plot β against an abscissa parameter

$$\lambda = \frac{\sigma}{\sigma_0} \qquad (9\text{-}30)$$

for various sample sizes n, where σ denotes the true value of the standard deviation. Charts VIk and VIl are for the one-sided alternative $H_1: \sigma^2 > \sigma_0^2$, while Charts VII$m$ and VIIn are for the other one-sided alternative $H_1: \sigma^2 < \sigma_0^2$. In using these charts, we think of σ as the value of the standard deviation that we want to detect.

These curves can be used to evaluate the β-error (or power) associated with a particular test. Alternatively, they can be used to **design** a test—that is, to determine what sample size is necessary to detect a particular value of σ that differs from the hypothesized value σ_0.

EXAMPLE 9-9
Automated Filling Sample Size

Consider the bottle-filling problem from Example 9-8. If the variance of the filling process exceeds 0.01 (fluid ounces)^2, too many bottles will be underfilled. Thus, the hypothesized value of the standard deviation is $\sigma_0 = 0.10$. Suppose that if the true standard deviation of the filling process exceeds this value by 25%, we would like to detect this with probability at least 0.8. Is the sample size of $n = 20$ adequate?

To solve this problem, note that we require

$$\lambda = \frac{\sigma}{\sigma_0} = \frac{0.125}{0.10} = 1.25$$

This is the abscissa parameter for Chart VIIk. From this chart, with $n = 20$ and $\lambda = 1.25$, we find that $\beta \simeq 0.6$. Therefore, there is only about a 40% chance that the null hypothesis will be rejected if the true standard deviation is really as large as $\sigma = 0.125$ fluid ounce.

To reduce the β-error, a larger sample size must be used. From the operating characteristic curve with $\beta = 0.20$ and $\lambda = 1.25$, we find that $n = 75$, approximately. Thus, if we want the test to perform as required above, the sample size must be at least 75 bottles.

EXERCISES FOR SECTION 9-4

9-63. Consider the test of $H_0: \sigma^2 = 7$ against $H_1: \sigma^2 \neq 7$. What are the critical values for the test statistic X_0^2 for the following significance levels and sample sizes?
(a) $\alpha = 0.01$ and $n = 20$
(b) $\alpha = 0.05$ and $n = 12$
(c) $\alpha = 0.10$ and $n = 15$

9-64. Consider the test of $H_0: \sigma^2 = 10$ against $H_1: \sigma^2 > 10$. What are the critical values for the test statistic X_0^2 for the following significance levels and sample sizes?
(a) $\alpha = 0.01$ and $n = 20$
(b) $\alpha = 0.05$ and $n = 12$
(c) $\alpha = 0.10$ and $n = 15$

9-65. Consider the test of $H_0: \sigma^2 = 5$ against $H_1: \sigma^2 < 5$. What are the critical values for the test statistic X_0^2 for the following significance levels and sample sizes?

(a) $\alpha = 0.01$ and $n = 20$
(b) $\alpha = 0.05$ and $n = 12$
(c) $\alpha = 0.10$ and $n = 15$

9-66. Consider the hypothesis test of $H_0: \sigma^2 = 7$ against $H_1: \sigma^2 \neq 7$. Approximate the P-value for each of the following test statistics.
(a) $x_0^2 = 25.2$ and $n = 20$
(b) $x_0^2 = 15.2$ and $n = 12$
(c) $x_0^2 = 23.0$ and $n = 15$

9-67. Consider the test of $H_0: \sigma^2 = 5$ against $H_1: \sigma^2 < 5$. Approximate the P-value for each of the following test statistics.
(a) $x_0^2 = 25.2$ and $n = 20$
(b) $x_0^2 = 15.2$ and $n = 12$
(c) $x_0^2 = 4.2$ and $n = 15$

9-68. Consider the hypothesis test of $H_0: \sigma^2 = 10$ against $H_1: \sigma^2 > 10$. Approximate the P-value for each of the following test statistics.
(a) $x_0^2 = 25.2$ and $n = 20$
(b) $x_0^2 = 15.2$ and $n = 12$
(c) $x_0^2 = 4.2$ and $n = 15$

9-69. The data from *Medicine and Science in Sports and Exercise* described in Exercise 8-42 considered ice hockey player performance after electrostimulation training. In summary, there were 17 players and the sample standard deviation of performance was 0.09 seconds.
(a) Is there strong evidence to conclude that the standard deviation of performance time exceeds the historical value of 0.75 seconds? Use $\alpha = 0.05$. Find the P-value for this test.
(b) Discuss how part (a) could be answered by constructing a 95% one-sided confidence interval for σ.

9-70. The data from *Technometrics* described in Exercise 8-45 considered the variability in repeated measurements of the weight of a sheet of paper. In summary, the sample standard deviation from 15 measurements was 0.0083 grams.
(a) Does the measurement standard deviation differ from 0.01 grams at $\alpha = 0.05$? Find the P-value for this test.
(b) Discuss how part (a) could be answered by constructing a confidence interval for σ.

9-71. Reconsider the percentage of titanium in an alloy used in aerospace castings from Exercise 8-41. Recall that $s = 0.37$ and $n = 51$.
(a) Test the hypothesis $H_0: \sigma = 0.25$ versus $H_1: \sigma \neq 0.25$ using $\alpha = 0.05$. State any necessary assumptions about the underlying distribution of the data. Find the P-value.
(b) Explain how you could answer the question in part (a) by constructing a 95% two-sided confidence interval for σ.

9-72. Data from an Izod impact test was described in Exercise 8-24. The sample standard deviation was 0.25 and $n = 20$ specimens were tested.
(a) Test the hypothesis that $\sigma = 0.10$ against an alternative specifying that $\sigma \neq 0.10$, using $\alpha = 0.01$, and draw a conclusion. State any necessary assumptions about the underlying distribution of the data.
(b) What is the P-value for this test?
(c) Could the question in part (a) have been answered by constructing a 99% two-sided confidence interval for σ^2?

9-73. Data for tire life was described in Exercise 8-23. The sample standard deviation was 3645.94 kilometers and $n = 16$.
(a) Can you conclude, using $\alpha = 0.05$, that the standard deviation of tire life is less than 4000 kilometers? State any necessary assumptions about the underlying distribution of the data. Find the P-value for this test.
(b) Explain how you could answer the question in part (a) by constructing a 95% one-sided confidence interval for σ.

9-74. If the standard deviation of hole diameter exceeds 0.01 millimeters, there is an unacceptably high probability that the rivet will not fit. Suppose that $n = 15$ and $s = 0.008$ millimeters.
(a) Is there strong evidence to indicate that the standard deviation of hole diameter exceeds 0.01 millimeters? Use $\alpha = 0.01$. State any necessary assumptions about the underlying distribution of the data. Find the P-value for this test.
(b) Suppose that the actual standard deviation of hole diameter exceeds the hypothesized value by 50%. What is the probability that this difference will be detected by the test described in part (a).
(c) If σ is really as large as 0.0125 millimeters, what sample size will be required to defect this with power of at least 0.8?

9-75. Recall the sugar content of the syrup in canned peaches from Exercise 8-40. Suppose that the variance is thought to be $\sigma^2 = 18$ (milligrams)^2. Recall that a random sample of $n = 10$ cans yields a sample standard deviation of $s = 4.8$ milligrams.
(a) Test the hypothesis $H_0: \sigma^2 = 18$ versus $H_1: \sigma^2 \neq 18$ using $\alpha = 0.05$. Find the P-value for this test.
(b) Suppose that the actual standard deviation is twice as large as the hypothesized value. What is the probability that this difference will be detected by the test described in part (a)?
(c) Suppose that the true variance is $\sigma^2 = 40$. How large a sample would be required to detect this difference with probability at least 0.90?

9-5 TESTS ON A POPULATION PROPORTION

It is often necessary to test hypotheses on a population proportion. For example, suppose that a random sample of size n has been taken from a large (possibly infinite) population and that $X(\leq n)$ observations in this sample belong to a class of interest. Then $\hat{P} = X/n$ is a point estimator of the proportion of the population p that belongs to this class. Note that n and p are the parameters of a binomial distribution. Furthermore, from Chapter 7 we know that the sampling distribution of $\hat{P}$ is approximately normal with mean p and variance $p(1-p)/n$, if p is not too close to either 0 or 1 and if n is relatively large. Typically, to apply this approximation we require that np and $n(1-p)$ be greater than or equal to 5. We will give a large-sample test that makes use of the normal approximation to the binomial distribution.

9-5.1 Large-Sample Tests on a Proportion

In many engineering problems, we are concerned with a random variable that follows the binomial distribution. For example, consider a production process that manufactures items that are classified as either acceptable or defective. It is usually reasonable to model the occurrence of defectives with the binomial distribution, where the binomial parameter p represents the proportion of defective items produced. Consequently, many engineering decision problems involve hypothesis testing about p.

We will consider testing

$$H_0: p = p_0$$
$$H_1: p \neq p_0$$
(9-31)

An approximate test based on the normal approximation to the binomial will be given. As noted above, this approximate procedure will be valid as long as p is not extremely close to zero or one, and if the sample size is relatively large. Let X be the number of observations in a random sample of size n that belongs to the class associated with p. Then, if the null hypothesis $H_0: p = p_0$ is true, we have $X \sim N[np_0, np_0(1 - p_0)]$, approximately. To test $H_0: p = p_0$, calculate the test statistic.

Test Statistic

$$Z_0 = \frac{X - np_0}{\sqrt{np_0(1 - p_0)}}$$
(9-32)

and reject $H_0: p = p_0$ if

$$z_0 > z_{\alpha/2} \quad \text{or} \quad z_0 < -z_{\alpha/2}$$

Note that the standard normal distribution is the reference distribution for this test statistic. Critical regions for the one-sided alternative hypotheses are constructed in the usual manner.

Approximate Tests on a Binomial Proportion

Null hypothesis: $H_0: p = p_0$

Test statistic: $Z_0 = \dfrac{X - np_0}{\sqrt{np_0(1 - p_0)}}$

Alternative hypothesis	Rejection criteria
$H_1: p \neq p_0$	$z_0 > z_{\alpha/2, n-1}$ or $z_0 < -z_{\alpha/2, n-1}$
$H_1: p > p_0$	$z_0 > z_{\alpha, n-1}$
$H_1: p < p_0$	$z_0 < -z_{\alpha, n-1}$

EXAMPLE 9-10
Automobile Engine Controller

A semiconductor manufacturer produces controllers used in automobile engine applications. The customer requires that the process fallout or fraction defective at a critical manufacturing step not exceed 0.05 and that the manufacturer demonstrate process capability at this level of quality using $\alpha = 0.05$. The semiconductor manufacturer takes a random sample of 200 devices and finds that four of them are defective. Can the manufacturer demonstrate process capability for the customer?

We may solve this problem using the eight-step hypothesis-testing procedure as follows:

1. The parameter of interest is the process fraction defective p.
2. $H_0: p = 0.05$
3. $H_1: p < 0.05$
   This formulation of the problem will allow the manufacturer to make a strong claim about process capability if the null hypothesis $H_0: p = 0.05$ is rejected.
4. $\alpha = 0.05$
5. The test statistic is (from Equation 9-32)

$$z_0 = \frac{x - np_0}{\sqrt{np_0(1 - p_0)}}$$

   where $x = 4$, $n = 200$, and $p_0 = 0.05$.
6. Reject $H_0: p = 0.05$ if $z_0 < -z_{0.05} = -1.645$
7. Computations: The test statistic is

$$z_0 = \frac{4 - 200(0.05)}{\sqrt{200(0.05)(0.95)}} = -1.95$$

8. Conclusions: Since $z_0 = -1.95 < -z_{0.05} = -1.645$, we reject H_0 and conclude that the process fraction defective p is less than 0.05. The P-value for this value of the test statistic z_0 is $P = 0.0256$, which is less than $\alpha = 0.05$. We conclude that the process is capable.

Another form of the test statistic Z_0 in Equation 9-32 is occasionally encountered. Note that if X is the number of observations in a random sample of size n that belongs to a class of interest, then $\hat{P} = X/n$ is the sample proportion that belongs to that class. Now divide both numerator and denominator of Z_0 in Equation 9-32 by n, giving

$$Z_0 = \frac{X/n - p_0}{\sqrt{p_0(1 - p_0)/n}}$$

or

$$Z_0 = \frac{\hat{P} - p_0}{\sqrt{p_0(1 - p_0)/n}} \tag{9-33}$$

This presents the test statistic in terms of the sample proportion instead of the number of items X in the sample that belongs to the class of interest.

Statistical software packages usually provide the one sample Z-test for a proportion. The Minitab output for Example 9-10 follows.

Minitab Computations

Test and CI for One Proportion

Test of $p = 0.05$ vs $p < 0.05$

Sample	X	N	Sample p	95.0% Upper Bound	Z-Value	P-Value
1	4	200	0.020000	0.036283	−1.95	0.026

* NOTE * The normal approximation may be inaccurate for small samples.

Notice that both the test statistic (and accompanying *P*-value) and the 95% one-sided upper confidence bound are displayed. The 95% upper confidence bound is 0.036283, which is less than 0.05. This is consistent with rejection of the null hypothesis $H_0: p = 0.05$.

9-5.2 Type II Error and Choice of Sample Size

It is possible to obtain closed-form equations for the approximate β-error for the tests in Section 9-5.1. Suppose that p is the true value of the population proportion. The approximate β-error for the two-sided alternative $H_1: p \neq p_0$ is

$$\beta = \Phi\left(\frac{p_0 - p + z_{\alpha/2}\sqrt{p_0(1-p_0)/n}}{\sqrt{p(1-p)/n}}\right) - \Phi\left(\frac{p_0 - p - z_{\alpha/2}\sqrt{p_0(1-p_0)/n}}{\sqrt{p(1-p)/n}}\right) \quad (9\text{-}34)$$

If the alternative is $H_1: p < p_0$,

$$\beta = 1 - \Phi\left(\frac{p_0 - p - z_\alpha\sqrt{p_0(1-p_0)/n}}{\sqrt{p(1-p)/n}}\right) \quad (9\text{-}35)$$

whereas if the alternative is $H_1: p > p_0$,

$$\beta = \Phi\left(\frac{p_0 - p + z_\alpha\sqrt{p_0(1-p_0)/n}}{\sqrt{p(1-p)/n}}\right) \quad (9\text{-}36)$$

These equations can be solved to find the approximate sample size n that gives a test of level α that has a specified β risk. The sample size equations are

Approximate Sample Size for a Two-Sided Test on a Binomial Proportion

$$n = \left[\frac{z_{\alpha/2}\sqrt{p_0(1-p_0)} + z_\beta\sqrt{p(1-p)}}{p - p_0}\right]^2 \quad (9\text{-}37)$$

for the two-sided alternative and

Approximate Sample Size for a One-Sided Test on a Binomial Proportion

$$n = \left[\frac{z_\alpha\sqrt{p_0(1-p_0)} + z_\beta\sqrt{p(1-p)}}{p - p_0}\right]^2 \quad (9\text{-}38)$$

for a one-sided alternative.

EXAMPLE 9-11
Automobile Engine Controller Type II Error

Consider the semiconductor manufacturer from Example 9-10. Suppose that its process fallout is really $p = 0.03$. What is the β-error for a test of process capability that uses $n = 200$ and $\alpha = 0.05$?

The β-error can be computed using Equation 9-35 as follows:

$$\beta = 1 - \Phi\left[\frac{0.05 - 0.03 - (1.645)\sqrt{0.05(0.95)/200}}{\sqrt{0.03(1-0.03)/200}}\right] = 1 - \Phi(-0.44) = 0.67$$

Thus, the probability is about 0.7 that the semiconductor manufacturer will fail to conclude that the process is capable if the true process fraction defective is $p = 0.03$ (3%). That is, the power of the test against this particular alternative is only about 0.3. This appears to be a large β-error (or small power), but the difference between $p = 0.05$ and $p = 0.03$ is fairly small, and the sample size $n = 200$ is not particularly large.

Suppose that the semiconductor manufacturer was willing to accept a β-error as large as 0.10 if the true value of the process fraction defective was $p = 0.03$. If the manufacturer continues to use $\alpha = 0.05$, what sample size would be required?

The required sample size can be computed from Equation 9-38 as follows:

$$n = \left[\frac{1.645\sqrt{0.05(0.95)} + 1.28\sqrt{0.03(0.97)}}{0.03 - 0.05}\right]^2$$
$$\approx 832$$

where we have used $p = 0.03$ in Equation 9-38. Note that $n = 832$ is a very large sample size. However, we are trying to detect a fairly small deviation from the null value $p_0 = 0.05$.

Minitab will also perform power and sample size calculations for the one-sample Z-test on a proportion. Output from Minitab for the engine controllers tested in Example 9-10 follows.

Power and Sample Size

Test for One Proportion
Testing proportion = 0.05 (versus < 0.05)
Alpha = 0.05

Alternative Proportion	Sample Size	Power
3.00E-02	200	0.3287

Power and Sample Size

Test for One Proportion
Testing proportion = 0.05 (versus < 0.05)
Alpha = 0.05

Alternative Proportion	Sample Size	Target Power	Actual Power
3.00E-02	833	0.9000	0.9001

Power and Sample Size

Test for One Proportion
Testing proportion = 0.05 (versus < 0.05)
Alpha = 0.05

Alternative Proportion	Sample Size	Target Power	Actual Power
3.00E-02	561	0.7500	0.7503

The first part of the output shows the power calculation based on the situation described in Example 9-11, where the true proportion is really 0.03. The power calculation from Minitab agrees with the results from Equation 9-35 in Example 9-11. The second part of the output computes the sample size necessary for a power of 0.9 ($\beta = 0.1$) if $p = 0.03$. Again, the results agree closely with those obtained from Equation 9-38. The final portion of the display shows the sample size that would be required if $p = 0.03$ and the power requirement is relaxed to 0.75. Notice that the sample size of $n = 561$ is still quite large because the difference between $p = 0.05$ and $p = 0.03$ is fairly small.

EXERCISES FOR SECTION 9-5

9-76. Suppose that 1000 customers are surveyed and 850 are satisfied or very satisfied with a corporation's products and services.
(a) Test the hypothesis $H_0: p = 0.9$ against $H_1: p \neq 0.9$ at $\alpha = 0.05$. Find the P-value.
(b) Explain how the question in part (a) could be answered by constructing a 95% two-sided confidence interval for p.

9-77. Suppose that 500 parts are tested in manufacturing and 10 are rejected.
(a) Test the hypothesis $H_0: p = 0.03$ against $H_1: p < 0.03$ at $\alpha = 0.05$. Find the P-value.
(b) Explain how the question in part (a) could be answered by constructing a 95% one-sided confidence interval for p.

9-78. A random sample of 300 circuits generated 13 defectives.
(a) Use the data to test $H_0: p = 0.05$ versus $H_1: p \neq 0.05$. Use $\alpha = 0.05$. Find the P-value for the test.
(b) Explain how the question in part (a) could be answered with a confidence interval.

9-79. An article in *British Medical Journal* (1986, Vol. 292, pp. 879–882) "Comparison of treatment of renal calculi by operative surgery, percutaneous nephrolithotomy, and extracorporeal shock wave lithotripsy," found that percutaneous nephrolithotomy (PN) had a success rate in removing kidney stones of 289 out of 350 patients. The traditional method was 78% effective.
(a) Is there evidence that the success rate for PN is greater than the historical success rate? Find the P-value.
(b) Explain how the question in part (a) could be answered with a confidence interval.

9-80. A manufacturer of interocular lenses is qualifying a new grinding machine and will qualify the machine if there is evidence that percentage of polished lenses that contain surface defects does not exceed 2%. A random sample of 250 lenses contains six defective lenses.
(a) Formulate and test an appropriate set of hypotheses to determine if the machine can be qualified. Use $\alpha = 0.05$. Find the P-value.
(b) Explain how the question in part (a) could be answered with a confidence interval.

9-81. A researcher claims that at least 10% of all football helmets have manufacturing flaws that could potentially cause injury to the wearer. A sample of 200 helmets revealed that 16 helmets contained such defects.

(a) Does this finding support the researcher's claim? Use $\alpha = 0.01$. Find the P-value.
(b) Explain how the question in part (a) could be answered with a confidence interval.

9-82. An article in *Fortune* (September 21, 1992) claimed that nearly one-half of all engineers continue academic studies beyond the B.S. degree, ultimately receiving either an M.S. or a Ph.D. degree. Data from an article in *Engineering Horizons* (Spring 1990) indicated that 117 of 484 new engineering graduates were planning graduate study.
(a) Are the data from *Engineering Horizons* consistent with the claim reported by *Fortune*? Use $\alpha = 0.05$ in reaching your conclusions. Find the P-value for this test.
(b) Discuss how you could have answered the question in part (a) by constructing a two-sided confidence interval on p.

9-83. The advertised claim for batteries for cell phones is set at 48 operating hours, with proper charging procedures. A study of 5000 batteries is carried out and 15 stop operating prior to 48 hours. Do these experimental results support the claim that less than 0.2 percent of the company's batteries will fail during the advertised time period, with proper charging procedures? Use a hypothesis-testing procedure with $\alpha = 0.01$.

9-84. A random sample of 500 registered voters in Phoenix is asked if they favor the use of oxygenated fuels year-round to reduce air pollution. If more than 315 voters respond positively, we will conclude that at least 60% of the voters favor the use of these fuels.
(a) Find the probability of type I error if exactly 60% of the voters favor the use of these fuels.
(b) What is the type II error probability β if 75% of the voters favor this action?

9-85. In a random sample of 85 automobile engine crankshaft bearings, 10 have a surface finish roughness that exceeds the specifications. Does this data present strong evidence that the proportion of crankshaft bearings exhibiting excess surface roughness exceeds 0.10?
(a) State and test the appropriate hypotheses using $\alpha = 0.05$.
(b) If it is really the situation that $p = 0.15$, how likely is it that the test procedure in part (a) will not reject the null hypothesis?
(c) If $p = 0.15$, how large would the sample size have to be for us to have a probability of correctly rejecting the null hypothesis of 0.9?

9-6 SUMMARY TABLE OF INFERENCE PROCEDURES FOR A SINGLE SAMPLE

The table in the end papers of this book (inside back cover) presents a summary of all the single-sample inference procedures from Chapters 8 and 9. The table contains the null hypothesis statement, the test statistic, the various alternative hypotheses and the criteria for rejecting H_0, and the formulas for constructing the $100(1 - \alpha)\%$ two-sided confidence interval. It would also be helpful to refer to the roadmap table on page 281 that provides guidance to match the problem type to the information inside the back cover.

9-7 TESTING FOR GOODNESS OF FIT

The hypothesis-testing procedures that we have discussed in previous sections are designed for problems in which the population or probability distribution is known and the hypotheses involve the parameters of the distribution. Another kind of hypothesis is often encountered: we do not know the underlying distribution of the population, and we wish to test the hypothesis that a particular distribution will be satisfactory as a population model. For example, we might wish to test the hypothesis that the population is normal.

We have previously discussed a very useful graphical technique for this problem called **probability plotting** and illustrated how it was applied in the case of a normal distribution. In this section, we describe a formal goodness-of-fit test procedure based on the chi-square distribution.

The test procedure requires a random sample of size n from the population whose probability distribution is unknown. These n observations are arranged in a frequency histogram, having k bins or class intervals. Let O_i be the observed frequency in the ith class interval. From the hypothesized probability distribution, we compute the expected frequency in the ith class interval, denoted E_i. The test statistic is

Goodness of Fit Test Statistic

$$X_0^2 = \sum_{i=1}^{k} \frac{(O_i - E_i)^2}{E_i} \qquad (9\text{-}39)$$

It can be shown that, if the population follows the hypothesized distribution, X_0^2 has, approximately, a chi-square distribution with $k - p - 1$ degrees of freedom, where p represents the number of parameters of the hypothesized distribution estimated by sample statistics. This approximation improves as n increases. We would reject the hypothesis that the distribution of the population is the hypothesized distribution if the calculated value of the test statistic $\chi_0^2 > \chi_{\alpha,k-p-1}^2$.

One point to be noted in the application of this test procedure concerns the magnitude of the expected frequencies. If these expected frequencies are too small, the test statistic X_0^2 will not reflect the departure of observed from expected, but only the small magnitude of the expected frequencies. There is no general agreement regarding the minimum value of expected frequencies, but values of 3, 4, and 5 are widely used as minimal. Some writers suggest that an expected frequency could be as small as 1 or 2, so long as most of them exceed 5. Should an expected frequency be too small, it can be combined with the expected frequency in an adjacent class interval. The corresponding observed frequencies would then also be combined, and k would be reduced by 1. Class intervals are not required to be of equal width.

EXAMPLE 9-12
Printed Circuit Board Defects

A Poisson Distribution

The number of defects in printed circuit boards is hypothesized to follow a Poisson distribution. A random sample of $n = 60$ printed boards has been collected, and the following number of defects observed.

Number of Defects	Observed Frequency
0	32
1	15
2	9
3	4

The mean of the assumed Poisson distribution in this example is unknown and must be estimated from the sample data. The estimate of the mean number of defects per board is the sample average, that is, $(32 \cdot 0 + 15 \cdot 1 + 9 \cdot 2 + 4 \cdot 3)/60 = 0.75$. From the Poisson distribution with parameter 0.75, we may compute p_i, the theoretical, hypothesized probability associated with the ith class interval. Since each class interval corresponds to a particular number of defects, we may find the p_i as follows:

$$p_1 = P(X = 0) = \frac{e^{-0.75}(0.75)^0}{0!} = 0.472$$

$$p_2 = P(X = 1) = \frac{e^{-0.75}(0.75)^1}{1!} = 0.354$$

$$p_3 = P(X = 2) = \frac{e^{-0.75}(0.75)^2}{2!} = 0.133$$

$$p_4 = P(X \geq 3) = 1 - (p_1 + p_2 + p_3) = 0.041$$

The expected frequencies are computed by multiplying the sample size $n = 60$ times the probabilities p_i. That is, $E_i = np_i$. The expected frequencies follow:

Number of Defects	Probability	Expected Frequency
0	0.472	28.32
1	0.354	21.24
2	0.133	7.98
3 (or more)	0.041	2.46

Since the expected frequency in the last cell is less than 3, we combine the last two cells:

Number of Defects	Observed Frequency	Expected Frequency
0	32	28.32
1	15	21.24
2 (or more)	13	10.44

The chi-square test statistic in Equation 9-39 will have $k - p - 1 = 3 - 1 - 1 = 1$ degree of freedom, because the mean of the Poisson distribution was estimated from the data.

The eight-step hypothesis-testing procedure may now be applied, using $\alpha = 0.05$, as follows:

1. The variable of interest is the form of the distribution of defects in printed circuit boards.
2. H_0: The form of the distribution of defects is Poisson.
3. H_1: The form of the distribution of defects is not Poisson.
4. $\alpha = 0.05$
5. The test statistic is

$$\chi_0^2 = \sum_{i=1}^{k} \frac{(o_i - E_i)^2}{E_i}$$

6. Reject H_0 if $\chi_0^2 > \chi_{0.05,1}^2 = 3.84$.
7. Computations:

$$\chi_0^2 = \frac{(32 - 28.32)^2}{28.32} + \frac{(15 - 21.24)^2}{21.24} + \frac{(13 - 10.44)^2}{10.44} = 2.94$$

8. Conclusions: Since $\chi_0^2 = 2.94 < \chi_{0.05,1}^2 = 3.84$, we are unable to reject the null hypothesis that the distribution of defects in printed circuit boards is Poisson. The P-value for the test is $P = 0.0864$. (This value was computed using an HP-48 calculator.)

EXAMPLE 9-13
Power Supply Distribution

Continuous Distribution

A manufacturing engineer is testing a power supply used in a notebook computer and, using $\alpha = 0.05$, wishes to determine whether output voltage is adequately described by a normal distribution. Sample estimates of the mean and standard deviation of $\bar{x} = 5.04$ V and $s = 0.08$ V are obtained from a random sample of $n = 100$ units.

A common practice in constructing the class intervals for the frequency distribution used in the chi-square goodness-of-fit test is to choose the cell boundaries so that the expected frequencies $E_i = np_i$ are equal for all cells. To use this method, we want to choose the cell boundaries $a_0, a_1, \ldots, a_k$ for the k cells so that all the probabilities

$$p_i = P(a_{i-1} \leq X \leq a_i) = \int_{a_{i-1}}^{a_i} f(x)\, dx$$

are equal. Suppose we decide to use $k = 8$ cells. For the standard normal distribution, the intervals that divide the scale into eight equally likely segments are $[0, 0.32)$, $[0.32, 0.675)$ $[0.675, 1.15)$, $[1.15, \infty)$ and their four "mirror image" intervals on the other side of zero. For each interval $p_i = 1/8 = 0.125$, so the expected cell frequencies are $E_i = np_i = 100(0.125) = 12.5$. The complete table of observed and expected frequencies is as follows:

Class Interval	Observed Frequency o_i	Expected Frequency E_i
$x < 4.948$	12	12.5
$4.948 \leq x < 4.986$	14	12.5
$4.986 \leq x < 5.014$	12	12.5
$5.014 \leq x < 5.040$	13	12.5
$5.040 \leq x < 5.066$	12	12.5
$5.066 \leq x < 5.094$	11	12.5
$5.094 \leq x < 5.132$	12	12.5
$5.132 \leq x$	14	12.5
Totals	100	100

334 CHAPTER 9 TESTS OF HYPOTHESES FOR A SINGLE SAMPLE

The boundary of the first class interval is $\bar{x} - 1.15s = 4.948$. The second class interval is $[\bar{x} - 1.15s, \bar{x} - 0.675s)$ and so forth. We may apply the eight-step hypothesis-testing procedure to this problem.

1. The variable of interest is the form of the distribution of power supply voltage.
2. H_0: The form of the distribution is normal.
3. H_1: The form of the distribution is nonnormal.
4. $\alpha = 0.05$
5. The test statistic is

$$\chi_0^2 = \sum_{i=1}^{k} \frac{(o_i - E_i)^2}{E_i}$$

6. Since two parameters in the normal distribution have been estimated, the chi-square statistic above will have $k - p - 1 = 8 - 2 - 1 = 5$ degrees of freedom. Therefore, we will reject H_0 if $\chi_0^2 > \chi_{0.05,5}^2 = 11.07$.
7. Computations:

$$\chi_0^2 = \sum_{i=1}^{8} \frac{(o_i - E_i)^2}{E_i}$$
$$= \frac{(12 - 12.5)^2}{12.5} + \frac{(14 - 12.5)^2}{12.5} + \cdots + \frac{(14 - 12.5)^2}{12.5}$$
$$= 0.64$$

8. Conclusions: Since $\chi_0^2 = 0.64 < \chi_{0.05,5}^2 = 11.07$, we are unable to reject H_0, and there is no strong evidence to indicate that output voltage is not normally distributed. The P-value for the chi-square statistic $\chi_0^2 = 0.64$ is $P = 0.9861$.

EXERCISES FOR SECTION 9-7

9-86. Consider the following frequency table of observations on the random variable X.

Values	0	1	2	3	4
Observed Frequency	24	30	31	11	4

(a) Based on these 100 observations, is a Poisson distribution with a mean of 1.2 an appropriate model? Perform a goodness-of-fit procedure with $\alpha = 0.05$.
(b) Calculate the P-value for this test.

9-87. Let X denote the number of flaws observed on a large coil of galvanized steel. Seventy-five coils are inspected and the following data were observed for the values of X:

Values	1	2	3	4	5	6	7	8
Observed Frequency	1	11	8	13	11	12	10	9

(a) Does the assumption of the Poisson distribution seem appropriate as a probability model for this data? Use $\alpha = 0.01$.
(b) Calculate the P-value for this test.

9-88. The number of calls arriving at a switchboard from noon to 1 PM during the business days Monday through Friday is monitored for six weeks (i.e., 30 days). Let X be defined as the number of calls during that one-hour period. The relative frequency of calls was recorded and reported as

Value	5	6	8	9	10
Relative Frequency	0.067	0.067	0.100	0.133	0.200

Value	11	12	13	14	15
Relative Frequency	0.133	0.133	0.067	0.033	0.067

(a) Does the assumption of a Poisson distribution seem appropriate as a probability model for this data? Use $\alpha = 0.05$.
(b) Calculate the P-value for this test.

9-89. Consider the following frequency table of observations on the random variable X:

Values	0	1	2	3	4
Frequency	4	21	10	13	2

(a) Based on these 50 observations, is a binomial distribution with $n = 6$ and $p = 0.25$ an appropriate model? Perform a goodness-of-fit procedure with $\alpha = 0.05$.
(b) Calculate the P-value for this test.

9-90. Define X as the number of underfilled bottles from a filling operation in a carton of 24 bottles. Seventy-five cartons are inspected and the following observations on X are recorded:

Values	0	1	2	3
Frequency	39	23	12	1

(a) Based on these 75 observations, is a binomial distribution an appropriate model? Perform a goodness-of-fit procedure with $\alpha = 0.05$.
(b) Calculate the P-value for this test.

9-91. The number of cars passing eastbound through the intersection of Mill and University Avenues has been tabulated by a group of civil engineering students. They have obtained the data in the adjacent table:
(a) Does the assumption of a Poisson distribution seem appropriate as a probability model for this process? Use $\alpha = 0.05$.
(b) Calculate the P-value for this test.

Vehicles per Minute	Observed Frequency	Vehicles per Minute	Observed Frequency
40	14	53	102
41	24	54	96
42	57	55	90
43	111	56	81
44	194	57	73
45	256	58	64
46	296	59	61
47	378	60	59
48	250	61	50
49	185	62	42
50	171	63	29
51	150	64	18
52	110	65	15

9-92. Reconsider Exercise 6-63. The data was the number of earthquakes per year of magnitude 7.0 and greater since 1900.

1900	13	1927	20	1954	17	1981	14
1901	14	1928	22	1955	19	1982	10
1902	8	1929	19	1956	15	1983	15
1903	10	1930	13	1957	34	1984	8
1904	16	1931	26	1958	10	1985	15
1905	26	1932	13	1959	15	1986	6
1906	32	1933	14	1960	22	1987	11
1907	27	1934	22	1961	18	1988	8
1908	18	1935	24	1962	15	1989	7
1909	32	1936	21	1963	20	1990	18
1910	36	1937	22	1964	15	1991	16
1911	24	1938	26	1965	22	1992	13
1912	22	1939	21	1966	19	1993	12
1913	23	1940	23	1967	16	1994	13
1914	22	1941	24	1968	30	1995	20
1915	18	1942	27	1969	27	1996	15
1916	25	1943	41	1970	29	1997	16
1917	21	1944	31	1971	23	1998	12
1918	21	1945	27	1972	20	1999	18
1919	14	1946	35	1973	16	2000	15
1920	8	1947	26	1974	21	2001	16
1921	11	1948	28	1975	21	2002	13
1922	14	1949	36	1976	25	2003	15
1923	23	1950	39	1977	16	2004	15
1924	18	1951	21	1978	18		
1925	17	1952	17	1979	15		
1926	19	1953	22	1980	18		

(a) Use computer software to summarize this data into a frequency distribution. Test the hypothesis that the number of earthquakes of magnitude 7.0 or greater each year follows a Poisson distribution at $\alpha = 0.05$.
(b) Calculate the P-value for the test.

9-8 CONTINGENCY TABLE TESTS

Many times, the n elements of a sample from a population may be classified according to two different criteria. It is then of interest to know whether the two methods of classification are statistically independent; for example, we may consider the population of graduating engineers, and we may wish to determine whether starting salary is independent of academic disciplines. Assume that the first method of classification has r levels and that the second method has c levels. We will let O_{ij} be the observed frequency for level i of the first classification

Table 9-2 An $r \times c$ Contingency Table

		Columns			
		1	2	...	c
Rows	1	O_{11}	O_{12}	...	O_{1c}
	2	O_{21}	O_{22}	...	O_{2c}
	⋮	⋮	⋮	⋮	⋮
	r	O_{r1}	O_{r2}	...	O_{rc}

method and level j on the second classification method. The data would, in general, appear as shown in Table 9-2. Such a table is usually called an $r \times c$ **contingency table**.

We are interested in testing the hypothesis that the row-and-column methods of classification are independent. If we reject this hypothesis, we conclude there is some interaction between the two criteria of classification. The exact test procedures are difficult to obtain, but an approximate test statistic is valid for large n. Let p_{ij} be the probability that a randomly selected element falls in the ijth cell, given that the two classifications are independent. Then $p_{ij} = u_i v_j$, where u_i is the probability that a randomly selected element falls in row class i and v_j is the probability that a randomly selected element falls in column class j. Now, assuming independence, the estimators of u_i and v_j are

$$\hat{u}_i = \frac{1}{n} \sum_{j=1}^{c} O_{ij}$$
$$\hat{v}_j = \frac{1}{n} \sum_{i=1}^{r} O_{ij} \qquad (9\text{-}40)$$

Therefore, the expected frequency of each cell is

$$E_{ij} = n\hat{u}_i\hat{v}_j = \frac{1}{n} \sum_{j=1}^{c} O_{ij} \sum_{i=1}^{r} O_{ij} \qquad (9\text{-}41)$$

Then, for large n, the statistic

$$\chi_0^2 = \sum_{i=1}^{r} \sum_{j=1}^{c} \frac{(O_{ij} - E_{ij})^2}{E_{ij}} \qquad (9\text{-}42)$$

has an approximate chi-square distribution with $(r-1)(c-1)$ degrees of freedom if the null hypothesis is true. Therefore, we would reject the hypothesis of independence if the observed value of the test statistic χ_0^2 exceeded $\chi_{\alpha,(r-1)(c-1)}^2$.

EXAMPLE 9-14
Pension Plan
Preference

A company has to choose among three pension plans. Management wishes to know whether the preference for plans is independent of job classification and wants to use $\alpha = 0.05$. The opinions of a random sample of 500 employees are shown in Table 9-3.

To find the expected frequencies, we must first compute $\hat{u}_1 = (340/500) = 0.68$, $\hat{u}_2 = (160/500) = 0.32$, $\hat{v}_1 = (200/500) = 0.40$, $\hat{v}_2 = (200/500) = 0.40$, and $\hat{v}_3 = (100/500) = 0.20$. The expected frequencies may now be computed from Equation 9-41. For example, the expected number of salaried workers favoring pension plan 1 is

$$E_{11} = n\hat{u}_1\hat{v}_1 = 500(0.68)(0.40) = 136$$

Table 9-3 Observed Data for Example 9-14

Job Classification	Pension Plan			Totals
	1	2	3	
Salaried workers	160	140	40	340
Hourly workers	40	60	60	160
Totals	200	200	100	500

Table 9-4 Expected Frequencies for Example 9-14

Job Classification	Pension Plan			Totals
	1	2	3	
Salaried workers	136	136	68	340
Hourly workers	64	64	32	160
Totals	200	200	100	500

The expected frequencies are shown in Table 9-4.

The eight-step hypothesis-testing procedure may now be applied to this problem.

1. The variable of interest is employee preference among pension plans.
2. H_0: Preference is independent of salaried versus hourly job classification.
3. H_1: Preference is not independent of salaried versus hourly job classification.
4. $\alpha = 0.05$
5. The test statistic is

$$\chi_0^2 = \sum_{i=1}^{r} \sum_{j=1}^{c} \frac{(o_{ij} - E_{ij})^2}{E_{ij}}$$

6. Since $r = 2$ and $c = 3$, the degrees of freedom for chi-square are $(r - 1)(c - 1) = (1)(2) = 2$, and we would reject H_0 if $\chi_0^2 > \chi_{0.05,2}^2 = 5.99$.
7. Computations:

$$\chi_0^2 = \sum_{i=1}^{2} \sum_{j=1}^{3} \frac{(o_{ij} - E_{ij})^2}{E_{ij}}$$

$$= \frac{(160 - 136)^2}{136} + \frac{(140 - 136)^2}{136} + \frac{(40 - 68)^2}{68} + \frac{(40 - 64)^2}{64}$$

$$+ \frac{(60 - 64)^2}{64} + \frac{(60 - 32)^2}{32} = 49.63$$

8. Conclusions: Since $\chi_0^2 = 49.63 > \chi_{0.05,2}^2 = 5.99$, we reject the hypothesis of independence and conclude that the preference for pension plans is not independent of job classification. The P-value for $\chi_0^2 = 49.63$ is $P = 1.671 \times 10^{-11}$. (This value was computed from computer software.) Further analysis would be necessary to explore the nature of the association between these factors. It might be helpful to examine the table of observed minus expected frequencies.

Using the two-way contingency table to test independence between two variables of classification in a sample from a single population of interest is only one application of contingency table methods. Another common situation occurs when there are r populations of interest and each population is divided into the same c categories. A sample is then taken from the ith population, and the counts are entered in the appropriate columns of the ith row. In this situation we want to investigate whether or not the proportions in the c categories are the same for all populations. The null hypothesis in this problem states that the populations are **homogeneous** with respect to the categories. For example, when there are only two categories, such as success and failure, defective and nondefective, and so on, the test for homogeneity is really

338　CHAPTER 9　TESTS OF HYPOTHESES FOR A SINGLE SAMPLE

a test of the equality of r binomial parameters. Calculation of expected frequencies, determination of degrees of freedom, and computation of the chi-square statistic for the test for homogeneity are identical to the test for independence.

EXERCISES FOR SECTION 9-8

9-93. A company operates four machines three shifts each day. From production records, the following data on the number of breakdowns are collected:

	Machines			
Shift	A	B	C	D
1	41	20	12	16
2	31	11	9	14
3	15	17	16	10

Test the hypothesis (using $\alpha = 0.05$) that breakdowns are independent of the shift. Find the P-value for this test.

9-94. Patients in a hospital are classified as surgical or medical. A record is kept of the number of times patients require nursing service during the night and whether or not these patients are on Medicare. The data are presented here:

	Patient Category	
Medicare	Surgical	Medical
Yes	46	52
No	36	43

Test the hypothesis (using $\alpha = 0.01$) that calls by surgical-medical patients are independent of whether the patients are receiving Medicare. Find the P-value for this test.

9-95. Grades in a statistics course and an operations research course taken simultaneously were as follows for a group of students.

	Operation Research Grade			
Statistics Grade	A	B	C	Other
A	25	6	17	13
B	17	16	15	6
C	18	4	18	10
Other	10	8	11	20

Are the grades in statistics and operations research related? Use $\alpha = 0.01$ in reaching your conclusion. What is the P-value for this test?

9-96. An experiment with artillery shells yields the following data on the characteristics of lateral deflections and ranges. Would you conclude that deflection and range are independent? Use $\alpha = 0.05$. What is the P-value for this test?

	Lateral Deflection		
Range (yards)	Left	Normal	Right
0–1,999	6	14	8
2,000–5,999	9	11	4
6,000–11,999	8	17	6

9-97. A study is being made of the failures of an electronic component. There are four types of failures possible and two mounting positions for the device. The following data have been taken:

	Failure Type			
Mounting Position	A	B	C	D
1	22	46	18	9
2	4	17	6	12

Would you conclude that the type of failure is independent of the mounting position? Use $\alpha = 0.01$. Find the P-value for this test.

9-98. A random sample of students is asked their opinions on a proposed core curriculum change. The results are as follows.

	Opinion	
Class	Favoring	Opposing
Freshman	120	80
Sophomore	70	130
Junior	60	70
Senior	40	60

Test the hypothesis that opinion on the change is independent of class standing. Use $\alpha = 0.05$. What is the P-value for this test?

9-99. An article in *British Medical Journal* (1986, Vol. 292, pp. 879–882) "Comparison of treatment of renal calculi by operative surgery, percutaneous nephrolithotomy, and extracorporeal shock wave lithotripsy," found that percutaneous nephrolithotomy (PN) had a success rate in removing kidney stones of 289 out of 350 (83%) patients. However, when the stone diameter was considered the results looked different. For stones of <2 cm, 87% (234/270) of cases were successful. For stones of ≥2 cm, a success rate of 69% (55/80) was observed for PN.

(a) Are the successes and size of stones independent? Use $\alpha = 0.05$.
(b) Find the P-value for this test.

Supplemental Exercises

9-100. Suppose we wish to test the hypothesis H_0: $\mu = 85$ versus the alternative H_1: $\mu > 85$ where $\sigma = 16$. Suppose that the true mean is $\mu = 86$ and that in the practical context of the problem this is not a departure from $\mu_0 = 85$ that has practical significance.
(a) For a test with $\alpha = 0.01$, compute β for the sample sizes $n = 25, 100, 400$, and 2500 assuming that $\mu = 86$.
(b) Suppose the sample average is $\bar{x} = 86$. Find the P-value for the test statistic for the different sample sizes specified in part (a). Would the data be statistically significant at $\alpha = 0.01$?
(c) Comment on the use of a large sample size in this problem.

9-101. A manufacturer of semiconductor devices takes a random sample of size n of chips and tests them, classifying each chip as defective or nondefective. Let $X_i = 0$ if the chip is nondefective and $X_i = 1$ if the chip is defective. The sample fraction defective is

$$\hat{p} = \frac{X_1 + X_2 + \cdots + X_n}{n}$$

What are the sampling distribution, the sample mean, and sample variance estimates of $\hat{p}$ when
(a) The sample size is $n = 50$?
(b) The sample size is $n = 80$?
(c) The sample size is $n = 100$?
(d) Compare your answers to parts (a)–(c) and comment on the effect of sample size on the variance of the sampling distribution.

9-102. Consider the situation of Exercise 9-101. After collecting a sample, we are interested in testing H_0: $p = 0.10$ versus H_1: $p \neq 0.10$ with $\alpha = 0.05$. For each of the following situations, compute the p-value for this test:
(a) $n = 50, \hat{p} = 0.095$
(b) $n = 100, \hat{p} = 0.095$
(c) $n = 500, \hat{p} = 0.095$
(d) $n = 1000, \hat{p} = 0.095$
(e) Comment on the effect of sample size on the observed P-value of the test.

9-103. An inspector of flow metering devices used to administer fluid intravenously will perform a hypothesis test to determine whether the mean flow rate is different from the flow rate setting of 200 milliliters per hour. Based on prior information the standard deviation of the flow rate is assumed to be known and equal to 12 milliliters per hour. For each of the following sample sizes, and a fixed $\alpha = 0.05$, find the probability of a type II error if the true mean is 205 milliliters per hour.

(a) $n = 20$
(b) $n = 50$
(c) $n = 100$
(d) Does the probability of a type II error increase or decrease as the sample size increases? Explain your answer.

9-104. Suppose that in Exercise 9-103, the experimenter had believed that $\sigma = 14$. For each of the following sample sizes, and a fixed $\alpha = 0.05$, find the probability of a type II error if the true mean is 205 milliliters per hour.
(a) $n = 20$
(b) $n = 50$
(c) $n = 100$
(d) Comparing your answers to those in Exercise 9-103, does the probability of a type II error increase or decrease with the increase in standard deviation? Explain your answer.

9-105. The marketers of shampoo products know that customers like their product to have a lot of foam. A manufacturer of shampoo claims that the foam height of his product exceeds 200 millimeters. It is known from prior experience that the standard deviation of foam height is 8 millimeters. For each of the following sample sizes, and a fixed $\alpha = 0.05$, find the power of the test if the true mean is 204 millimeters.
(a) $n = 20$
(b) $n = 50$
(c) $n = 100$
(d) Does the power of the test increase or decrease as the sample size increases? Explain your answer.

9-106. Suppose we are testing H_0: $p = 0.5$ versus H_0: $p \neq 0.5$. Suppose that p is the true value of the population proportion.
(a) Using $\alpha = 0.05$, find the power of the test for $n = 100$, 150, and 300 assuming that $p = 0.6$. Comment on the effect of sample size on the power of the test.
(b) Using $\alpha = 0.01$, find the power of the test for $n = 100$, 150, and 300 assuming that $p = 0.6$. Compare your answers to those from part (a) and comment on the effect of α on the power of the test for different sample sizes.
(c) Using $\alpha = 0.05$, find the power of the test for $n = 100$, assuming $p = 0.08$. Compare your answer to part (a) and comment on the effect of the true value of p on the power of the test for the same sample size and α level.
(d) Using $\alpha = 0.01$, what sample size is required if $p = 0.6$ and we want $\beta = 0.05$? What sample is required if $p = 0.8$ and we want $\beta = 0.05$? Compare the two sample sizes and comment on the effect of the true value of p on sample size required when β is held approximately constant.

9-107. The cooling system in a nuclear submarine consists of an assembly of welded pipes through which a coolant is circulated. Specifications require that weld strength must meet or exceed 150 psi.
(a) Suppose that the design engineers decide to test the hypothesis H_0: $\mu = 150$ versus H_1: $\mu > 150$. Explain why this choice of alternative hypothesis is better than H_1: $\mu < 150$.

(b) A random sample of 20 welds results in $\bar{x} = 153.7$ psi and $s = 11.3$ psi. What conclusions can you draw about the hypothesis in part (a)? State any necessary assumptions about the underlying distribution of the data.

9-108. The mean pull-off force of an adhesive used in manufacturing a connector for an automotive engine application should be at least 75 pounds. This adhesive will be used unless there is strong evidence that the pull-off force does not meet this requirement. A test of an appropriate hypothesis is to be conducted with sample size $n = 10$ and $\alpha = 0.05$. Assume that the pull-off force is normally distributed, and σ is not known.
(a) If the true standard deviation is $\sigma = 1$, what is the risk that the adhesive will be judged acceptable when the true mean pull-off force is only 73 pounds? Only 72 pounds?
(b) What sample size is required to give a 90% chance of detecting that the true mean is only 72 pounds when $\sigma = 1$?
(c) Rework parts (a) and (b) assuming that $\sigma = 2$. How much impact does increasing the value of σ have on the answers you obtain?

9-109. A manufacturer of precision measuring instruments claims that the standard deviation in the use of the instruments is at most 0.00002 millimeter. An analyst, who is unaware of the claim, uses the instrument eight times and obtains a sample standard deviation of 0.00001 millimeter.
(a) Confirm using a test procedure and an α level of 0.01 that there is insufficient evidence to support the claim that the standard deviation of the instruments is at most 0.00002. State any necessary assumptions about the underlying distribution of the data.
(b) Explain why the sample standard deviation, $s = 0.00001$, is less than 0.00002, yet the statistical test procedure results do not support the claim.

9-110. A biotechnology company produces a therapeutic drug whose concentration has a standard deviation of 4 grams per liter. A new method of producing this drug has been proposed, although some additional cost is involved. Management will authorize a change in production technique only if the standard deviation of the concentration in the new process is less than 4 grams per liter. The researchers chose $n = 10$ and obtained the following data in grams per liter. Perform the necessary analysis to determine whether a change in production technique should be implemented.

16.628	16.630
16.622	16.631
16.627	16.624
16.623	16.622
16.618	16.626

9-111. Consider the 40 observations collected on the number of nonconforming coil springs in production batches of size 50 given in Exercise 6-85.

(a) Based on the description of the random variable and these 40 observations, is a binomial distribution an appropriate model? Perform a goodness-of-fit procedure with $\alpha = 0.05$.
(b) Calculate the P-value for this test.

9-112. Consider the 20 observations collected on the number of errors in a string of 1000 bits of a communication channel given in Exercise 6-86.
(a) Based on the description of the random variable and these 20 observations, is a binomial distribution an appropriate model? Perform a goodness-of-fit procedure with $\alpha = 0.05$.
(b) Calculate the P-value for this test.

9-113. Consider the spot weld shear strength data in Exercise 6-23. Does the normal distribution seem to be a reasonable model for these data? Perform an appropriate goodness-of-fit test to answer this question.

9-114. Consider the water quality data in Exercise 6-24.
(a) Do these data support the claim that mean concentration of suspended solids does not exceed 50 parts per million? Use $\alpha = 0.05$.
(b) What is the P-value for the test in part (a)?
(c) Does the normal distribution seem to be a reasonable model for these data? Perform an appropriate goodness-of-fit test to answer this question.

9-115. Consider the golf ball overall distance data in Exercise 6-25.
(a) Do these data support the claim that the mean overall distance for this brand of ball does not exceed 270 yards? Use $\alpha = 0.05$.
(b) What is the P-value for the test in part (a)?
(c) Do these data appear to be well modeled by a normal distribution? Use a formal goodness-of-fit test in answering this question.

9-116. Consider the baseball coefficient of restitution data in Exercise 8-86. If the mean coefficient of restitution exceeds 0.635, the population of balls from which the sample has been taken will be too "lively" and considered unacceptable for play.
(a) Formulate an appropriate hypothesis testing procedure to answer this question.
(b) Test these hypotheses and draw conclusions, using $\alpha = 0.01$.
(c) Find the P-value for this test.
(d) In Exercise 8-86(b), you found a 99% confidence interval on the mean coefficient of restitution. Does this interval, or a one-sided CI, provide additional useful information to the decision maker? Explain why or why not.

9-117. Consider the dissolved oxygen data in Exercise 8-88. Water quality engineers are interested in knowing whether these data support a claim that mean dissolved oxygen concentration is 2.5 milligrams per liter.
(a) Formulate an appropriate hypothesis testing procedure to investigate this claim.
(b) Test these hypotheses and draw conclusions, using $\alpha = 0.05$.
(c) Find the P-value for this test.

(d) In Exercise 8-88(b) you found a 95% CI on the mean dissolved oxygen concentration. Does this interval provide useful additional information beyond that of the hypothesis testing results? Explain your answer.

9-118. An article in *Food Testing and Analysis*, "Improving Reproducibility of Refractometry Measurements of Fruit Juices" (1999, Vol. 4, No. 4, pp. 13–17) measured the sugar concentration (Brix) in clear apple juice. All readings were taken at 20°C:

11.48	11.45	11.48	11.47	11.48
11.50	11.42	11.49	11.45	11.44
11.45	11.47	11.46	11.47	11.43
11.50	11.49	11.45	11.46	11.47

(a) Test the hypothesis $H_0: \mu = 11.5$ versus $H_1: \mu \neq 11.5$ using $\alpha = 0.05$. Find the P-value.
(b) Compute the power of the test if the true mean is 11.4.
(c) What sample size would be required to detect a true mean sugar concentration of 11.45 if we wanted the power of test to be at least 0.9?
(d) Explain how the question in part (a) could be answered by constructing a two-sided confidence interval on the mean sugar concentration.
(e) Is there evidence to support the assumption that the sugar concentration is normally distributed?

9-119. An article in *Growth: A Journal Devoted to Problems of Normal and Abnormal Growth*, "Comparison of Measured and Estimated Fat-Free Weight, Fat, Potassium and Nitrogen of Growing Guinea Pigs" (1982, Vol. 46, No. 4, pp. 306–321) measured the body weight (grams) for guinea pigs at their birth.

421.0	452.6	456.1	494.6	373.8
90.5	110.7	96.4	81.7	102.4
241.0	296.0	317.0	290.9	256.5
447.8	687.6	705.7	879.0	88.8
296.0	273.0	268.0	227.5	279.3
258.5	296.0			

(a) Test the hypothesis that mean body weight is 300 grams. Use $\alpha = 0.05$.
(b) What is the smallest level of significance at which you would be willing to reject the null hypothesis?
(c) Explain how you could answer the question in part (a) with a two-sided confidence interval on mean body weight.

9-120. An article in *Food Chemistry*, "A Study of Factors Affecting Extraction of Peanut (*Arachis Hypgaea* L.) Solids with Water" (1991, Vol. 42, No. 2, pp. 153–165) found the percent protein extracted from peanut milk as follows:

78.3	77.1	71.3	84.5	87.8	75.7	64.8	72.5
78.2	91.2	86.2	80.9	82.1	89.3	89.4	81.6

(a) Can you support a claim that mean percent protein extracted exceeds 80 percent? Use $\alpha = 0.05$.
(b) Is there evidence that percent protein extracted is normally distributed?
(c) What is the P-value of the test statistic computed in part (a)?

9-121. An article in *Biological Trace Element Research*, "Interaction of Dietary Calcium, Manganese, and Manganese Source (Mn Oxide or Mn Methionine Complex) or Chick Performance and Manganese Utilization" (1991, Vol. 29, No. 3, pp. 217–228) showed the following results of tissue assay for liver manganese (ppm) in chicks fed high-Ca diets.

6.02	6.08	7.11	5.73	5.32	7.10
5.29	5.84	6.03	5.99	4.53	6.81

(a) Test the hypothesis $H_0: \sigma^2 = 0.6$ versus $H_1: \sigma^2 \neq 0.6$ using $\alpha = 0.01$.
(b) What is the P-value for this test?
(c) Discuss how part (a) could be answered by constructing a 99% two-sided confidence interval for σ.

9-122. An article in *Experimental Brain Research*, "Synapses in the Granule Cell Layer of the Rat Dentate Gyrus: Serial-Sectionin Study" (1996, Vol. 112, No. 2, pp. 237–243) showed the ratio between the numbers of symmetrical and total synapses on somata and azon initial segments of reconstructed granule cells in the dentate gyrus of a 12-week-old rat:

0.65	0.90	0.78	0.94	0.40	0.94
0.91	0.86	0.53	0.84	0.42	0.50
0.50	0.68	1.00	0.57	1.00	1.00
0.84	0.9	0.91	0.92	0.96	
0.96	0.56	0.67	0.96	0.52	
0.89	0.60	0.54			

(a) Use the data to test $H_0: \sigma^2 = 0.02$ versus $H_1: \sigma^2 \neq 0.02$ using $\alpha = 0.05$.
(b) Find the P-value for the test.

9-123. An article in the *Journal of Electronic Material*, "Progress in CdZnTe Substrate Producibility and Critical Drive of IRFPA Yield Originating with CdZnTe Substrates" (1998, Vol. 27, No. 6, pp. 564–572) improved the quality of CdZnTe substrates used to produce the HgCdTe infrared focal plane arrays (IRFPAs), also defined as sensor chip assemblies (SCAs). The cut-on wave length (μm) on 11 wafers was measured and shown below:

6.06 6.16 6.57 6.67 6.98 6.17 6.17 6.93 6.73 6.87 6.76

(a) Is there evidence that the mean of cut-on wave length is not 6.50 μm?
(b) What is the P-value for this test?
(c) What sample size would be required to detect a true mean cut-on wave length of 6.25 μm with probability 95%?

(d) What is the type II error probability if the true mean cut-on wave length is 6.95 μm?

9-124. Consider the fatty acid measurements for the diet margarine described in Exercise 8-32.
(a) For the sample size $n = 6$, using a two-sided alternative hypothesis and $\alpha = 0.01$, test $H_0: \sigma^2 = 1.0$.
(b) Suppose instead of $n = 6$, the sample size was $n = 51$. Repeat the analysis performed in part (a) using $n = 51$.
(c) Compare your answers and comment on how sample size affects your conclusions drawn in parts (a) and (b).

9-125. Consider the television picture tube brightness experiment described in Exercise 8-31.
(a) For the sample size $n = 10$, do the data support the claim that the standard deviation of current is less than 20 microamps?
(b) Suppose instead of $n = 10$, the sample size was 51. Repeat the analysis performed in part (a) using $n = 51$.
(c) Compare your answers and comment on how sample size affects your conclusions drawn in parts (a) and (b).

MIND-EXPANDING EXERCISES

9-126. Suppose that we wish to test $H_0: \mu = \mu_0$ versus $H_1: \mu \neq \mu_0$, where the population is normal with known σ. Let $0 < \epsilon < \alpha$, and define the critical region so that we will reject H_0 if $z_0 > z_\epsilon$ or if $z_0 < -z_{\alpha-\epsilon}$, where z_0 is the value of the usual test statistic for these hypotheses.
(a) Show that the probability of type I error for this test is α.
(b) Suppose that the true mean is $\mu_1 = \mu_0 + \delta$. Derive an expression for β for the above test.

9-127. Derive an expression for β for the test on the variance of a normal distribution. Assume that the two-sided alternative is specified.

9-128. When $X_1, X_2, \ldots, X_n$ are independent Poisson random variables, each with parameter λ, and n is large, the sample mean $\overline{X}$ has an approximate normal distribution with mean λ and variance λ/n. Therefore,

$$Z = \frac{\overline{X} - \lambda}{\sqrt{\lambda/n}}$$

has approximately a standard normal distribution. Thus we can test $H_0: \lambda = \lambda_0$ by replacing λ in Z by λ_0. When X_i are Poisson variables, this test is preferable to the large-sample test of Section 9-2.3, which would use $S/\sqrt{n}$ in the denominator, because it is designed just for the Poisson distribution. Suppose that the number of open circuits on a semiconductor wafer has a Poisson distribution. Test data for 500 wafers indicate a total of 1038 opens. Using $\alpha = 0.05$, does this suggest that the mean number of open circuits per wafer exceeds 2.0?

9-129. When $X_1, X_2, \ldots, X_n$ is a random sample from a normal distribution and n is large, the sample standard deviation has approximately a normal distribution with mean σ and variance $\sigma^2/(2n)$. Therefore, a large-sample test for $H_0: \sigma = \sigma_0$ can be based on the statistic

$$Z = \frac{S - \sigma_0}{\sqrt{\sigma_0^2/(2n)}}$$

(a) Use this result to test $H_0: \sigma = 10$ versus $H_1: \sigma < 10$ for the golf ball overall distance data in Exercise 6-25.
(b) Find an approximately unbiased estimator of the 95 percentile $\theta = \mu + 1.645\sigma$. From the fact that $\overline{X}$ and S are independent random variables, find the standard error of the estimator of θ. How would you estimate the standard error?
(c) Consider the golf ball overall distance data in Exercise 6-25. We wish to investigate a claim that the 95 percentile of overall distance does not exceed 285 yards. Construct a test statistic that can be used for testing the appropriate hypotheses. Apply this procedure to the data from Exercise 6-25. What are your conclusions?

9-130. Let $X_1, X_2, \ldots, X_n$ be a sample from an exponential distribution with parameter λ. It can be shown that $2\lambda \sum_{i=1}^{n} X_i$ has a chi-square distribution with $2n$ degrees of freedom. Use this fact to devise a test statistic and critical region for $H_0: \lambda = \lambda_0$ versus the three usual alternatives.

IMPORTANT TERMS AND CONCEPTS

α and β
Connection between hypothesis tests and confidence intervals
Critical region for a test statistic
Hypothesis test
Inference
Null distribution
Null hypothesis
One- and two-sided alternative hypotheses
Operating characteristic (OC) curves
Power of a test
P-value
Reference distribution for a test statistic
Sample size determination for hypothesis tests
Significance level of a test
Statistical hypotheses
Statistical versus practical significance
Test for goodness of fit
Test for homogeneity
Test for independence
Test statistic
Type I and type II errors

10 Statistical Inference for Two Samples

CHAPTER OUTLINE

10-1 INTRODUCTION

10-2 INFERENCE ON THE DIFFERENCE IN MEANS OF TWO NORMAL DISTRIBUTIONS, VARIANCES KNOWN

    10-2.1 Hypothesis Tests on the Difference in Means, Variances Known

    10-2.2 Type II Error and Choice of Sample Size

    10-2.3 Confidence Interval on the Difference in Means, Variances Known

10-3 INFERENCE ON THE DIFFERENCE IN MEANS OF TWO NORMAL DISTRIBUTIONS, VARIANCES UNKNOWN

    10-3.1 Hypothesis Tests on the Difference in Means, Variances Unknown

    10-3.2 Type II Error and Choice of Sample Size

    10-3.3 Confidence Interval on the Difference in Means, Variances Unknown

10-4 PAIRED t-TEST

10-5 INFERENCE ON THE VARIANCES OF TWO NORMAL DISTRIBUTIONS

    10-5.1 F Distribution

    10-5.2 Hypothesis Tests on the Ratio of Two Variances

    10-5.3 Type II Error and Choice of Sample Size

    10-5.4 Confidence Interval on the Ratio of Two Variances

10-6 INFERENCE ON TWO POPULATION PROPORTIONS

    10-6.1 Large-Sample Tests on the Difference in Population Proportions

    10-6.2 Type II Error and Choice of Sample Size

    10-6.3 Confidence Interval on the Difference in Population Proportions

10-7 SUMMARY TABLE AND ROADMAP FOR INFERENCE PROCEDURES FOR TWO SAMPLES

LEARNING OBJECTIVES

After careful study of this chapter, you should be able to do the following:
1. Structure comparative experiments involving two samples as hypothesis tests
2. Test hypotheses and construct confidence intervals on the difference in means of two normal distributions
3. Test hypotheses and construct confidence intervals on the ratio of the variances or standard deviations of two normal distributions
4. Test hypotheses and construct confidence intervals on the difference in two population proportions
5. Use the P-value approach for making decisions in hypotheses tests
6. Compute power, type II error probability, and make sample size decisions for two-sample tests on means, variances, and proportions
7. Explain and use the relationship between confidence intervals and hypothesis tests

10-1 INTRODUCTION

The previous two chapters presented hypothesis tests and confidence intervals for a single population parameter (the mean μ, the variance σ^2, or a proportion p). This chapter extends those results to the case of two independent populations.

The general situation is shown in Fig. 10-1. Population 1 has mean μ_1 and variance σ_1^2, while population 2 has mean μ_2 and variance σ_2^2. Inferences will be based on two random samples of sizes n_1 and n_2, respectively. That is, $X_{11}, X_{12}, p, X_{1n_1}$ is a random sample of n_1 observations from population 1, and $X_{21}, X_{22}, p, X_{2n_2}$ is a random sample of n_2 observations from population 2. Most of the practical applications of the procedures in this chapter arise in the context of **simple comparative experiments** in which the objective is to study the difference in the parameters of the two populations.

Engineers and scientists are often interested in comparing two different conditions to determine whether either condition produces a significant effect on the response that is observed. These conditions are sometimes called **treatments**. Example 10-1 illustrates such an experiment; the two different treatments are two paint formulations, and the response is the drying time. The purpose of the study is to determine whether the new formulation results in a significant effect—reducing drying time. In this situation, the product developer (the experimenter) randomly assigned 10 test specimens to one formulation and 10 test specimens to the other formulation. Then the paints were applied to the test specimens in random order until all 20 specimens were painted. This is an example of a **completely randomized experiment**.

When statistical significance is observed in a randomized experiment, the experimenter can be confident in the conclusion that it was the difference in treatments that resulted in the difference in response. That is, we can be confident that a **cause-and-effect** relationship has been found.

Sometimes the objects to be used in the comparison are not assigned at random to the treatments. For example, the September 1992 issue of *Circulation* (a medical journal

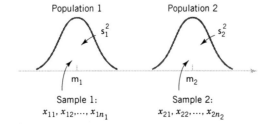

Figure 10-1 Two independent populations.

published by the American Heart Association) reports a study linking high iron levels in the body with increased risk of heart attack. The study, done in Finland, tracked 1931 men for five years and showed a statistically significant effect of increasing iron levels on the incidence of heart attacks. In this study, the comparison was not performed by randomly selecting a sample of men and then assigning some to a "low iron level" treatment and the others to a "high iron level" treatment. The researchers just tracked the subjects over time. Recall from Chapter 1 that this type of study is called an **observational study.**

It is difficult to identify causality in observational studies, because the observed statistically significant difference in response between the two groups may be due to some other underlying factor (or group of factors) that was not equalized by randomization and not due to the treatments. For example, the difference in heart attack risk could be attributable to the difference in iron levels, or to other underlying factors that form a reasonable explanation for the observed results—such as cholesterol levels or hypertension.

10-2 INFERENCE ON THE DIFFERENCE IN MEANS OF TWO NORMAL DISTRIBUTIONS, VARIANCES KNOWN

In this section we consider statistical inferences on the difference in means $\mu_1 - \mu_2$ of two normal distributions, where the variances σ_1^2 and σ_2^2 are known. The assumptions for this section are summarized as follows.

Assumptions for Two-Sample Inference

1. $X_{11}, X_{12}, \ldots, X_{1n_1}$ is a random sample from population 1.
2. $X_{21}, X_{22}, \ldots, X_{2n_2}$ is a random sample from population 2.
3. The two populations represented by X_1 and X_2 are independent.
4. Both populations are normal.

A logical point estimator of $\mu_1 - \mu_2$ is the difference in sample means $\overline{X}_1 - \overline{X}_2$. Based on the properties of expected values

$$E(\overline{X}_1 - \overline{X}_2) = E(\overline{X}_1) - E(\overline{X}_2) = \mu_1 - \mu_2$$

and the variance of $\overline{X}_1 - \overline{X}_2$ is

$$V(\overline{X}_1 - \overline{X}_2) = V(\overline{X}_1) + V(\overline{X}_2) = \frac{\sigma_1^2}{n_1} + \frac{\sigma_2^2}{n_2}$$

Based on the assumptions and the preceding results, we may state the following.

The quantity

$$Z = \frac{\overline{X}_1 - \overline{X}_2 - (\mu_1 - \mu_2)}{\sqrt{\dfrac{\sigma_1^2}{n_1} + \dfrac{\sigma_2^2}{n_2}}} \tag{10-1}$$

has a $N(0, 1)$ distribution.

10-2 INFERENCE ON THE DIFFERENCE IN MEANS OF TWO NORMAL DISTRIBUTIONS, VARIANCES KNOWN

This result will be used to form tests of hypotheses and confidence intervals on $\mu_1 - \mu_2$. Essentially, we may think of $\mu_1 - \mu_2$ as a parameter θ, and its estimator is $\hat{\Theta} = \bar{X}_1 - \bar{X}_2$ with variance $\sigma_{\hat{\Theta}}^2 = \sigma_1^2/n_1 + \sigma_2^2/n_2$. If θ_0 is the null hypothesis value specified for θ, the test statistic will be $(\hat{\Theta} - \theta_0)/\sigma_{\hat{\Theta}}$. Notice how similar this is to the test statistic for a single mean used in Equation 9-8 of Chapter 9.

10-2.1 Hypothesis Tests on the Difference in Means, Variances Known

We now consider hypothesis testing on the difference in the means $\mu_1 - \mu_2$ of two normal populations. Suppose that we are interested in testing that the difference in means $\mu_1 - \mu_2$ is equal to a specified value Δ_0. Thus, the null hypothesis will be stated as $H_0: \mu_1 - \mu_2 = \Delta_0$. Obviously, in many cases, we will specify $\Delta_0 = 0$ so that we are testing the equality of two means (i.e., $H_0: \mu_1 = \mu_2$). The appropriate test statistic would be found by replacing $\mu_1 - \mu_2$ in Equation 10-1 by Δ_0, and this test statistic would have a standard normal distribution under H_0. That is, the standard normal distribution is the reference distribution for the test statistic. Suppose that the alternative hypothesis is $H_1: \mu_1 - \mu_2 \neq \Delta_0$. Now, a sample value of $\bar{x}_1 - \bar{x}_2$ that is considerably different from Δ_0 is evidence that H_1 is true. Because Z_0 has the $N(0, 1)$ distribution when H_0 is true, we would take $-z_{\alpha/2}$ and $z_{\alpha/2}$ as the boundaries of the critical region just as we did in the single-sample hypothesis-testing problem of Section 9-2.1. This would give a test with level of significance α. Critical regions for the one-sided alternatives would be located similarly. Formally, we summarize these results below.

Tests on the Difference in Means, Variances Known

Null hypothesis: $H_0: \mu_1 - \mu_2 = \Delta_0$

Test statistic: $$Z_0 = \frac{\bar{X}_1 - \bar{X}_2 - \Delta_0}{\sqrt{\dfrac{\sigma_1^2}{n_1} + \dfrac{\sigma_2^2}{n_2}}} \qquad (10\text{-}2)$$

Alternative Hypotheses	Rejection Criterion
$H_1: \mu_1 - \mu_2 \neq \Delta_0$	$z_0 > z_{\alpha/2}$ or $z_0 < -z_{\alpha/2}$
$H_1: \mu_1 - \mu_2 > \Delta_0$	$z_0 > z_\alpha$
$H_1: \mu_1 - \mu_2 < \Delta_0$	$z_0 < -z_\alpha$

EXAMPLE 10-1
Paint Drying Time

A product developer is interested in reducing the drying time of a primer paint. Two formulations of the paint are tested; formulation 1 is the standard chemistry, and formulation 2 has a new drying ingredient that should reduce the drying time. From experience, it is known that the standard deviation of drying time is 8 minutes, and this inherent variability should be unaffected by the addition of the new ingredient. Ten specimens are painted with formulation 1, and another 10 specimens are painted with formulation 2; the 20 specimens are painted in random order. The two sample average drying times are $\bar{x}_1 = 121$ minutes and $\bar{x}_2 = 112$ minutes, respectively. What conclusions can the product developer draw about the effectiveness of the new ingredient, using $\alpha = 0.05$?

We apply the eight-step procedure to this problem as follows:

1. The quantity of interest is the difference in mean drying times, $\mu_1 - \mu_2$, and $\Delta_0 = 0$.
2. $H_0: \mu_1 - \mu_2 = 0$, or $H_0: \mu_1 = \mu_2$.
3. $H_1: \mu_1 > \mu_2$. We want to reject H_0 if the new ingredient reduces mean drying time.

348 CHAPTER 10 STATISTICAL INFERENCE FOR TWO SAMPLES

4. $\alpha = 0.05$
5. The test statistic is

$$z_0 = \frac{\bar{x}_1 - \bar{x}_2 - 0}{\sqrt{\dfrac{\sigma_1^2}{n_1} + \dfrac{\sigma_2^2}{n_2}}}$$

where $\sigma_1^2 = \sigma_2^2 = (8)^2 = 64$ and $n_1 = n_2 = 10$.
6. Reject H_0: $\mu_1 = \mu_2$ if $z_0 > 1.645 = z_{0.05}$.
7. Computations: Since $\bar{x}_1 = 121$ minutes and $\bar{x}_2 = 112$ minutes, the test statistic is

$$z_0 = \frac{121 - 112}{\sqrt{\dfrac{(8)^2}{10} + \dfrac{(8)^2}{10}}} = 2.52$$

8. Conclusion: Since $z_0 = 2.52 > 1.645$, we reject H_0: $\mu_1 = \mu_2$ at the $\alpha = 0.05$ level and conclude that adding the new ingredient to the paint significantly reduces the drying time. Alternatively, we can find the *P*-value for this test as

$$P\text{-value} = 1 - \Phi(2.52) = 0.0059$$

Therefore, H_0: $\mu_1 = \mu_2$ would be rejected at any significance level $\alpha \geq 0.0059$.

When the population variances are unknown, the sample variances s_1^2 and s_2^2 can be substituted into the test statistic Equation 10-2 to produce a **large-sample test** for the difference in means. This procedure will also work well when the populations are not necessarily normally distributed. However, both n_1 and n_2 should exceed 40 for this large-sample test to be valid.

10-2.2 Type II Error and Choice of Sample Size

Use of Operating Characteristic Curves
The operating characteristic curves (OC) in Appendix Charts VIIa, VIIb, VIIc, and VIId may be used to evaluate the type II error probability for the hypotheses in the display (10-2). These curves are also useful in determining sample size. Curves are provided for $\alpha = 0.05$ and $\alpha = 0.01$. For the two-sided alternative hypothesis, the abscissa scale of the operating characteristic curve in charts VIIa and VIIb is d, where

$$d = \frac{|\mu_1 - \mu_2 - \Delta_0|}{\sqrt{\sigma_1^2 + \sigma_2^2}} = \frac{|\Delta - \Delta_0|}{\sqrt{\sigma_1^2 + \sigma_2^2}} \tag{10-3}$$

and one must choose equal sample sizes, say, $n = n_1 = n_2$. The one-sided alternative hypotheses require the use of Charts VIIc and VIId. For the one-sided alternatives H_1: $\mu_1 - \mu_2 > \Delta_0$ or H_1: $\mu_1 - \mu_2 < \Delta_0$, the abscissa scale is also given by

$$d = \frac{|\mu_1 - \mu_2 - \Delta_0|}{\sqrt{\sigma_1^2 + \sigma_2^2}} = \frac{|\Delta - \Delta_0|}{\sqrt{\sigma_1^2 + \sigma_2^2}}$$

It is not unusual to encounter problems where the costs of collecting data differ substantially between the two populations, or where one population variance is much greater than the other.

10-2 INFERENCE ON THE DIFFERENCE IN MEANS OF TWO NORMAL DISTRIBUTIONS, VARIANCES KNOWN

In those cases, we often use unequal sample sizes. If $n_1 \neq n_2$, the operating characteristic curves may be entered with an *equivalent* value of n computed from

$$n = \frac{\sigma_1^2 + \sigma_2^2}{\sigma_1^2/n_1 + \sigma_2^2/n_2} \quad (10\text{-}4)$$

If $n_1 \neq n_2$, and their values are fixed in advance, Equation 10-4 is used directly to calculate n, and the operating characteristic curves are entered with a specified d to obtain β. If we are given d and it is necessary to determine n_1 and n_2 to obtain a specified β, say, β^*, we guess at trial values of n_1 and n_2, calculate n in Equation 10-4, and enter the curves with the specified value of d to find β. If $\beta = \beta^*$, the trial values of n_1 and n_2 are satisfactory. If $\beta \neq \beta^*$, adjustments to n_1 and n_2 are made and the process is repeated.

EXAMPLE 10-2
Paint Drying Time, Sample Size from OC Curves

Consider the paint drying time experiment from Example 10-1. If the true difference in mean drying times is as much as 10 minutes, find the sample sizes required to detect this difference with probability at least 0.90.

The appropriate value of the abscissa parameter is (since $\Delta_0 = 0$, and $\Delta = 10$)

$$d = \frac{|\mu_1 - \mu_2|}{\sqrt{\sigma_1^2 + \sigma_2^2}} = \frac{10}{\sqrt{8^2 + 8^2}} = 0.88$$

and since the detection probability or power of the test must be at least 0.9, with $\alpha = 0.05$, we find from Appendix Chart VIIc that $n = n_1 = n_2 \simeq 11$.

Sample Size Formulas

It is also possible to obtain formulas for calculating the sample sizes directly. Suppose that the null hypothesis H_0: $\mu_1 - \mu_2 = \Delta_0$ is false and that the true difference in means is $\mu_1 - \mu_2 = \Delta$, where $\Delta > \Delta_0$. One may find formulas for the sample size required to obtain a specific value of the type II error probability β for a given difference in means Δ and level of significance α.

For example, we first write the expression for the β-error for the two-sided alternative, which is

$$\beta = \Phi\left(z_{\alpha/2} - \frac{\Delta - \Delta_0}{\sqrt{\frac{\sigma_1^2}{n_1} + \frac{\sigma_2^2}{n_2}}}\right) - \Phi\left(-z_{\alpha/2} - \frac{\Delta - \Delta_0}{\sqrt{\frac{\sigma_1^2}{n_1} + \frac{\sigma_2^2}{n_2}}}\right)$$

The derivation for sample size closely follows the single-sample case in Section 9-2.2.

Sample Size for a Two-Sided Test on the Difference in Means with $n_1 = n_2$, Variances Known

For the two-sided alternative hypothesis with significance level α, the sample size $n_1 = n_2 = n$ required to detect a true difference in means of Δ with power at least $1 - \beta$ is

$$n \simeq \frac{(z_{\alpha/2} + z_\beta)^2(\sigma_1^2 + \sigma_2^2)}{(\Delta - \Delta_0)^2} \quad (10\text{-}5)$$

350 CHAPTER 10 STATISTICAL INFERENCE FOR TWO SAMPLES

This approximation is valid when $\Phi(-z_{\alpha/2} - (\Delta - \Delta_0)\sqrt{n}/\sqrt{\sigma_1^2 + \sigma_2^2})$ is small compared to β.

Sample Size for a One-Sided Test on the Difference in Means with $n_1 = n_2$, Variances Known

For a one-sided alternative hypothesis with significance level α, the sample size $n_1 = n_2 = n$ required to detect a true difference in means of $\Delta(\neq \Delta_0)$ with power at least $1 - \beta$ is

$$n = \frac{(z_\alpha + z_\beta)^2(\sigma_1^2 + \sigma_2^2)}{(\Delta - \Delta_0)^2} \quad (10\text{-}6)$$

where Δ is the true difference in means of interest. Then by following a procedure similar to that used to obtain Equation 9-17, the expression for β can be obtained for the case where $n = n_1 = n_2$.

EXAMPLE 10-3
Paint Drying Time Sample Size

To illustrate the use of these sample size equations, consider the situation described in Example 10-1, and suppose that if the true difference in drying times is as much as 10 minutes, we want to detect this with probability at least 0.90. Under the null hypothesis, $\Delta_0 = 0$. We have a one-sided alternative hypothesis with $\Delta = 10$, $\alpha = 0.05$ (so $z_\alpha = z_{0.05} = 1.645$), and since the power is 0.9, $\beta = 0.10$ (so $z_\beta = z_{0.10} = 1.28$). Therefore we may find the required sample size from Equation 10-6 as follows:

$$n = \frac{(z_\alpha + z_\beta)^2(\sigma_1^2 + \sigma_2^2)}{(\Delta - \Delta_0)^2} = \frac{(1.645 + 1.28)^2[(8)^2 + (8)^2]}{(10 - 0)^2} = 11$$

This is exactly the same as the result obtained from using the OC curves.

10-2.3 Confidence Interval on the Difference in Means, Variances Known

The $100(1 - \alpha)\%$ confidence interval on the difference in two means $\mu_1 - \mu_2$ when the variances are known can be found directly from results given previously in this section. Recall that $X_{11}, X_{12}, \ldots, X_{1n_1}$ is a random sample of n_1 observations from the first population and $X_{21}, X_{22}, \ldots, X_{2n_2}$ is a random sample of n_2 observations from the second population. The difference in sample means $\overline{X}_1 - \overline{X}_2$ is a point estimator of $\mu_1 - \mu_2$, and

$$Z = \frac{\overline{X}_1 - \overline{X}_2 - (\mu_1 - \mu_2)}{\sqrt{\frac{\sigma_1^2}{n_1} + \frac{\sigma_2^2}{n_2}}}$$

has a standard normal distribution if the two populations are normal or is approximately standard normal if the conditions of the central limit theorem apply, respectively. This implies that $P(-z_{\alpha/2} \leq Z \leq z_{\alpha/2}) = 1 - \alpha$, or

$$P\left[-z_{\alpha/2} \leq \frac{\overline{X}_1 - \overline{X}_2 - (\mu_1 - \mu_2)}{\sqrt{\frac{\sigma_1^2}{n_1} + \frac{\sigma_2^2}{n_2}}} \leq z_{\alpha/2}\right] = 1 - \alpha$$

10-2 INFERENCE ON THE DIFFERENCE IN MEANS OF TWO NORMAL DISTRIBUTIONS, VARIANCES KNOWN

This can be rearranged as

$$P\left(\overline{X}_1 - \overline{X}_2 - z_{\alpha/2}\sqrt{\frac{\sigma_1^2}{n_1} + \frac{\sigma_2^2}{n_2}} \leq \mu_1 - \mu_2 \leq \overline{X}_1 - \overline{X}_2 + z_{\alpha/2}\sqrt{\frac{\sigma_1^2}{n_1} + \frac{\sigma_2^2}{n_2}}\right) = 1 - \alpha$$

Therefore, the $100(1 - \alpha)\%$ confidence interval for $\mu_1 - \mu_2$ is defined as follows.

Confidence Interval on the Difference in Means, Variances Known

If $\bar{x}_1$ and $\bar{x}_2$ are the means of independent random samples of sizes n_1 and n_2 from two independent normal populations with known variances σ_1^2 and σ_2^2, respectively, a **$100(1 - \alpha)\%$ confidence interval for $\mu_1 - \mu_2$** is

$$\bar{x}_1 - \bar{x}_2 - z_{\alpha/2}\sqrt{\frac{\sigma_1^2}{n_1} + \frac{\sigma_2^2}{n_2}} \leq \mu_1 - \mu_2 \leq \bar{x}_1 - \bar{x}_2 + z_{\alpha/2}\sqrt{\frac{\sigma_1^2}{n_1} + \frac{\sigma_2^2}{n_2}} \quad (10\text{-}7)$$

where $z_{\alpha/2}$ is the upper $\alpha/2$ percentage point of the standard normal distribution.

The confidence level $1 - \alpha$ is exact when the populations are normal. For nonnormal populations, the confidence level is approximately valid for large sample sizes.

EXAMPLE 10-4
Aluminum Tensile Strength

Tensile strength tests were performed on two different grades of aluminum spars used in manufacturing the wing of a commercial transport aircraft. From past experience with the spar manufacturing process and the testing procedure, the standard deviations of tensile strengths are assumed to be known. The data obtained are as follows: $n_1 = 10$, $\bar{x}_1 = 87.6$, $\sigma_1 = 1$, $n_2 = 12$, $\bar{x}_2 = 74.5$, and $\sigma_2 = 1.5$. If μ_1 and μ_2 denote the true mean tensile strengths for the two grades of spars, we may find a 90% confidence interval on the difference in mean strength $\mu_1 - \mu_2$ as follows:

$$\bar{x}_1 - \bar{x}_2 - z_{\alpha/2}\sqrt{\frac{\sigma_1^2}{n_1} + \frac{\sigma_2^2}{n_2}} \leq \mu_1 - \mu_2 \leq \bar{x}_1 - \bar{x}_2 + z_{\alpha/2}\sqrt{\frac{\sigma_1^2}{n_1} + \frac{\sigma_2^2}{n_2}}$$

$$87.6 - 74.5 - 1.645\sqrt{\frac{(1)^2}{10} + \frac{(1.5)^2}{12}} \leq \mu_1 - \mu_2 \leq 87.6 - 74.5 + 1.645\sqrt{\frac{(1^2)}{10} + \frac{(1.5)^2}{12}}$$

Therefore, the 90% confidence interval on the difference in mean tensile strength (in kilograms per square millimeter) is

$$12.22 \leq \mu_1 - \mu_2 \leq 13.98 \text{ (in kilograms per square millimeter)}$$

Notice that the confidence interval does not include zero, implying that the mean strength of aluminum grade 1 (μ_1) exceeds the mean strength of aluminum grade 2 (μ_2). In fact, we can state that we are 90% confident that the mean tensile strength of aluminum grade 1 exceeds that of aluminum grade 2 by between 12.22 and 13.98 kilograms per square millimeter.

Choice of Sample Size

If the standard deviations σ_1 and σ_2 are known (at least approximately) and the two sample sizes n_1 and n_2 are equal ($n_1 = n_2 = n$, say), we can determine the sample size required so that the error in estimating $\mu_1 - \mu_2$ by $\bar{x}_1 - \bar{x}_2$ will be less than E at $100(1 - \alpha)\%$ confidence. The required sample size from each population is

Sample Size for a Confidence Interval on the Difference in Means, Variances Known

$$n = \left(\frac{z_{\alpha/2}}{E}\right)^2 (\sigma_1^2 + \sigma_2^2) \qquad (10\text{-}8)$$

Remember to round up if n is not an integer. This will ensure that the level of confidence does not drop below $100(1 - \alpha)\%$.

One-Sided Confidence Bounds

One-sided confidence bounds on $\mu_1 - \mu_2$ may also be obtained. A $100(1 - \alpha)\%$ upper-confidence bound on $\mu_1 - \mu_2$ is

One-Sided Upper Confidence Bound

$$\mu_1 - \mu_2 \leq \bar{x}_1 - \bar{x}_2 + z_\alpha \sqrt{\frac{\sigma_1^2}{n_1} + \frac{\sigma_2^2}{n_2}} \qquad (10\text{-}9)$$

and a $100(1 - \alpha)\%$ lower-confidence bound is

One-Sided Lower Confidence Bound

$$\bar{x}_1 - \bar{x}_2 - z_\alpha \sqrt{\frac{\sigma_1^2}{n_1} + \frac{\sigma_2^2}{n_2}} \leq \mu_1 - \mu_2 \qquad (10\text{-}10)$$

EXERCISES FOR SECTION 10-2

10-1. Consider the hypothesis test $H_0: \mu_1 = \mu_2$ against $H_1: \mu_1 \neq \mu_2$ with known variances $\sigma_1 = 10$ and $\sigma_2 = 5$. Suppose that sample sizes $n_1 = 10$ and $n_2 = 15$ and that $\bar{x}_1 = 4.7$ and $\bar{x}_2 = 7.8$. Use $\alpha = 0.05$.
(a) Test the hypothesis and find the P-value.
(b) Explain how the test could be conducted with a confidence interval.
(c) What is the power of the test in part (a) for a true difference in means of 3?
(d) Assuming equal sample sizes, what sample size should be used to obtain $\beta = 0.05$ if the true difference in means is 3? Assume that $\alpha = 0.05$.

10-2. Consider the hypothesis test $H_0: \mu_1 = \mu_2$ against $H_1: \mu_1 < \mu_2$ with known variances $\sigma_1 = 10$ and $\sigma_2 = 5$. Suppose that sample sizes $n_1 = 10$ and $n_2 = 15$ and that $\bar{x}_1 = 14.2$ and $\bar{x}_2 = 19.7$. Use $\alpha = 0.05$.
(a) Test the hypothesis and find the P-value.

(b) Explain how the test could be conducted with a confidence interval.
(c) What is the power of the test in part (a) if μ_1 is 4 units less than μ_2?
(d) Assuming equal sample sizes, what sample size should be used to obtain $\beta = 0.05$ if μ_1 is 4 units less than μ_2? Assume that $\alpha = 0.05$.

10-3. Consider the hypothesis test $H_0: \mu_1 = \mu_2$ against $H_1: \mu_1 > \mu_2$ with known variances $\sigma_1 = 10$ and $\sigma_2 = 5$. Suppose that sample sizes $n_1 = 10$ and $n_2 = 15$ and that $\bar{x}_1 = 24.5$ and $\bar{x}_2 = 21.3$. Use $\alpha = 0.01$.
(a) Test the hypothesis and find the P-value.
(b) Explain how the test could be conducted with a confidence interval.
(c) What is the power of the test in part (a) if μ_1 is 2 units greater than μ_2?

(d) Assuming equal sample sizes, what sample size should be used to obtain $\beta = 0.05$ if μ_1 is 2 units greater than μ_2? Assume that $\alpha = 0.05$.

10-4. Two machines are used for filling plastic bottles with a net volume of 16.0 ounces. The fill volume can be assumed normal, with standard deviation $\sigma_1 = 0.020$ and $\sigma_2 = 0.025$ ounces. A member of the quality engineering staff suspects that both machines fill to the same mean net volume, whether or not this volume is 16.0 ounces. A random sample of 10 bottles is taken from the output of each machine.

Machine 1		Machine 2	
16.03	16.01	16.02	16.03
16.04	15.96	15.97	16.04
16.05	15.98	15.96	16.02
16.05	16.02	16.01	16.01
16.02	15.99	15.99	16.00

(a) Do you think the engineer is correct? Use $\alpha = 0.05$. What is the P-value for this test?
(b) Calculate a 95% confidence interval on the difference in means. Provide a practical interpretation of this interval.
(c) What is the power of the test in part (a) for a true difference in means of 0.04?
(d) Assuming equal sample sizes, what sample size should be used to assure that $\beta = 0.05$ if the true difference in means is 0.04? Assume that $\alpha = 0.05$.

10-5. Two types of plastic are suitable for use by an electronics component manufacturer. The breaking strength of this plastic is important. It is known that $\sigma_1 = \sigma_2 = 1.0$ psi. From a random sample of size $n_1 = 10$ and $n_2 = 12$, we obtain $\bar{x}_1 = 162.5$ and $\bar{x}_2 = 155.0$. The company will not adopt plastic 1 unless its mean breaking strength exceeds that of plastic 2 by at least 10 psi.
(a) Based on the sample information, should it use plastic 1? Use $\alpha = 0.05$ in reaching a decision. Find the P-value.
(b) Calculate a 95% confidence interval on the difference in means. Suppose that the true difference in means is really 12 psi.
(c) Find the power of the test assuming that $\alpha = 0.05$.
(d) If it is really important to detect a difference of 12 psi, are the sample sizes employed in part (a) adequate, in your opinion?

10-6. The burning rates of two different solid-fuel propellants used in aircrew escape systems are being studied. It is known that both propellants have approximately the same standard deviation of burning rate; that is $\sigma_1 = \sigma_2 = 3$ centimeters per second. Two random samples of $n_1 = 20$ and $n_2 = 20$ specimens are tested; the sample mean burning rates are $\bar{x}_1 = 18$ centimeters per second and $\bar{x}_2 = 24$ centimeters per second.

(a) Test the hypothesis that both propellants have the same mean burning rate. Use $\alpha = 0.05$. What is the P-value?
(b) Construct a 95% confidence interval on the difference in means $\mu_1 - \mu_2$. What is the practical meaning of this interval?
(c) What is the β-error of the test in part (a) if the true difference in mean burning rate is 2.5 centimeters per second?
(d) Assuming equal sample sizes, what sample size is needed to obtain power of 0.9 at a true difference in means of 14 cm/s?

10-7. Two different formulations of an oxygenated motor fuel are being tested to study their road octane numbers. The variance of road octane number for formulation 1 is $\sigma_1^2 = 1.5$, and for formulation 2 it is $\sigma_2^2 = 1.2$. Two random samples of size $n_1 = 15$ and $n_2 = 20$ are tested, and the mean road octane numbers observed are $\bar{x}_1 = 89.6$ and $\bar{x}_2 = 92.5$. Assume normality.
(a) If formulation 2 produces a higher road octane number than formulation 1, the manufacturer would like to detect it. Formulate and test an appropriate hypothesis, using $\alpha = 0.05$. What is the P-value?
(b) Explain how the question in part (a) could be answered with a 95% confidence interval on the difference in mean road octane number.
(c) What sample size would be required in each population if we wanted to be 95% confident that the error in estimating the difference in mean road octane number is less than 1?

10-8. A polymer is manufactured in a batch chemical process. Viscosity measurements are normally made on each batch, and long experience with the process has indicated that the variability in the process is fairly stable with $\sigma = 20$. Fifteen batch viscosity measurements are given as follows:

724, 718, 776, 760, 745, 759, 795, 756, 742, 740, 761, 749, 739, 747, 742

A process change is made which involves switching the type of catalyst used in the process. Following the process change, eight batch viscosity measurements are taken:

735, 775, 729, 755, 783, 760, 738, 780

Assume that process variability is unaffected by the catalyst change. If the difference in mean batch viscosity is 10 or less, the manufacturer would like to detect it with a high probability.
(a) Formulate and test an appropriate hypothesis using $\alpha = 0.10$. What are your conclusions? Find the P-value.
(b) Find a 90% confidence interval on the difference in mean batch viscosity resulting from the process change.
(c) Compare the results of parts (a) and (b) and discuss your findings.

10-9. The concentration of active ingredient in a liquid laundry detergent is thought to be affected by the type of catalyst used in the process. The standard deviation of active concentration is known to be 3 grams per liter, regardless of the catalyst type. Ten observations on concentration are taken with each catalyst, and the data follow:

Catalyst 1: 57.9, 66.2, 65.4, 65.4, 65.2, 62.6, 67.6, 63.7, 67.2, 71.0

Catalyst 2: 66.4, 71.7, 70.3, 69.3, 64.8, 69.6, 68.6, 69.4, 65.3, 68.8

(a) Find a 95% confidence interval on the difference in mean active concentrations for the two catalysts. Find the P-value.

(b) Is there any evidence to indicate that the mean active concentrations depend on the choice of catalyst? Base your answer on the results of part (a).

(c) Suppose that the true mean difference in active concentration is 5 grams per liter. What is the power of the test to detect this difference if $\alpha = 0.05$?

(d) If this difference of 5 grams per liter is really important, do you consider the sample sizes used by the experimenter to be adequate? Does the assumption of normality seem reasonable for both samples?

(e) Does the assumption of normality of the differences seem reasonable?

10-3 INFERENCE ON THE DIFFERENCE IN MEANS OF TWO NORMAL DISTRIBUTIONS, VARIANCES UNKNOWN

We now extend the results of the previous section to the difference in means of the two distributions in Fig. 10-1 when the variances of both distributions σ_1^2 and σ_2^2 are unknown. If the sample sizes n_1 and n_2 exceed 40, the normal distribution procedures in Section 10-2 could be used. However, when small samples are taken, we will assume that the populations are normally distributed and base our hypotheses tests and confidence intervals on the t distribution. This nicely parallels the case of inference on the mean of a single sample with unknown variance.

10-3.1 Hypotheses Tests on the Difference in Means, Variances Unknown

We now consider tests of hypotheses on the difference in means $\mu_1 - \mu_2$ of two normal distributions where the variances σ_1^2 and σ_2^2 are unknown. A t-statistic will be used to test these hypotheses. As noted above and in Section 9-3, the normality assumption is required to develop the test procedure, but moderate departures from normality do not adversely affect the procedure. Two different situations must be treated. In the first case, we assume that the variances of the two normal distributions are unknown but equal; that is, $\sigma_1^2 = \sigma_2^2 = \sigma^2$. In the second, we assume that σ_1^2 and σ_2^2 are unknown and not necessarily equal.

Case 1: $\sigma_1^2 = \sigma_2^2 = \sigma^2$

Suppose we have two independent normal populations with unknown means μ_1 and μ_2, and unknown but equal variances, $\sigma_1^2 = \sigma_2^2 = \sigma^2$. We wish to test

$$H_0: \mu_1 - \mu_2 = \Delta_0$$
$$H_1: \mu_1 - \mu_2 \neq \Delta_0 \quad (10\text{-}11)$$

Let $X_{11}, X_{12}, \ldots, X_{1n_1}$ be a random sample of n_1 observations from the first population and $X_{21}, X_{22}, \ldots, X_{2n_2}$ be a random sample of n_2 observations from the second population. Let $\overline{X}_1, \overline{X}_2, S_1^2$, and S_2^2 be the sample means and sample variances, respectively. Now the expected value of the difference in sample means $\overline{X}_1 - \overline{X}_2$ is $E(\overline{X}_1 - \overline{X}_2) = \mu_1 - \mu_2$, so $\overline{X}_1 - \overline{X}_2$ is an unbiased estimator of the difference in means. The variance of $\overline{X}_1 - \overline{X}_2$ is

$$V(\overline{X}_1 - \overline{X}_2) = \frac{\sigma^2}{n_1} + \frac{\sigma^2}{n_2} = \sigma^2 \left(\frac{1}{n_1} + \frac{1}{n_2} \right)$$

10-3 INFERENCE ON THE DIFFERENCE IN MEANS OF TWO NORMAL DISTRIBUTIONS, VARIANCES UNKNOWN

It seems reasonable to combine the two sample variances S_1^2 and S_2^2 to form an estimator of σ^2. The **pooled estimator** of σ^2 is defined as follows.

Pooled Estimator of Variance

The **pooled estimator** of σ^2, denoted by S_p^2, is defined by

$$S_p^2 = \frac{(n_1 - 1)S_1^2 + (n_2 - 1)S_2^2}{n_1 + n_2 - 2} \qquad (10\text{-}12)$$

It is easy to see that the pooled estimator S_p^2 can be written as

$$S_p^2 = \frac{n_1 - 1}{n_1 + n_2 - 2}S_1^2 + \frac{n_2 - 1}{n_1 + n_2 - 2}S_2^2 = wS_1^2 + (1-w)S_2^2$$

where $0 < w \le 1$. Thus S_p^2 is a **weighted average** of the two sample variances S_1^2 and S_2^2, where the weights w and $1 - w$ depend on the two sample sizes n_1 and n_2. Obviously, if $n_1 = n_2 = n$, $w = 0.5$ and S_p^2 is just the arithmetic average of S_1^2 and S_2^2. If $n_1 = 10$ and $n_2 = 20$ (say), $w = 0.32$ and $1 - w = 0.68$. The first sample contributes $n_1 - 1$ degrees of freedom to S_p^2 and the second sample contributes $n_2 - 1$ degrees of freedom. Therefore, S_p^2 has $n_1 + n_2 - 2$ degrees of freedom.

Now we know that

$$Z = \frac{\bar{X}_1 - \bar{X}_2 - (\mu_1 - \mu_2)}{\sigma\sqrt{\frac{1}{n_1} + \frac{1}{n_2}}}$$

has a $N(0, 1)$ distribution. Replacing σ by S_p gives the following.

Given the assumptions of this section, the quantity

$$T = \frac{\bar{X}_1 - \bar{X}_2 - (\mu_1 - \mu_2)}{S_p\sqrt{\frac{1}{n_1} + \frac{1}{n_2}}} \qquad (10\text{-}13)$$

has a t distribution with $n_1 + n_2 - 2$ degrees of freedom.

The use of this information to test the hypotheses in Equation 10-11 is now straightforward: simply replace $\mu_1 - \mu_2$ by Δ_0, and the resulting test statistic has a t distribution with $n_1 + n_2 - 2$ degrees of freedom under $H_0: \mu_1 - \mu_2 = \Delta_0$. Therefore, the reference distribution for the test statistic is the t distribution with $n_1 + n_2 - 2$ degrees of freedom. The location of the critical region for both two- and one-sided alternatives parallels those in the one-sample case. Because a pooled estimate of variance is used, the procedure is often called the **pooled t-test.**

Case 1: Tests on the Difference in Means, Variances Unknown and Equal*

Null hypothesis: $H_0: \mu_1 - \mu_2 = \Delta_0$

Test statistic:
$$T_0 = \frac{\bar{X}_1 - \bar{X}_2 - \Delta_0}{S_p\sqrt{\frac{1}{n_1} + \frac{1}{n_2}}} \quad (10\text{-}14)$$

Alternative Hypothesis	Rejection Criterion
$H_1: \mu_1 - \mu_2 \neq \Delta_0$	$t_0 > t_{\alpha/2, n_1+n_2-2}$ or $t_0 < -t_{\alpha/2, n_1+n_2-2}$
$H_1: \mu_1 - \mu_2 > \Delta_0$	$t_0 > t_{\alpha, n_1+n_2-2}$
$H_1: \mu_1 - \mu_2 < \Delta_0$	$t_0 < -t_{\alpha, n_1+n_2-2}$

EXAMPLE 10-5
Yield from a Catalyst

Two catalysts are being analyzed to determine how they affect the mean yield of a chemical process. Specifically, catalyst 1 is currently in use, but catalyst 2 is acceptable. Since catalyst 2 is cheaper, it should be adopted, providing it does not change the process yield. A test is run in the pilot plant and results in the data shown in Table 10-1. Is there any difference between the mean yields? Use $\alpha = 0.05$, and assume equal variances.

The solution using the eight-step hypothesis-testing procedure is as follows:

1. The parameters of interest are μ_1 and μ_2, the mean process yield using catalysts 1 and 2, respectively, and we want to know if $\mu_1 - \mu_2 = 0$.
2. $H_0: \mu_1 - \mu_2 = 0$, or $H_0: \mu_1 = \mu_2$
3. $H_1: \mu_1 \neq \mu_2$
4. $\alpha = 0.05$
5. The test statistic is

$$t_0 = \frac{\bar{x}_1 - \bar{x}_2 - 0}{s_p\sqrt{\frac{1}{n_1} + \frac{1}{n_2}}}$$

Table 10-1 Catalyst Yield Data, Example 10-5

Observation Number	Catalyst 1	Catalyst 2
1	91.50	89.19
2	94.18	90.95
3	92.18	90.46
4	95.39	93.21
5	91.79	97.19
6	89.07	97.04
7	94.72	91.07
8	89.21	92.75
	$\bar{x}_1 = 92.255$	$\bar{x}_2 = 92.733$
	$s_1 = 2.39$	$s_2 = 2.98$

*While we have given the development of this procedure for the case where the sample sizes could be different, there is an advantage to using equal sample sizes $n_1 = n_2 = n$. When the sample sizes are the same from both populations, the t-test is more robust to the assumption of equal variances.

6. Reject H_0 if $t_0 > t_{0.025,14} = 2.145$ or if $t_0 < -t_{0.025,14} = -2.145$.
7. Computations: From Table 10-1 we have $\bar{x}_1 = 92.255$, $s_1 = 2.39$, $n_1 = 8$, $\bar{x}_2 = 92.733$, $s_2 = 2.98$, and $n_2 = 8$. Therefore

$$s_p^2 = \frac{(n_1 - 1)s_1^2 + (n_2 - 1)s_2^2}{n_1 + n_2 - 2} = \frac{(7)(2.39)^2 + 7(2.98)^2}{8 + 8 - 2} = 7.30$$

$$s_p = \sqrt{7.30} = 2.70$$

and

$$t_0 = \frac{\bar{x}_1 - \bar{x}_2}{2.70\sqrt{\frac{1}{n_1} + \frac{1}{n_2}}} = \frac{92.255 - 92.733}{2.70\sqrt{\frac{1}{8} + \frac{1}{8}}} = -0.35$$

8. Conclusions: Since $-2.145 < t_0 = -0.35 < 2.145$, the null hypothesis cannot be rejected. That is, at the 0.05 level of significance, we do not have strong evidence to conclude that catalyst 2 results in a mean yield that differs from the mean yield when catalyst 1 is used.

A *P*-value could also be used for decision making in this example. From Appendix Table V we find that $t_{0.40,14} = 0.258$ and $t_{0.25,14} = 0.692$. Therefore, since $0.258 < 0.35 < 0.692$, we conclude that lower and upper bounds on the *P*-value are $0.50 < P < 0.80$. Therefore, since the *P*-value exceeds $\alpha = 0.05$, the null hypothesis cannot be rejected.

The Minitab two-sample *t*-test and confidence interval procedure for Example 10-5 follows:

Minitab Computations

Two-Sample T-Test and CI: Cat 1, Cat 2

Two-sample T for Cat 1 vs Cat 2

	N	Mean	StDev	SE Mean
Cat 1	8	92.26	2.39	0.84
Cat 2	8	92.73	2.99	1.1

Difference = mu Cat 1 − mu Cat 2
Estimate for difference: −0.48
95% CI for difference: (−3.37, 2.42)
T-Test of difference = 0 (vs not =): T-Value = −0.35 P-Value = 0.730 DF = 14
Both use Pooled StDev = 2.70

Notice that the numerical results are essentially the same as the manual computations in Example 10-5. The *P*-value is reported as $P = 0.73$. The two-sided CI on $\mu_1 - \mu_2$ is also reported. We will give the computing formula for the CI in Section 10-3.3. Figure 10-2 shows the normal probability plot of the two samples of yield data and comparative box plots. The normal probability plots indicate that there is no problem with the normality assumption. Furthermore, both straight lines have similar slopes, providing some verification of the assumption of equal variances. The comparative box plots indicate that there is no obvious difference in the two catalysts, although catalyst 2 has slightly greater sample variability.

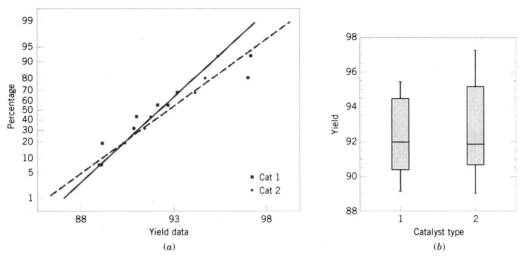

Figure 10-2 Normal probability plot and comparative box plot for the catalyst yield data in Example 10-5. (a) Normal probability plot, (b) Box plots.

Case 2: $\sigma_1^2 \neq \sigma_2^2$

In some situations, we cannot reasonably assume that the unknown variances σ_1^2 and σ_2^2 are equal. There is not an exact t-statistic available for testing $H_0: \mu_1 - \mu_2 = \Delta_0$ in this case. However, an approximate result can be applied.

Case 2: Test Statistic for the Difference in Means, Variances Unknown and Not Assumed Equal

If $H_0: \mu_1 - \mu_2 = \Delta_0$ is true, the statistic

$$T_0^* = \frac{\bar{X}_1 - \bar{X}_2 - \Delta_0}{\sqrt{\dfrac{S_1^2}{n_1} + \dfrac{S_2^2}{n_2}}} \tag{10-15}$$

is distributed approximately as t with degrees of freedom given by

$$v = \frac{\left(\dfrac{s_1^2}{n_1} + \dfrac{s_2^2}{n_2}\right)^2}{\dfrac{(s_1^2/n_1)^2}{n_1 - 1} + \dfrac{(s_2^2/n_2)^2}{n_2 - 1}} \tag{10-16}$$

If v is not an integer, round down to the nearest integer.

Therefore, if $\sigma_1^2 \neq \sigma_2^2$, the hypotheses on differences in the means of two normal distributions are tested as in the equal variances case, except that T_0^* is used as the test statistic and $n_1 + n_2 - 2$ is replaced by v in determining the degrees of freedom for the test.

EXAMPLE 10-6
Arsenic in Water

Arsenic concentration in public drinking water supplies is a potential health risk. An article in the *Arizona Republic* (Sunday, May 27, 2001) reported drinking water arsenic concentrations in parts per billion (ppb) for 10 metropolitan Phoenix communities and 10 communities in rural Arizona. The data follow:

Metro Phoenix ($\bar{x}_1 = 12.5$, $s_1 = 7.63$)	Rural Arizona ($\bar{x}_2 = 27.5$, $s_2 = 15.3$)
Phoenix, 3	Rimrock, 48
Chandler, 7	Goodyear, 44
Gilbert, 25	New River, 40
Glendale, 10	Apache Junction, 38
Mesa, 15	Buckeye, 33
Paradise Valley, 6	Nogales, 21
Peoria, 12	Black Canyon City, 20
Scottsdale, 25	Sedona, 12
Tempe, 15	Payson, 1
Sun City, 7	Casa Grande, 18

We wish to determine it there is any difference in mean arsenic concentrations between metropolitan Phoenix communities and communities in rural Arizona. Figure 10-3 shows a normal probability plot for the two samples of arsenic concentration. The assumption of normality appears quite reasonable, but since the slopes of the two straight lines are very different, it is unlikely that the population variances are the same.

Applying the eight-step procedure gives the following:

1. The parameters of interest are the mean arsenic concentrations for the two geographic regions, say, μ_1 and μ_2, and we are interested in determining whether $\mu_1 - \mu_2 = 0$.
2. $H_0: \mu_1 - \mu_2 = 0$, or $H_0: \mu_1 = \mu_2$
3. $H_1: \mu_1 \neq \mu_2$
4. $\alpha = 0.05$ (say)
5. The test statistic is

$$t_0^* = \frac{\bar{x}_1 - \bar{x}_2 - 0}{\sqrt{\dfrac{s_1^2}{n_1} + \dfrac{s_2^2}{n_2}}}$$

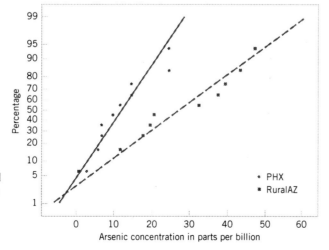

Figure 10-3 Normal probability plot of the arsenic concentration data from Example 10-6.

6. The degrees of freedom on t_0^* are found from Equation 10-16 as

$$v = \frac{\left(\frac{s_1^2}{n_1} + \frac{s_2^2}{n_2}\right)^2}{\frac{(s_1^2/n_1)^2}{n_1-1} + \frac{(s_2^2/n_2)^2}{n_2-1}} = \frac{\left[\frac{(7.63)^2}{10} + \frac{(15.3)^2}{10}\right]^2}{\frac{[(7.63)^2/10]^2}{9} + \frac{[(15.3)^2/10]^2}{9}} = 13.2 \simeq 13$$

Therefore, using $\alpha = 0.05$, we would reject $H_0: \mu_1 = \mu_2$ if $t_0^* > t_{0.025,13} = 2.160$ or if $t_0^* < -t_{0.025,13} = -2.160$

7. Computations: Using the sample data we find

$$t_0^* = \frac{\bar{x}_1 - \bar{x}_2}{\sqrt{\frac{s_1^2}{n_1} + \frac{s_2^2}{n_2}}} = \frac{12.5 - 27.5}{\sqrt{\frac{(7.63)^2}{10} + \frac{(15.3)^2}{10}}} = -2.77$$

8. Conclusions: Because $t_0^* = -2.77 < t_{0.025,13} = -2.160$, we reject the null hypothesis. Therefore, there is evidence to conclude that mean arsenic concentration in the drinking water in rural Arizona is different from the mean arsenic concentration in metropolitan Phoenix drinking water. Furthermore, the mean arsenic concentration is higher in rural Arizona communities. The P-value for this test is approximately $P = 0.016$.

The Minitab output for this example follows:

Minitab Computations

Two-Sample T-Test and CI: PHX, RuralAZ

Two-sample T for PHX vs RuralAZ

	N	Mean	StDev	SE Mean
PHX	10	12.50	7.63	2.4
RuralAZ	10	27.5	15.3	4.9

Difference = mu PHX − mu RuralAZ
Estimate for difference: −15.00
95% CI for difference: (−26.71, −3.29)
T-Test of difference = 0 (vs not =): T-Value = −2.77   P-Value = 0.016 DF = 13

The numerical results from Minitab exactly match the calculations from Example 10-6. Note that a two-sided 95% CI on $\mu_1 - \mu_2$ is also reported. We will discuss its computation in Section 10-3.3; however, note that the interval does not include zero. Indeed, the upper 95% of confidence limit is −3.29 ppb, well below zero, and the mean observed difference is $\bar{x}_1 - \bar{x}_2 = 12.5 - 27.5 = -15$ ppb.

10-3.2 Type II Error and Choice of Sample Size

The operating characteristic curves in Appendix Charts VIIe, VIIf, VIIg, and VIIh are used to evaluate the type II error for the case where $\sigma_1^2 = \sigma_2^2 = \sigma^2$. Unfortunately, when $\sigma_1^2 \neq \sigma_2^2$, the distribution of T_0^* is unknown if the null hypothesis is false, and no operating characteristic curves are available for this case.

For the two-sided alternative $H_1: \mu_1 - \mu_2 = \Delta \neq \Delta_0$, when $\sigma_1^2 = \sigma_2^2 = \sigma^2$ and $n_1 = n_2 = n$, Charts VIIe and VIIf are used with

$$d = \frac{|\Delta - \Delta_0|}{2\sigma} \quad (10\text{-}17)$$

where Δ is the true difference in means that is of interest. To use these curves, they must be entered with the sample size $n^* = 2n - 1$. For the one-sided alternative hypothesis, we use Charts VIIg and VIIh and define d and Δ as in Equation 10-17. It is noted that the parameter d is a function of σ, which is unknown. As in the single-sample t-test, we may have to rely on a prior estimate of σ or use a subjective estimate. Alternatively, we could define the differences in the mean that we wish to detect relative to σ.

EXAMPLE 10-7
Yield from Catalyst Sample Size

Consider the catalyst experiment in Example 10-5. Suppose that, if catalyst 2 produces a mean yield that differs from the mean yield of catalyst 1 by 4.0%, we would like to reject the null hypothesis with probability at least 0.85. What sample size is required?

Using $s_p = 2.70$ as a rough estimate of the common standard deviation σ, we have $d = |\Delta|/2\sigma = |4.0|/[(2)(2.70)] = 0.74$. From Appendix Chart VIIe with $d = 0.74$ and $\beta = 0.15$, we find $n^* = 20$, approximately. Therefore, since $n^* = 2n - 1$,

$$n = \frac{n^* + 1}{2} = \frac{20 + 1}{2} = 10.5 \simeq 11(\text{say})$$

and we would use sample sizes of $n_1 = n_2 = n = 11$.

Minitab will also perform power and sample size calculations for the two-sample t-test (equal variances). The output from Example 10-7 is as follows:

Minitab Computations

Power and Sample Size

2-Sample t Test
Testing mean 1 = mean 2 (versus not =)
Calculating power for mean 1 = mean 2 + difference
Alpha = 0.05 Sigma = 2.7

Difference	Sample Size	Target Power	Actual Power
4	10	0.8500	0.8793

The results agree fairly closely with the results obtained from the O.C. curve.

10-3.3 Confidence Interval on the Difference in Means, Variances Unknown

Case 1: $\sigma_1^2 = \sigma_2^2 = \sigma^2$
To develop the confidence interval for the difference in means $\mu_1 - \mu_2$ when both variances are equal, note that the distribution of the statistic

$$T = \frac{\bar{X}_1 - \bar{X}_2 - (\mu_1 - \mu_2)}{S_p\sqrt{\frac{1}{n_1} + \frac{1}{n_2}}} \quad (10\text{-}18)$$

is the t distribution with $n_1 + n_2 - 2$ degrees of freedom. Therefore $P(-t_{\alpha/2, n_1+n_2-2} \le T \le t_{\alpha/2, n_1+n_2-2}) = 1 - \alpha$. Now substituting Equation 10-18 for T and manipulating the quantities inside the probability statement will lead to the $100(1 - \alpha)\%$ confidence interval on $\mu_1 - \mu_2$.

Case 1: Confidence Interval on the Difference in Means, Variances Unknowns and Equal

If $\bar{x}_1, \bar{x}_2, s_1^2$ and s_2^2 are the sample means and variances of two random samples of sizes n_1 and n_2, respectively, from two independent normal populations with unknown but equal variances, then a **$100(1 - \alpha)\%$ confidence interval on the difference in means $\mu_1 - \mu_2$ is**

$$\bar{x}_1 - \bar{x}_2 - t_{\alpha/2, n_1+n_2-2} s_p \sqrt{\frac{1}{n_1} + \frac{1}{n_2}}$$

$$\le \mu_1 - \mu_2 \le \bar{x}_1 - \bar{x}_2 + t_{\alpha/2, n_1+n_2-2} s_p \sqrt{\frac{1}{n_1} + \frac{1}{n_2}} \quad (10\text{-}19)$$

where $s_p = \sqrt{[(n_1 - 1)s_1^2 + (n_2 - 1)s_2^2]/(n_1 + n_2 - 2)}$ is the pooled estimate of the common population standard deviation, and $t_{\alpha/2, n_1+n_2-2}$ is the upper $\alpha/2$ percentage point of the t distribution with $n_1 + n_2 - 2$ degrees of freedom.

EXAMPLE 10-8
Cement Hydration

An article in the journal *Hazardous Waste and Hazardous Materials* (Vol. 6, 1989) reported the results of an analysis of the weight of calcium in standard cement and cement doped with lead. Reduced levels of calcium would indicate that the hydration mechanism in the cement is blocked and would allow water to attack various locations in the cement structure. Ten samples of standard cement had an average weight percent calcium of $\bar{x}_1 = 90.0$, with a sample standard deviation of $s_1 = 5.0$, while 15 samples of the lead-doped cement had an average weight percent calcium of $\bar{x}_2 = 87.0$, with a sample standard deviation of $s_2 = 4.0$.

We will assume that weight percent calcium is normally distributed and find a 95% confidence interval on the difference in means, $\mu_1 - \mu_2$, for the two types of cement. Furthermore, we will assume that both normal populations have the same standard deviation.

The pooled estimate of the common standard deviation is found using Equation 10-12 as follows:

$$s_p^2 = \frac{(n_1 - 1)s_1^2 + (n_2 - 1)s_2^2}{n_1 + n_2 - 2}$$

$$= \frac{9(5.0)^2 + 14(4.0)^2}{10 + 15 - 2}$$

$$= 19.52$$

Therefore, the pooled standard deviation estimate is $s_p = \sqrt{19.52} = 4.4$. The 95% confidence interval is found using Equation 10-19:

$$\bar{x}_1 - \bar{x}_2 - t_{0.025, 23} s_p \sqrt{\frac{1}{n_1} + \frac{1}{n_2}} \le \mu_1 - \mu_2 \le \bar{x}_1 - \bar{x}_2 + t_{0.025, 23} s_p \sqrt{\frac{1}{n_1} + \frac{1}{n_2}}$$

or upon substituting the sample values and using $t_{0.025,23} = 2.069$,

$$90.0 - 87.0 - 2.069(4.4)\sqrt{\frac{1}{10} + \frac{1}{15}} \leq \mu_1 - \mu_2$$

$$\leq 90.0 - 87.0 + 2.069(4.4)\sqrt{\frac{1}{10} + \frac{1}{15}}$$

which reduces to

$$-0.72 \leq \mu_1 - \mu_2 \leq 6.72$$

Notice that the 95% confidence interval includes zero; therefore, at this level of confidence we cannot conclude that there is a difference in the means. Put another way, there is no evidence that doping the cement with lead affected the mean weight percent of calcium; therefore, we cannot claim that the presence of lead affects this aspect of the hydration mechanism at the 95% level of confidence.

Case 2: $\sigma_1^2 \neq \sigma_2^2$

In many situations it is not reasonable to assume that $\sigma_1^2 = \sigma_2^2$. When this assumption is unwarranted, we may still find a $100(1 - \alpha)\%$ confidence interval on $\mu_1 - \mu_2$ using the fact that $T^* = [\overline{X}_1 - \overline{X}_2 - (\mu_1 - \mu_2)]/\sqrt{S_1^2/n_1 + S_2^2/n_2}$ is distributed approximately as t with degrees of freedom v given by Equation 10-16. The CI expression follows.

Case 2: Approximate Confidence Interval on the Difference in Means, Variances Unknown Are Not Assumed Equal

If $\overline{x}_1, \overline{x}_2, s_1^2$, and s_2^2 are the means and variances of two random samples of sizes n_1 and n_2, respectively, from two independent normal populations with unknown and unequal variances, an approximate $100(1 - \alpha)\%$ confidence interval on the difference in means $\mu_1 - \mu_2$ is

$$\overline{x}_1 - \overline{x}_2 - t_{\alpha/2,v}\sqrt{\frac{s_1^2}{n_1} + \frac{s_2^2}{n_2}} \leq \mu_1 - \mu_2 \leq \overline{x}_1 - \overline{x}_2 + t_{\alpha/2,v}\sqrt{\frac{s_1^2}{n_1} + \frac{s_2^2}{n_2}} \quad (10\text{-}20)$$

where v is given by Equation 10-16 and $t_{\alpha/2,v}$ is the upper $\alpha/2$ percentage point of the t distribution with v degrees of freedom.

EXERCISES FOR SECTION 10-3

10-10. Consider the hypothesis test $H_0: \mu_1 = \mu_2$ against $H_1: \mu_1 \neq \mu_2$. Suppose that sample sizes are $n_1 = 15$ and $n_2 = 15$, that $\overline{x}_1 = 4.7$ and $\overline{x}_2 = 7.8$ and that $s_1^2 = 4$ and $s_2^2 = 6.25$. Assume that $\sigma_1^2 = \sigma_2^2$ and that the data are drawn from normal distributions. Use $\alpha = 0.05$.
(a) Test the hypothesis and find the P-value.
(b) Explain how the test could be conducted with a confidence interval.
(c) What is the power of the test in part (a) for a true difference in means of 3?
(d) Assuming equal sample sizes, what sample size should be used to obtain $\beta = 0.05$ if the true difference in means is -2? Assume that $\alpha = 0.05$.

10-11. Consider the hypothesis test $H_0: \mu_1 = \mu_2$ against $H_1: \mu_1 < \mu_2$. Suppose that sample sizes $n_1 = 15$ and $n_2 = 15$, that $\overline{x}_1 = 6.2$ and $\overline{x}_2 = 7.8$ and that $s_1^2 = 4$ and $s_2^2 = 6.25$. Assume that $\sigma_1^2 = \sigma_2^2$ and that the data are drawn from normal distributions. Use $\alpha = 0.05$.
(a) Test the hypothesis and find the P-value.
(b) Explain how the test could be conducted with a confidence interval.
(c) What is the power of the test in part (a) if μ_1 is 3 units less than μ_2?
(d) Assuming equal sample sizes, what sample size should be used to obtain $\beta = 0.05$ if μ_1 is 2.5 units less than μ_2? Assume that $\alpha = 0.05$.

10-12. Consider the hypothesis test $H_0: \mu_1 = \mu_2$ against $H_1: \mu_1 > \mu_2$. Suppose that sample sizes $n_1 = 10$ and $n_2 = 10$, that $\bar{x}_1 = 7.8$ and $\bar{x}_2 = 5.6$ and that $s_1^2 = 4$ and $s_2^2 = 9$. Assume that $\sigma_1^2 = \sigma_2^2$ and that the data are drawn from normal distributions. Use $\alpha = 0.05$.
(a) Test the hypothesis and find the P-value.
(b) Explain how the test could be conducted with a confidence interval.
(c) What is the power of the test in part (a) if μ_1 is 3 units greater than μ_2?
(d) Assuming equal sample sizes, what sample size should be used to obtain $\beta = 0.05$ if μ_1 is 3 units greater than μ_2? Assume that $\alpha = 0.05$.

10-13. The diameter of steel rods manufactured on two different extrusion machines is being investigated. Two random samples of sizes $n_1 = 15$ and $n_2 = 17$ are selected, and the sample means and sample variances are $\bar{x}_1 = 8.73$, $s_1^2 = 0.35$, $\bar{x}_2 = 8.68$, and $s_2^2 = 0.40$, respectively. Assume that $\sigma_1^2 = \sigma_2^2$ and that the data are drawn from a normal distribution.
(a) Is there evidence to support the claim that the two machines produce rods with different mean diameters? Use $\alpha = 0.05$ in arriving at this conclusion. Find the P-value.
(b) Construct a 95% confidence interval for the difference in mean rod diameter. Interpret this interval.

10-14. An article in *Fire Technology* investigated two different foam expanding agents that can be used in the nozzles of fire-fighting spray equipment. A random sample of five observations with an aqueous film-forming foam (AFFF) had a sample mean of 4.7 and a standard deviation of 0.6. A random sample of five observations with alcohol-type concentrates (ATC) had a sample mean of 6.9 and a standard deviation 0.8.
(a) Can you draw any conclusions about differences in mean foam expansion? Assume that both populations are well represented by normal distributions with the same standard deviations.
(b) Find a 95% confidence interval on the difference in mean foam expansion of these two agents.

10-15. Two catalysts may be used in a batch chemical process. Twelve batches were prepared using catalyst 1, resulting in an average yield of 86 and a sample standard deviation of 3. Fifteen batches were prepared using catalyst 2, and they resulted in an average yield of 89 with a standard deviation of 2. Assume that yield measurements are approximately normally distributed with the same standard deviation.
(a) Is there evidence to support a claim that catalyst 2 produces a higher mean yield than catalyst 1? Use $\alpha = 0.01$.
(b) Find a 95% confidence interval on the difference in mean yields that can be used to test the claim in part (a).

10-16. The deflection temperature under load for two different types of plastic pipe is being investigated. Two random samples of 15 pipe specimens are tested, and the deflection temperatures observed are as follows (in °F):

Type 1: 206, 188, 205, 187, 194, 193, 207, 185, 189, 213, 192, 210, 194, 178, 205.

Type 2: 177, 197, 206, 201, 180, 176, 185, 200, 197, 192, 198, 188, 189, 203, 192.

(a) Construct box plots and normal probability plots for the two samples. Do these plots provide support of the assumptions of normality and equal variances? Write a practical interpretation for these plots.
(b) Do the data support the claim that the deflection temperature under load for type 1 pipe exceeds that of type 2? In reaching your conclusions, use $\alpha = 0.05$. Calculate a P-value.
(c) If the mean deflection temperature for type 2 pipe exceeds that of type 1 by as much as 5°F, it is important to detect this difference with probability at least 0.90. Is the choice of $n_1 = n_2 = 15$ adequate? Use $\alpha = 0.05$.

10-17. In semiconductor manufacturing, wet chemical etching is often used to remove silicon from the backs of wafers prior to metallization. The etch rate is an important characteristic in this process and known to follow a normal distribution. Two different etching solutions have been compared, using two random samples of 10 wafers for each solution. The observed etch rates are as follows (in mils per minute):

Solution 1		Solution 2	
9.9	10.6	10.2	10.0
9.4	10.3	10.6	10.2
9.3	10.0	10.7	10.7
9.6	10.3	10.4	10.4
10.2	10.1	10.5	10.3

(a) Construct normal probability plots for the two samples. Do these plots provide support for the assumptions of normality and equal variances? Write a practical interpretation for these plots.
(b) Do the data support the claim that the mean etch rate is the same for both solutions? In reaching your conclusions, use $\alpha = 0.05$ and assume that both population variances are equal. Calculate a P-value.
(c) Find a 95% confidence interval on the difference in mean etch rates.

10-18. Two suppliers manufacture a plastic gear used in a laser printer. The impact strength of these gears measured in foot-pounds is an important characteristic. A random sample of 10 gears from supplier 1 results in $\bar{x}_1 = 290$ and $s_1 = 12$, while another random sample of 16 gears from the second supplier results in $\bar{x}_2 = 321$ and $s_2 = 22$.
(a) Is there evidence to support the claim that supplier 2 provides gears with higher mean impact strength? Use $\alpha = 0.05$, and assume that both populations are normally distributed but the variances are not equal. What is the P-value for this test?

(b) Do the data support the claim that the mean impact strength of gears from supplier 2 is at least 25 foot-pounds higher than that of supplier 1? Make the same assumptions as in part (a).

(c) Construct a confidence interval estimate for the difference in mean impact strength, and explain how this interval could be used to answer the question posed regarding supplier-to-supplier differences.

10-19. The melting points of two alloys used in formulating solder were investigated by melting 21 samples of each material. The sample mean and standard deviation for alloy 1 was $\bar{x}_1 = 420°F$ and $s_1 = 4°F$, while for alloy 2 they were $\bar{x}_2 = 426°F$, and $s_2 = 3°F$.

(a) Do the sample data support the claim that both alloys have the same melting point? Use $\alpha = 0.05$ and assume that both populations are normally distributed and have the same standard deviation. Find the P-value for the test.

(b) Suppose that the true mean difference in melting points is 3°F. How large a sample would be required to detect this difference using an $\alpha = 0.05$ level test with probability at least 0.9? Use $\sigma_1 = \sigma_2 = 4$ as an initial estimate of the common standard deviation.

10-20. A photoconductor film is manufactured at a nominal thickness of 25 mils. The product engineer wishes to increase the mean speed of the film, and believes that this can be achieved by reducing the thickness of the film to 20 mils. Eight samples of each film thickness are manufactured in a pilot production process, and the film speed (in microjoules per square inch) is measured. For the 25-mil film the sample data result is $\bar{x}_1 = 1.15$ and $s_1 = 0.11$, while for the 20-mil film, the data yield $\bar{x}_2 = 1.06$ and $s_2 = 0.09$. Note that an increase in film speed would lower the value of the observation in microjoules per square inch.

(a) Do the data support the claim that reducing the film thickness increases the mean speed of the film? Use $\alpha = 0.10$ and assume that the two population variances are equal and the underlying population of film speed is normally distributed. What is the P-value for this test?

(b) Find a 95% confidence interval on the difference in the two means that can be used to test the claim in part (a).

10-21. Two companies manufacture a rubber material intended for use in an automotive application. The part will be subjected to abrasive wear in the field application, so we decide to compare the material produced by each company in a test. Twenty-five samples of material from each company are tested in an abrasion test, and the amount of wear after 1000 cycles is observed. For company 1, the sample mean and standard deviation of wear are $\bar{x}_1 = 20$ milligrams/1000 cycles and $s_1 = 2$ milligrams/1000 cycles, while for company 2 we obtain $\bar{x}_2 = 15$ milligrams/1000 cycles and $s_2 = 8$ milligrams/1000 cycles.

(a) Do the data support the claim that the two companies produce material with different mean wear? Use $\alpha = 0.05$, and assume each population is normally distributed but that their variances are not equal. What is the P-value for this test?

(b) Do the data support a claim that the material from company 1 has higher mean wear than the material from company 2? Use the same assumptions as in part (a).

(c) Construct confidence intervals that will address the questions in parts (a) and (b) above.

10-22. The thickness of a plastic film (in mils) on a substrate material is thought to be influenced by the temperature at which the coating is applied. A completely randomized experiment is carried out. Eleven substrates are coated at 125°F, resulting in a sample mean coating thickness of $\bar{x}_1 = 103.5$ and a sample standard deviation of $s_1 = 10.2$. Another 13 substrates are coated at 150°F, for which $\bar{x}_2 = 99.7$ and $s_2 = 20.1$ are observed. It was originally suspected that raising the process temperature would reduce mean coating thickness.

(a) Do the data support this claim? Use $\alpha = 0.01$ and assume that the two population standard deviations are not equal. Calculate an approximate P-value for this test.

(b) How could you have answered the question posed regarding the effect of temperature on coating thickness by using a confidence interval? Explain your answer.

10-23. An article in *Electronic Components and Technology Conference* (2001, Vol. 52, pp. 1167–1171) compared single versus dual spindle saw processes for copper metallized wafers. A total of 15 devices of each type were measured for the width of the backside chipouts, $\bar{x}_{single} = 66.385, s_{single} = 7.895$ and $\bar{x}_{double} = 45.278, s_{double} = 8.612$.

(a) Do the sample data support the claim that both processes have the same chip outputs? Use $\alpha = 0.05$ and assume that both populations are normally distributed and have the same variance. Find the P-value for the test.

(b) Construct a 95% two-sided confidence interval on the mean difference in spindle saw process. Compare this interval to the results in part (a).

(c) If the β-error of the test when the true difference in chip outputs is 15 should not exceed 0.1, what sample sizes must be used? Use $\alpha = 0.05$.

10-24. An article in *IEEE International Symposium on Electromagnetic Compatibility* (2002, Vol. 2, pp. 667–670) quantified the absorption of electromagnetic energy and the resulting thermal effect from cellular phones. The experimental results were obtained from in vivo experiments conducted on rats. The arterial blood pressure values (mmHg) for the control group (8 rats) during the experiment are $\bar{x}_1 = 90$, $s_1 = 5$ and for the test group (9 rats) are $\bar{x}_2 = 115, s_2 = 10$.

(a) Is there evidence to support the claim that the test group has higher mean blood pressure? Use $\alpha = 0.05$, and assume that both populations are normally distributed but the variances are not equal. What is the P-value for this test?

(b) Calculate a confidence interval to answer the question in part (a).

(c) Do the data support the claim that the mean blood pressure from the test group is at least 15 mmHg higher than the control group? Make the same assumptions as in part (a).

(d) Explain how the question in part (c) could be answered with a confidence interval.

10-25. An article in *Radio Engineering and Electronic Physics* (1984, Vol. 29, No. 3, pp. 63–66) investigated the behavior of a stochastic generator in the presence of external noise. The number of periods was measured in a sample of 100 trains for each of two different levels of noise voltage, 100 and 150 mV. For 100 mV, the mean number of periods in a train was 7.9 with $s = 2.6$. For 150 mV, the mean was 6.9 with $s = 2.4$.

(a) It was originally suspected that raising noise voltage would reduce mean number of periods. Do the data support this claim? Use $\alpha = 0.01$ and assume that each population is normally distributed and the two population variances are equal. What is the *P*-value for this test?

(b) Calculate a confidence interval to answer the question in part (a).

10-26. An article in *Technometrics* (1999, Vol. 41, pp. 202–211) studied the capability of a gauge by measuring the weights of two sheets of paper. The data are shown below.

Paper	Observations				
1	3.481	3.448	3.485	3.475	3.472
	3.477	3.472	3.464	3.472	3.470
	3.470	3.470	3.477	3.473	3.474
2	3.258	3.254	3.256	3.249	3.241
	3.254	3.247	3.257	3.239	3.250
	3.258	3.239	3.245	3.240	3.254

(a) Check the assumption that the data from each sheet are from normal distributions.

(b) Test the hypothesis that the mean weight of the two sheets are equal against the alternative that they are not (and assume equal variances). Use $\alpha = 0.05$ and assume equal variances. Find the *P*-value.

(c) Repeat the previous test with $\alpha = 0.10$.

(d) Compare your answers for parts (b) and (c) and explain why they are the same or different.

(e) Explain how the questions in parts (b) and (c) could be answered with confidence intervals.

10-27. The overall distance traveled by a golf ball is tested by hitting the ball with Iron Byron, a mechanical golfer with a swing that is said to emulate the legendary champion, Byron Nelson. Ten randomly selected balls of two different brands are tested and the overall distance measured. The data follow:

Brand 1: 275, 286, 287, 271, 283, 271, 279, 275, 263, 267

Brand 2: 258, 244, 260, 265, 273, 281, 271, 270, 263, 268

(a) Is there evidence that overall distance is approximately normally distributed? Is an assumption of equal variances justified?

(b) Test the hypothesis that both brands of ball have equal mean overall distance. Use $\alpha = 0.05$. What is the *P*-value?

(c) Construct a 95% two-sided CI on the mean difference in overall distance between the two brands of golf balls.

(d) What is the power of the statistical test in part (b) to detect a true difference in mean overall distance of 5 yards?

(e) What sample size would be required to detect a true difference in mean overall distance of 3 yards with power of approximately 0.75?

10-28. The "spring-like effect" in a golf club could be determined by measuring the coefficient of restitution (the ratio of the outbound velocity to the inbound velocity of a golf ball fired at the clubhead). Twelve randomly selected drivers produced by two clubmakers are tested and the coefficient of restitution measured. The data follow:

Club 1: 0.8406, 0.8104, 0.8234, 0.8198, 0.8235, 0.8562, 0.8123, 0.7976, 0.8184, 0.8265, 0.7773, 0.7871

Club 2: 0.8305, 0.7905, 0.8352, 0.8380, 0.8145, 0.8465, 0.8244, 0.8014, 0.8309, 0.8405, 0.8256, 0.8476

(a) Is there evidence that coefficient of restitution is approximately normally distributed? Is an assumption of equal variances justified?

(b) Test the hypothesis that both brands of clubs have equal mean coefficient of restitution. Use $\alpha = 0.05$. What is the *P*-value of the test?

(c) Construct a 95% two-sided CI on the mean difference in coefficient of restitution between the two brands of golf clubs.

(d) What is the power of the statistical test in part (b) to detect a true difference in mean coefficient of restitution of 0.2?

(e) What sample size would be required to detect a true difference in mean coefficient of restitution of 0.1 with power of approximately 0.8?

10-4 PAIRED *t*-TEST

A special case of the two-sample *t*-tests of Section 10-3 occurs when the observations on the two populations of interest are collected in **pairs**. Each pair of observations, say (X_{1j}, X_{2j}), is taken under homogeneous conditions, but these conditions may change from one pair to another. For example, suppose that we are interested in comparing two different types of tips for a hardness-testing machine. This machine presses the tip into a metal specimen with a known force. By measuring the depth of the depression caused by the tip, the hardness of the specimen can be determined. If several specimens were selected at random, half tested with tip 1, half tested with tip 2, and the pooled or independent *t*-test in Section 10-3 was applied, the results of the test could be erroneous. The metal specimens could have been cut from bar stock that was produced in different heats, or they might not be homogeneous in some other way that might affect hardness. Then the observed difference between mean hardness readings for the two tip types also includes hardness differences between specimens.

A more powerful experimental procedure is to collect the data in pairs—that is, to make two hardness readings on each specimen, one with each tip. The test procedure would then consist of analyzing the *differences* between hardness readings on each specimen. If there is no difference between tips, the mean of the differences should be zero. This test procedure is called the **paired *t*-test.**

Let $(X_{11}, X_{21}), (X_{12}, X_{22}), \ldots, (X_{1n}, X_{2n})$ be a set of n paired observations where we assume that the mean and variance of the population represented by X_1 are μ_1 and σ_1^2, and the mean and variance of the population represented by X_2 are μ_2 and σ_2^2. Define the differences between each pair of observations as $D_j = X_{1j} - X_{2j}$, $j = 1, 2, \ldots, n$. The D_j's are assumed to be normally distributed with mean

$$\mu_D = E(X_1 - X_2) = E(X_1) - E(X_2) = \mu_1 - \mu_2$$

and variance σ_D^2, so testing hypotheses about the difference between μ_1 and μ_2 can be accomplished by performing a one-sample *t*-test on μ_D. Specifically, testing $H_0: \mu_1 - \mu_2 = \Delta_0$ against $H_1: \mu_1 - \mu_2 \neq \Delta_0$ is equivalent to testing

$$H_0: \mu_D = \Delta_0$$
$$H_1: \mu_D \neq \Delta_0 \tag{10-21}$$

The test statistic is given below.

Paired *t*-Test

Null hypothesis: $H_0: \mu_D = \Delta_0$

Test statistic: $$T_0 = \frac{\overline{D} - \Delta_0}{S_D/\sqrt{n}} \tag{10-22}$$

Alternative Hypothesis	Rejection Region
$H_1: \mu_D \neq \Delta_0$	$t_0 > t_{\alpha/2,n-1}$ or $t_0 < -t_{\alpha/2,n-1}$
$H_1: \mu_D > \Delta_0$	$t_0 > t_{\alpha,n-1}$
$H_1: \mu_D < \Delta_0$	$t_0 < -t_{\alpha,n-1}$

In Equation 10-22, $\overline{D}$ is the sample average of the n differences $D_1, D_2, \ldots, D_n$, and S_D is the sample standard deviation of these differences.

Table 10-2 Strength Predictions for Nine Steel Plate Girders (Predicted Load/Observed Load)

Girder	Karlsruhe Method	Lehigh Method	Difference d_j
S1/1	1.186	1.061	0.125
S2/1	1.151	0.992	0.159
S3/1	1.322	1.063	0.259
S4/1	1.339	1.062	0.277
S5/1	1.200	1.065	0.135
S2/1	1.402	1.178	0.224
S2/2	1.365	1.037	0.328
S2/3	1.537	1.086	0.451
S2/4	1.559	1.052	0.507

EXAMPLE 10-9
Strength for Steel Girders

An article in the *Journal of Strain Analysis* (1983, Vol. 18, No. 2) compares several methods for predicting the shear strength for steel plate girders. Data for two of these methods, the Karlsruhe and Lehigh procedures, when applied to nine specific girders, are shown in Table 10-2. We wish to determine whether there is any difference (on the average) between the two methods.

The eight-step procedure is applied as follows:

1. The parameter of interest is the difference in mean shear strength between the two methods, say, $\mu_D = \mu_1 - \mu_2 = 0$.
2. $H_0: \mu_D = 0$
3. $H_1: \mu_D \neq 0$
4. $\alpha = 0.05$
5. The test statistic is

$$t_0 = \frac{\bar{d}}{s_D/\sqrt{n}}$$

6. Reject H_0 if $t_0 > t_{0.025,8} = 2.306$ or if $t_0 < -t_{0.025,8} = -2.306$.
7. Computations: The sample average and standard deviation of the differences d_j are $\bar{d} = 0.2739$ and $s_D = 0.1351$, so the test statistic is

$$t_0 = \frac{\bar{d}}{s_D/\sqrt{n}} = \frac{0.2739}{0.1351/\sqrt{9}} = 6.08$$

8. Conclusions: Since $t_0 = 6.08 > 2.306$, we conclude that the strength prediction methods yield different results. Specifically, the data indicate that the Karlsruhe method produces, on the average, higher strength predictions than does the Lehigh method. The *P*-value for $t_0 = 6.08$ is $P = 0.0003$, so the test statistic is well into the critical region.

Paired Versus Unpaired Comparisons[3]
In performing a comparative experiment, the investigator can sometimes choose between the paired experiment and the two-sample (or unpaired) experiment. If n measurements are to be made on each population, the two-sample *t*-statistic is

$$T_0 = \frac{\bar{X}_1 - \bar{X}_2 - \Delta_0}{S_p\sqrt{\frac{1}{n} + \frac{1}{n}}}$$

which would be compared to t_{2n-2}, and of course, the paired t-statistic is

$$T_0 = \frac{\overline{D} - \Delta_0}{S_D/\sqrt{n}}$$

which is compared to t_{n-1}. Notice that since

$$\overline{D} = \sum_{j=1}^{n} \frac{D_j}{n} = \sum_{j=1}^{n} \frac{(X_{1j} - X_{2j})}{n} = \sum_{j=1}^{n} \frac{X_{1j}}{n} - \sum_{j=1}^{n} \frac{X_{2j}}{n} = \overline{X}_1 - \overline{X}_2$$

the numerators of both statistics are identical. However, the denominator of the two-sample t-test is based on the assumption that X_1 and X_2 are *independent*. In many paired experiments, a strong positive correlation ρ exists between X_1 and X_2. Then it can be shown that

$$\begin{aligned} V(\overline{D}) &= V(\overline{X}_1 - \overline{X}_2 - \Delta_0) \\ &= V(\overline{X}_1) + V(\overline{X}_2) - 2\operatorname{cov}(\overline{X}_1, \overline{X}_2) \\ &= \frac{2\sigma^2(1-\rho)}{n} \end{aligned}$$

assuming that both populations X_1 and X_2 have identical variances σ^2. Furthermore, S_D^2/n estimates the variance of $\overline{D}$. Whenever there is positive correlation within the pairs, the denominator for the paired t-test will be smaller than the denominator of the two-sample t-test. This can cause the two-sample t-test to considerably understate the significance of the data if it is incorrectly applied to paired samples.

Although pairing will often lead to a smaller value of the variance of $\overline{X}_1 - \overline{X}_2$, it does have a disadvantage—namely, the paired t-test leads to a loss of $n - 1$ degrees of freedom in comparison to the two-sample t-test. Generally, we know that increasing the degrees of freedom of a test increases the power against any fixed alternative values of the parameter.

So how do we decide to conduct the experiment? Should we pair the observations or not? Although there is no general answer to this question, we can give some guidelines based on the above discussion.

1. If the experimental units are relatively homogeneous (small σ) and the correlation within pairs is small, the gain in precision attributable to pairing will be offset by the loss of degrees of freedom, so an independent-sample experiment should be used.

2. If the experimental units are relatively heterogeneous (large σ) and there is large positive correlation within pairs, the paired experiment should be used. Typically, this case occurs when the experimental units are the *same* for both treatments; as in Example 10-9, the same girders were used to test the two methods.

Implementing the rules still requires judgment, because σ and ρ are never known precisely. Furthermore, if the number of degrees of freedom is large (say, 40 or 50), the loss of $n - 1$ of them for pairing may not be serious. However, if the number of degrees of freedom is small (say, 10 or 20), losing half of them is potentially serious if not compensated for by increased precision from pairing.

Confidence Interval for μ_D

To construct the confidence interval for $\mu_D = \mu_1 - \mu_2$, note that

$$T = \frac{\overline{D} - \mu_D}{S_D/\sqrt{n}}$$

follows a t distribution with $n - 1$ degrees of freedom. Then, since $P(-t_{\alpha/2,n-1} \leq T \leq t_{\alpha/2,n-1}) = 1 - \alpha$, we can substitute for T in the above expression and perform the necessary steps to isolate $\mu_D = \mu_1 - \mu_2$ between the inequalities. This leads to the following $100(1 - \alpha)\%$ confidence interval on $\mu_1 - \mu_2$.

Confidence Interval for μ_D from Paired Samples

If $\bar{d}$ and s_D are the sample mean and standard deviation of the difference of n random pairs of normally distributed measurements, a **$100(1 - \alpha)\%$ confidence interval on the difference in means $\mu_D = \mu_1 - \mu_2$** is

$$\bar{d} - t_{\alpha/2,n-1} s_D/\sqrt{n} \leq \mu_D \leq \bar{d} + t_{\alpha/2,n-1} s_D/\sqrt{n} \qquad (10\text{-}23)$$

where $t_{\alpha/2,n-1}$ is the upper $\alpha/2\%$ point of the t-distribution with $n - 1$ degrees of freedom.

This confidence interval is also valid for the case where $\sigma_1^2 \neq \sigma_2^2$, because s_D^2 estimates $\sigma_D^2 = V(X_1 - X_2)$. Also, for large samples (say, $n \geq 30$ pairs), the explicit assumption of normality is unnecessary because of the central limit theorem.

EXAMPLE 10-10
Parallel Park Cars

The journal *Human Factors* (1962, pp. 375–380) reports a study in which $n = 14$ subjects were asked to parallel park two cars having very different wheel bases and turning radii. The time in seconds for each subject was recorded and is given in Table 10-3. From the column of observed differences we calculate $\bar{d} = 1.21$ and $s_D = 12.68$. The 90% confidence interval for $\mu_D = \mu_1 - \mu_2$ is found from Equation 10-23 as follows:

$$\bar{d} - t_{0.05,13} s_D/\sqrt{n} \leq \mu_D \leq \bar{d} + t_{0.05,13} s_D/\sqrt{n}$$
$$1.21 - 1.771(12.68)/\sqrt{14} \leq \mu_D \leq 1.21 + 1.771(12.68)/\sqrt{14}$$
$$-4.79 \leq \mu_D \leq 7.21$$

Table 10-3 Time in Seconds to Parallel Park Two Automobiles

	Automobile		Difference
Subject	1(x_{1j})	2(x_{2j})	(d_j)
1	37.0	17.8	19.2
2	25.8	20.2	5.6
3	16.2	16.8	−0.6
4	24.2	41.4	−17.2
5	22.0	21.4	0.6
6	33.4	38.4	−5.0
7	23.8	16.8	7.0
8	58.2	32.2	26.0
9	33.6	27.8	5.8
10	24.4	23.2	1.2
11	23.4	29.6	−6.2
12	21.2	20.6	0.6
13	36.2	32.2	4.0
14	29.8	53.8	−24.0

EXERCISES FOR SECTION 10-4

10-29. Consider the shear strength experiment described in Example 10-9.
(a) Construct a 95% confidence interval on the difference in mean shear strength for the two methods. Is the result you obtained consistent with the findings in Example 10-9? Explain why.
(b) Do each of the individual shear strengths have to be normally distributed for the paired t-test to be appropriate, or is it only the difference in shear strengths that must be normal? Use a normal probability plot to investigate the normality assumption.

10-30. Consider the parking data in Example 10-10.
(a) Use the paired t-test to investigate the claim that the two types of cars have different levels of difficulty to parallel park. Use $\alpha = 0.10$.
(b) Compare your results with the confidence interval constructed in Example 10-10 and comment on why they are the same or different.
(c) Investigate the assumption that the differences in parking times are normally distributed.

10-31. The manager of a fleet of automobiles is testing two brands of radial tires and assigns one tire of each brand at random to the two rear wheels of eight cars and runs the cars until the tires wear out. The data (in kilometers) follow. Find a 99% confidence interval on the difference in mean life. Which brand would you prefer, based on this calculation?

Car	Brand 1	Brand 2
1	36,925	34,318
2	45,300	42,280
3	36,240	35,500
4	32,100	31,950
5	37,210	38,015
6	48,360	47,800
7	38,200	37,810
8	33,500	33,215

10-32. A computer scientist is investigating the usefulness of two different design languages in improving programming tasks. Twelve expert programmers, familiar with both languages, are asked to code a standard function in both languages, and the time (in minutes) is recorded. The data follow:
(a) Is the assumption that the difference in coding time is normally distributed reasonable?

	Time	
Programmer	Design Language 1	Design Language 2
1	17	18
2	16	14
3	21	19
4	14	11
5	18	23
6	24	21
7	16	10
8	14	13
9	21	19
10	23	24
11	13	15
12	18	20

(b) Find a 95% confidence interval on the difference in mean coding times. Is there any indication that one design language is preferable?

10-33. Fifteen adult males between the ages of 35 and 50 participated in a study to evaluate the effect of diet and exercise on blood cholesterol levels. The total cholesterol was measured in each subject initially and then three months after

Blood Cholesterol Level		
Subject	Before	After
1	265	229
2	240	231
3	258	227
4	295	240
5	251	238
6	245	241
7	287	234
8	314	256
9	260	247
10	279	239
11	283	246
12	240	218
13	238	219
14	225	226
15	247	233

participating in an aerobic exercise program and switching to a low-fat diet. The data are shown in the accompanying table
(a) Do the data support the claim that low-fat diet and aerobic exercise are of value in producing a mean reduction in blood cholesterol levels? Use $\alpha = 0.05$. Find the P-value.
(b) Calculate a one-sided confidence limit that can be used to answer the question in part (a).

10-34. An article in the *Journal of Aircraft* (Vol. 23, 1986, pp. 859–864) described a new equivalent plate analysis method formulation that is capable of modeling aircraft structures such as cranked wing boxes and that produces results similar to the more computationally intensive finite element analysis method. Natural vibration frequencies for the cranked wing box structure are calculated using both methods, and results for the first seven natural frequencies follow:

Freq.	Finite Element Cycle/s	Equivalent Plate, Cycle/s
1	14.58	14.76
2	48.52	49.10
3	97.22	99.99
4	113.99	117.53
5	174.73	181.22
6	212.72	220.14
7	277.38	294.80

(a) Do the data suggest that the two methods provide the same mean value for natural vibration frequency? Use $\alpha = 0.05$. Find the P-value.
(b) Find a 95% confidence interval on the mean difference between the two methods.

10-35. Ten individuals have participated in a diet-modification program to stimulate weight loss. Their weight both before and after participation in the program is shown in the following list.

Subject	Before	After
1	195	187
2	213	195
3	247	221
4	201	190
5	187	175
6	210	197
7	215	199
8	246	221
9	294	278
10	310	285

(a) Is there evidence to support the claim that this particular diet-modification program is effective in producing a mean weight reduction? Use $\alpha = 0.05$.
(b) Is there evidence to support the claim that this particular diet-modification program will result in a mean weight loss of at least 10 pounds? Use $\alpha = 0.05$.
(c) Suppose that, if the diet-modification program results in mean weight loss of at least 10 pounds, it is important to detect this with probability of at least 0.90. Was the use of 10 subjects an adequate sample size? If not, how many subjects should have been used?

10-36. Two different analytical tests can be used to determine the impurity level in steel alloys. Eight specimens are tested using both procedures, and the results are shown in the following tabulation.

Specimen	Test 1	Test 2
1	1.2	1.4
2	1.3	1.7
3	1.5	1.5
4	1.4	1.3
5	1.7	2.0
6	1.8	2.1
7	1.4	1.7
8	1.3	1.6

(a) Is there sufficient evidence to conclude that tests differ in the mean impurity level, using $\alpha = 0.01$?
(b) Is there evidence to support the claim that Test 1 generates a mean difference 0.1 units lower than Test 2? Use $\alpha = 0.05$.
(c) If the mean from Test 1 is 0.1 less than the mean from Test 2, it is important to detect this with probability at least 0.90. Was the use of 8 alloys an adequate sample size? If not, how many alloys should have been used?

10-37. An article in *Neurology* (1998, Vol. 50, pp. 1246–1252) discussed that monozygotic twins share numerous physical, psychological, and pathological traits. The investigators measured an intelligence score of 10 pairs of twins and the data are as follows:

Pair	Birth Order: 1	Birth Order: 2
1	6.08	5.73
2	6.22	5.80
3	7.99	8.42
4	7.44	6.84
5	6.48	6.43
6	7.99	8.76
7	6.32	6.32
8	7.60	7.62
9	6.03	6.59
10	7.52	7.67

(a) Is the assumption that the difference in score is normally distributed reasonable? Show results to support your answer.
(b) Find a 95% confidence interval on the difference in mean score. Is there any evidence that mean score depends on birth order?
(c) It is important to detect a mean difference in score of one point, with a probability of at least 0.90. Was the use of 10 pairs an adequate sample size? If not, how many pairs should have been used?

10-38. In *Biometrics* (1990, Vol. 46, pp. 673–87), the authors analyzed the circumference of five orange trees (labeled as A–E) measured on seven occasions (x_i).

Tree	x_1	x_2	x_3	x_4	x_5	x_6	x_7
A	30	58	87	115	120	142	145
B	33	69	111	156	172	203	203
C	30	51	75	108	115	139	140
D	32	62	112	167	179	209	214
E	30	49	81	125	142	174	177

(a) Compare the mean increase in circumference in periods 1 to 2 to the mean increase in periods 2 to 3. The increase is the difference in circumference in the two periods. Are these means significantly different at $\alpha = 0.10$?
(b) Is there evidence that the mean increase in period 1 to period 2 is greater than the mean increase in period 6 to period 7 at $\alpha = 0.05$?
(c) Are the assumptions of the test in part (a) violated because the same data (period 2 circumference) is used to calculate both mean increases?

10-5 INFERENCE ON THE VARIANCES OF TWO NORMAL DISTRIBUTIONS

We now introduce tests and confidence intervals for the two population variances shown in Fig. 10-1. We will assume that both populations are normal. Both the hypothesis-testing and confidence interval procedures are relatively sensitive to the normality assumption.

10-5.1 F Distribution

Suppose that two independent normal populations are of interest, where the population means and variances, say, μ_1, σ_1^2, μ_2, and σ_2^2, are unknown. We wish to test hypotheses about the equality of the two variances, say, $H_0: \sigma_1^2 = \sigma_2^2$. Assume that two random samples of size n_1 from population 1 and of size n_2 from population 2 are available, and let S_1^2 and S_2^2 be the sample variances. We wish to test the hypotheses

$$H_0: \sigma_1^2 = \sigma_2^2$$
$$H_1: \sigma_1^2 \neq \sigma_2^2 \qquad (10\text{-}24)$$

The development of a test procedure for these hypotheses requires a new probability distribution, the F distribution. The random variable F is defined to be the ratio of two independent chi-square random variables, each divided by its number of degrees of freedom. That is,

$$F = \frac{W/u}{Y/v} \qquad (10\text{-}25)$$

where W and Y are independent chi-square random variables with u and v degrees of freedom, respectively. We now formally state the sampling distribution of F.

F Distribution

> Let W and Y be independent chi-square random variables with u and v degrees of freedom, respectively. Then the ratio
>
> $$F = \frac{W/u}{Y/v} \qquad (10\text{-}26)$$
>
> has the probability density function
>
> $$f(x) = \frac{\Gamma\left(\dfrac{u+v}{2}\right)\left(\dfrac{u}{v}\right)^{u/2} x^{(u/2)-1}}{\Gamma\left(\dfrac{u}{2}\right)\Gamma\left(\dfrac{v}{2}\right)\left[\left(\dfrac{u}{v}\right)x + 1\right]^{(u+v)/2}}, \qquad 0 < x < \infty \qquad (10\text{-}27)$$
>
> and is said to follow the F distribution with u degrees of freedom in the numerator and v degrees of freedom in the denominator. It is usually abbreviated as $F_{u,v}$.

The mean and variance of the F distribution are $\mu = v/(v-2)$ for $v > 2$, and

$$\sigma^2 = \frac{2v^2(u+v-2)}{u(v-2)^2(v-4)}, \qquad v > 4$$

Two F distributions are shown in Fig. 10-4. The F random variable is nonnegative, and the distribution is skewed to the right. The F distribution looks very similar to the chi-square distribution; however, the two parameters u and v provide extra flexibility regarding shape.

The percentage points of the F distribution are given in Table VI of the Appendix. Let $f_{\alpha,u,v}$ be the percentage point of the F distribution, with numerator degrees of freedom u and denominator degrees of freedom v such that the probability that the random variable F exceeds this value is

$$P(F > f_{\alpha,u,v}) = \int_{f_{\alpha,u,v}}^{\infty} f(x)\,dx = \alpha$$

This is illustrated in Fig. 10-5. For example, if $u = 5$ and $v = 10$, we find from Table V of the Appendix that

$$P(F > f_{0.05,5,10}) = P(F_{5,10} > 3.33) = 0.05$$

That is, the upper 5 percentage point of $F_{5,10}$ is $f_{0.05,5,10} = 3.33$.

Table V contains only upper-tail percentage points (for selected values of $f_{\alpha,u,v}$ for $\alpha \leq 0.25$) of the F distribution. The lower-tail percentage points $f_{1-\alpha,u,v}$ can be found as follows.

$$f_{1-\alpha,u,v} = \frac{1}{f_{\alpha,v,u}} \qquad (10\text{-}28)$$

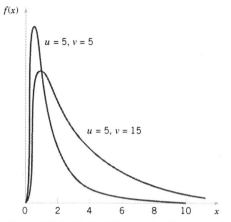
Figure 10-4 Probability density functions of two F distributions.

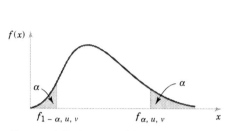
Figure 10-5 Upper and lower percentage points of the F distribution.

For example, to find the lower-tail percentage point $f_{0.95,5,10}$, note that

$$f_{0.95,5,10} = \frac{1}{f_{0.05,10,5}} = \frac{1}{4.74} = 0.211$$

10-5.2 Hypothesis Tests on the Ratio of Two Variances

A hypothesis-testing procedure for the equality of two variances is based on the following result.

Distribution of the Ratio of Sample Variances from Two Normal Distributions

Let $X_{11}, X_{12}, \ldots, X_{1n_1}$ be a random sample from a normal population with mean μ_1 and variance σ_1^2, and let $X_{21}, X_{22}, \ldots, X_{2n_2}$ be a random sample from a second normal population with mean μ_2 and variance σ_2^2. Assume that both normal populations are independent. Let S_1^2 and S_2^2 be the sample variances. Then the ratio

$$F = \frac{S_1^2/\sigma_1^2}{S_2^2/\sigma_2^2}$$

has an F distribution with $n_1 - 1$ numerator degrees of freedom and $n_2 - 1$ denominator degrees of freedom.

This result is based on the fact that $(n_1 - 1)S_1^2/\sigma_1^2$ is a chi-square random variable with $n_1 - 1$ degrees of freedom, that $(n_2 - 1)S_2^2/\sigma_2^2$ is a chi-square random variable with $n_2 - 1$ degrees of freedom, and that the two normal populations are independent. Clearly under the null hypothesis $H_0: \sigma_1^2 = \sigma_2^2$ the ratio $F_0 = S_1^2/S_2^2$ has an F_{n_1-1,n_2-1} distribution. This is the basis of the following test procedure.

Tests on the Ratio of Variances from Two Normal Distributions

Null hypothesis: $H_0: \sigma_1^2 = \sigma_2^2$

Test statistic: $F_0 = \dfrac{S_1^2}{S_2^2}$ (10-29)

Alternative Hypotheses	Rejection Criterion
$H_1: \sigma_1^2 \neq \sigma_2^2$	$f_0 > f_{\alpha/2, n_1-1, n_2-1}$ or $f_0 < f_{1-\alpha/2, n_1-1, n_2-1}$
$H_1: \sigma_1^2 > \sigma_2^2$	$f_0 > f_{\alpha, n_1-1, n_2-1}$
$H_1: \sigma_1^2 < \sigma_2^2$	$f_0 < f_{1-\alpha, n_1-1, n_2-1}$

EXAMPLE 10-11
Semiconductor Etch Variability

Oxide layers on semiconductor wafers are etched in a mixture of gases to achieve the proper thickness. The variability in the thickness of these oxide layers is a critical characteristic of the wafer, and low variability is desirable for subsequent processing steps. Two different mixtures of gases are being studied to determine whether one is superior in reducing the variability of the oxide thickness. Twenty wafers are etched in each gas. The sample standard deviations of oxide thickness are $s_1 = 1.96$ angstroms and $s_2 = 2.13$ angstroms, respectively. Is there any evidence to indicate that either gas is preferable? Use $\alpha = 0.05$.

The eight-step hypothesis-testing procedure may be applied to this problem as follows:

1. The parameters of interest are the variances of oxide thickness σ_1^2 and σ_2^2. We will assume that oxide thickness is a normal random variable for both gas mixtures.
2. $H_0: \sigma_1^2 = \sigma_2^2$
3. $H_1: \sigma_1^2 \neq \sigma_2^2$
4. $\alpha = 0.05$
5. The test statistic is given by Equation 10-29:

$$f_0 = \frac{s_1^2}{s_2^2}$$

6. Since $n_1 = n_2 = 20$, we will reject $H_0: \sigma_1^2 = \sigma_2^2$ if $f_0 > f_{0.025, 19, 19} = 2.53$ or if $f_0 < f_{0.975, 19, 19} = 1/f_{0.025, 19, 19} = 1/2.53 = 0.40$.
7. Computations: Since $s_1^2 = (1.96)^2 = 3.84$ and $s_2^2 = (2.13)^2 = 4.54$, the test statistic is

$$f_0 = \frac{s_1^2}{s_2^2} = \frac{3.84}{4.54} = 0.85$$

8. Conclusions: Since $f_{0.975, 19, 19} = 0.40 < f_0 = 0.85 < f_{0.025, 19, 19} = 2.53$, we cannot reject the null hypothesis $H_0: \sigma_1^2 = \sigma_2^2$ at the 0.05 level of significance. Therefore, there is no strong evidence to indicate that either gas results in a smaller variance of oxide thickness.

We may also find a P-value for the F-statistic in Example 10-11. Since $f_{0.50, 19, 19} = 1.00$, the computed value of the test statistic $f_0 = s_1^2/s_2^2 = 3.84/4.54 = 0.85$ is nearer the lower tail of the F distribution than the upper tail. The probability that an F-random variable with 19 numerator and denominator degrees of freedom is less than 0.85 is 0.3634. Since it is arbitrary which population is identified as "one," we could have computed the test statistic as $f_0 = 4.54/3.84 = 1.18$. The probability that an F-random variable with 19 numerator and denominator degrees of freedom exceeds 1.18 is 0.3610. Therefore, the P-value for the test statistic $f_0 = 0.85$ is the sum of these two probabilities, or $P = 0.3634 + 0.3610 = 0.7244$. Since the P-value exceeds 0.05, the null hypothesis $H_0: \sigma_1^2 = \sigma_2^2$ cannot be rejected. (The probabilities given above were computed using a hand-held calculator.)

10-5.3 Type II Error and Choice of Sample Size

Appendix Charts VIIo, VIIp, VIIq, and VIIr provide operating characteristic curves for the F-test given in Section 10-5.1 for $\alpha = 0.05$ and $\alpha = 0.01$, assuming that $n_1 = n_2 = n$. Charts VIIo and VIIp are used with the two-sided alternate hypothesis. They plot β against the abscissa parameter

$$\lambda = \frac{\sigma_1}{\sigma_2} \tag{10-30}$$

for various $n_1 = n_2 = n$. Charts VIIq and VIIr are used for the one-sided alternative hypotheses.

EXAMPLE 10-12
Semiconductor Etch Variability Sample Size

For the semiconductor wafer oxide etching problem in Example 10-11, suppose that one gas resulted in a standard deviation of oxide thickness that is half the standard deviation of oxide thickness of the other gas. If we wish to detect such a situation with probability at least 0.80, is the sample size $n_1 = n_2 = 20$ adequate?

Note that if one standard deviation is half the other,

$$\lambda = \frac{\sigma_1}{\sigma_2} = 2$$

By referring to Appendix Chart VIIo with $n_1 = n_2 = n = 20$ and $\lambda = 2$, we find that $\beta \simeq 0.20$. Therefore, if $\beta = 0.20$, the power of the test (which is the probability that the difference in standard deviations will be detected by the test) is 0.80, and we conclude that the sample sizes $n_1 = n_2 = 20$ are adequate.

10-5.4 Confidence Interval on the Ratio of Two Variances

To find the confidence interval on σ_1^2/σ_2^2, recall that the sampling distribution of

$$F = \frac{S_2^2/\sigma_2^2}{S_1^2/\sigma_1^2}$$

is an F with $n_2 - 1$ and $n_1 - 1$ degrees of freedom. Therefore, $P(f_{1-\alpha/2, n_2-1, n_1-1} \leq F \leq f_{\alpha/2, n_2-1, n_1-1}) = 1 - \alpha$. Substitution for F and manipulation of the inequalities will lead to the $100(1 - \alpha)\%$ confidence interval for σ_1^2/σ_2^2.

Confidence Interval on the Ratio of Variances from Two Normal Distributions

If s_1^2 and s_2^2 are the sample variances of random samples of sizes n_1 and n_2, respectively, from two independent normal populations with unknown variances σ_1^2 and σ_2^2, then a **$100(1 - \alpha)\%$ confidence interval on the ratio σ_1^2/σ_2^2** is

$$\frac{s_1^2}{s_2^2} f_{1-\alpha/2, n_2-1, n_1-1} \leq \frac{\sigma_1^2}{\sigma_2^2} \leq \frac{s_1^2}{s_2^2} f_{\alpha/2, n_2-1, n_1-1} \tag{10-31}$$

where $f_{\alpha/2, n_2-1, n_1-1}$ and $f_{1-\alpha/2, n_2-1, n_1-1}$ are the upper and lower $\alpha/2$ percentage points of the F distribution with $n_2 - 1$ numerator and $n_1 - 1$ denominator degrees of freedom, respectively. A confidence interval on the ratio of the standard deviations can be obtained by taking square roots in Equation 10-31.

EXAMPLE 10-13
Surface Finish for Titanium Alloy

A company manufactures impellers for use in jet-turbine engines. One of the operations involves grinding a particular surface finish on a titanium alloy component. Two different grinding processes can be used, and both processes can produce parts at identical mean surface roughness. The manufacturing engineer would like to select the process having the least variability in surface roughness. A random sample of $n_1 = 11$ parts from the first process results in a sample standard deviation $s_1 = 5.1$ microinches, and a random sample of $n_2 = 16$ parts from the second process results in a sample standard deviation of $s_2 = 4.7$ microinches. We will find a 90% confidence interval on the ratio of the two standard deviations, σ_1/σ_2.

Assuming that the two processes are independent and that surface roughness is normally distributed, we can use Equation 10-31 as follows:

$$\frac{s_1^2}{s_2^2} f_{0.95,15,10} \leq \frac{\sigma_1^2}{\sigma_2^2} \leq \frac{s_1^2}{s_2^2} f_{0.05,15,10}$$

$$\frac{(5.1)^2}{(4.7)^2} 0.39 \leq \frac{\sigma_1^2}{\sigma_2^2} \leq \frac{(5.1)^2}{(4.7)^2} 2.85$$

or upon completing the implied calculations and taking square roots,

$$0.678 \leq \frac{\sigma_1}{\sigma_2} \leq 1.832$$

Notice that we have used Equation 10-28 to find $f_{0.95,15,10} = 1/f_{0.05,10,15} = 1/2.54 = 0.39$. Since this confidence interval includes unity, we cannot claim that the standard deviations of surface roughness for the two processes are different at the 90% level of confidence.

EXERCISES FOR SECTION 10-5

10-39. For an F distribution, find the following:
(a) $f_{0.25,5,10}$ (b) $f_{0.10,24,9}$
(c) $f_{0.05,8,15}$ (d) $f_{0.75,5,10}$
(e) $f_{0.90,24,9}$ (f) $f_{0.95,8,15}$

10-40. For an F distribution, find the following:
(a) $f_{0.25,7,15}$ (b) $f_{0.10,10,12}$
(c) $f_{0.01,20,10}$ (d) $f_{0.75,7,15}$
(e) $f_{0.90,10,12}$ (f) $f_{0.99,20,10}$

10-41. Consider the hypothesis test $H_0: \sigma_1^2 = \sigma_2^2$ against $H_1: \sigma_1^2 < \sigma_2^2$. Suppose that the sample sizes are $n_1 = 5$ and $n_2 = 10$ and that $s_1^2 = 23.2$ and $s_2^2 = 28.8$. Use $\alpha = 0.05$. Test the hypothesis and explain how the test could be conducted with a confidence interval on σ_1/σ_2.

10-42. Consider the hypothesis test $H_0: \sigma_1^2 = \sigma_2^2$ against $H_0: \sigma_1^2 > \sigma_2^2$. Suppose that the sample sizes are $n_1 = 20$ and $n_2 = 8$ and that $s_1^2 = 4.5$ and $s_2^2 = 2.3$. Use $\alpha = 0.01$. Test the hypothesis and explain how the test could be conducted with a confidence interval on σ_1/σ_2.

10-43. Consider the hypothesis test $H_0: \sigma_1^2 = \sigma_2^2$ against $H_1: \sigma_1^2 \neq \sigma_2^2$. Suppose that the sample sizes are $n_1 = 15$ and $n_2 = 15$ and the sample variances are $s_1^2 = 2.3$ and $s_2^2 = 1.9$. Use $\alpha = 0.05$.
(a) Test the hypothesis and explain how the test could be conducted with a confidence interval on σ_1/σ_2.

(b) What is the power of the test in part (a) if σ_1 is twice as large as σ_2?
(c) Assuming equal sample sizes, what sample size should be used to obtain $\beta = 0.05$ if the σ_2 is half of σ_1?

10-44. Two chemical companies can supply a raw material. The concentration of a particular element in this material is important. The mean concentration for both suppliers is the same, but we suspect that the variability in concentration may differ between the two companies. The standard deviation of concentration in a random sample of $n_1 = 10$ batches produced by company 1 is $s_1 = 4.7$ grams per liter, while for company 2, a random sample of $n_2 = 16$ batches yields $s_2 = 5.8$ grams per liter. Is there sufficient evidence to conclude that the two population variances differ? Use $\alpha = 0.05$.

10-45. A study was performed to determine whether men and women differ in their repeatability in assembling components on printed circuit boards. Random samples of 25 men and 21 women were selected, and each subject assembled the units. The two sample standard deviations of assembly time were $s_{men} = 0.98$ minutes and $s_{women} = 1.02$ minutes.
(a) Is there evidence to support the claim that men and women differ in repeatability for this assembly task? Use $\alpha = 0.02$ and state any necessary assumptions about the underlying distribution of the data.

(b) Find a 98% confidence interval on the ratio of the two variances. Provide an interpretation of the interval.

10-46. Consider the foam data in Exercise 10-14. Construct the following:
(a) A 90% two-sided confidence interval on σ_1^2/σ_2^2.
(b) A 95% two-sided confidence interval on σ_1^2/σ_2^2. Comment on the comparison of the width of this interval with the width of the interval in part (a).
(c) A 90% lower-confidence bound on σ_1/σ_2.

10-47. Consider the diameter data in Exercise 10-13. Construct the following:
(a) A 90% two-sided confidence interval on σ_1/σ_2.
(b) A 95% two-sided confidence interval on σ_1/σ_2. Comment on the comparison of the width of this interval with the width of the interval in part (a).
(c) A 90% lower-confidence bound on σ_1/σ_2.

10-48. Consider the gear impact strength data in Exercise 10-18. Is there sufficient evidence to conclude that the variance of impact strength is different for the two suppliers? Use $\alpha = 0.05$.

10-49. Consider the melting point data in Exercise 10-19. Do the sample data support a claim that both alloys have the same variance of melting point? Use $\alpha = 0.05$ in reaching your conclusion.

10-50. Exercise 10-22 presented measurements of plastic coating thickness at two different application temperatures. Test $H_0: \sigma_1^2 = \sigma_2^2$ against $H_1: \sigma_1^2 \neq \sigma_2^2$ using $\alpha = 0.01$.

10-51. Reconsider the overall distance data for golf balls in Exercise 10-27. Is there evidence to support the claim that the standard deviation of overall distance is the same for both brands of balls (use $\alpha = 0.05$)? Explain how this question can be answered with a 95% confidence interval on σ_1/σ_2.

10-52. Reconsider the coefficient of restitution data in Exercise 10-28. Do the data suggest that the standard deviation is the same for both brands of drivers (use $\alpha = 0.05$)? Explain how to answer this question with a confidence interval on σ_1/σ_2.

10-53. Consider the weight of paper data from Technometrics In Exercise 10-26. Is there evidence that the variance of the weight measurement differs between the sheets of paper? Use $\alpha = 0.05$. Explain how this test can be conducted with a confidence interval.

10-54. Consider the film speed data in Exercise 10-20.
(a) Test $H_0: \sigma_1^2 = \sigma_2^2$ versus $H_1: \sigma_1^2 \neq \sigma_2^2$ using $\alpha = 0.02$.
(b) Suppose that one population standard deviation is 50% larger than the other. Is the sample size $n_1 = n_2 = 8$ adequate to detect this difference with high probability? Use $\alpha = 0.01$ in answering this question.

10-55. Consider the etch rate data in Exercise 10-17.
(a) Test the hypothesis $H_0: \sigma_1^2 = \sigma_2^2$ against $H_1: \sigma_1^2 \neq \sigma_2^2$ using $\alpha = 0.05$, and draw conclusions.
(b) Suppose that if one population variance is twice as large as the other, we want to detect this with probability at least 0.90 (using $\alpha = 0.05$). Are the sample sizes $n_1 = n_2 = 10$ adequate?

10-6 INFERENCE ON TWO POPULATION PROPORTIONS

We now consider the case where there are two binomial parameters of interest, say, p_1 and p_2, and we wish to draw inferences about these proportions. We will present large-sample hypothesis testing and confidence interval procedures based on the normal approximation to the binomial.

10-6.1 Large-Sample Test on the Difference in Population Proportions

Suppose that two independent random samples of sizes n_1 and n_2 are taken from two populations, and let X_1 and X_2 represent the number of observations that belong to the class of interest in samples 1 and 2, respectively. Furthermore, suppose that the normal approximation to the binomial is applied to each population, so the estimators of the population proportions $\hat{P}_1 = X_1/n_1$ and $\hat{P}_2 = X_2/n_2$ have approximate normal distributions. We are interested in testing the hypotheses

$$H_0: p_1 = p_2$$
$$H_1: p_1 \neq p_2$$

380   CHAPTER 10 STATISTICAL INFERENCE FOR TWO SAMPLES

The statistic

Test Statistic for the Difference of Two Population Proportions

$$Z = \frac{\hat{P}_1 - \hat{P}_2 - (p_1 - p_2)}{\sqrt{\frac{p_1(1 - p_1)}{n_1} + \frac{p_2(1 - p_2)}{n_2}}} \quad (10\text{-}32)$$

is distributed approximately as standard normal and is the basis of a test for $H_0: p_1 = p_2$. Specifically, if the null hypothesis $H_0: p_1 = p_2$ is true, using the fact that $p_1 = p_2 = p$, the random variable

$$Z = \frac{\hat{P}_1 - \hat{P}_2}{\sqrt{p(1 - p)\left(\frac{1}{n_1} + \frac{1}{n_2}\right)}}$$

is distributed approximately $N(0, 1)$. An estimator of the common parameter p is

$$\hat{P} = \frac{X_1 + X_2}{n_1 + n_2}$$

The **test statistic** for $H_0: p_1 = p_2$ is then

$$Z_0 = \frac{\hat{P}_1 - \hat{P}_2}{\sqrt{\hat{P}(1 - \hat{P})\left(\frac{1}{n_1} + \frac{1}{n_2}\right)}}$$

This leads to the test procedures described below.

Approximate Tests on the Difference of Two Population Proportions

Null hypothesis: $H_0: p_1 = p_2$

Test statistic: $Z_0 = \dfrac{\hat{P}_1 - \hat{P}_2}{\sqrt{\hat{P}(1 - \hat{P})\left(\dfrac{1}{n_1} + \dfrac{1}{n_2}\right)}}$ (10-33)

Alternative Hypotheses	Rejection Criterion
$H_1: p_1 \neq p_2$	$z_0 > z_{\alpha/2}$ or $z_0 < -z_{\alpha/2}$
$H_1: p_1 > p_2$	$z_0 > z_{\alpha}$
$H_1: p_1 < p_2$	$z_0 < -z_{\alpha}$

EXAMPLE 10-14
St. John's Wort

Extracts of St. John's Wort are widely used to treat depression. An article in the April 18, 2001 issue of the *Journal of the American Medical Association* ("Effectiveness of St. John's Wort on Major Depression: A Randomized Controlled Trial") compared the efficacy of a standard extract of St. John's Wort with a placebo in 200 outpatients diagnosed with major depression. Patients were randomly assigned to two groups; one group received the St. John's Wort, and the other received the placebo. After eight weeks, 19 of the placebo-treated patients showed improvement, whereas 27 of those treated with St. John's Wort improved. Is there any reason to believe that St. John's Wort is effective in treating major depression? Use $\alpha = 0.05$.

The eight-step hypothesis testing procedure leads to the following results:

1. The parameters of interest are p_1 and p_2, the proportion of patients who improve following treatment with St. John's Wort (p_1) or the placebo (p_2).
2. $H_0: p_1 = p_2$
3. $H_1: p_1 \neq p_2$
4. $\alpha = 0.05$
5. The test statistic is

$$z_0 = \frac{\hat{p}_1 - \hat{p}_2}{\sqrt{\hat{p}(1-\hat{p})\left(\frac{1}{n_1} + \frac{1}{n_2}\right)}}$$

where $\hat{p}_1 = 27/100 = 0.27$, $\hat{p}_2 = 19/100 = 0.19$, $n_1 = n_2 = 100$, and

$$\hat{p} = \frac{x_1 + x_2}{n_1 + n_2} = \frac{19 + 27}{100 + 100} = 0.23$$

6. Reject $H_0: p_1 = p_2$ if $z_0 > z_{0.025} = 1.96$ or if $z_0 < -z_{0.025} = -1.96$.
7. Computations: The value of the test statistic is

$$z_0 = \frac{0.27 - 0.19}{\sqrt{0.23(0.77)\left(\frac{1}{100} + \frac{1}{100}\right)}} = 1.34$$

8. Conclusions: Since $z_0 = 1.34$ does not exceed $z_{0.025}$, we cannot reject the null hypothesis. Note that the P-value is $P \simeq 0.179$. There is insufficient evidence to support the claim that St. John's Wort is effective in treating major depression.

The following box shows the Minitab two-sample hypothesis test and CI procedure for proportions. Notice that the 95% CI on $p_1 - p_2$ includes zero. The equation for constructing the CI will be given in Section 10-6.3.

Minitab Computations

Test and CI for Two Proportions

Sample	X	N	Sample p
1	27	100	0.270000
2	19	100	0.190000

Estimate for p(1) − p(2): 0.08
95% CI for p(1) − p(2): (−0.0361186, 0.196119)
Test for p(1) − p(2) = 0 (vs not = 0): Z = 1.35 P-Value = 0.177

10-6.2 Type II Error and Choice of Sample Size

The computation of the β-error for the large-sample test of $H_0: p_1 = p_2$ is somewhat more involved than in the single-sample case. The problem is that the denominator of the test statistic Z_0 is an estimate of the standard deviation of $\hat{P}_1 - \hat{P}_2$ under the assumption that $p_1 = p_2 = p$. When $H_0: p_1 = p_2$ is false, the standard deviation of $\hat{P}_1 - \hat{P}_2$ is

$$\sigma_{\hat{P}_1 - \hat{P}_2} = \sqrt{\frac{p_1(1-p_1)}{n_1} + \frac{p_2(1-p_2)}{n_2}} \tag{10-34}$$

Approximate Type II Error for a Two-Sided Test on the Difference of Two Population Proportions

If the alternative hypothesis is two sided, the β-error is

$$\beta = \Phi\left[\frac{z_{\alpha/2}\sqrt{\bar{p}\bar{q}(1/n_1 + 1/n_2)} - (p_1 - p_2)}{\sigma_{\hat{P}_1 - \hat{P}_2}}\right]$$

$$- \Phi\left[\frac{-z_{\alpha/2}\sqrt{\bar{p}\bar{q}(1/n_1 + 1/n_2)} - (p_1 - p_2)}{\sigma_{\hat{P}_1 - \hat{P}_2}}\right] \tag{10-35}$$

where

$$\bar{p} = \frac{n_1 p_1 + n_2 p_2}{n_1 + n_2} \quad \text{and} \quad \bar{q} = \frac{n_1(1-p_1) + n_2(1-p_2)}{n_1 + n_2}$$

and $\sigma_{\hat{P}_1 - \hat{P}_2}$ is given by Equation 10-34.

Approximate Type II Error for a One-Sided Test on the Difference of Two Population Proportions

If the alternative hypothesis is $H_1: p_1 > p_2$,

$$\beta = \Phi\left[\frac{z_{\alpha}\sqrt{\bar{p}\bar{q}(1/n_1 + 1/n_2)} - (p_1 - p_2)}{\sigma_{\hat{P}_1 - \hat{P}_2}}\right] \tag{10-36}$$

and if the alternative hypothesis is $H_1: p_1 < p_2$,

$$\beta = 1 - \Phi\left[\frac{-z_{\alpha}\sqrt{\bar{p}\bar{q}(1/n_1 + 1/n_2)} - (p_1 - p_2)}{\sigma_{\hat{P}_1 - \hat{P}_2}}\right] \tag{10-37}$$

For a specified pair of values p_1 and p_2, we can find the sample sizes $n_1 = n_2 = n$ required to give the test of size α that has specified type II error β.

Approximate Sample Size for a Two-Sided Test on the Difference in Population Proportions

For the two-sided alternative, the common sample size is

$$n = \frac{[z_{\alpha/2}\sqrt{(p_1 + p_2)(q_1 + q_2)/2} + z_{\beta}\sqrt{p_1 q_1 + p_2 q_2}]^2}{(p_1 - p_2)^2} \quad (10\text{-}38)$$

where $q_1 = 1 - p_1$ and $q_2 = 1 - p_2$.

For a one-sided alternative, replace $z_{\alpha/2}$ in Equation 10-38 by z_α.

10-6.3 Confidence Interval on the Difference in Population Proportions

The confidence interval for $p_1 - p_2$ can be found directly, since we know that

$$Z = \frac{\hat{P}_1 - \hat{P}_2 - (p_1 - p_2)}{\sqrt{\dfrac{p_1(1 - p_1)}{n_1} + \dfrac{p_2(1 - p_2)}{n_2}}}$$

is a standard normal random variable. Thus $P(-z_{\alpha/2} \leq Z \leq z_{\alpha/2}) \simeq 1 - \alpha$, so we can substitute for Z in this last expression and use an approach similar to the one employed previously to find an approximate $100(1 - \alpha)\%$ two-sided confidence interval for $p_1 - p_2$.

Approximate Confidence Interval on the Difference in Population Proportions

If $\hat{p}_1$ and $\hat{p}_2$ are the sample proportions of observations in two independent random samples of sizes n_1 and n_2 that belong to a class of interest, an **approximate two-sided $100(1 - \alpha)\%$ confidence interval on the difference in the true proportions $p_1 - p_2$** is

$$\hat{p}_1 - \hat{p}_2 - z_{\alpha/2}\sqrt{\frac{\hat{p}_1(1 - \hat{p}_1)}{n_1} + \frac{\hat{p}_2(1 - \hat{p}_2)}{n_2}}$$

$$\leq p_1 - p_2 \leq \hat{p}_1 - \hat{p}_2 + z_{\alpha/2}\sqrt{\frac{\hat{p}_1(1 - \hat{p}_1)}{n_1} + \frac{\hat{p}_2(1 - \hat{p}_2)}{n_2}} \quad (10\text{-}39)$$

where $z_{\alpha/2}$ is the upper $\alpha/2$ percentage point of the standard normal distribution.

EXAMPLE 10-15
Defective Bearings

Consider the process manufacturing crankshaft bearings described in Example 8-7. Suppose that a modification is made in the surface finishing process and that, subsequently, a second random sample of 85 bearings is obtained. The number of defective bearings in this second sample is 8. Therefore, since $n_1 = 85$, $\hat{p}_1 = 0.12$, $n_2 = 85$, and $\hat{p}_2 = 8/85 = 0.09$, we can obtain an approximate 95% confidence interval on the difference in the proportion of defective bearings produced under the two processes from Equation 10-39 as follows:

$$\hat{p}_1 - \hat{p}_2 - z_{0.025}\sqrt{\frac{\hat{p}_1(1-\hat{p}_1)}{n_1} + \frac{\hat{p}_2(1-\hat{p}_2)}{n_2}}$$
$$\leq p_1 - p_2 \leq \hat{p}_1 - \hat{p}_2 + z_{0.025}\sqrt{\frac{\hat{p}_1(1-\hat{p}_1)}{n_1} + \frac{\hat{p}_2(1-\hat{p}_2)}{n_2}}$$

or

$$0.12 - 0.09 - 1.96\sqrt{\frac{0.12(0.88)}{85} + \frac{0.09(0.91)}{85}}$$
$$\leq p_1 - p_2 \leq 0.12 - 0.09 + 1.96\sqrt{\frac{0.12(0.88)}{85} + \frac{0.09(0.91)}{85}}$$

This simplifies to

$$-0.06 \leq p_1 - p_2 \leq 0.12$$

This confidence interval includes zero, so, based on the sample data, it seems unlikely that the changes made in the surface finish process have reduced the proportion of defective crankshaft bearings being produced.

EXERCISES FOR SECTION 10-6

10-56. An article in *Knee Surgery, Sports Traumatology, Arthroscopy* (2005, Vol. 13, pp. 273–279), considered arthroscopic meniscal repair with an absorbable screw. Results showed that for tears greater than 25 millimeters, 14 of 18 (78%) repairs were successful while for shorter tears, 22 of 30 (73%) repairs were successful.
(a) Is there evidence that the success rate is greater for longer tears? Use $\alpha = 0.05$. What is the *P*-value?
(b) Calculate a one-sided 95% confidence bound on the difference in proportions that can be used to answer the question in part (a).

10-57. In the 2004 presidential election, exit polls from the critical state of Ohio provided the following results: for respondents with college degrees, 53% voted for Bush and 46% voted for Kerry. There were 2020 respondents.
(a) Is there a significant difference in these proportions? Use $\alpha = 0.05$. What is the *P*-value?
(b) Calculate a 95% confidence interval for the difference in the two proportions and comment on the use of this interval to answer the question in part (a).

10-58. Two different types of injection-molding machines are used to form plastic parts. A part is considered defective if it has excessive shrinkage or is discolored. Two random samples, each of size 300, are selected, and 15 defective parts are found in the sample from machine 1 while 8 defective parts are found in the sample from machine 2.
(a) Is it reasonable to conclude that both machines produce the same fraction of defective parts, using $\alpha = 0.05$? Find the *P*-value for this test.

(b) Construct a 95% confidence interval on the difference in the two fractions defective.
(c) Suppose that $p_1 = 0.05$ and $p_2 = 0.01$. With the sample sizes given here, what is the power of the test for this two-sided alternate?
(d) Supoose that $p_1 = 0.05$ and $p_2 = 0.01$. Determine the sample size needed to detect this difference with a probability of at least 0.9.
(e) Suppose that $p_1 = 0.05$ and $p_2 = 0.02$. With the sample sizes given here, what is the power of the test for this two-sided alternate?
(f) Suppose that $p_1 = 0.05$ and $p_2 = 0.02$. Determine the sample size needed to detect this difference with a probability of at least 0.9.

10-59. Two different types of polishing solutions are being evaluated for possible use in a tumble-polish operation for manufacturing interocular lenses used in the human eye following cataract surgery. Three hundred lenses were tumble polished using the first polishing solution, and of this number 253 had no polishing-induced defects. Another 300 lenses were tumble-polished using the second polishing solution, and 196 lenses were satisfactory upon completion.
(a) Is there any reason to believe that the two polishing solutions differ? Use $\alpha = 0.01$. What is the *P*-value for this test?
(b) Discuss how this question could be answered with a confidence interval on $p_1 - p_2$.

10-60. A random sample of 500 adult residents of Maricopa County found that 385 were in favor of increasing the highway speed limit to 75 mph, while another sample of 400 adult residents of Pima County found that 267 were in favor of the increased speed limit.

(a) Do these data indicate that there is a difference in the support for increasing the speed limit between the residents of the two counties? Use $\alpha = 0.05$. What is the P-value for this test?

(b) Construct a 95% confidence interval on the difference in the two proportions. Provide a practical interpretation of this interval.

10-7 SUMMARY TABLE AND ROADMAPS FOR INFERENCE PROCEDURES FOR TWO SAMPLES

The table in the end papers of the book summarizes all of the two-sample inference procedures given in this chapter. The table contains the null hypothesis statements, the test statistics, the criteria for rejection of the various alternative hypotheses, and the formulas for constructing the $100(1 - \alpha)\%$ confidence intervals.

The roadmap to select the appropriate confidence interval formula or hypothesis test method for one-sample problems was presented in Table 8-1. In Table 10-4, we extend the road

Table 10-4 Roadmap to Construct Confidence Intervals and Hypothesis Tests, Two-Sample Case

Function of the Parameters to be Bounded by the Confidence Interval or Tested with a Hypothesis	Symbol	Other Parameters?	Confidence Interval Section	Hypothesis Test Section	Comments
Difference in means from two normal distributions	$\mu_1 - \mu_2$	Standard deviations σ_1 and σ_2 known	10-2.3	10-2.1	
Difference in means from two arbitrary distributions with large sample sizes	$\mu_1 - \mu_2$	Sample sizes large enough that σ_1 and σ_2 are essentially known	10-2.3	10-2.1	Large sample size is often taken to be n_1 and $n_2 \geq 40$
Difference in means from two normal distributions	$\mu_1 - \mu_2$	Standard deviations σ_1 and σ_2 are unknown, and assumed equal	10-3.3	10-3.1	Case 1: $\sigma_1 = \sigma_2$
Difference in means from two normal distributions	$\mu_1 - \mu_2$	Standard deviations σ_1 and σ_2 are unknown, and NOT assumed equal	10-3.3	10-3.1	Case 2: $\sigma_1 \neq \sigma_2$
Difference in means from two normal distributions in a paired analysis	$\mu_D = \mu_1 - \mu_2$	Standard deviation of differences are unknown	10-4	10-4	Paired analysis calculates differences and uses a one-sample method for inference on the mean difference
Ratio of variances of two normal distributions	σ_1^2/σ_2^2	Means μ_1 and μ_2 unknown and estimated	10-5.4	10-5.2	
Difference in Two Population Proportions	$p_1 - p_2$	None	10-6.3	10-6.1	Normal approximation to the binomial distribution used for the tests and confidence intervals

map to two-sample problems. The primary comments stated previously also apply here (except we usually apply conclusions to a function of the parameters from each sample, such as the difference in means):

1. Determine the function of the parameters (and the distribution of the data) that is to be bounded by the confidence interval or tested by the hypothesis.
2. Check if other parameters are known or need to be estimated (and if any assumptions are made).

Supplemental Exercises

10-61. An article in the *Journal of Materials Engineering* (1989, Vol. 11, No. 4, pp. 275–282) reported the results of an experiment to determine failure mechanisms for plasma-sprayed thermal barrier coatings. The failure stress for one particular coating (NiCrAlZr) under two different test conditions is as follows:

Failure stress ($\times 10^6$ Pa) after nine 1-hour cycles: 19.8, 18.5, 17.6, 16.7, 16.7, 14.8, 15.4, 14.1, 13.6

Failure stress ($\times 10^6$ Pa) after six 1-hour cycles: 14.9, 12.7, 11.9, 11.4, 10.1, 7.9

(a) What assumptions are needed to construct confidence intervals for the difference in mean failure stress under the two different test conditions? Use normal probability plots of the data to check these assumptions.
(b) Find a 99% confidence interval on the difference in mean failure stress under the two different test conditions.
(c) Using the confidence interval constructed in part (b), does the evidence support the claim that the first test conditions yield higher results, on the average, than the second? Explain your answer.
(d) Construct a 95% confidence interval on the ratio of the variances, σ_1^2/σ_2^2, of failure stress under the two different test conditions.
(e) Use your answer in part (b) to determine whether there is a significant difference in variances of the two different test conditions. Explain your answer.

10-62. A procurement specialist has purchased 25 resistors from vendor 1 and 35 resistors from vendor 2. Each resistor's resistance is measured with the following results:

Vendor 1					
96.8	100.0	100.3	98.5	98.3	98.2
99.6	99.4	99.9	101.1	103.7	97.7
99.7	101.1	97.7	98.6	101.9	101.0
99.4	99.8	99.1	99.6	101.2	98.2
98.6					

Vendor 2					
106.8	106.8	104.7	104.7	108.0	102.2
103.2	103.7	106.8	105.1	104.0	106.2
102.6	100.3	104.0	107.0	104.3	105.8
104.0	106.3	102.2	102.8	104.2	103.4
104.6	103.5	106.3	109.2	107.2	105.4
106.4	106.8	104.1	107.1	107.7	

(a) What distributional assumption is needed to test the claim that the variance of resistance of product from vendor 1 is not significantly different from the variance of resistance of product from vendor 2? Perform a graphical procedure to check this assumption.
(b) Perform an appropriate statistical hypothesis-testing procedure to determine whether the procurement specialist can claim that the variance of resistance of product from vendor 1 is significantly different from the variance of resistance of product from vendor 2.

10-63. A liquid dietary product implies in its advertising that use of the product for one month results in an average weight loss of at least 3 pounds. Eight subjects use the product for one month, and the resulting weight loss data are reported below. Use hypothesis-testing procedures to answer the following questions.

Subject	Initial Weight (lb)	Final Weight (lb)
1	165	161
2	201	195
3	195	192
4	198	193
5	155	150
6	143	141
7	150	146
8	187	183

(a) Do the data support the claim of the producer of the dietary product with the probability of a type I error set to 0.05?

(b) Do the data support the claim of the producer of the dietary product with the probability of a type I error set to 0.01?

(c) In an effort to improve sales, the producer is considering changing its claim from "at least 3 pounds" to "at least 5 pounds." Repeat parts (a) and (b) to test this new claim.

10-64. The breaking strength of yarn supplied by two manufacturers is being investigated. We know from experience with the manufacturers' processes that $\sigma_1 = 5$ psi and $\sigma_2 = 4$ psi. A random sample of 20 test specimens from each manufacturer results in $\bar{x}_1 = 88$ psi and $\bar{x}_2 = 91$ psi, respectively.

(a) Using a 90% confidence interval on the difference in mean breaking strength, comment on whether or not there is evidence to support the claim that manufacturer 2 produces yarn with higher mean breaking strength.

(b) Using a 98% confidence interval on the difference in mean breaking strength, comment on whether or not there is evidence to support the claim that manufacturer 2 produces yarn with higher mean breaking strength.

(c) Comment on why the results from parts (a) and (b) are different or the same. Which would you choose to make your decision and why?

10-65. The Salk polio vaccine experiment in 1954 focused on the effectiveness of the vaccine in combating paralytic polio. Because it was felt that without a control group of children there would be no sound basis for evaluating the efficacy of the Salk vaccine, the vaccine was administered to one group, and a placebo (visually identical to the vaccine but known to have no effect) was administered to a second group. For ethical reasons, and because it was suspected that knowledge of vaccine administration would affect subsequent diagnoses, the experiment was conducted in a double-blind fashion. That is, neither the subjects nor the administrators knew who received the vaccine and who received the placebo. The actual data for this experiment are as follows:

Placebo group: $n = 201,299$: 110 cases of polio observed

Vaccine group: $n = 200,745$: 33 cases of polio observed

(a) Use a hypothesis-testing procedure to determine if the proportion of children in the two groups who contracted paralytic polio is statistically different. Use a probability of a type I error equal to 0.05.

(b) Repeat part (a) using a probability of a type I error equal to 0.01.

(c) Compare your conclusions from parts (a) and (b) and explain why they are the same or different.

10-66. Consider Supplemental Exercise 10-64. Suppose that prior to collecting the data, you decide that you want the error in estimating $\mu_1 - \mu_2$ by $\bar{x}_1 - \bar{x}_2$ to be less than 1.5 psi. Specify the sample size for the following percentage confidence:

(a) 90%

(b) 98%

(c) Comment on the effect of increasing the percentage confidence on the sample size needed.

(d) Repeat parts (a)–(c) with an error of less than 0.75 psi instead of 1.5 psi.

(e) Comment on the effect of decreasing the error on the sample size needed.

10-67. A random sample of 1500 residential telephones in Phoenix in 1990 found that 387 of the numbers were unlisted. A random sample in the same year of 1200 telephones in Scottsdale found that 310 were unlisted.

(a) Find a 95% confidence interval on the difference in the two proportions and use this confidence interval to determine if there is a statistically significant difference in proportions of unlisted numbers between the two cities.

(b) Find a 90% confidence interval on the difference in the two proportions and use this confidence interval to determine if there is a statistically significant difference in proportions of unlisted numbers between the two cities.

(c) Suppose that all the numbers in the problem description were doubled. That is, 774 residents out of 3000 sampled in Phoenix and 620 residents out of 2400 in Scottsdale had unlisted phone numbers. Repeat parts (a) and (b) and comment on the effect of increasing the sample size without changing the proportions on your results.

10-68. In a random sample of 200 Phoenix residents who drive a domestic car, 165 reported wearing their seat belt regularly, while another sample of 250 Phoenix residents who drive a foreign car revealed 198 who regularly wore their seat belt.

(a) Perform a hypothesis-testing procedure to determine if there is a statistically significant difference in seat belt usage between domestic and foreign car drivers. Set your probability of a type I error to 0.05.

(b) Perform a hypothesis-testing procedure to determine if there is a statistically significant difference in seat belt usage between domestic and foreign car drivers. Set your probability of a type I error to 0.1.

(c) Compare your answers for parts (a) and (b) and explain why they are the same or different.

(d) Suppose that all the numbers in the problem description were doubled. That is, in a random sample of 400 Phoenix residents who drive a domestic car, 330 reported wearing their seat belt regularly, while another sample of 500 Phoenix residents who drive a foreign car revealed 396 who regularly wore their seat belt. Repeat parts (a) and (b) and comment on the effect of increasing the sample size without changing the proportions on your results.

10-69. Consider the previous exercise, which summarized data collected from drivers about their seat belt usage.

(a) Do you think there is a reason not to believe these data? Explain your answer.

(b) Is it reasonable to use the hypothesis-testing results from the previous problem to draw an inference about the difference in proportion of seat belt usage

(i) of the spouses of these drivers of domestic and foreign cars? Explain your answer.

(ii) of the children of these drivers of domestic and foreign cars? Explain your answer.
(iii) of all drivers of domestic and foreign cars? Explain your answer.
(iv) of all drivers of domestic and foreign trucks? Explain your answer.

10-70. A manufacturer of a new pain relief tablet would like to demonstrate that its product works twice as fast as the competitor's product. Specifically, the manufacturer would like to test

$$H_0: \mu_1 = 2\mu_2$$
$$H_1: \mu_1 > 2\mu_2$$

where μ_1 is the mean absorption time of the competitive product and μ_2 is the mean absorption time of the new product. Assuming that the variances σ_1^2 and σ_2^2 are known, develop a procedure for testing this hypothesis.

10-71. Two machines are used to fill plastic bottles with dishwashing detergent. The standard deviations of fill volume are known to be $\sigma_1 = 0.10$ fluid ounces and $\sigma_2 = 0.15$ fluid ounces for the two machines, respectively. Two random samples of $n_1 = 12$ bottles from machine 1 and $n_2 = 10$ bottles from machine 2 are selected, and the sample mean fill volumes are $\bar{x}_1 = 30.87$ fluid ounces and $\bar{x}_2 = 30.68$ fluid ounces. Assume normality.
(a) Construct a 90% two-sided confidence interval on the mean difference in fill volume. Interpret this interval.
(b) Construct a 95% two-sided confidence interval on the mean difference in fill volume. Compare and comment on the width of this interval to the width of the interval in part (a).
(c) Construct a 95% upper-confidence interval on the mean difference in fill volume. Interpret this interval.
(d) Test the hypothesis that both machines fill to the same mean volume. Use $\alpha = 0.05$. What is the P-value?
(e) If the β-error of the test when the true difference in fill volume is 0.2 fluid ounces should not exceed 0.1, what sample sizes must be used? Use $\alpha = 0.05$.

10-72. Suppose that we are testing $H_0: \mu_1 = \mu_2$ versus $H_1: \mu_1 \neq \mu_2$, and we plan to use equal sample sizes from the two populations. Both populations are assumed to be normal with unknown but equal variances. If we use $\alpha = 0.05$ and if the true mean $\mu_1 = \mu_2 + \sigma$, what sample size must be used for the power of this test to be at least 0.90?

10-73. Consider the situation described in Exercise 10-59.
(a) Redefine the parameters of interest to be the proportion of lenses that are unsatisfactory following tumble polishing with polishing fluids 1 or 2. Test the hypothesis that the two polishing solutions give different results using $\alpha = 0.01$.
(b) Compare your answer in part (a) with that for Exercise 10-59. Explain why they are the same or different.
(c) We wish to use $\alpha = 0.01$. Suppose that if $p_1 = 0.9$ and $p_2 = 0.6$, we wish to detect this with a high probability, say, at least 0.9. What sample sizes are required to meet this objective?

10-74. Consider the fire-fighting foam expanding agents investigated in Exercise 10-14, in which five observations of each agent were recorded. Suppose that, if agent 1 produces a mean expansion that differs from the mean expansion of agent 1 by 1.5, we would like to reject the null hypothesis with probability at least 0.95.
(a) What sample size is required?
(b) Do you think that the original sample size in Exercise 10-14 was appropriate to detect this difference? Explain your answer.

10-75. A fuel-economy study was conducted for two German automobiles, Mercedes and Volkswagen. One vehicle of each brand was selected, and the mileage performance was observed for 10 tanks of fuel in each car. The data are as follows (in miles per gallon):

Mercedes		Volkswagen	
24.7	24.9	41.7	42.8
24.8	24.6	42.3	42.4
24.9	23.9	41.6	39.9
24.7	24.9	39.5	40.8
24.5	24.8	41.9	29.6

(a) Construct a normal probability plot of each of the data sets. Based on these plots, is it reasonable to assume that they are each drawn from a normal population?
(b) Suppose that it was determined that the lowest observation of the Mercedes data was erroneously recorded and should be 24.6. Furthermore, the lowest observation of the Volkswagen data was also mistaken and should be 39.6. Again construct normal probability plots of each of the data sets with the corrected values. Based on these new plots, is it reasonable to assume that they are each drawn from a normal population?
(c) Compare your answers from parts (a) and (b) and comment on the effect of these mistaken observations on the normality assumption.
(d) Using the corrected data from part (b) and a 95% confidence interval, is there evidence to support the claim that the variability in mileage performance is greater for a Volkswagen than for a Mercedes?
(e) Rework part (d) of this problem using an appropriate hypothesis-testing procedure. Did you get the same answer as you did originally? Why?

10-76. An experiment was conducted to compare the filling capability of packaging equipment at two different wineries. Ten bottles of pinot noir from Ridgecrest Vineyards were randomly selected and measured, along with 10 bottles of pinot noir from Valley View Vineyards. The data are as follows (fill volume is in milliliters):

Ridgecrest				Valley View			
755	751	752	753	756	754	757	756
753	753	753	754	755	756	756	755
752	751			755	756		

(a) What assumptions are necessary to perform a hypothesis-testing procedure for equality of means of these data? Check these assumptions.
(b) Perform the appropriate hypothesis-testing procedure to determine whether the data support the claim that both wineries will fill bottles to the same mean volume.
(c) Suppose that the true difference in mean fill volume is as much as 2 fluid ounces; did the sample sizes of 10 from each vineyard provide good detection capability when $\alpha = 0.05$? Explain your answer.

10-77. A Rockwell hardness-testing machine presses a tip into a test coupon and uses the depth of the resulting depression to indicate hardness. Two different tips are being compared to determine whether they provide the same Rockwell C-scale hardness readings. Nine coupons are tested, with both tips being tested on each coupon. The data are shown in the accompanying table.

Coupon	Tip 1	Tip 2	Coupon	Tip 1	Tip 2
1	47	46	6	41	41
2	42	40	7	45	46
3	43	45	8	45	46
4	40	41	9	49	48
5	42	43			

(a) State any assumptions necessary to test the claim that both tips produce the same Rockwell C-scale hardness readings. Check those assumptions for which you have the information.
(b) Apply an appropriate statistical method to determine if the data support the claim that the difference in Rockwell C-scale hardness readings of the two tips is significantly different from zero
(c) Suppose that if the two tips differ in mean hardness readings by as much as 1.0, we want the power of the test to be at least 0.9. For an $\alpha = 0.01$, how many coupons should have been used in the test?

10-78. Two different gauges can be used to measure the depth of bath material in a Hall cell used in smelting aluminum. Each gauge is used once in 15 cells by the same operator.

Cell	Gauge 1	Gauge 2	Cell	Gauge 1	Gauge 2
1	46 in.	47 in.	9	52	51
2	50	53	10	47	45
3	47	45	11	49	51
4	53	50	12	45	45
5	49	51	13	47	49
6	48	48	14	46	43
7	53	54	15	50	51
8	56	53			

(a) State any assumptions necessary to test the claim that both gauges produce the same mean bath depth readings. Check those assumptions for which you have the information.
(b) Apply an appropriate statistical procedure to determine if the data support the claim that the two gauges produce different mean bath depth readings.
(c) Suppose that if the two gauges differ in mean bath depth readings by as much as 1.65 inch, we want the power of the test to be at least 0.8. For $\alpha = 0.01$, how many cells should have been used?

10-79. An article in the *Journal of the Environmental Engineering Division* ("Distribution of Toxic Substances in Rivers," 1982, Vol. 108, pp. 639–649) investigated the concentration of several hydrophobic organic substances in the Wolf River in Tennessee. Measurements on hexachlorobenzene (HCB) in nanograms per liter were taken at different depths downstream of an abandoned dump site. Data for two depths follow:

Surface: 3.74, 4.61, 4.00, 4.67, 4.87, 5.12, 4.52, 5.29, 5.74, 5.48
Bottom: 5.44, 6.88, 5.37, 5.44, 5.03, 6.48, 3.89, 5.85, 6.85, 7.16

(a) What assumptions are required to test the claim that mean HCB concentration is the same at both depths? Check those assumptions for which you have the information.
(b) Apply an appropriate procedure to determine if the data support the claim in part a.
(c) Suppose that the true difference in mean concentrations is 2.0 nanograms per liter. For $\alpha = 0.05$, what is the power of a statistical test for $H_0: \mu_1 = \mu_2$ versus $H_1: \mu_1 \neq \mu_2$?
(d) What sample size would be required to detect a difference of 1.0 nanograms per liter at $\alpha = 0.05$ if the power must be at least 0.9?

MIND-EXPANDING EXERCISES

10-80. Three different pesticides can be used to control infestation of grapes. It is suspected that pesticide 3 is more effective than the other two. In a particular vineyard, three different plantings of pinot noir grapes are selected for study. The following results on yield are obtained:

Pesticide	$\bar{x}_i$ (Bushels/Plant)	s_i	n_i (Number of Plants)
1	4.6	0.7	100
2	5.2	0.6	120
3	6.1	0.8	130

If μ_i is the true mean yield after treatment with the ith pesticide, we are interested in the quantity

$$\mu = \frac{1}{2}(\mu_1 + \mu_2) - \mu_3$$

which measures the difference in mean yields between pesticides 1 and 2 and pesticide 3. If the sample sizes n_i are large, the estimator (say, $\hat{\mu}$) obtained by replacing each individual μ_i by $\bar{X}_i$ is approximately normal.
(a) Find an approximate $100(1 - \alpha)\%$ large-sample confidence interval for μ.
(b) Do these data support the claim that pesticide 3 is more effective than the other two? Use $\alpha = 0.05$ in determining your answer.

10-81. Suppose that we wish to test $H_0: \mu_1 = \mu_2$ versus $H_1: \mu_1 \neq \mu_2$, where σ_1^2 and σ_2^2 are known. The total sample size N is to be determined, and the allocation of observations to the two populations such that $n_1 + n_2 = N$ is to be made on the basis of cost. If the cost of sampling for populations 1 and 2 are C_1 and C_2, respectively, find the minimum cost sample sizes that provide a specified variance for the difference in sample means.

10-82. Suppose that we wish to test the hypothesis $H_0: \mu_1 = \mu_2$ versus $H_1: \mu_1 \neq \mu_2$, where both variances σ_1^2 and σ_2^2 are known. A total of $n_1 + n_2 = N$ observations can be taken. How should these observations be allocated to the two populations to maximize the probability that H_0 will be rejected if H_1 is true and $\mu_1 - \mu_2 = \Delta \neq 0$?

10-83. Suppose that we wish to test $H_0: \mu = \mu_0$ versus $H_1: \mu \neq \mu_0$, where the population is normal with known σ. Let $0 < \epsilon < \alpha$, and define the critical region so that we will reject H_0 if $z_0 > z_\epsilon$ or if $z_0 < -z_{\alpha-\epsilon}$, where z_0 is the value of the usual test statistic for these hypotheses.
(a) Show that the probability of type I error for this test is α.
(b) Suppose that the true mean is $\mu_1 = \mu_0 + \Delta$. Derive an expression for β for the above test.

10-84. Construct a data set for which the paired t-test statistic is very large, indicating that when this analysis is used the two population means are different, but t_0 for the two-sample t-test is very small so that the incorrect analysis would indicate that there is no significant difference between the means.

10-85. In some situations involving proportions, we are interested in the ratio $\theta = p_1/p_2$ rather than the difference $p_1 - p_2$. Let $\hat{\theta} = \hat{p}_1/\hat{p}_2$. We can show that $\ln(\hat{\theta})$ has an approximate normal distribution with the mean $\ln(\theta)$ and variance $[(n_1 - x_1)/(n_1 x_1) + (n_2 - x_2)/(n_2 x_2)]^{1/2}$.
(a) Use the information above to derive a large-sample confidence interval for $\ln \theta$.
(b) Show how to find a large-sample CI for θ.
(c) Use the data from the St. John's Wort study in Example 10-14, and find a 95% CI on $\theta = p_1/p_2$. Provide a practical interpretation for this CI.

10-86. Derive an expression for β for the test of the equality of the variances of two normal distributions. Assume that the two-sided alternative is specified.

IMPORTANT TERMS AND CONCEPTS

Comparative experiments
Critical region for a test statistic
Identifying cause and effect
Null and alternative hypotheses
One-sided and two-sided alternative hypotheses
Operating characteristic curves
Paired t-test
Pooled t-test
P-value
Reference distribution for a test statistic
Sample size determination for hypothesis tests and confidence intervals
Statistical hypotheses
Test statistic

Simple Linear Regression and Correlation

CHAPTER OUTLINE

- 11-1 EMPIRICAL MODELS
- 11-2 SIMPLE LINEAR REGRESSION
- 11-3 PROPERTIES OF THE LEAST SQUARES ESTIMATORS
- 11-4 HYPOTHESIS TESTS IN SIMPLE LINEAR REGRESSION
  - 11-4.1 Use of t-Tests
  - 11-4.2 Analysis of Variance Approach to Test Significance of Regression
- 11-5 CONFIDENCE INTERVALS
  - 11-5.1 Confidence Intervals on the Slope and Intercept
  - 11-5.2 Confidence Interval on the Mean Response
- 11-6 PREDICTION OF NEW OBSERVATIONS
- 11-7 ADEQUACY OF THE REGRESSION MODEL
  - 11-7.1 Residual Analysis
  - 11-7.2 Coefficient of Determination (R^2)
- 11-8 CORRELATION
- 11-9 TRANSFORMATIONS
  - 11-9.1 Logistic Regression available at www.wiley.com/college/montgomery

LEARNING OBJECTIVES

After careful study of this chapter, you should be able to do the following:

1. Use simple linear regression for building empirical models to engineering and scientific data
2. Understand how the method of least squares is used to estimate the parameters in a linear regression model
3. Analyze residuals to determine if the regression model is an adequate fit to the data or to see if any underlying assumptions are violated
4. Test statistical hypotheses and construct confidence intervals on regression model parameters
5. Use the regression model to make a prediction of a future observation and construct an appropriate prediction interval on the future observation
6. Apply the correlation model
7. Use simple transformations to achieve a linear regression model

11-1 EMPIRICAL MODELS

Many problems in engineering and science involve exploring the relationships between two or more variables. **Regression analysis** is a statistical technique that is very useful for these types of problems. For example, in a chemical process, suppose that the yield of the product is related to the process-operating temperature. Regression analysis can be used to build a model to predict yield at a given temperature level. This model can also be used for process optimization, such as finding the level of temperature that maximizes yield, or for process control purposes.

As an illustration, consider the data in Table 11-1. In this table y is the purity of oxygen produced in a chemical distillation process, and x is the percentage of hydrocarbons that are present in the main condenser of the distillation unit. Figure 11-1 presents a **scatter diagram** of the data in Table 11-1. This is just a graph on which each (x_i, y_i) pair is represented as a point plotted in a two-dimensional coordinate system. This scatter diagram was produced by Minitab, and we selected an option that shows dot diagrams of the x and y variables along the top and right margins of the graph, respectively, making it easy to see the distributions of the individual variables (box plots or histograms could also be selected). Inspection of this scatter diagram indicates that, although no simple curve will pass exactly through all the points, there is a strong indication that the points lie scattered randomly around a straight line. Therefore, it is probably reasonable to assume that the mean of the random variable Y is related to x by the following straight-line relationship:

$$E(Y|x) = \mu_{Y|x} = \beta_0 + \beta_1 x$$

Table 11-1 Oxygen and Hydrocarbon Levels

Observation Number	Hydrocarbon Level $x(\%)$	Purity $y(\%)$
1	0.99	90.01
2	1.02	89.05
3	1.15	91.43
4	1.29	93.74
5	1.46	96.73
6	1.36	94.45
7	0.87	87.59
8	1.23	91.77
9	1.55	99.42
10	1.40	93.65
11	1.19	93.54
12	1.15	92.52
13	0.98	90.56
14	1.01	89.54
15	1.11	89.85
16	1.20	90.39
17	1.26	93.25
18	1.32	93.41
19	1.43	94.98
20	0.95	87.33

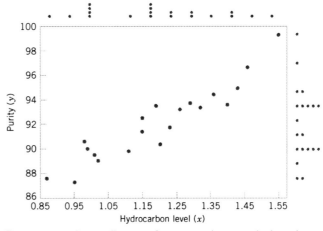

Figure 11-1 Scatter diagram of oxygen purity versus hydrocarbon level from Table 11-1.

where the slope and intercept of the line are called **regression coefficients**. While the mean of Y is a linear function of x, the actual observed value y does not fall exactly on a straight line. The appropriate way to generalize this to a probabilistic linear model is to assume that the expected value of Y is a linear function of x, but that for a fixed value of x the actual value of Y is determined by the mean value function (the linear model) plus a random error term, say,

$$Y = \beta_0 + \beta_1 x + \epsilon \tag{11-1}$$

where ϵ is the random error term. We will call this model the **simple linear regression model**, because it has only one independent variable or **regressor**. Sometimes a model like this will arise from a theoretical relationship. At other times, we will have no theoretical knowledge of the relationship between x and y, and the choice of the model is based on inspection of a scatter diagram, such as we did with the oxygen purity data. We then think of the regression model as an **empirical model.**

To gain more insight into this model, suppose that we can fix the value of x and observe the value of the random variable Y. Now if x is fixed, the random component ϵ on the right-hand side of the model in Equation 11-1 determines the properties of Y. Suppose that the mean and variance of ϵ are 0 and σ^2, respectively. Then

$$E(Y|x) = E(\beta_0 + \beta_1 x + \epsilon) = \beta_0 + \beta_1 x + E(\epsilon) = \beta_0 + \beta_1 x$$

Notice that this is the same relationship that we initially wrote down empirically from inspection of the scatter diagram in Fig. 11-1. The variance of Y given x is

$$V(Y|x) = V(\beta_0 + \beta_1 x + \epsilon) = V(\beta_0 + \beta_1 x) + V(\epsilon) = 0 + \sigma^2 = \sigma^2$$

Thus, the true regression model $\mu_{Y|x} = \beta_0 + \beta_1 x$ is a line of mean values; that is, the height of the regression line at any value of x is just the expected value of Y for that x. The slope, β_1, can be interpreted as the change in the mean of Y for a unit change in x. Furthermore, the variability of Y at a particular value of x is determined by the error variance σ^2. This implies that there is a distribution of Y-values at each x and that the variance of this distribution is the same at each x.

For example, suppose that the true regression model relating oxygen purity to hydrocarbon level is $\mu_{Y|x} = 75 + 15x$, and suppose that the variance is $\sigma^2 = 2$. Figure 11-2 illustrates this situation. Notice that we have used a normal distribution to describe the random variation in ϵ. Since Y is the sum of a constant $\beta_0 + \beta_1 x$ (the mean) and a normally distributed

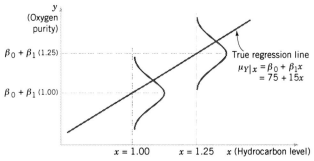

Figure 11-2 The distribution of Y for a given value of x for the oxygen purity-hydrocarbon data.

random variable, Y is a normally distributed random variable. The variance σ^2 determines the variability in the observations Y on oxygen purity. Thus, when σ^2 is small, the observed values of Y will fall close to the line, and when σ^2 is large, the observed values of Y may deviate considerably from the line. Because σ^2 is constant, the variability in Y at any value of x is the same.

The regression model describes the relationship between oxygen purity Y and hydrocarbon level x. Thus, for any value of hydrocarbon level, oxygen purity has a normal distribution with mean $75 + 15x$ and variance 2. For example, if $x = 1.25$, Y has mean value $\mu_{Y|x} = 75 + 15(1.25) = 93.75$ and variance 2.

In most real-world problems, the values of the intercept and slope (β_0, β_1) and the error variance σ^2 will not be known, and they must be estimated from sample data. Then this fitted regression equation or model is typically used in prediction of future observations of Y, or for estimating the mean response at a particular level of x. To illustrate, a chemical engineer might be interested in estimating the mean purity of oxygen produced when the hydrocarbon level is $x = 1.25\%$. This chapter discusses such procedures and applications for the simple linear regression model. Chapter 12 will discuss multiple linear regression models that involve more than one regressor.

Historical Note

Sir Francis Galton first used the term regression **analysis** in a study of the heights of fathers (x) and sons (y). Galton fit a least squares line and used it to predict the son's height from the father's height. He found that if a father's height was above average, the son's height would also be above average, but not by as much as the father's height was. A similar effect was observed for below average heights. That is, the son's height "regressed" toward the average. Consequently, Galton referred to the least squares line as a regression line.

Abuses of Regression

Regression is widely used and frequently misused; several common abuses of regression are briefly mentioned here. Care should be taken in selecting variables with which to construct regression equations and in determining the form of the model. It is possible to develop statistically significant relationships among variables that are completely unrelated in a **causal** sense. For example, we might attempt to relate the shear strength of spot welds with the number of empty parking spaces in the visitor parking lot. A straight line may even appear to provide a good fit to the data, but the relationship is an unreasonable one on which to rely. You can't increase the weld strength by blocking off parking spaces. A strong observed association between variables does not necessarily imply that a causal relationship exists between those variables. This type of effect is encountered fairly often in retrospective data analysis, and even in observational studies. **Designed experiments** are the only way to determine cause-and-effect relationships.

Regression relationships are valid only for values of the regressor variable within the range of the original data. The linear relationship that we have tentatively assumed may be valid over the original range of x, but it may be unlikely to remain so as we extrapolate—that is, if we use values of x beyond that range. In other words, as we move beyond the range of values of x for which data were collected, we become less certain about the validity of the assumed model. Regression models are not necessarily valid for extrapolation purposes.

Now this does not mean *don't ever extrapolate*. There are many problem situations in science and engineering where extrapolation of a regression model is the only way to even approach the problem. However, there is a strong warning to **be careful**. A modest extrapolation may be perfectly all right in many cases, but a large extrapolation will almost never produce acceptable results.

11-2 SIMPLE LINEAR REGRESSION

The case of **simple linear regression** considers a single regressor variable or predictor variable x and a dependent or response variable Y. Suppose that the true relationship between Y and x is a straight line and that the observation Y at each level of x is a random variable. As noted previously, the expected value of Y for each value of x is

$$E(Y|x) = \beta_0 + \beta_1 x$$

where the intercept β_0 and the slope β_1 are unknown regression coefficients. We assume that each observation, Y, can be described by the model

$$Y = \beta_0 + \beta_1 x + \epsilon \tag{11-2}$$

where ϵ is a random error with mean zero and (unknown) variance σ^2. The random errors corresponding to different observations are also assumed to be uncorrelated random variables.

Suppose that we have n pairs of observations $(x_1, y_1), (x_2, y_2), \ldots, (x_n, y_n)$. Figure 11-3 shows a typical scatter plot of observed data and a candidate for the estimated regression line. The estimates of β_0 and β_1 should result in a line that is (in some sense) a "best fit" to the data. The German scientist Karl Gauss (1777–1855) proposed estimating the parameters β_0 and β_1 in Equation 11-2 to minimize the sum of the squares of the vertical deviations in Fig. 11-3.

We call this criterion for estimating the regression coefficients the method of least squares. Using Equation 11-2, we may express the n observations in the sample as

$$y_i = \beta_0 + \beta_1 x_i + \epsilon_i, \quad i = 1, 2, \ldots, n \tag{11-3}$$

and the sum of the squares of the deviations of the observations from the true regression line is

$$L = \sum_{i=1}^{n} \epsilon_i^2 = \sum_{i=1}^{n} (y_i - \beta_0 - \beta_1 x_i)^2 \tag{11-4}$$

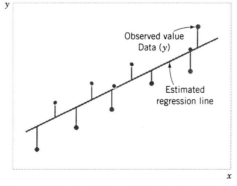

Figure 11-3 Deviations of the data from the estimated regression model.

The least squares estimators of β_0 and β_1, say, $\hat{\beta}_0$ and $\hat{\beta}_1$, must satisfy

$$\left.\frac{\partial L}{\partial \beta_0}\right|_{\hat{\beta}_0,\hat{\beta}_1} = -2\sum_{i=1}^{n}(y_i - \hat{\beta}_0 - \hat{\beta}_1 x_i) = 0$$

$$\left.\frac{\partial L}{\partial \beta_1}\right|_{\hat{\beta}_0,\hat{\beta}_1} = -2\sum_{i=1}^{n}(y_i - \hat{\beta}_0 - \hat{\beta}_1 x_i)x_i = 0 \quad (11\text{-}5)$$

Simplifying these two equations yields

$$n\hat{\beta}_0 + \hat{\beta}_1 \sum_{i=1}^{n} x_i = \sum_{i=1}^{n} y_i$$

$$\hat{\beta}_0 \sum_{i=1}^{n} x_i + \hat{\beta}_1 \sum_{i=1}^{n} x_i^2 = \sum_{i=1}^{n} y_i x_i \quad (11\text{-}6)$$

Equations 11-6 are called the **least squares normal equations**. The solution to the normal equations results in the least squares estimators $\hat{\beta}_0$ and $\hat{\beta}_1$.

Least Squares Estimates

The **least squares estimates** of the intercept and slope in the simple linear regression model are

$$\hat{\beta}_0 = \bar{y} - \hat{\beta}_1 \bar{x} \quad (11\text{-}7)$$

$$\hat{\beta}_1 = \frac{\sum_{i=1}^{n} y_i x_i - \dfrac{\left(\sum_{i=1}^{n} y_i\right)\left(\sum_{i=1}^{n} x_i\right)}{n}}{\sum_{i=1}^{n} x_i^2 - \dfrac{\left(\sum_{i=1}^{n} x_i\right)^2}{n}} \quad (11\text{-}8)$$

where $\bar{y} = (1/n)\sum_{i=1}^{n} y_i$ and $\bar{x} = (1/n)\sum_{i=1}^{n} x_i$.

The **fitted** or **estimated regression line** is therefore

$$\hat{y} = \hat{\beta}_0 + \hat{\beta}_1 x \quad (11\text{-}9)$$

Note that each pair of observations satisfies the relationship

$$y_i = \hat{\beta}_0 + \hat{\beta}_1 x_i + e_i, \quad i = 1, 2, \ldots, n$$

where $e_i = y_i - \hat{y}_i$ is called the **residual**. The residual describes the error in the fit of the model to the ith observation y_i. Later in this chapter we will use the residuals to provide information about the adequacy of the fitted model.

Notationally, it is occasionally convenient to give special symbols to the numerator and denominator of Equation 11-8. Given data $(x_1, y_1), (x_2, y_2), \ldots, (x_n, y_n)$, let

$$S_{xx} = \sum_{i=1}^{n} (x_i - \bar{x})^2 = \sum_{i=1}^{n} x_i^2 - \frac{\left(\sum_{i=1}^{n} x_i\right)^2}{n} \tag{11-10}$$

and

$$S_{xy} = \sum_{i=1}^{n} (y_i - \bar{y})(x_i - \bar{x}) = \sum_{i=1}^{n} x_i y_i - \frac{\left(\sum_{i=1}^{n} x_i\right)\left(\sum_{i=1}^{n} y_i\right)}{n} \tag{11-11}$$

EXAMPLE 11-1
Oxygen Purity

We will fit a simple linear regression model to the oxygen purity data in Table 11-1. The following quantities may be computed:

$$n = 20 \quad \sum_{i=1}^{20} x_i = 23.92 \quad \sum_{i=1}^{20} y_i = 1{,}843.21 \quad \bar{x} = 1.1960 \quad \bar{y} = 92.1605$$

$$\sum_{i=1}^{20} y_i^2 = 170{,}044.5321 \quad \sum_{i=1}^{20} x_i^2 = 29.2892 \quad \sum_{i=1}^{20} x_i y_i = 2{,}214.6566$$

$$S_{xx} = \sum_{i=1}^{20} x_i^2 - \frac{\left(\sum_{i=1}^{20} x_i\right)^2}{20} = 29.2892 - \frac{(23.92)^2}{20} = 0.68088$$

and

$$S_{xy} = \sum_{i=1}^{20} x_i y_i - \frac{\left(\sum_{i=1}^{20} x_i\right)\left(\sum_{i=1}^{20} y_i\right)}{20} = 2{,}214.6566 - \frac{(23.92)(1{,}843.21)}{20} = 10.17744$$

Therefore, the least squares estimates of the slope and intercept are

$$\hat{\beta}_1 = \frac{S_{xy}}{S_{xx}} = \frac{10.17744}{0.68088} = 14.94748$$

and

$$\hat{\beta}_0 = \bar{y} - \hat{\beta}_1 \bar{x} = 92.1605 - (14.94748)1.196 = 74.28331$$

The fitted simple linear regression model (with the coefficients reported to three decimal places) is

$$\hat{y} = 74.283 + 14.947x$$

This model is plotted in Fig. 11-4, along with the sample data.

Computer software programs are widely used in regression modeling. These programs typically carry more decimal places in the calculations. Table 11-2 shows a portion of the output from Minitab for this problem. The estimates $\hat{\beta}_0$ and $\hat{\beta}_1$ are highlighted. In subsequent sections we will provide explanations for the information provided in this computer output.

398 CHAPTER 11 SIMPLE LINEAR REGRESSION AND CORRELATION

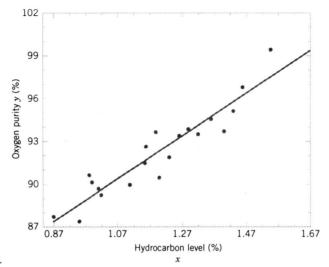

Figure 11-4 Scatter plot of oxygen purity y versus hydrocarbon level x and regression model $\hat{y} = 74.283 + 14.947x$.

Using the regression model of Example 11-1, we would predict oxygen purity of $\hat{y} = 89.23\%$ when the hydrocarbon level is $x = 1.00\%$. The purity 89.23% may be interpreted as an estimate of the true population mean purity when $x = 1.00\%$, or as an estimate of a new observation when $x = 1.00\%$. These estimates are, of course, subject to error; that is, it is unlikely that a future observation on purity would be exactly 89.23% when the hydrocarbon level is 1.00%. In subsequent sections we will see how to use confidence intervals and prediction intervals to describe the error in estimation from a regression model.

Table 11-2 Minitab Output for the Oxygen Purity Data in Example 11-1

Regression Analysis

The regression equation is

Purity = 74.3 + 14.9 HC Level

Predictor	Coef	SE Coef	T	P
Constant	74.283 ← $\hat{\beta}_0$	1.593	46.62	0.000
HC Level	14.947 ← $\hat{\beta}_1$	1.317	11.35	0.000

S = 1.087   R-Sq = 87.7%   R-Sq (adj) = 87.1%

Analysis of Variance

Source	DF	SS	MS	F	P
Regression	1	152.13	152.13	128.86	0.000
Residual Error	18	21.25 ← SS_E	1.18 ← $\hat{\sigma}^2$		
Total	19	173.38			

Predicted Values for New Observations

New Obs	Fit	SE Fit	95.0% CI	95.0% PI
1	89.231	0.354	(88.486, 89.975)	(86.830, 91.632)

Values of Predictors for New Observations

New Obs	HC Level
1	1.00

Estimating σ^2

There is actually another unknown parameter in our regression model, σ^2 (the variance of the error term ϵ). The residuals $e_i = y_i - \hat{y}_i$ are used to obtain an estimate of σ^2. The sum of squares of the residuals, often called the **error sum of squares**, is

$$SS_E = \sum_{i=1}^{n} e_i^2 = \sum_{i=1}^{n} (y_i - \hat{y}_i)^2 \qquad (11\text{-}12)$$

We can show that the expected value of the error sum of squares is $E(SS_E) = (n-2)\sigma^2$. Therefore an unbiased estimator of σ^2 is

Estimator of Variance

$$\hat{\sigma}^2 = \frac{SS_E}{n-2} \qquad (11\text{-}13)$$

Computing SS_E using Equation 11-12 would be fairly tedious. A more convenient computing formula can be obtained by substituting $\hat{y}_i = \hat{\beta}_0 + \hat{\beta}_1 x_i$ into Equation 11-12 and simplifying. The resulting computing formula is

$$SS_E = SS_T - \hat{\beta}_1 S_{xy} \qquad (11\text{-}14)$$

where $SS_T = \sum_{i=1}^{n} (y_i - \bar{y})^2 = \sum_{i=1}^{n} y_i^2 - n\bar{y}^2$ is the total sum of squares of the response variable y. Formulas such as this are presented in Section 11-4. The error sum of squares and the estimate of σ^2 for the oxygen purity data, $\hat{\sigma}^2 = 1.18$, are highlighted in the Minitab output in Table 11-2.

EXERCISES FOR SECTION 11-2

11-1. An article in *Concrete Research* ("Near Surface Characteristics of Concrete: Intrinsic Permeability," Vol. 41, 1989), presented data on compressive strength x and intrinsic permeability y of various concrete mixes and cures. Summary quantities are $n = 14$, $\sum y_i = 572$, $\sum y_i^2 = 23{,}530$, $\sum x_i = 43$, $\sum x_i^2 = 157.42$, and $\sum x_i y_i = 1697.80$. Assume that the two variables are related according to the simple linear regression model.
(a) Calculate the least squares estimates of the slope and intercept. Estimate σ^2. Graph the regression line.
(b) Use the equation of the fitted line to predict what permeability would be observed when the compressive strength is $x = 4.3$.
(c) Give a point estimate of the mean permeability when compressive strength is $x = 3.7$.
(d) Suppose that the observed value of permeability at $x = 3.7$ is $y = 46.1$. Calculate the value of the corresponding residual.

11-2. Regression methods were used to analyze the data from a study investigating the relationship between roadway surface temperature (x) and pavement deflection (y). Summary quantities were $n = 20$, $\sum y_i = 12.75$, $\sum y_i^2 = 8.86$, $\sum x_i = 1478$, $\sum x_i^2 = 143{,}215.8$, and $\sum x_i y_i = 1083.67$.
(a) Calculate the least squares estimates of the slope and intercept. Graph the regression line. Estimate σ^2.
(b) Use the equation of the fitted line to predict what pavement deflection would be observed when the surface temperature is 85°F.
(c) What is the mean pavement deflection when the surface temperature is 90°F?
(d) What change in mean pavement deflection would be expected for a 1°F change in surface temperature?

11-3. The following table presents data on the ratings of quarterbacks for the 2004 National Football League season (source: *The Sports Network*). It is suspected that the rating (y)

Player	Yards per Attempt	Rating Points
D. Culpepper, MIN	8.61	110.9
D. McNabb, PHI	8.26	104.7
B. Griese, TAM	7.83	97.5
M. Bulger, STL	8.17	93.7
B. Favre, GBP	7.57	92.4
J. Delhomme, CAR	7.29	87.3
K. Warner, NYG	7.42	86.5
M. Hasselbeck, SEA	7.14	83.1
A. Brooks, NOS	7.03	79.5
T. Rattay, SFX	6.67	78.1
M. Vick, ATL	7.21	78.1
J. Harrington, DET	6.23	77.5
V. Testaverde, DAL	7.14	76.4
P. Ramsey, WAS	6.12	74.8
J. McCown, ARI	6.15	74.1
P. Manning, IND	9.17	121.1
D. Brees, SDC	7.9	104.8
B. Roethlisberger, PIT	8.88	98.1
T. Green, KAN	8.26	95.2
T. Brady, NEP	7.79	92.6
C. Pennington, NYJ	7.22	91
B. Volek, TEN	6.96	87.1
J. Plummer, DEN	7.85	84.5
D. Carr, HOU	7.58	83.5
B. Leftwich, JAC	6.67	82.2
C. Palmer, CIN	6.71	77.3
J. Garcia, CLE	6.87	76.7
D. Bledsoe, BUF	6.52	76.6
K. Collins, OAK	6.81	74.8
K. Boller, BAL	5.52	70.9

is related to the average number of yards gained per pass attempt (x).

(a) Calculate the least squares estimates of the slope and intercept. What is the estimate of σ^2? Graph the regression model.
(b) Find an estimate of the mean rating if a quarterback averages 7.5 yards per attempt.
(c) What change in the mean rating is associated with a decrease of one yard per attempt?
(d) To increase the mean rating by 10 points, how much increase in the average yards per attempt must be generated?
(e) Given that $x = 7.21$ yards (M. Vick), find the fitted value of y and the corresponding residual.

11-4. An article in *Technometrics* by S. C. Narula and J. F. Wellington ("Prediction, Linear Regression, and a Minimum Sum of Relative Errors," Vol. 19, 1977) presents data on the selling price and annual taxes for 24 houses. The data are shown in the following table.

Sale Price/1000	Taxes (Local, School, County)/1000	Sale Price/1000	Taxes (Local, School, County)/1000
25.9	4.9176	30.0	5.0500
29.5	5.0208	36.9	8.2464
27.9	4.5429	41.9	6.6969
25.9	4.5573	40.5	7.7841
29.9	5.0597	43.9	9.0384
29.9	3.8910	37.5	5.9894
30.9	5.8980	37.9	7.5422
28.9	5.6039	44.5	8.7951
35.9	5.8282	37.9	6.0831
31.5	5.3003	38.9	8.3607
31.0	6.2712	36.9	8.1400
30.9	5.9592	45.8	9.1416

(a) Assuming that a simple linear regression model is appropriate, obtain the least squares fit relating selling price to taxes paid. What is the estimate of σ^2?
(b) Find the mean selling price given that the taxes paid are $x = 7.50$.
(c) Calculate the fitted value of y corresponding to $x = 5.8980$. Find the corresponding residual.
(d) Calculate the fitted $\hat{y}_i$ for each value of x_i used to fit the model. Then construct a graph of $\hat{y}_i$ versus the corresponding observed value y_i and comment on what this plot would look like if the relationship between y and x was a deterministic (no random error) straight line. Does the plot actually obtained indicate that taxes paid is an effective regressor variable in predicting selling price?

11-5. The number of pounds of steam used per month by a chemical plant is thought to be related to the average ambient temperature (in° F) for that month. The past year's usage and temperature are shown in the following table:

Month	Temp.	Usage/1000	Month	Temp.	Usage/1000
Jan.	21	185.79	July	68	621.55
Feb.	24	214.47	Aug.	74	675.06
Mar.	32	288.03	Sept.	62	562.03
Apr.	47	424.84	Oct.	50	452.93
May	50	454.58	Nov.	41	369.95
June	59	539.03	Dec.	30	273.98

(a) Assuming that a simple linear regression model is appropriate, fit the regression model relating steam usage (y) to the average temperature (x). What is the estimate of σ^2? Graph the regression line.
(b) What is the estimate of expected steam usage when the average temperature is 55°F?
(c) What change in mean steam usage is expected when the monthly average temperature changes by 1°F?
(d) Suppose the monthly average temperature is 47°F. Calculate the fitted value of y and the corresponding residual.

11-6. The following table presents the highway gasoline mileage performance and engine displacement for Damler-Chrysler vehicles for model year 2005 (source: U.S. Environmental Protection Agency).
(a) Fit a simple linear model relating highway miles per gallon (y) to engine displacement (x) in cubic inches using least squares.
(b) Find an estimate of the mean highway gasoline mileage performance for a car with 150 cubic inches engine displacement.
(c) Obtain the fitted value of y and the corresponding residual for a car, the Neon, with an engine displacement of 122 cubic inches.

Carline	Engine Displacement (in³)	MPG (highway)
300C/SRT-8	215	30.8
CARAVAN 2WD	201	32.5
CROSSFIRE ROADSTER	196	35.4
DAKOTA PICKUP 2WD	226	28.1
DAKOTA PICKUP 4WD	226	24.4
DURANGO 2WD	348	24.1
GRAND CHEROKEE 2WD	226	28.5
GRAND CHEROKEE 4WD	348	24.2
LIBERTY/CHEROKEE 2WD	148	32.8
LIBERTY/CHEROKEE 4WD	226	28
NEON/SRT-4/SX 2.0	122	41.3
PACIFICA 2WD	215	30.0
PACIFICA AWD	215	28.2
PT CRUISER	148	34.1
RAM 1500 PICKUP 2WD	500	18.7
RAM 1500 PICKUP 4WD	348	20.3
SEBRING 4-DR	165	35.1
STRATUS 4-DR	148	37.9
TOWN & COUNTRY 2WD	148	33.8
VIPER CONVERTIBLE	500	25.9
WRANGLER/TJ 4WD	148	26.4

11-7. An article in the *Tappi Journal* (March, 1986) presented data on green liquor Na_2S concentration (in grams per liter) and paper machine production (in tons per day). The data (read from a graph) are shown as follows:

y	40	42	49	46	44	48
x	825	830	890	895	890	910

y	46	43	53	52	54	57	58
x	915	960	990	1010	1012	1030	1050

(a) Fit a simple linear regression model with y = green liquor Na_2S concentration and x = production. Find an estimate of σ^2. Draw a scatter diagram of the data and the resulting least squares fitted model.
(b) Find the fitted value of y corresponding to x = 910 and the associated residual.
(c) Find the mean green liquor Na_2S concentration when the production rate is 950 tons per day.

11-8. An article in the *Journal of Sound and Vibration* (Vol. 151, 1991, pp. 383–394) described a study investigating the relationship between noise exposure and hypertension. The following data are representative of those reported in the article.

y	1	0	1	2	5	1	4	6	2	3
x	60	63	65	70	70	70	80	90	80	80

y	5	4	6	8	4	5	7	9	7	6
x	85	89	90	90	90	90	94	100	100	100

(a) Draw a scatter diagram of y (blood pressure rise in millimeters of mercury) versus x (sound pressure level in decibels). Does a simple linear regression model seem reasonable in this situation?
(b) Fit the simple linear regression model using least squares. Find an estimate of σ^2.
(c) Find the predicted mean rise in blood pressure level associated with a sound pressure level of 85 decibels.

11-9. An article in *Wear* (Vol. 152, 1992, pp. 171–181) presents data on the fretting wear of mild steel and oil viscosity. Representative data follow, with x = oil viscosity and y = wear volume (10^{-4} cubic millimeters).

y	240	181	193	155	172
x	1.6	9.4	15.5	20.0	22.0

y	110	113	75	94
x	35.5	43.0	40.5	33.0

(a) Construct a scatter plot of the data. Does a simple linear regression model appear to be plausible?
(b) Fit the simple linear regression model using least squares. Find an estimate of σ^2.
(c) Predict fretting wear when viscosity $x = 30$.
(d) Obtain the fitted value of y when $x = 22.0$ and calculate the corresponding residual.

11-10. An article in the *Journal of Environmental Engineering* (Vol. 115, No. 3, 1989, pp. 608–619) reported the results of a study on the occurrence of sodium and chloride in surface streams in central Rhode Island. The following data are chloride concentration y (in milligrams per liter) and roadway area in the watershed x (in percentage).

y	4.4	6.6	9.7	10.6	10.8	10.9
x	0.19	0.15	0.57	0.70	0.67	0.63

y	11.8	12.1	14.3	14.7	15.0	17.3
x	0.47	0.70	0.60	0.78	0.81	0.78

y	19.2	23.1	27.4	27.7	31.8	39.5
x	0.69	1.30	1.05	1.06	1.74	1.62

(a) Draw a scatter diagram of the data. Does a simple linear regression model seem appropriate here?
(b) Fit the simple linear regression model using the method of least squares. Find an estimate of σ^2.
(c) Estimate the mean chloride concentration for a watershed that has 1% roadway area.
(d) Find the fitted value corresponding to $x = 0.47$ and the associated residual.

11-11. A rocket motor is manufactured by bonding together two types of propellants, an igniter and a sustainer. The shear strength of the bond y is thought to be a linear function of the age of the propellant x when the motor is cast. Twenty observations are shown in the following table.

(a) Draw a scatter diagram of the data. Does the straight-line regression model seem to be plausible?
(b) Find the least squares estimates of the slope and intercept in the simple linear regression model. Find an estimate of σ^2.
(c) Estimate the mean shear strength of a motor made from propellant that is 20 weeks old.
(d) Obtain the fitted values $\hat{y}_i$ that correspond to each observed value y_i. Plot $\hat{y}_i$ versus y_i, and comment on what this plot would look like if the linear relationship between shear strength and age were perfectly deterministic (no error). Does this plot indicate that age is a reasonable choice of regressor variable in this model?

Observation Number	Strength y (psi)	Age x (weeks)
1	2158.70	15.50
2	1678.15	23.75
3	2316.00	8.00
4	2061.30	17.00
5	2207.50	5.00
6	1708.30	19.00
7	1784.70	24.00
8	2575.00	2.50
9	2357.90	7.50
10	2277.70	11.00
11	2165.20	13.00
12	2399.55	3.75
13	1779.80	25.00
14	2336.75	9.75
15	1765.30	22.00
16	2053.50	18.00
17	2414.40	6.00
18	2200.50	12.50
19	2654.20	2.00
20	1753.70	21.50

11-12. An article in the *Journal of the American Ceramic Society*, "Rapid Hot-Pressing of Ultrafine PSZ Powders" (1991, Vol. 74, pp. 1547–1553) considered the microstructure of the ultrafine powder of partially stabilized zirconia as a function of temperature. The data are shown below:

x = Temperature (°C): 1100 1200 1300 1100 1500
                        1200 1300

y = Porosity (%): 30.8 19.2 6.0 13.5 11.4
                     7.7   3.6

(a) Fit the simple linear regression model using the method of least squares. Find an estimate of σ^2.
(b) Estimate the mean porosity for a temperature of 1400°C.
(c) Find the fitted value corresponding to $y = 11.4$ and the associated residual.
(d) Draw a scatter diagram of the data. Does a simple linear regression model seem appropriate here? Explain.

11-13. An article in the *Journal of the Environmental Engineering Division*, "Least Squares Estimates of BOD Parameters" (1980, Vol. 106, pp. 1197–1202) took a sample from the Holston River below Kingport, TN, during August 1977. The biochemical oxygen demand (BOD) test is conducted over a period of time in days. The resulting data is shown below:

Time (days):   1   2   4   6   8   10   12   14   16   18   20

BOD (mg/liter): 0.6 0.7 1.5 1.9 2.1 2.6 2.9 3.7 3.5 3.7 3.8

(a) Assuming that a simple linear regression model is appropriate, fit the regression model relating BOD (y) to the time (x). What is the estimate of σ^2?
(b) What is the estimate of expected BOD level when the time is 15 days?
(c) What change in mean BOD is expected when the time changes by 3 days?
(d) Suppose the time used is 6 days. Calculate the fitted value of y and the corresponding residual.
(e) Calculate the fitted $\hat{y}_i$ for each value of x_i used to fit the model. Then construct a graph of $\hat{y}_i$ versus the corresponding observed values y_i and comment on what this plot would look like if the relationship between y and x was a deterministic (no random error) straight line. Does the plot actually obtained indicate that time is an effective regressor variable in predicting BOD?

11-14. An article in *Wood Science and Technology*, "Creep in Chipboard, Part 3: Initial Assessment of the Influence of Moisture Content and Level of Stressing on Rate of Creep and Time to Failure" (1981, Vol. 15, pp. 125–144) studied the deflection (mm) of particleboard from stress levels of relative humidity. Assume that the two variables are related according to the simple linear regression model. The data are shown below:

$x =$ Stress level (%):   54   54   61   61   68
$y =$ Deflection (mm): 16.473 18.693 14.305 15.121 13.505

$x =$ Stress level (%):   68   75   75   75
$y =$ Deflection (mm): 11.640 11.168 12.534 11.224

(a) Calculate the least square estimates of the slope and intercept. What is the estimate of σ^2? Graph the regression model and the data.
(b) Find the estimate of the mean deflection if the stress level can be limited to 65%.
(c) Estimate the change in the mean deflection associated with a 5% increment in stress level.
(d) To decrease the mean deflection by one millimeter, how much increase in stress level must be generated?
(e) Given that the stress level is 68%, find the fitted value of deflection and the corresponding residual.

11-15. In an article in *Statistics and Computing*, "An iterative Monte Carlo method for nonconjugate Bayesian analysis" (1991, pp. 119–128) Carlin and Gelfand investigated the age (x) and length (y) of 27 captured dugongs (sea cows).

$x =$ 1.0, 1.5, 1.5, 1.5, 2.5, 4.0, 5.0, 5.0, 7.0, 8.0, 8.5, 9.0, 9.5, 9.5, 10.0, 12.0, 12.0, 13.0, 13.0, 14.5, 15.5, 15.5, 16.5, 17.0, 22.5, 29.0, 31.5

$y =$ 1.80, 1.85, 1.87, 1.77, 2.02, 2.27, 2.15, 2.26, 2.47, 2.19, 2.26, 2.40, 2.39, 2.41, 2.50, 2.32, 2.32, 2.43, 2.47, 2.56, 2.65, 2.47, 2.64, 2.56, 2.70, 2.72, 2.57

(a) Find the least squares estimates of the slope and the intercept in the simple linear regression model. Find an estimate of σ^2.
(b) Estimate the mean length of dugongs at age 11.
(c) Obtain the fitted values $\hat{y}_i$ that correspond to each observed value y_i. Plot $\hat{y}_i$ versus y_i, and comment on what this plot would look like if the linear relationship between length and age were perfectly deterministic (no error). Does this plot indicate that age is a reasonable choice of regressor variable in this model?

11-16. Consider the regression model developed in Exercise 11-2.
(a) Suppose that temperature is measured in °C rather than °F. Write the new regression model.
(b) What change in expected pavement deflection is associated with a 1°C change in surface temperature?

11-17. Consider the regression model developed in Exercise 11-6. Suppose that engine displacement is measured in cubic centimeters instead of cubic inches.
(a) Write the new regression model.
(b) What change in gasoline mileage is associated with a 1 cm^3 change is engine displacement?

11-18. Show that in a simple linear regression model the point $(\bar{x}, \bar{y})$ lies exactly on the least squares regression line.

11-19. Consider the simple linear regression model $Y = \beta_0 + \beta_1 x + \epsilon$. Suppose that the analyst wants to use $z = x - \bar{x}$ as the regressor variable.
(a) Using the data in Exercise 11-11, construct one scatter plot of the (x_i, y_i) points and then another of the $(z_i = x_i - \bar{x}, y_i)$ points. Use the two plots to intuitively explain how the two models, $Y = \beta_0 + \beta_1 x + \epsilon$ and $Y = \beta_0^* + \beta_1^* z + \epsilon$, are related.
(b) Find the least squares estimates of β_0^* and β_1^* in the model $Y = \beta_0^* + \beta_1^* z + \epsilon$. How do they relate to the least squares estimates $\hat{\beta}_0$ and $\hat{\beta}_1$?

11-20. Suppose we wish to fit a regression model for which the true regression line passes through the point (0, 0). The appropriate model is $Y = \beta x + \epsilon$. Assume that we have n pairs of data $(x_1, y_1), (x_2, y_2), \ldots, (x_n, y_n)$.
(a) Find the least squares estimate of β.
(b) Fit the model $Y = \beta x + \epsilon$ to the chloride concentration-roadway area data in Exercise 11-10. Plot the fitted model on a scatter diagram of the data and comment on the appropriateness of the model.

11-3 PROPERTIES OF THE LEAST SQUARES ESTIMATORS

The statistical properties of the least squares estimators $\hat{\beta}_0$ and $\hat{\beta}_1$ may be easily described. Recall that we have assumed that the error term ϵ in the model $Y = \beta_0 + \beta_1 x + \epsilon$ is a random variable with mean zero and variance σ^2. Since the values of x are fixed, Y is a random variable with mean $\mu_{Y|x} = \beta_0 + \beta_1 x$ and variance σ^2. Therefore, the values of $\hat{\beta}_0$ and $\hat{\beta}_1$ depend on the observed y's; thus, the least squares estimators of the regression coefficients may be viewed as random variables. We will investigate the bias and variance properties of the least squares estimators $\hat{\beta}_0$ and $\hat{\beta}_1$.

Consider first $\hat{\beta}_1$. Because $\hat{\beta}_1$ is a linear combination of the observations Y_i, we can use properties of expectation to show that the expected value of $\hat{\beta}_1$ is

$$E(\hat{\beta}_1) = \beta_1 \qquad (11\text{-}15)$$

Thus, $\hat{\beta}_1$ is an unbiased estimator of the true slope β_1.

Now consider the variance of $\hat{\beta}_1$. Since we have assumed that $V(\epsilon_i) = \sigma^2$, it follows that $V(Y_i) = \sigma^2$. Because $\hat{\beta}_1$ is a linear combination of the observations Y_i, the results in Section 5-5 can be applied to show that

$$V(\hat{\beta}_1) = \frac{\sigma^2}{S_{xx}} \qquad (11\text{-}16)$$

For the intercept, we can show in a similar manner that

$$E(\hat{\beta}_0) = \beta_0 \quad \text{and} \quad V(\hat{\beta}_0) = \sigma^2 \left[\frac{1}{n} + \frac{\bar{x}^2}{S_{xx}}\right] \qquad (11\text{-}17)$$

Thus, $\hat{\beta}_0$ is an unbiased estimator of the intercept β_0. The covariance of the random variables $\hat{\beta}_0$ and $\hat{\beta}_1$ is not zero. It can be shown (see Exercise 11-92) that $\text{cov}(\hat{\beta}_0, \hat{\beta}_1) = -\sigma^2 \bar{x}/S_{xx}$.

The estimate of σ^2 could be used in Equations 11-16 and 11-17 to provide estimates of the variance of the slope and the intercept. We call the square roots of the resulting variance estimators the **estimated standard errors** of the slope and intercept, respectively.

Estimated Standard Errors

In simple linear regression the **estimated standard error of the slope** and the **estimated standard error of the intercept** are

$$se(\hat{\beta}_1) = \sqrt{\frac{\hat{\sigma}^2}{S_{xx}}} \quad \text{and} \quad se(\hat{\beta}_0) = \sqrt{\hat{\sigma}^2 \left[\frac{1}{n} + \frac{\bar{x}^2}{S_{xx}}\right]}$$

respectively, where $\hat{\sigma}^2$ is computed from Equation 11-13.

The Minitab computer output in Table 11-2 reports the estimated standard errors of the slope and intercept under the column heading "SE coeff."

11-4 HYPOTHESIS TESTS IN SIMPLE LINEAR REGRESSION

An important part of assessing the adequacy of a linear regression model is testing statistical hypotheses about the model parameters and constructing certain confidence intervals. Hypothesis testing in simple linear regression is discussed in this section, and Section 11-5 presents methods for constructing confidence intervals. To test hypotheses about the slope and intercept of the regression model, we must make the additional assumption that the error component in the model, ϵ, is normally distributed. Thus, the complete assumptions are that the errors are normally and independently distributed with mean zero and variance σ^2, abbreviated NID$(0, \sigma^2)$.

11-4.1 Use of t-Tests

Suppose we wish to test the hypothesis that the slope equals a constant, say, $\beta_{1,0}$. The appropriate hypotheses are

$$H_0: \beta_1 = \beta_{1,0}$$
$$H_1: \beta_1 \neq \beta_{1,0} \tag{11-18}$$

where we have assumed a two-sided alternative. Since the errors ϵ_i are NID$(0, \sigma^2)$, it follows directly that the observations Y_i are NID$(\beta_0 + \beta_1 x_i, \sigma^2)$. Now $\hat{\beta}_1$ is a linear combination of independent normal random variables, and consequently, $\hat{\beta}_1$ is $N(\beta_1, \sigma^2/S_{xx})$, using the bias and variance properties of the slope discussed in Section 11-3. In addition, $(n-2)\hat{\sigma}^2/\sigma^2$ has a chi-square distribution with $n-2$ degrees of freedom, and $\hat{\beta}_1$ is independent of $\hat{\sigma}^2$. As a result of those properties, the statistic

$$T_0 = \frac{\hat{\beta}_1 - \beta_{1,0}}{\sqrt{\hat{\sigma}^2/S_{xx}}} \tag{11-19}$$

follows the t distribution with $n-2$ degrees of freedom under $H_0: \beta_1 = \beta_{1,0}$. We would reject $H_0: \beta_1 = \beta_{1,0}$ if

$$|t_0| > t_{\alpha/2, n-2} \tag{11-20}$$

where t_0 is computed from Equation 11-19. The denominator of Equation 11-19 is the standard error of the slope, so we could write the test statistic as

$$T_0 = \frac{\hat{\beta}_1 - \beta_{1,0}}{se(\hat{\beta}_1)}$$

A similar procedure can be used to test hypotheses about the intercept. To test

$$H_0: \beta_0 = \beta_{0,0}$$
$$H_1: \beta_0 \neq \beta_{0,0} \tag{11-21}$$

we would use the statistic

$$T_0 = \frac{\hat{\beta}_0 - \beta_{0,0}}{\sqrt{\hat{\sigma}^2 \left[\frac{1}{n} + \frac{\bar{x}^2}{S_{xx}}\right]}} = \frac{\hat{\beta}_0 - \beta_{0,0}}{se(\hat{\beta}_0)} \tag{11-22}$$

Figure 11-5 The hypothesis H_0: $\beta_1 = 0$ is not rejected.

and reject the null hypothesis if the computed value of this test statistic, t_0, is such that $|t_0| > t_{\alpha/2, n-2}$. Note that the denominator of the test statistic in Equation 11-22 is just the standard error of the intercept.

A very important special case of the hypotheses of Equation 11-18 is

$$H_0: \beta_1 = 0$$
$$H_1: \beta_1 \neq 0 \qquad (11\text{-}23)$$

These hypotheses relate to the **significance of regression**. Failure to reject H_0: $\beta_1 = 0$ is equivalent to concluding that there is no linear relationship between x and Y. This situation is illustrated in Fig. 11-5. Note that this may imply either that x is of little value in explaining the variation in Y and that the best estimator of Y for any x is $\hat{y} = \overline{Y}$ (Fig. 11-5a) or that the true relationship between x and Y is not linear (Fig. 11-5b). Alternatively, if H_0: $\beta_1 = 0$ is rejected, this implies that x is of value in explaining the variability in Y (see Fig. 11-6). Rejecting H_0: $\beta_1 = 0$ could mean either that the straight-line model is adequate (Fig. 11-6a) or that, although there is a linear effect of x, better results could be obtained with the addition of higher order polynomial terms in x (Fig. 11-6b).

EXAMPLE 11-2
Oxygen Purity
Tests of
Coefficients

We will test for significance of regression using the model for the oxygen purity data from Example 11-1. The hypotheses are

$$H_0: \beta_1 = 0$$
$$H_1: \beta_1 \neq 0$$

and we will use $\alpha = 0.01$. From Example 11-1 and Table 11-2 we have

$$\hat{\beta}_1 = 14.947 \quad n = 20, \quad S_{xx} = 0.68088, \quad \hat{\sigma}^2 = 1.18$$

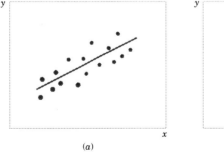
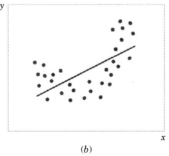

Figure 11-6 The hypothesis H_0: $\beta_1 = 0$ is rejected.

so the *t*-statistic in Equation 10-20 becomes

$$t_0 = \frac{\hat{\beta}_1}{\sqrt{\hat{\sigma}^2/S_{xx}}} = \frac{\hat{\beta}_1}{se(\hat{\beta}_1)} = \frac{14.947}{\sqrt{1.18/0.68088}} = 11.35$$

Since the reference value of t is $t_{0.005,18} = 2.88$, the value of the test statistic is very far into the critical region, implying that H_0: $\beta_1 = 0$ should be rejected. The *P*-value for this test is $P \simeq 1.23 \times 10^{-9}$. This was obtained manually with a calculator.

Table 11-2 presents the Minitab output for this problem. Notice that the *t*-statistic value for the slope is computed as 11.35 and that the reported *P*-value is $P = 0.000$. Minitab also reports the *t*-statistic for testing the hypothesis H_0: $\beta_0 = 0$. This statistic is computed from Equation 11-22, with $\beta_{0,0} = 0$, as $t_0 = 46.62$. Clearly, then, the hypothesis that the intercept is zero is rejected.

11-4.2 Analysis of Variance Approach to Test Significance of Regression

A method called the **analysis of variance** can be used to test for significance of regression. The procedure partitions the total variability in the response variable into meaningful components as the basis for the test. The **analysis of variance identity** is as follows:

Analysis of Variance Identity

$$\sum_{i=1}^{n}(y_i - \bar{y})^2 = \sum_{i=1}^{n}(\hat{y}_i - \bar{y})^2 + \sum_{i=1}^{n}(y_i - \hat{y}_i)^2 \qquad (11\text{-}24)$$

The two components on the right-hand-side of Equation 11-24 measure, respectively, the amount of variability in y_i accounted for by the regression line and the residual variation left unexplained by the regression line. We usually call $SS_E = \sum_{i=1}^{n}(y_i - \hat{y}_i)^2$ the **error sum of squares** and $SS_R = \sum_{i=1}^{n}(\hat{y}_i - \bar{y})^2$ the **regression sum of squares**. Symbolically, Equation 11-24 may be written as

$$SS_T = SS_R + SS_E \qquad (11\text{-}25)$$

where $SS_T = \sum_{i=1}^{n}(y_i - \bar{y})^2$ is the **total corrected sum of squares** of y. In Section 11-2 we noted that $SS_E = SS_T - \hat{\beta}_1 S_{xy}$ (see Equation 11-14), so since $SS_T = \hat{\beta}_1 S_{xy} + SS_E$, we note that the regression sum of squares in Equation 11-25 is $SS_R = \hat{\beta}_1 S_{xy}$. The total sum of squares SS_T has $n - 1$ degrees of freedom, and SS_R and SS_E have 1 and $n - 2$ degrees of freedom, respectively.

We may show that $E[SS_E/(n - 2)] = \sigma^2$, $E(SS_R) = \sigma^2 + \beta_1^2 S_{xx}$ and that SS_E/σ^2 and SS_R/σ^2 are independent chi-square random variables with $n - 2$ and 1 degrees of freedom, respectively. Thus, if the null hypothesis H_0: $\beta_1 = 0$ is true, the statistic

Test for Significance of Regression

$$F_0 = \frac{SS_R/1}{SS_E/(n-2)} = \frac{MS_R}{MS_E} \qquad (11\text{-}26)$$

Table 11-3 Analysis of Variance for Testing Significance of Regression

Source of Variation	Sum of Squares	Degrees of Freedom	Mean Square	F_0
Regression	$SS_R = \hat{\beta}_1 S_{xy}$	1	MS_R	MS_R/MS_E
Error	$SS_E = SS_T - \hat{\beta}_1 S_{xy}$	$n-2$	MS_E	
Total	SS_T	$n-1$		

Note that $MS_E = \hat{\sigma}^2$.

follows the $F_{1,n-2}$ distribution, and we would reject H_0 if $f_0 > f_{\alpha,1,n-2}$. The quantities $MS_R = SS_R/1$ and $MS_E = SS_E/(n-2)$ are called **mean squares**. In general, a mean square is always computed by dividing a sum of squares by its number of degrees of freedom. The test procedure is usually arranged in an **analysis of variance table**, such as Table 11-3.

EXAMPLE 11-3
Oxygen Purity ANOVA

We will use the analysis of variance approach to test for significance of regression using the oxygen purity data model from Example 11-1. Recall that $SS_T = 173.38$, $\hat{\beta}_1 = 14.947$, $S_{xy} = 10.17744$, and $n = 20$. The regression sum of squares is

$$SS_R = \hat{\beta}_1 S_{xy} = (14.947)10.17744 = 152.13$$

and the error sum of squares is

$$SS_E = SS_T - SS_R = 173.38 - 152.13 = 21.25$$

The analysis of variance for testing $H_0: \beta_1 = 0$ is summarized in the Minitab output in Table 11-2. The test statistic is $f_0 = MS_R/MS_E = 152.13/1.18 = 128.86$, for which we find that the P-value is $P \simeq 1.23 \times 10^{-9}$, so we conclude that β_1 is not zero.

There are frequently minor differences in terminology among computer packages. For example, sometimes the regression sum of squares is called the "model" sum of squares, and the error sum of squares is called the "residual" sum of squares.

Note that the analysis of variance procedure for testing for significance of regression is equivalent to the t-test in Section 11-4.1. That is, either procedure will lead to the same conclusions. This is easy to demonstrate by starting with the t-test statistic in Equation 11-19 with $\beta_{1,0} = 0$, say

$$T_0 = \frac{\hat{\beta}_1}{\sqrt{\hat{\sigma}^2/S_{xx}}} \qquad (11\text{-}27)$$

Squaring both sides of Equation 11-27 and using the fact that $\hat{\sigma}^2 = MS_E$ results in

$$T_0^2 = \frac{\hat{\beta}_1^2 S_{xx}}{MS_E} = \frac{\hat{\beta}_1 S_{xy}}{MS_E} = \frac{MS_R}{MS_E} \qquad (11\text{-}28)$$

Note that T_0^2 in Equation 11-28 is identical to F_0 in Equation 11-26. It is true, in general, that the square of a t random variable with v degrees of freedom is an F random variable, with one and v degrees of freedom in the numerator and denominator, respectively. Thus, the test using T_0 is equivalent to the test based on F_0. Note, however, that the t-test is somewhat more flexible in that it would allow testing against a one-sided alternative hypothesis, while the F-test is restricted to a two-sided alternative.

EXERCISES FOR SECTION 11-4

11-21. Consider the data from Exercise 11-1 on $x =$ compressive strength and $y =$ intrinsic permeability of concrete.
(a) Test for significance of regression using $\alpha = 0.05$. Find the P-value for this test. Can you conclude that the model specifies a useful linear relationship between these two variables?
(b) Estimate σ^2 and the standard deviation of $\hat{\beta}_1$.
(c) What is the standard error of the intercept in this model?

11-22. Consider the data from Exercise 11-2 on $x =$ roadway surface temperature and $y =$ pavement deflection.
(a) Test for significance of regression using $\alpha = 0.05$. Find the P-value for this test. What conclusions can you draw?
(b) Estimate the standard errors of the slope and intercept.

11-23. Consider the National Football League data in Exercise 11-3.
(a) Test for significance of regression using $\alpha = 0.01$. Find the P-value for this test. What conclusions can you draw?
(b) Estimate the standard errors of the slope and intercept.
(c) Test H_0: $\beta_1 = 10$ versus H_1: $\beta_1 \neq 10$ with $\alpha = 0.01$. Would you agree with the statement that this is a test of the hypothesis that a one-yard increase in the average yards per attempt results in a mean increase of 10 rating points?

11-24. Consider the data from Exercise 11-4 on $y =$ sales price and $x =$ taxes paid.

(a) Test H_0: $\beta_1 = 0$ using the t-test; use $\alpha = 0.05$.
(b) Test H_0: $\beta_1 = 0$ using the analysis of variance with $\alpha = 0.05$. Discuss the relationship of this test to the test from part (a).
(c) Estimate the standard errors of the slope and intercept.
(d) Test the hypothesis that $\beta_0 = 0$.

11-25. Consider the data from Exercise 11-5 on $y =$ steam usage and $x =$ average temperature.

(a) Test for significance of regression using $\alpha = 0.01$. What is the P-value for this test? State the conclusions that result from this test.
(b) Estimate the standard errors of the slope and intercept.
(c) Test the hypothesis H_0: $\beta_1 = 10$ versus H_1: $\beta_1 \neq 10$ using $\alpha = 0.01$. Find the P-value for this test.
(d) Test H_0: $\beta_0 = 0$ versus H_1: $\beta_0 \neq 0$ using $\alpha = 0.01$. Find the P-value for this test and draw conclusions.

11-26. Consider the data from Exercise 11-6 on $y =$ highway gasoline mileage and $x =$ engine displacement.
(a) Test for significance of regression using $\alpha = 0.01$. Find the P-value for this test. What conclusions can you reach?
(b) Estimate the standard errors of the slope and intercept.
(c) Test H_0: $\beta_1 = -0.05$ versus H_1: $\beta_1 < -0.05$ using $\alpha = 0.01$ and draw conclusions. What is the P-value for this test?
(d) Test the hypothesis H_0: $\beta_0 = 0$ versus H_1: $\beta_0 \neq 0$ using $\alpha = 0.01$. What is the P-value for this test?

11-27. Consider the data from Exercise 11-7 on $y =$ green liquor Na$_2$S concentration and $x =$ production in a paper mill.
(a) Test for significance of regression using $\alpha = 0.05$. Find the P-value for this test.
(b) Estimate the standard errors of the slope and intercept.
(c) Test H_0: $\beta_0 = 0$ versus H_1: $\beta_0 \neq 0$ using $\alpha = 0.05$. What is the P-value for this test?

11-28. Consider the data from Exercise 11-8 on $y =$ blood pressure rise and $x =$ sound pressure level.
(a) Test for significance of regression using $\alpha = 0.05$. What is the P-value for this test?
(b) Estimate the standard errors of the slope and intercept.
(c) Test H_0: $\beta_0 = 0$ versus H_1: $\beta_0 \neq 0$ using $\alpha = 0.05$. Find the P-value for this test.

11-29. Consider the data from Exercise 11-11, on $y =$ shear strength of a propellant and $x =$ propellant age.
(a) Test for significance of regression with $\alpha = 0.01$. Find the P-value for this test.
(b) Estimate the standard errors of $\hat{\beta}_0$ and $\hat{\beta}_1$.
(c) Test H_0: $\beta_1 = -30$ versus H_1: $\beta_1 \neq -30$ using $\alpha = 0.01$. What is the P-value for this test?
(d) Test H_0: $\beta_0 = 0$ versus H_1: $\beta_0 \neq 0$ using $\alpha = 0.01$. What is the P-value for this test?
(e) Test H_0: $\beta_0 = 2500$ versus H_1: $\beta_0 > 2500$ using $\alpha = 0.01$. What is the P-value for this test?

11-30. Consider the data from Exercise 11-10 on $y =$ chloride concentration in surface streams and $x =$ roadway area.

(a) Test the hypothesis H_0: $\beta_1 = 0$ versus H_1: $\beta_1 \neq 0$ using the analysis of variance procedure with $\alpha = 0.01$.
(b) Find the P-value for the test in part (a).
(c) Estimate the standard errors of $\hat{\beta}_1$ and $\hat{\beta}_0$.
(d) Test H_0: $\beta_0 = 0$ versus H_1: $\beta_0 \neq 0$ using $\alpha = 0.01$. What conclusions can you draw? Does it seem that the model might be a better fit to the data if the intercept were removed?

11-31. Consider the data in Exercise 11-13 on $y =$ oxygen demand and $x =$ time.
(a) Test for significance of regression using $\alpha = 0.01$. Find the P-value for this test. What conclusions can you draw?
(b) Estimate the standard errors of the slope and intercept.
(c) Test the hypothesis that $\beta_0 = 0$.

11-32. Consider the data in Exercise 11-14 on $y =$ deflection and $x =$ stress level.
(a) Test for significance of regression using $\alpha = 0.01$. What is the P-value for this test? State the conclusions that result from this test.
(b) Does this model appear to be adequate?
(c) Estimate the standard errors of the slope and intercept.

11-33. An article in *The Journal of Clinical Endocrinology and Metabolism*, "Simultaneous and Continuous 24-Hour Plasma and Cerebrospinal Fluid Leptin Measurements: Dissociation of Concentrations in Central and Peripheral Compartments" (2004, Vol. 89, pp. 258–265) studied the demographics of simultaneous and continuous 24-hour plasma and cerebrospinal fluid leptin measurements. The data follow:

y = BMI (kg/m^2): 19.92 20.59 29.02 20.78 25.97
20.39 23.29 17.27 35.24

x = Age (yr): 45.5 34.6 40.6 32.9 28.2 30.1
52.1 33.3 47.0

(a) Test for significance of regression using $\alpha = 0.05$. Find the P-value for this test. Can you conclude that the model specifies a useful linear relationship between these two variables?
(b) Estimate σ^2 and the standard deviation of $\hat{\beta}_1$.
(c) What is the standard error of the intercept in this model?

11-34. Suppose that each value of x_i is multiplied by a positive constant a, and each value of y_i is multiplied by another positive constant b. Show that the t-statistic for testing $H_0: \beta_1 = 0$ versus $H_1: \beta_1 \neq 0$ is unchanged in value.

11-35. The type II error probability for the t-test for $H_0: \beta_1 = \beta_{1,0}$ can be computed in a similar manner to the t-tests of Chapter 9. If the true value of β_1 is β_1', the value $d = |\beta_{1,0} - \beta_1'|/(\sigma\sqrt{(n-1)/S_{xx}}$ is calculated and used as the horizontal scale factor on the operating characteristic curves for the t-test (Appendix Charts VIIe through VIIh) and the type II error probability is read from the vertical scale using the curve for $n - 2$ degrees of freedom. Apply this procedure to the football data of Exercise 11-3, using $\sigma = 5.5$ and $\beta_1' = 12.5$, where the hypotheses are $H_0: \beta_1 = 10$ versus $H_1: \beta_1 \neq 10$.

11-36. Consider the no-intercept model $Y = \beta x + \epsilon$ with the ϵ's NID(0, σ^2). The estimate of σ^2 is $s^2 = \sum_{i=1}^{n}(y_i - \hat{\beta}x_i)^2/(n-1)$ and $V(\hat{\beta}) = \sigma^2/\sum_{i=1}^{n}x_i^2$.
(a) Devise a test statistic for $H_0: \beta = 0$ versus $H_1: \beta \neq 0$.
(b) Apply the test in (a) to the model from Exercise 11-20.

11-5 CONFIDENCE INTERVALS

11-5.1 Confidence Intervals on the Slope and Intercept

In addition to point estimates of the slope and intercept, it is possible to obtain confidence interval estimates of these parameters. The width of these confidence intervals is a measure of the overall quality of the regression line. If the error terms, ϵ_i, in the regression model are normally and independently distributed,

$$(\hat{\beta}_1 - \beta_1)/\sqrt{\hat{\sigma}^2/S_{xx}} \quad \text{and} \quad (\hat{\beta}_0 - \beta_0)/\sqrt{\hat{\sigma}^2\left[\frac{1}{n} + \frac{\bar{x}^2}{S_{xx}}\right]}$$

are both distributed as t random variables with $n - 2$ degrees of freedom. This leads to the following definition of $100(1 - \alpha)\%$ confidence intervals on the slope and intercept.

Confidence Intervals on Parameters

Under the assumption that the observations are normally and independently distributed, a $100(1 - \alpha)\%$ confidence interval **on the slope** β_1 in simple linear regression is

$$\hat{\beta}_1 - t_{\alpha/2, n-2}\sqrt{\frac{\hat{\sigma}^2}{S_{xx}}} \leq \beta_1 \leq \hat{\beta}_1 + t_{\alpha/2, n-2}\sqrt{\frac{\hat{\sigma}^2}{S_{xx}}} \qquad (11\text{-}29)$$

Similarly, a $100(1 - \alpha)\%$ confidence interval **on the intercept** β_0 is

$$\hat{\beta}_0 - t_{\alpha/2, n-2}\sqrt{\hat{\sigma}^2\left[\frac{1}{n} + \frac{\bar{x}^2}{S_{xx}}\right]}$$
$$\leq \beta_0 \leq \hat{\beta}_0 + t_{\alpha/2, n-2}\sqrt{\hat{\sigma}^2\left[\frac{1}{n} + \frac{\bar{x}^2}{S_{xx}}\right]} \qquad (11\text{-}30)$$

EXAMPLE 11-4

Oxygen Purity Confidence Interval on the Slope

We will find a 95% confidence interval on the slope of the regression line using the data in Example 11-1. Recall that $\hat{\beta}_1 = 14.947$, $S_{xx} = 0.68088$, and $\hat{\sigma}^2 = 1.18$ (see Table 11-2). Then, from Equation 11-29 we find

$$\hat{\beta}_1 - t_{0.025,18}\sqrt{\frac{\hat{\sigma}^2}{S_{xx}}} \leq \beta_1 \leq \hat{\beta}_1 + t_{0.025,18}\sqrt{\frac{\hat{\sigma}^2}{S_{xx}}}$$

or

$$14.947 - 2.101\sqrt{\frac{1.18}{0.68088}} \leq \beta_1 \leq 14.947 + 2.101\sqrt{\frac{1.18}{0.68088}}$$

This simplifies to

$$12.181 \leq \beta_1 \leq 17.713$$

11-5.2 Confidence Interval on the Mean Response

A confidence interval may be constructed on the mean response at a specified value of x, say, x_0. This is a confidence interval about $E(Y|x_0) = \mu_{Y|x_0}$ and is often called a confidence interval about the regression line. Since $E(Y|x_0) = \mu_{Y|x_0} = \beta_0 + \beta_1 x_0$, we may obtain a point estimate of the mean of Y at $x = x_0$ ($\mu_{Y|x_0}$) from the fitted model as

$$\hat{\mu}_{Y|x_0} = \hat{\beta}_0 + \hat{\beta}_1 x_0$$

Now $\hat{\mu}_{Y|x_0}$ is an unbiased point estimator of $\mu_{Y|x_0}$, since $\hat{\beta}_0$ and $\hat{\beta}_1$ are unbiased estimators of β_0 and β_1. The variance of $\hat{\mu}_{Y|x_0}$ is

$$V(\hat{\mu}_{Y|x_0}) = \sigma^2\left[\frac{1}{n} + \frac{(x_0 - \bar{x})^2}{S_{xx}}\right]$$

This last result follows from the fact that $\hat{\mu}_{Y|x_0} = \bar{y} + \hat{\beta}_1(x_0 - \bar{x})$ and cov $(\bar{Y}, \hat{\beta}_1) = 0$. The zero covariance result is left as a Mind-Expanding exercise. Also, $\hat{\mu}_{Y|x_0}$ is normally distributed, because $\hat{\beta}_1$ and $\hat{\beta}_0$ are normally distributed, and if we $\hat{\sigma}^2$ use as an estimate of σ^2, it is easy to show that

$$\frac{\hat{\mu}_{Y|x_0} - \mu_{Y|x_0}}{\sqrt{\hat{\sigma}^2\left[\frac{1}{n} + \frac{(x_0 - \bar{x})^2}{S_{xx}}\right]}}$$

has a t distribution with $n - 2$ degrees of freedom. This leads to the following confidence interval definition.

Confidence Interval on the Mean Response

A $100(1 - \alpha)\%$ confidence interval **about the mean response** at the value of $x = x_0$, say $\mu_{Y|x_0}$, is given by

$$\hat{\mu}_{Y|x_0} - t_{\alpha/2, n-2}\sqrt{\hat{\sigma}^2\left[\frac{1}{n} + \frac{(x_0 - \bar{x})^2}{S_{xx}}\right]}$$

$$\leq \mu_{Y|x_0} \leq \hat{\mu}_{Y|x_0} + t_{\alpha/2, n-2}\sqrt{\hat{\sigma}^2\left[\frac{1}{n} + \frac{(x_0 - \bar{x})^2}{S_{xx}}\right]} \quad (11\text{-}31)$$

where $\hat{\mu}_{Y|x_0} = \hat{\beta}_0 + \hat{\beta}_1 x_0$ is computed from the fitted regression model.

Note that the width of the confidence interval for $\mu_{Y|x_0}$ is a function of the value specified for x_0. The interval width is a minimum for $x_0 = \bar{x}$ and widens as $|x_0 - \bar{x}|$ increases.

EXAMPLE 11-5 Oxygen Purity Confidence Interval on the Mean Response

We will construct a 95% confidence interval about the mean response for the data in Example 11-1. The fitted model is $\hat{\mu}_{Y|x_0} = 74.283 + 14.947 x_0$, and the 95% confidence interval on $\mu_{Y|x_0}$ is found from Equation 11-31 as

$$\hat{\mu}_{Y|x_0} \pm 2.101\sqrt{1.18\left[\frac{1}{20} + \frac{(x_0 - 1.1960)^2}{0.68088}\right]}$$

Suppose that we are interested in predicting mean oxygen purity when $x_0 = 1.00\%$. Then

$$\hat{\mu}_{Y|x_{1.00}} = 74.283 + 14.947(1.00) = 89.23$$

and the 95% confidence interval is

$$\left\{89.23 \pm 2.101\sqrt{1.18\left[\frac{1}{20} + \frac{(1.00 - 1.1960)^2}{0.68088}\right]}\right\}$$

or

$$89.23 \pm 0.75$$

Therefore, the 95% confidence interval on $\mu_{Y|1.00}$ is

$$88.48 \leq \mu_{Y|1.00} \leq 89.98$$

Minitab will also perform these calculations. Refer to Table 11-2. The predicted value of y at $x = 1.00$ is shown along with the 95% CI on the mean of y at this level of x.

By repeating these calculations for several different values for x_0 we can obtain confidence limits for each corresponding value of $\mu_{Y|x_0}$. Figure 11-7 displays the scatter diagram with the fitted model and the corresponding 95% confidence limits plotted as the upper and lower lines. The 95% confidence level applies only to the interval obtained at one value of x and not to the entire set of x-levels. Notice that the width of the confidence interval on $\mu_{Y|x_0}$ increases as $|x_0 - \bar{x}|$ increases.

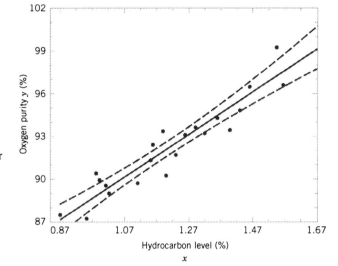

Figure 11-7 Scatter diagram of oxygen purity data from Example 11-1 with fitted regression line and 95 percent confidence limits on $\mu_{Y|x_0}$.

11-6 PREDICTION OF NEW OBSERVATIONS

An important application of a regression model is predicting new or future observations Y corresponding to a specified level of the regressor variable x. If x_0 is the value of the regressor variable of interest,

$$\hat{Y}_0 = \hat{\beta}_0 + \hat{\beta}_1 x_0 \tag{11-32}$$

is the point estimator of the new or future value of the response Y_0.

Now consider obtaining an interval estimate for this future observation Y_0. This new observation is independent of the observations used to develop the regression model. Therefore, the confidence interval for $\mu_{Y|x_0}$ in Equation 11-31 is inappropriate, since it is based only on the data used to fit the regression model. The confidence interval about $\mu_{Y|x_0}$ refers to the true mean response at $x = x_0$ (that is, a population parameter), not to future observations.

Let Y_0 be the future observation at $x = x_0$, and let $\hat{Y}_0$ given by Equation 11-32 be the estimator of Y_0. Note that the error in prediction

$$e_{\hat{p}} = Y_0 - \hat{Y}_0$$

is a normally distributed random variable with mean zero and variance

$$V(e_{\hat{p}}) = V(Y_0 - \hat{Y}_0) = \sigma^2 \left[1 + \frac{1}{n} + \frac{(x_0 - \bar{x})^2}{S_{xx}} \right]$$

because Y_0 is independent of $\hat{Y}_0$. If we use $\hat{\sigma}^2$ to estimate σ^2, we can show that

$$\frac{Y_0 - \hat{Y}_0}{\sqrt{\hat{\sigma}^2 \left[1 + \frac{1}{n} + \frac{(x_0 - \bar{x})^2}{S_{xx}} \right]}}$$

Prediction Interval

A $100(1 - \alpha)\%$ **prediction interval on a future observation** Y_0 at the value x_0 is given by

$$\hat{y}_0 - t_{\alpha/2, n-2} \sqrt{\hat{\sigma}^2 \left[1 + \frac{1}{n} + \frac{(x_0 - \bar{x})^2}{S_{xx}}\right]}$$

$$\leq Y_0 \leq \hat{y}_0 + t_{\alpha/2, n-2} \sqrt{\hat{\sigma}^2 \left[1 + \frac{1}{n} + \frac{(x_0 - \bar{x})^2}{S_{xx}}\right]} \quad (11\text{-}33)$$

The value $\hat{y}_0$ is computed from the regression model $\hat{y}_0 = \hat{\beta}_0 + \hat{\beta}_1 x_0$.

Notice that the prediction interval is of minimum width at $x_0 = \bar{x}$ and widens as $|x_0 - \bar{x}|$ increases. By comparing Equation 11-33 with Equation 11-31, we observe that the prediction interval at the point x_0 is always wider than the confidence interval at x_0. This results because the prediction interval depends on both the error from the fitted model and the error associated with future observations.

EXAMPLE 11-6 Oxygen Purity Prediction Interval

To illustrate the construction of a prediction interval, suppose we use the data in Example 11-1 and find a 95% prediction interval on the next observation of oxygen purity at $x_0 = 1.00\%$. Using Equation 11-33 and recalling from Example 11-5 that $\hat{y}_0 = 89.23$, we find that the prediction interval is

$$89.23 - 2.101 \sqrt{1.18 \left[1 + \frac{1}{20} + \frac{(1.00 - 1.1960)^2}{0.68088}\right]}$$

$$\leq Y_0 \leq 89.23 + 2.101 \sqrt{1.18 \left[1 + \frac{1}{20} + \frac{(1.00 - 1.1960)^2}{0.68088}\right]}$$

which simplifies to

$$86.83 \leq y_0 \leq 91.63$$

Minitab will also calculate prediction intervals. Refer to the output in Table 11-2. The 95% PI on the future observation at $x_0 = 1.00$ is shown in the display.

By repeating the foregoing calculations at different levels of x_0, we may obtain the 95% prediction intervals shown graphically as the lower and upper lines about the fitted regression model in Fig. 11-8. Notice that this graph also shows the 95% confidence limits on $\mu_{Y|x_0}$ calculated in Example 11-5. It illustrates that the prediction limits are always wider than the confidence limits.

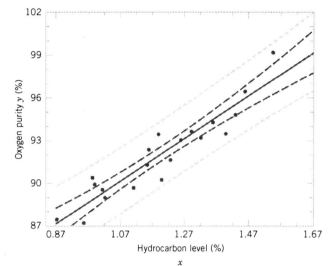

Figure 11-8 Scatter diagram of oxygen purity data from Example 11-1 with fitted regression line, 95% prediction limits (outer lines) and 95% confidence limits on $\mu_{Y|x_0}$.

EXERCISES FOR SECTIONS 11-5 AND 11-4

11-37. Refer to the data in Exercise 11-1 on y = intrinsic permeability of concrete and x = compressive strength. Find a 95% confidence interval on each of the following:
(a) Slope (b) Intercept
(c) Mean permeability when $x = 2.5$
(d) Find a 95% prediction interval on permeability when $x = 2.5$. Explain why this interval is wider than the interval in part (c).

11-38. Exercise 11-2 presented data on roadway surface temperature x and pavement deflection y. Find a 99% confidence interval on each of the following:
(a) Slope (b) Intercept
(c) Mean deflection when temperature $x = 85°F$
(d) Find a 99% prediction interval on pavement deflection when the temperature is $90°F$.

11-39. Refer to the NFL quarterback ratings data in Exercise 11-3. Find a 95% confidence interval on each of the following:
(a) Slope
(b) Intercept
(c) Mean rating when the average yards per attempt is 8.0
(d) Find a 95% prediction interval on the rating when the average yards per attempt is 8.0.

11-40. Refer to the data on y = house selling price and x = taxes paid in Exercise 11-4. Find a 95% confidence interval on each of the following:
(a) β_1 (b) β_0
(c) Mean selling price when the taxes paid are $x = 7.50$
(d) Compute the 95% prediction interval for selling price when the taxes paid are $x = 7.50$.

11-41. Exercise 11-5 presented data on y = steam usage and x = monthly average temperature.
(a) Find a 99% confidence interval for β_1.
(b) Find a 99% confidence interval for β_0.
(c) Find a 95% confidence interval on mean steam usage when the average temperature is $55°F$.
(d) Find a 95% prediction interval on steam usage when temperature is $55°F$. Explain why this interval is wider than the interval in part (c).

11-42. Exercise 11-6 presented gasoline mileage performance for 20 cars, along with information about the engine displacement. Find a 95% confidence interval on each of the following:
(a) Slope (b) Intercept
(c) Mean highway gasoline mileage when the engine displacement is $x = 150$ in^3
(d) Construct a 95% prediction interval on highway gasoline mileage when the engine displacement is $x = 150$ in^3.

11-43. Consider the data in Exercise 11-7 on y = green liquor Na$_2$S concentration and x = production in a paper mill. Find a 99% confidence interval on each of the following:
(a) β_1 (b) β_0
(c) Mean Na$_2$S concentration when production $x = 910$ tons/day
(d) Find a 99% prediction interval on Na$_2$S concentration when $x = 910$ tons/day.

11-44. Exercise 11-8 presented data on y = blood pressure rise and x = sound pressure level. Find a 95% confidence interval on each of the following:
(a) β_1 (b) β_0
(c) Mean blood pressure rise when the sound pressure level is 85 decibels
(d) Find a 95% prediction interval on blood pressure rise when the sound pressure level is 85 decibels.

11-45. Refer to the data in Exercise 11-9 on y = wear volume of mild steel and x = oil viscosity. Find a 95% confidence interval on each of the following:
(a) Intercept   (b) Slope
(c) Mean wear when oil viscosity $x = 30$

11-46. Exercise 11-10 presented data on chloride concentration y and roadway area x on watersheds in central Rhode Island. Find a 99% confidence interval on each of the following:
(a) β_1 (b) β_0
(c) Mean chloride concentration when roadway area $x = 1.0\%$
(d) Find a 99% prediction interval on chloride concentration when roadway area $x = 1.0\%$.

11-47. Refer to the data in Exercise 11-11 on rocket motor shear strength y and propellant age x. Find a 95% confidence interval on each of the following:
(a) Slope β_1 (b) Intercept β_0
(c) Mean shear strength when age $x = 20$ weeks

(d) Find a 95% prediction interval on shear strength when age $x = 20$ weeks.

11-48. Refer to the data in Exercise 11-12 on the microstructure of zirconia. Find a 95% confidence interval on each of the following:
(a) Slope   (b) Intercept
(c) Mean length when $x = 1500$
(d) Find a 95% prediction interval on length when $x = 1500$. Explain why this interval is wider than the interval in part (c).

11-49. Refer to the data in Exercise 11-13 on oxygen demand. Find a 99% confidence interval on each of the following:
(a) β_1
(b) β_0
(c) Find a 95% confidence interval on mean BOD when the time is 8 days.

11-7 ADEQUACY OF THE REGRESSION MODEL

Fitting a regression model requires several **assumptions.** Estimation of the model parameters requires the assumption that the errors are uncorrelated random variables with mean zero and constant variance. Tests of hypotheses and interval estimation require that the errors be normally distributed. In addition, we assume that the order of the model is correct; that is, if we fit a simple linear regression model, we are assuming that the phenomenon actually behaves in a linear or first-order manner.

The analyst should always consider the validity of these assumptions to be doubtful and conduct analyses to examine the adequacy of the model that has been tentatively entertained. In this section we discuss methods useful in this respect.

11-7.1 Residual Analysis

The **residuals** from a regression model are $e_i = y_i - \hat{y}_i$, $i = 1, 2, \ldots, n$, where y_i is an actual observation and $\hat{y}_i$ is the corresponding fitted value from the regression model. Analysis of the residuals is frequently helpful in checking the assumption that the errors are approximately normally distributed with constant variance, and in determining whether additional terms in the model would be useful.

As an approximate check of normality, the experimenter can construct a frequency histogram of the residuals or a **normal probability plot of residuals.** Many computer programs will produce a normal probability plot of residuals, and since the sample sizes in regression are often too small for a histogram to be meaningful, the normal probability plotting method is preferred. It requires judgment to assess the abnormality of such plots. (Refer to the discussion of the "fat pencil" method in Section 6-6).

We may also **standardize** the residuals by computing $d_i = e_i/\sqrt{\hat{\sigma}^2}$, $i = 1, 2, \ldots, n$. If the errors are normally distributed, approximately 95% of the standardized residuals should fall in the interval $(-2, +2)$. Residuals that are far outside this interval may indicate the presence of an **outlier,** that is, an observation that is not typical of the rest of the data. Various rules have been proposed for discarding outliers. However, outliers sometimes provide

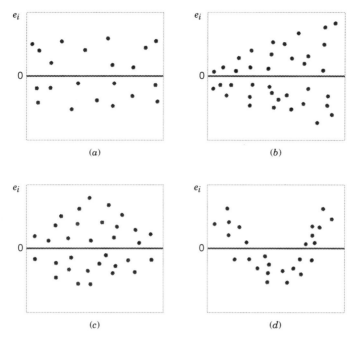

Figure 11-9 Patterns for residual plots. (a) satisfactory, (b) funnel, (c) double bow, (d) nonlinear. [Adapted from Montgomery, Peck, and Vining (2001).]

important information about unusual circumstances of interest to experimenters and should not be automatically discarded. For further discussion of outliers, see Montgomery, Peck and Vining (2001).

It is frequently helpful to plot the residuals (1) in time sequence (if known), (2), against the $\hat{y}_i$, and (3) against the independent variable x. These graphs will usually look like one of the four general patterns shown in Fig. 11-9. Pattern (a) in Fig. 11-9 represents the ideal situation, while patterns (b), (c), and (d) represent anomalies. If the residuals appear as in (b), the variance of the observations may be increasing with time or with the magnitude of y_i or x_i. Data transformation on the response y is often used to eliminate this problem. Widely used variance-stabilizing transformations include the use of $\sqrt{y}$, $\ln y$, or $1/y$ as the response. See Montgomery, Peck, and Vining (2001) for more details regarding methods for selecting an appropriate transformation. Plots of residuals against $\hat{y}_i$ and x_i that look like (c) also indicate inequality of variance. Residual plots that look like (d) indicate model inadequacy; that is, higher order terms should be added to the model, a transformation on the x-variable or the y-variable (or both) should be considered, or other regressors should be considered.

EXAMPLE 11-7
Oxygen Purity
Residuals

The regression model for the oxygen purity data in Example 11-1 is $\hat{y} = 74.283 + 14.947x$. Table 11-4 presents the observed and predicted values of y at each value of x from this data set, along with the corresponding residual. These values were computed using Minitab and show the number of decimal places typical of computer output. A normal probability plot of the residuals is shown in Fig. 11-10. Since the residuals fall approximately along a straight line in the figure, we conclude that there is no severe departure from normality. The residuals are also plotted against the predicted value $\hat{y}_i$ in Fig. 11-11 and against the hydrocarbon levels x_i in Fig. 11-12. These plots do not indicate any serious model inadequacies.

Table 11-4 Oxygen Purity Data from Example 11-1, Predicted Values, and Residuals

	Hydrocarbon Level, x	Oxygen Purity, y	Predicted Value, $\hat{y}$	Residual $e = y - \hat{y}$		Hydrocarbon Level, x	Oxygen Purity, y	Predicted Value, $\hat{y}$	Residual $e = y - \hat{y}$
1	0.99	90.01	89.081	0.929	11	1.19	93.54	92.071	1.469
2	1.02	89.05	89.530	−0.480	12	1.15	92.52	91.473	1.047
3	1.15	91.43	91.473	−0.043	13	0.98	90.56	88.932	1.628
4	1.29	93.74	93.566	0.174	14	1.01	89.54	89.380	0.160
5	1.46	96.73	96.107	0.623	15	1.11	89.85	90.875	−1.025
6	1.36	94.45	94.612	−0.162	16	1.20	90.39	92.220	−1.830
7	0.87	87.59	87.288	0.302	17	1.26	93.25	93.117	0.133
8	1.23	91.77	92.669	−0.899	18	1.32	93.41	94.014	−0.604
9	1.55	99.42	97.452	1.968	19	1.43	94.98	95.658	−0.678
10	1.40	93.65	95.210	−1.560	20	0.95	87.33	88.483	−1.153

11-7.2 Coefficient of Determination (R^2)

A widely used measure for a regression model is a ratio of sum of squares.

R^2

The coefficient of determination is

$$R^2 = \frac{SS_R}{SS_T} = 1 - \frac{SS_E}{SS_T} \qquad (11\text{-}34)$$

The coefficient is often used to judge the adequacy of a regression model. Subsequently, we will see that in the case where X and Y are jointly distributed random variables, R^2 is the square of the correlation coefficient between X and Y. From the analysis of variance identity in Equations 11-24 and 11-25, $0 \leq R^2 \leq 1$. We often refer loosely to R^2 as the amount of variability in the data explained or accounted for by the regression model. For the oxygen purity regression model, we have $R^2 = SS_R/SS_T = 152.13/173.38 = 0.877$; that is, the model accounts for 87.7% of the variability in the data.

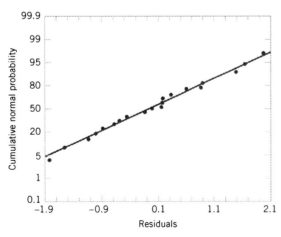

Figure 11-10 Normal probability plot of residuals, Example 11-7.

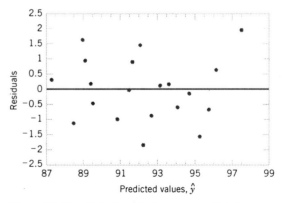

Figure 11-11 Plot of residuals versus predicted oxygen purity $\hat{y}$, Example 11-7.

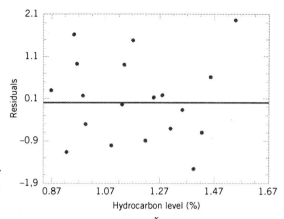

Figure 11-12 Plot of residuals versus hydrocarbon level x, Example 11-8.

The statistic R^2 should be used with caution, because it is always possible to make R^2 unity by simply adding enough terms to the model. For example, we can obtain a "perfect" fit to n data points with a polynomial of degree $n-1$. In addition, R^2 will always increase if we add a variable to the model, but this does not necessarily imply that the new model is superior to the old one. Unless the error sum of squares in the new model is reduced by an amount equal to the original error mean square, the new model will have a larger error mean square than the old one, because of the loss of one error degree of freedom. Thus, the new model will actually be worse than the old one.

There are several misconceptions about R^2. In general, R^2 does not measure the magnitude of the slope of the regression line. A large value of R^2 does not imply a steep slope. Furthermore, R^2 does not measure the appropriateness of the model, since it can be artificially inflated by adding higher order polynomial terms in x to the model. Even if y and x are related in a nonlinear fashion, R^2 will often be large. For example, R^2 for the regression equation in Fig. 11-6(b) will be relatively large, even though the linear approximation is poor. Finally, even though R^2 is large, this does not necessarily imply that the regression model will provide accurate predictions of future observations.

EXERCISES FOR SECTION 11-7

11-50. Refer to the compressive strength data in Exercise 11-1. Use the summary statistics provided to calculate R^2 and provide a practical interpretation of this quantity.

11-51. Refer to the NFL quarterback ratings data in Exercise 11-3.
(a) Calculate R^2 for this model and provide a practical interpretation of this quantity.
(b) Prepare a normal probability plot of the residuals from the least squares model. Does the normality assumption seem to be satisfied?
(c) Plot the residuals versus the fitted values and against x. Interpret these graphs.

11-52. Refer to the data in Exercise 11-4 on house selling price y and taxes paid x.
(a) Find the residuals for the least squares model.
(b) Prepare a normal probability plot of the residuals and interpret this display.

(c) Plot the residuals versus $\hat{y}$ and versus x. Does the assumption of constant variance seem to be satisfied?
(d) What proportion of total variability is explained by the regression model?

11-53. Refer to the data in Exercise 11-5 on $y=$ steam usage and $x=$ average monthly temperature.
(a) What proportion of total variability is accounted for by the simple linear regression model?
(b) Prepare a normal probability plot of the residuals and interpret this graph.
(c) Plot residuals versus $\hat{y}$ and x. Do the regression assumptions appear to be satisfied?

11-54. Refer to the gasoline mileage data in Exercise 11-6.
(a) What proportion of total variability in highway gasoline mileage performance is accounted for by engine displacement?
(b) Plot the residuals versus $\hat{y}$ and x, and comment on the graphs.

(c) Prepare a normal probability plot of the residuals. Does the normality assumption appear to be satisfied?

11-55. Exercise 11-9 presents data on wear volume y and oil viscosity x.

(a) Calculate R^2 for this model. Provide an interpretation of this quantity.
(b) Plot the residuals from this model versus $\hat{y}$ and versus x. Interpret these plots.
(c) Prepare a normal probability plot of the residuals. Does the normality assumption appear to be satisfied?

11-56. Refer to Exercise 11-8, which presented data on blood pressure rise y and sound pressure level x.

(a) What proportion of total variability in blood pressure rise is accounted for by sound pressure level?
(b) Prepare a normal probability plot of the residuals from this least squares model. Interpret this plot.
(c) Plot residuals versus $\hat{y}$ and versus x. Comment on these plots.

11-57. Refer to Exercise 11-10, which presented data on chloride concentration y and roadway area x.

(a) What proportion of the total variability in chloride concentration is accounted for by the regression model?
(b) Plot the residuals versus $\hat{y}$ and versus x. Interpret these plots.
(c) Prepare a normal probability plot of the residuals. Does the normality assumption appear to be satisfied?

11-58. An article in the *Journal of the American Statistical Association*, "Markov Chain Monte Carlo Methods for Computing Bayes Factors: A Comparative Review" (2001, Vol. 96, pp. 1122–1132) analyzed the tabulated data on compressive strength parallel to the grain versus resin-adjusted density for specimens of radiata pine.

(a) Fit a regression model relating compressive strength to density.

(b) Test for significance of regression with $\alpha = 0.05$.
(c) Estimate σ^2 for this model.
(d) Calculate R^2 for this model. Provide an interpretation of this quantity.
(e) Prepare a normal probability plot of the residuals and interpret this display.
(f) Plot the residuals versus $\hat{y}$ and versus x. Does the assumption of constant variance seem to be satisfied?

11-59. Consider the rocket propellant data in Exercise 11-11.

(a) Calculate R^2 for this model. Provide an interpretation of this quantity.
(b) Plot the residuals on a normal probability scale. Do any points seem unusual on this plot?
(c) Delete the two points identified in part (b) from the sample and fit the simple linear regression model to the remaining 18 points. Calculate the value of R^2 for the new model. Is it larger or smaller than the value of R^2 computed in part (a)? Why?
(d) Did the value of $\hat{\sigma}^2$ change dramatically when the two points identified above were deleted and the model fit to the remaining points? Why?

11-60. Consider the data in Exercise 11-7 on $y =$ green liquor Na_2S concentration and $x =$ paper machine production. Suppose that a 14th sample point is added to the original data, where $y_{14} = 59$ and $x_{14} = 855$.

(a) Prepare a scatter diagram of y versus x. Fit the simple linear regression model to all 14 observations.
(b) Test for significance of regression with $\alpha = 0.05$.
(c) Estimate σ^2 for this model.
(d) Compare the estimate of σ^2 obtained in part (c) above with the estimate of σ^2 obtained from the original 13 points. Which estimate is larger and why?
(e) Compute the residuals for this model. Does the value of e_{14} appear unusual?
(f) Prepare and interpret a normal probability plot of the residuals.
(g) Plot the residuals versus $\hat{y}$ and versus x. Comment on these graphs.

11-61. Consider the rocket propellant data in Exercise 11-11. Calculate the standardized residuals for these data. Does this provide any helpful information about the magnitude of the residuals?

11-62. **Studentized Residuals.** Show that the variance of the ith residual is

$$V(e_i) = \sigma^2 \left[1 - \left(\frac{1}{n} + \frac{(x_i - \bar{x})^2}{S_{xx}}\right)\right]$$

Compressive Strength	Density	Compressive Strength	Density
3040	29.2	3840	30.7
2470	24.7	3800	32.7
3610	32.3	4600	32.6
3480	31.3	1900	22.1
3810	31.5	2530	25.3
2330	24.5	2920	30.8
1800	19.9	4990	38.9
3110	27.3	1670	22.1
3160	27.1	3310	29.2
2310	24.0	3450	30.1
4360	33.8	3600	31.4
1880	21.5	2850	26.7
3670	32.2	1590	22.1
1740	22.5	3770	30.3
2250	27.5	3850	32.0
2650	25.6	2480	23.2
4970	34.5	3570	30.3
2620	26.2	2620	29.9
2900	26.7	1890	20.8
1670	21.1	3030	33.2
2540	24.1	3030	28.2

Hint:

$$\text{cov}(Y_i, \hat{Y}_i) = \sigma^2 \left[\frac{1}{n} + \frac{(x_i - \bar{x})^2}{S_{xx}} \right].$$

The ith studentized residual is defined as

$$r_i = \frac{e_i}{\sqrt{\hat{\sigma}^2 \left[1 - \left(\frac{1}{n} + \frac{(x_i - \bar{x})^2}{S_{xx}} \right) \right]}}$$

(a) Explain why r_i has unit standard deviation.
(b) Do the standardized residuals have unit standard deviation?
(c) Discuss the behavior of the studentized residual when the sample value x_i is very close to the middle of the range of x.
(d) Discuss the behavior of the studentized residual when the sample value x_i is very near one end of the range of x.

11-63. Show that an equivalent way to define the test for significance of regression in simple linear regression is to base the test on R^2 as follows: to test $H_0: \beta_1 = 0$ versus $H_1: \beta_1 \neq 0$, calculate

$$F_0 = \frac{R^2(n-2)}{1 - R^2}$$

and to reject $H_0: \beta_1 = 0$ if the computed value $f_0 > f_{\alpha,1,n-2}$. Suppose that a simple linear regression model has been fit to $n = 25$ observations and $R^2 = 0.90$.

(a) Test for significance of regression at $\alpha = 0.05$.
(b) What is the smallest value of R^2 that would lead to the conclusion of a significant regression if $\alpha = 0.05$?

11-8 CORRELATION

Our development of regression analysis has assumed that x is a mathematical variable, measured with negligible error, and that Y is a random variable. Many applications of regression analysis involve situations in which both X and Y are random variables. In these situations, it is usually assumed that the observations (X_i, Y_i), $i = 1, 2, \ldots, n$ are jointly distributed random variables obtained from the distribution $f(x, y)$.

For example, suppose we wish to develop a regression model relating the shear strength of spot welds to the weld diameter. In this example, weld diameter cannot be controlled. We would randomly select n spot welds and observe a diameter (X_i) and a shear strength (Y_i) for each. Therefore (X_i, Y_i) are jointly distributed random variables.

We assume that the joint distribution of X_i and Y_i is the bivariate normal distribution presented in Chapter 5, and μ_Y and σ_Y^2 are the mean and variance of Y, μ_X and σ_X^2 are the mean and variance of X, and ρ is the correlation coefficient between Y and X. Recall that the correlation coefficient is defined as

$$\rho = \frac{\sigma_{XY}}{\sigma_X \sigma_Y} \qquad (11\text{-}35)$$

where σ_{XY} is the covariance between Y and X.

The conditional distribution of Y for a given value of $X = x$ is

$$f_{Y|x}(y) = \frac{1}{\sqrt{2\pi}\sigma_{Y|x}} \exp\left[-\frac{1}{2} \left(\frac{y - \beta_0 - \beta_1 x}{\sigma_{Y|x}} \right)^2 \right] \qquad (11\text{-}36)$$

where

$$\beta_0 = \mu_Y - \mu_X \rho \frac{\sigma_Y}{\sigma_X} \qquad (11\text{-}37)$$

$$\beta_1 = \frac{\sigma_Y}{\sigma_X} \rho \qquad (11\text{-}38)$$

and the variance of the conditional distribution of Y given $X = x$ is

$$\sigma_{Y|x}^2 = \sigma_Y^2 (1 - \rho^2) \qquad (11\text{-}39)$$

That is, the conditional distribution of Y given $X = x$ is normal with mean

$$E(Y|x) = \beta_0 + \beta_1 x \qquad (11\text{-}40)$$

and variance $\sigma^2_{Y|x}$. Thus, the mean of the conditional distribution of Y given $X = x$ is a simple linear regression model. Furthermore, there is a relationship between the correlation coefficient ρ and the slope β_1. From Equation 11-38 we see that if $\rho = 0$, then $\beta_1 = 0$, which implies that there is no regression of Y on X. That is, knowledge of X does not assist us in predicting Y.

The method of maximum likelihood may be used to estimate the parameters β_0 and β_1. It can be shown that the maximum likelihood estimators of those parameters are

$$\hat{\beta}_0 = \overline{Y} - \hat{\beta}_1 \overline{X} \qquad (11\text{-}41)$$

and

$$\hat{\beta}_1 = \frac{\sum_{i=1}^{n} Y_i(X_i - \overline{X})}{\sum_{i=1}^{n}(X_i - \overline{X})^2} = \frac{S_{XY}}{S_{XX}} \qquad (11\text{-}42)$$

We note that the estimators of the intercept and slope in Equations 11-41 and 11-42 are identical to those given by the method of least squares in the case where X was assumed to be a mathematical variable. That is, the regression model with Y and X jointly normally distributed is equivalent to the model with X considered as a mathematical variable. This follows because the random variables Y given $X = x$ are independently and normally distributed with mean $\beta_0 + \beta_1 x$ and constant variance $\sigma^2_{Y|x}$. These results will also hold for any joint distribution of Y and X such that the conditional distribution of Y given X is normal.

It is possible to draw inferences about the correlation coefficient ρ in this model. The estimator of ρ is the **sample correlation coefficient**

$$R = \frac{\sum_{i=1}^{n} Y_i(X_i - \overline{X})}{\left[\sum_{i=1}^{n}(X_i - \overline{X})^2 \sum_{i=1}^{n}(Y_i - \overline{Y})^2\right]^{1/2}} = \frac{S_{XY}}{(S_{XX}SS_T)^{1/2}} \qquad (11\text{-}43)$$

Note that

$$\hat{\beta}_1 = \left(\frac{SS_T}{S_{XX}}\right)^{1/2} R \qquad (11\text{-}44)$$

so the slope $\hat{\beta}_1$ is just the sample correlation coefficient R multiplied by a scale factor that is the square root of the "spread" of the Y values divided by the "spread" of the X values. Thus, $\hat{\beta}_1$ and R are closely related, although they provide somewhat different information. The sample correlation coefficient R measures the linear association between Y and X, while $\hat{\beta}_1$ measures the predicted change in the mean of Y for a unit change in X. In the case of a mathematical variable x, R has no meaning because the magnitude of R depends on the choice of spacing of x. We may also write, from Equation 11-44,

$$R^2 = \hat{\beta}_1^2 \frac{S_{XX}}{SS_T} = \frac{\hat{\beta}_1 S_{XY}}{SS_T} = \frac{SS_R}{SS_T}$$

which is just the coefficient of determination. That is, the coefficient of determination R^2 is just the square of the correlation coefficient between Y and X.

It is often useful to test the hypotheses

$$H_0: \rho = 0$$
$$H_1: \rho \neq 0 \tag{11-45}$$

The appropriate test statistic for these hypotheses is

Test Statistic for Zero Correlation

$$T_0 = \frac{R\sqrt{n-2}}{\sqrt{1-R^2}} \tag{11-46}$$

which has the t distribution with $n-2$ degrees of freedom if $H_0: \rho = 0$ is true. Therefore, we would reject the null hypothesis if $|t_0| > t_{\alpha/2, n-2}$. This test is equivalent to the test of the hypothesis $H_0: \beta_1 = 0$ given in Section 11-5.1. This equivalence follows directly from Equation 11-46.

The test procedure for the hypothesis

$$H_0: \rho = \rho_0$$
$$H_1: \rho \neq \rho_0 \tag{11-47}$$

where $\rho_0 \neq 0$ is somewhat more complicated. For moderately large samples (say, $n \geq 25$) the statistic

$$Z = \text{arctanh } R = \frac{1}{2} \ln \frac{1+R}{1-R} \tag{11-48}$$

is approximately normally distributed with mean and variance

$$\mu_Z = \text{arctanh } \rho = \frac{1}{2} \ln \frac{1+\rho}{1-\rho} \quad \text{and} \quad \sigma_Z^2 = \frac{1}{n-3}$$

respectively. Therefore, to test the hypothesis $H_0: \rho = \rho_0$, we may use the test statistic

$$Z_0 = (\text{arctanh } R - \text{arctanh } \rho_0)(n-3)^{1/2} \tag{11-49}$$

and reject $H_0: \rho = \rho_0$ if the value of the test statistic in Equation 11-49 is such that $|z_0| > z_{\alpha/2}$.

It is also possible to construct an approximate $100(1-\alpha)\%$ confidence interval for ρ, using the transformation in Equation 11-48. The approximate $100(1-\alpha)\%$ confidence interval is

Confidence Interval for a Correlation Coefficient

$$\tanh\left(\text{arctanh } r - \frac{z_{\alpha/2}}{\sqrt{n-3}}\right) \leq \rho \leq \tanh\left(\text{arctanh } r + \frac{z_{\alpha/2}}{\sqrt{n-3}}\right) \tag{11-50}$$

where $\tanh u = (e^u - e^{-u})/(e^u + e^{-u})$.

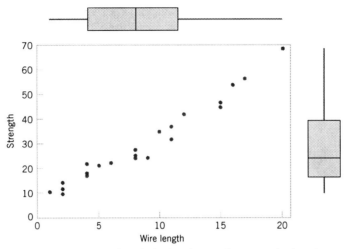

Figure 11-13 Scatter plot of wire bond strength versus wire length, Example 11-8.

EXAMPLE 11-8
Wire Bond
Pull Strength

In Chapter 1 (Section 1-3) an application of regression analysis is described in which an engineer at a semiconductor assembly plant is investigating the relationship between pull strength of a wire bond and two factors: wire length and die height. In this example, we will consider only one of the factors, the wire length. A random sample of 25 units is selected and tested, and the wire bond pull strength and wire length are observed for each unit. The data are shown in Table 1-2. We assume that pull strength and wire length are jointly normally distributed.

Figure 11-13 shows a scatter diagram of wire bond strength versus wire length. We have used the Minitab option of displaying box plots of each individual variable on the scatter diagram. There is evidence of a linear relationship between the two variables.

The Minitab output for fitting a simple linear regression model to the data is shown below. Now $S_{xx} = 698.56$ and $S_{xy} = 2027.7132$, and the sample correlation coefficient is

$$r = \frac{S_{xy}}{[S_{xx}SS_T]^{1/2}} = \frac{2027.7132}{[(698.560)(6105.9)]^{1/2}} = 0.9818$$

Regression Analysis: Strength versus Length

The regression equation is
Strength = 5.11 + 2.90 Length

Predictor	Coef	SE Coef	T	P
Constant	5.115	1.146	4.46	0.000
Length	2.9027	0.1170	24.80	0.000

S = 3.093   R-Sq = 96.4%   R-Sq(adj) = 96.2%
PRESS = 272.144   R-Sq(pred) = 95.54%

Analysis of Variance

Source	DF	SS	MS	F	P
Regression	1	5885.9	5885.9	615.08	0.000
Residual Error	23	220.1	9.6		
Total	24	6105.9			

Note that $r^2 = (0.9818)^2 = 0.9640$ (which is reported in the Minitab output), or that approximately 96.40% of the variability in pull strength is explained by the linear relationship to wire length.

Now suppose that we wish to test the hypothesis

$$H_0: \rho = 0$$
$$H_1: \rho \neq 0$$

with $\alpha = 0.05$. We can compute the t-statistic of Equation 11-46 as

$$t_0 = \frac{r\sqrt{n-2}}{\sqrt{1-r^2}} = \frac{0.9818\sqrt{23}}{\sqrt{1-0.9640}} = 24.8$$

This statistic is also reported in the Minitab output as a test of $H_0: \beta_1 = 0$. Because $t_{0.025,23} = 2.069$, we reject H_0 and conclude that the correlation coefficient $\rho \neq 0$.

Finally, we may construct an approximate 95% confidence interval on ρ from Equation 11-50. Since arctanh r = arctanh $0.9818 = 2.3452$, Equation 11-50 becomes

$$\tanh\left(2.3452 - \frac{1.96}{\sqrt{22}}\right) \leq \rho \leq \tanh\left(2.3452 + \frac{1.96}{\sqrt{22}}\right)$$

which reduces to

$$0.9585 \leq \rho \leq 0.9921$$

EXERCISES FOR SECTION 11-8

11-64. Suppose data is obtained from 20 pairs of (x, y) and the sample correlation coefficient is 0.8.
(a) Test the hypothesis that $H_0: \rho = 0$ against $H_1: \rho \neq 0$ with $\alpha = 0.05$. Calculate the P-value.
(b) Test the hypothesis that $H_1: \rho = 0.5$ against $H_1: \rho \neq 0.5$ with $\alpha = 0.05$. Calculate the P-value.
(c) Construct a 95% two-sided confidence interval for the correlation coefficient. Explain how the questions in parts (a) and (b) could be answered with a confidence interval.

11-65. Suppose data is obtained from 20 pairs of (x, y) and the sample correlation coefficient is 0.75.
(a) Test the hypothesis that $H_0: \rho = 0$ against $H_1: \rho > 0$ with $\alpha = 0.05$. Calculate the P-value.
(b) Test the hypothesis that $H_1: \rho = 0.5$ against $H_1: \rho > 0.5$ with $\alpha = 0.05$. Calculate the P-value.
(c) Construct a 95% one-sided confidence interval for the correlation coefficient. Explain how the questions in parts (a) and (b) could be answered with a confidence interval.

11-66. A random sample of $n = 25$ observations was made on the time to failure of an electronic component and the temperature in the application environment in which the component was used.
(a) Given that $r = 0.83$, test the hypothesis that $\rho = 0$, using $\alpha = 0.05$. What is the P-value for this test?
(b) Find a 95% confidence interval on ρ.
(c) Test the hypothesis $H_0: \rho = 0.8$ versus $H_1: \rho \neq 0.8$, using $\alpha = 0.05$. Find the P-value for this test.

11-67. A random sample of 50 observations was made on the diameter of spot welds and the corresponding weld shear strength.
(a) Given that $r = 0.62$, test the hypothesis that $\rho = 0$, using $\alpha = 0.01$. What is the P-value for this test?
(b) Find a 99% confidence interval for ρ.
(c) Based on the confidence interval in part (b), can you conclude that $\rho = 0.5$ at the 0.01 level of significance?

11-68. The following data gave X = the water content of snow on April 1 and Y = the yield from April to July (in inches) on the Snake River watershed in Wyoming for 1919 to 1935. (The data were taken from an article in *Research Notes*, Vol. 61, 1950, Pacific Northwest Forest Range Experiment Station, Oregon)

x	y	x	y
23.1	10.5	37.9	22.8
32.8	16.7	30.5	14.1
31.8	18.2	25.1	12.9
32.0	17.0	12.4	8.8
30.4	16.3	35.1	17.4
24.0	10.5	31.5	14.9
39.5	23.1	21.1	10.5
24.2	12.4	27.6	16.1
52.5	24.9		

(a) Estimate the correlation between Y and X.
(b) Test the hypothesis that $\rho = 0$, using $\alpha = 0.05$.
(c) Fit a simple linear regression model and test for significance of regression using $\alpha = 0.05$. What conclusions can you draw? How is the test for significance of regression related to the test on ρ in part (b)?
(d) Analyze the residuals and comment on model adequacy.

11-69. The final test and exam averages for 20 randomly selected students taking a course in engineering statistics and a course in operations research follow. Assume that the final averages are jointly normally distributed.
(a) Find the regression line relating the statistics final average to the OR final average. Graph the data.
(b) Test for significance of regression using $\alpha = 0.05$.
(c) Estimate the correlation coefficient.
(d) Test the hypothesis that $\rho = 0$, using $\alpha = 0.05$.
(e) Test the hypothesis that $\rho = 0.5$, using $\alpha = 0.05$.
(f) Construct a 95% confidence interval for the correlation coefficient.

Statistics	OR	Statistics	OR	Statistics	OR
86	80	86	81	83	81
75	81	71	76	75	70
69	75	65	72	71	73
75	81	84	85	76	72
90	92	71	72	84	80
94	95	62	65	97	98
83	80	90	93		

11-70. The weight and systolic blood pressure of 26 randomly selected males in the age group 25 to 30 are shown in the following table. Assume that weight and blood pressure are jointly normally distributed.

Subject	Weight	Systolic BP	Subject	Weight	Systolic BP
1	165	130	14	172	153
2	167	133	15	159	128
3	180	150	16	168	132
4	155	128	17	174	149
5	212	151	18	183	158
6	175	146	19	215	150
7	190	150	20	195	163
8	210	140	21	180	156
9	200	148	22	143	124
10	149	125	23	240	170
11	158	133	24	235	165
12	169	135	25	192	160
13	170	150	26	187	159

(a) Find a regression line relating systolic blood pressure to weight.
(b) Test for significance of regression using $\alpha = 0.05$.
(c) Estimate the correlation coefficient.
(d) Test the hypothesis that $\rho = 0$, using $\alpha = 0.05$.
(e) Test the hypothesis that $\rho = 0.6$, using $\alpha = 0.05$.
(f) Construct a 95% confidence interval for the correlation coefficient.

11-71. In an article in *IEEE Transactions on Instrumentation and Measurement* (2001, Vol. 50, pp. 986–990) researchers studied the effects of reducing current draw in a magnetic core by electronic means. They measured the current in a magnetic winding with and without the electronics in a paired experiment and the data for the case without electronics is provided in the following table.

Supply Voltage	Current Without Electronics (mA)
0.66	7.32
1.32	12.22
1.98	16.34
2.64	23.66
3.3	28.06
3.96	33.39
4.62	34.12
3.28	39.21
5.94	44.21
6.6	47.48

(a) Graph the data and fit a regression line to predict current without electronics to supply voltage. Is there a significant regression at $\alpha = 0.05$? What is the P-value?
(b) Estimate the correlation coefficient.
(c) Test the hypothesis that $\rho = 0$ against the alternative $\rho \neq 0$ with $\alpha = 0.05$. What is the P-value?
(d) Compute a 95% confidence interval for the correlation coefficient.

11-72. The monthly absolute estimate of global (land and ocean combined) temperature indexes (degrees C) in 2000 and 2001 are (source: (http://www.ncdc.noaa.gov/oa/climate/)

2000: 12.28, 12.63, 13.22, 14.21, 15.13, 15.82, 16.05, 16.02, 15.29, 14.29, 13.16, 12.47

2001: 12.44, 12.55, 13.35, 14.22, 15.28, 15.99, 16.23, 16.17, 15.44, 14.52, 13.52, 12.61

(a) Graph the data and fit a regression line to predict 2001 temperatures from those in 2000. Is there a significant regression at $\alpha = 0.05$? What is the P-value?
(b) Estimate the correlation coefficient.
(c) Test the hypothesis that $\rho = 0.9$ against the alternative $\rho \neq 0.9$ with $\alpha = 0.05$. What is the P-value?
(d) Compute a 95% confidence interval for the correlation coefficient.

11-73 Refer to the NFL quarterback ratings data in Exercise 11-3.
(a) Estimate the correlation coefficient between the ratings and the average yards per attempt.
(b) Test the hypothesis $H_0: \rho = 0$ versus $H_1: \rho \neq 0$ using $\alpha = 0.05$. What is the P-value for this test?
(c) Construct a 95% confidence interval for ρ.
(d) Test the hypothesis $H_0: \rho = 0.7$ versus $H_1: \rho \neq 0.7$ using $\alpha = 0.05$. Find the P-value for this test.

11-74. Consider the following (x, y) data. Calculate the correlation coefficient. Graph the data and comment on the relationship between x and y. Explain why the correlation coefficient does not detect the relationship between x and y.

x	y	x	y
−4	0	0	−4
−3	−2.65	1	3.87
−3	2.65	1	−3.87
−2	−3.46	2	3.46
−2	3.46	2	−3.46
−1	−3.87	3	2.65
−1	3.87	3	−2.65
0	4	4	0

11-9 TRANSFORMATIONS

We occasionally find that the straight-line regression model $Y = \beta_0 + \beta_1 x + \epsilon$ is inappropriate because the true regression function is nonlinear. Sometimes nonlinearity is visually determined from the scatter diagram, and sometimes, because of prior experience or underlying theory, we know in advance that the model is nonlinear. Occasionally, a scatter diagram will exhibit an apparent nonlinear relationship between Y and x. In some of these situations, a nonlinear function can be expressed as a straight line by using a suitable transformation. Such nonlinear models are called **intrinsically linear.**

As an example of a nonlinear model that is intrinsically linear, consider the exponential function

$$Y = \beta_0 e^{\beta_1 x} \epsilon$$

This function is intrinsically linear, since it can be transformed to a straight line by a logarithmic transformation

$$\ln Y = \ln \beta_0 + \beta_1 x + \ln \epsilon$$

This transformation requires that the transformed error terms $\ln \epsilon$ are normally and independently distributed with mean 0 and variance σ^2.

Another intrinsically linear function is

$$Y = \beta_0 + \beta_1 \left(\frac{1}{x}\right) + \epsilon$$

By using the reciprocal transformation $z = 1/x$, the model is linearized to

$$Y = \beta_0 + \beta_1 z + \epsilon$$

Sometimes several transformations can be employed jointly to linearize a function. For example, consider the function

$$Y = \frac{1}{\exp(\beta_0 + \beta_1 x + \epsilon)}$$

letting $Y^* = 1/Y$, we have the linearized form

$$\ln Y^* = \beta_0 + \beta_1 x + \epsilon$$

For examples of fitting these models, refer to Montgomery, Peck, and Vining (2001) or Myers (1990).

Transformations can be very useful in many situations where the true relationship between the response Y and the regressor x is not well approximated by a straight line. The utility of a transformation is illustrated in the following example.

EXAMPLE 11-9
Windmill Power

A research engineer is investigating the use of a windmill to generate electricity and has collected data on the DC output from this windmill and the corresponding wind velocity. The data are plotted in Figure 11-14 and listed in Table 11-5.

Inspection of the scatter diagram indicates that the relationship between DC output Y and wind velocity (x) may be nonlinear. However, we initially fit a straight-line model to the data. The regression model is

$$\hat{y} = 0.1309 + 0.2411x$$

The summary statistics for this model are $R^2 = 0.8745$, $MS_E = \hat{\sigma}^2 = 0.0557$ and $F_0 = 160.26$ (the P value is <0.0001).

A plot of the residuals versus $\hat{y}_i$ is shown in Figure 11-15. This residual plot indicates model inadequacy and implies that the linear relationship has not captured all of the information in the wind speed

Table 11-5 Observed Values y_i and Regressor Variable x_i for Example 11-9

Observation Number, i	Wind Velocity (mph), x_i	DC Output, y_i
1	5.00	1.582
2	6.00	1.822
3	3.40	1.057
4	2.70	0.500
5	10.00	2.236
6	9.70	2.386
7	9.55	2.294
8	3.05	0.558
9	8.15	2.166
10	6.20	1.866
11	2.90	0.653
12	6.35	1.930
13	4.60	1.562
14	5.80	1.737
15	7.40	2.088
16	3.60	1.137
17	7.85	2.179
18	8.80	2.112
19	7.00	1.800
20	5.45	1.501
21	9.10	2.303
22	10.20	2.310
23	4.10	1.194
24	3.95	1.144
25	2.45	0.123

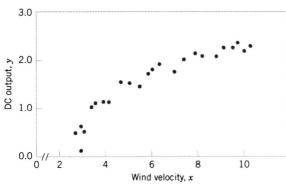

Figure 11-14 Plot of DC output y versus wind velocity x for the windmill data.

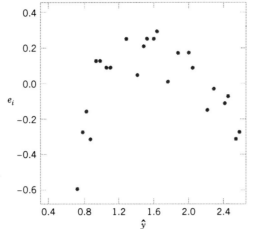

Figure 11-15 Plot of residuals e_i versus fitted values $\hat{y}_i$ for the windmill data.

variable. Note that the curvature that was apparent in the scatter diagram of Figure 11-14 is greatly amplified in the residual plots. Clearly some other model form must be considered.

We might initially consider using a quadratic model such as

$$y = \beta_0 + \beta_1 x + \beta_2 x^2 + \epsilon$$

to account for the apparent curvature. However, the scatter diagram Figure 11-14 suggests that as wind speed increases, DC output approaches an upper limit of approximately 2.5. This is also consistent with the theory of windmill operation. Since the quadratic model will eventually bend downward as wind speed increases, it would not be appropriate for these data. A more reasonable model for the windmill data that incorporates an upper asymptote would be

$$y = \beta_0 + \beta_1\left(\frac{1}{x}\right) + \epsilon$$

Figure 11-16 is a scatter diagram with the transformed variable $x' = 1/x$. This plot appears linear, indicating that the reciprocal transformation is appropriate. The fitted regression model is

$$\hat{y} = 2.9789 - 6.9345 x'$$

The summary statistics for this model are $R^2 = 0.9800$, $MS_E = \hat{\sigma}^2 = 0.0089$, and $F_0 = 1128.43$ (the P value is <0.0001).

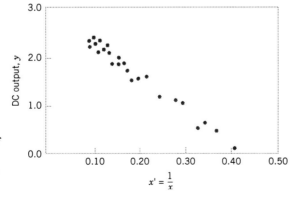

Figure 11-16 Plot of DC output versus $x' = 1/x$ for the windmill data.

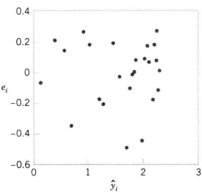

Figure 11-17 Plot of residuals versus fitted values $\hat{y}_i$ for the transformed model for the windmill data.

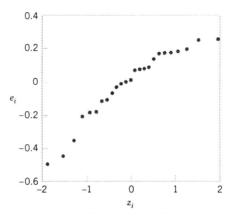

Figure 11-18 Normal probability plot of the residuals for the transformed model for the windmill data.

A plot of the residuals from the transformed model versus $\hat{y}$ is shown in Figure 11-17. This plot does not reveal any serious problem with inequality of variance. The normal probability plot, shown in Figure 11-18, gives a mild indication that the errors come from a distribution with heavier tails than the normal (notice the slight upward and downward curve at the extremes). This normal probability plot has the z-score value plotted on the horizontal axis. Since there is no strong signal of model inadequacy, we conclude that the transformed model is satisfactory.

11-9.1 Logistic Regression available at www.wiley.com/college/montgomery

EXERCISES FOR SECTION 11-9

11-75. Determine if the following models are intrinsically linear. If yes, determine the appropriate transformation to generate the linear model.

(a) $Y = \beta_0 x^{\beta_1} \epsilon$ (b) $Y = \dfrac{3 + 5x}{x} + \epsilon$

(c) $Y = \beta_0 \beta_1^x \epsilon$ (d) $Y = \dfrac{x}{\beta_0 x + \beta_1 + x\epsilon}$

11-76. The vapor pressure of water at various temperatures follows:

Observation Number, i	Temperature (K)	Vapor pressure (mm Hg)
1	273	4.6
2	283	9.2
3	293	17.5
4	303	31.8
5	313	55.3
6	323	92.5
7	333	149.4
8	343	233.7
9	353	355.1
10	363	525.8
11	373	760.0

(a) Draw a scatter diagram of these data. What type of relationship seems appropriate in relating y to x?
(b) Fit a simple linear regression model to these data.
(c) Test for significance of regression using $\alpha = 0.05$. What conclusions can you draw?
(d) Plot the residuals from the simple linear regression model versus $\hat{y}_i$. What do you conclude about model adequacy?
(e) The Clausis-Clapeyron relationship states that $\ln(P_v) \propto -\frac{1}{T}$, where P_v is the vapor pressure of water. Repeat parts (a)–(d), using an appropriate transformation.

11-77. An electric utility is interested in developing a model relating peak hour demand (y in kilowatts) to total monthly energy usage during the month (x, in kilowatt hours). Data for 50 residential customers are shown in the following table.

Customer	x	y	Customer	x	y
1	679	0.79	26	1434	0.31
2	292	0.44	27	837	4.20
3	1012	0.56	28	1748	4.88
4	493	0.79	29	1381	3.48
5	582	2.70	30	1428	7.58
6	1156	3.64	31	1255	2.63

continued

Customer	x	y	Customer	x	y
7	997	4.73	32	1777	4.99
8	2189	9.50	33	370	0.59
9	1097	5.34	34	2316	8.19
10	2078	6.85	35	1130	4.79
11	1818	5.84	36	463	0.51
12	1700	5.21	37	770	1.74
13	747	3.25	38	724	4.10
14	2030	4.43	39	808	3.94
15	1643	3.16	40	790	0.96
16	414	0.50	41	783	3.29
17	354	0.17	42	406	0.44
18	1276	1.88	43	1242	3.24
19	745	0.77	44	658	2.14
20	795	3.70	45	1746	5.71
21	540	0.56	46	895	4.12
22	874	1.56	47	1114	1.90
23	1543	5.28	48	413	0.51
24	1029	0.64	49	1787	8.33
25	710	4.00	50	3560	14.94

(a) Draw a scatter diagram of y versus x.
(b) Fit the simple linear regression model.
(c) Test for significance of regression using $\alpha = 0.05$.
(d) Plot the residuals versus $\hat{y}_i$ and comment on the underlying regression assumptions. Specifically, does it seem that the equality of variance assumption is satisfied?
(e) Find a simple linear regression model using $\sqrt{y}$ as the response. Does this transformation on y stabilize the inequality of variance problem noted in part (d) above?

Supplemental Exercises

11-78. Show that, for the simple linear regression model, the following statements are true:

(a) $\sum_{i=1}^{n}(y_i - \hat{y}_i) = 0$ (b) $\sum_{i=1}^{n}(y_i - \hat{y}_i)x_i = 0$

(c) $\frac{1}{n}\sum_{i=1}^{n}\hat{y}_i = \bar{y}$

11-79. An article in the *IEEE Transactions on Instrumentation and Measurement* ("Direct, Fast, and Accurate Measurement of V_T and K of MOS Transistor Using V_T-Sift Circuit," 1991, Vol. 40, pp. 951–955) described the use of a simple linear regression model to express drain current y (in milliamperes) as a function of ground-to-source voltage x (in volts). The data are as follows:

y	x	y	x
0.734	1.1	1.50	1.6
0.886	1.2	1.66	1.7
1.04	1.3	1.81	1.8
1.19	1.4	1.97	1.9
1.35	1.5	2.12	2.0

(a) Draw a scatter diagram of these data. Does a straight-line relationship seem plausible?
(b) Fit a simple linear regression model to these data.
(c) Test for significance of regression using $\alpha = 0.05$. What is the *P*-value for this test?
(d) Find a 95% confidence interval estimate on the slope.
(e) Test the hypothesis $H_0: \beta_0 = 0$ versus $H_1: \beta_0 \neq 0$ using $\alpha = 0.05$. What conclusions can you draw?

11-80. The strength of paper used in the manufacture of cardboard boxes (y) is related to the percentage of hardwood concentration in the original pulp (x). Under controlled conditions, a pilot plant manufactures 16 samples, each from a different batch of pulp, and measures the tensile strength. The data are shown in the table that follows:

y	101.4	117.4	117.1	106.2
x	1.0	1.5	1.5	1.5
y	131.9	146.9	146.8	133.9
x	2.0	2.0	2.2	2.4
y	111.0	123.0	125.1	145.2
x	2.5	2.5	2.8	2.8
y	134.3	144.5	143.7	146.9
x	3.0	3.0	3.2	3.3

(a) Fit a simple linear regression model to the data.
(b) Test for significance of regression using $\alpha = 0.05$.
(c) Construct a 90% confidence interval on the slope β_1.
(d) Construct a 90% confidence interval on the intercept β_0.
(e) Construct a 95% confidence interval on the mean strength at $x = 2.5$.
(f) Analyze the residuals and comment on model adequacy.

11-81. Consider the following data. Suppose that the relationship between Y and x is hypothesized to be $Y = (\beta_0 + \beta_1 x + \epsilon)^{-1}$. Fit an appropriate model to the data. Does the assumed model form seem reasonable?

x	10	15	18	12	9	8	11	6
y	0.1	0.13	0.09	0.15	0.20	0.21	0.18	0.24

11-82. The following data, adapted from Montgomery, Peck, and Vining (2001), present the number of certified mental defectives per 10,000 of estimated population in the United Kingdom (y) and the number of radio receiver licenses issued (x) by the BBC (in millions) for the years 1924 through 1937. Fit a regression model relating y and x. Comment on the model. Specifically, does the existence of a strong correlation imply a cause-and-effect relationship?

Year	y	x	Year	y	x
1924	8	1.350	1931	16	4.620
1925	8	1.960	1932	18	5.497
1926	9	2.270	1933	19	6.260
1927	10	2.483	1934	20	7.012
1928	11	2.730	1935	21	7.618
1929	11	3.091	1936	22	8.131
1930	12	3.674	1937	23	8.593

11-83. Consider the weight and blood pressure data in Exercise 11-70. Fit a no-intercept model to the data, and compare it to the model obtained in Exercise 11-70. Which model is superior?

11-84. An article in *Air and Waste* ("Update on Ozone Trends in California's South Coast Air Basin," Vol. 43, 1993) studied the ozone levels on the South Coast air basin of California for the years 1976–1991. The author believes that the number of days that the ozone level exceeds 0.20 parts per million depends on the seasonal meteorological index (the seasonal average 850 millibar temperature). The data follow:

Year	Days	Index	Year	Days	Index
1976	91	16.7	1984	81	18.0
1977	105	17.1	1985	65	17.2
1978	106	18.2	1986	61	16.9
1979	108	18.1	1987	48	17.1
1980	88	17.2	1988	61	18.2
1981	91	18.2	1989	43	17.3
1982	58	16.0	1990	33	17.5
1983	82	17.2	1991	36	16.6

(a) Construct a scatter diagram of the data.
(b) Fit a simple linear regression model to the data. Test for significance of regression.
(c) Find a 95% CI on the slope β_1.
(d) Analyze the residuals and comment on model adequacy.

11-85. An article in the *Journal of Applied Polymer Science* (Vol. 56, pp. 471–476, 1995) studied the effect of the mole ratio of sebacic acid on the intrinsic viscosity of copolyesters. The data follow:

Mole ratio x	1.0	0.9	0.8	0.7	0.6	0.5	0.4	0.3
Viscosity y	0.45	0.20	0.34	0.58	0.70	0.57	0.55	0.44

(a) Construct a scatter diagram of the data.
(b) Fit a simple linear repression model.
(c) Test for significance of regression. Calculate R^2 for the model.
(d) Analyze the residuals and comment on model adequacy.

11-86. Two different methods can be used for measuring the temperature of the solution in a Hall cell used in aluminum smelting, a thermocouple implanted in the cell and an indirect measurement produced from an IR device. The indirect method is preferable because the thermocouples are eventually destroyed by the solution. Consider the following 10 measurements:

Thermocouple	921	935	916	920	940
IR	918	934	924	921	945
Thermocouple	936	925	940	933	927
IR	930	919	943	932	935

(a) Construct a scatter diagram for these data, letting x = thermocouple measurement and y = IR measurement.
(b) Fit a simple linear regression model.
(c) Test for significance a regression and calculate R^2. What conclusions can you draw?
(d) Is there evidence to support a claim that both devices produce equivalent temperature measurements? Formulate and test an appropriate hypothesis to support this claim.
(e) Analyze the residuals and comment on model adequacy.

11-87. The grams of solids removed from a material (y) is thought to be related to the drying time. Ten observations obtained from an experimental study follow:

y	4.3	1.5	1.8	4.9	4.2	4.8	5.8	6.2	7.0	7.9
x	2.5	3.0	3.5	4.0	4.5	5.0	5.5	6.0	6.5	7.0

(a) Construct a scatter diagram for these data.
(b) Fit a simple linear regression model.
(c) Test for significance of regression.
(d) Based on these data, what is your estimate of the mean grams of solids removed at 4.25 hours? Find a 95% confidence interval on the mean.
(e) Analyze the residuals and comment on model adequacy.

11-88. Cesium atoms cooled by laser light could be used to build inexpensive atomic clocks. In a study in *IEEE Transactions on Instrumentation and Measurement* (2001, Vol. 50, pp. 1224–1228) the number of atoms cooled by lasers of various powers were counted.

Power (mW)	Number of Atoms ($\times 10E9$)
11	0
12	0.02
18	0.08
21	0.13
22	0.15
24	0.18
28	0.31
32	0.4
37	0.49
39	0.57
41	0.64
46	0.71
48	0.79
50	0.82
51	0.83

(a) Graph the data and fit a regression line to predict the number of atoms from laser power. Comment on the adequacy of a linear model.
(b) Is there a significant regression at $\alpha = 0.05$? What is the P-value?
(c) Estimate the correlation coefficient.
(d) Test the hypothesis that $\rho = 0$ against the alternative $\rho \neq 0$ with $\alpha = 0.05$. What is the P-value?
(e) Compute a 95% confidence interval for the slope coefficient.

11-89. The following data related diamond carats to purchase prices. It appeared in Singapore's *Business Times*, edition of February 18, 2000.

Carat	Price	Carat	Price
0.3	1302	0.33	1327
0.3	1510	0.33	1098
0.3	1510	0.34	1693
0.3	1260	0.34	1551
0.31	1641	0.34	1410
0.31	1555	0.34	1269
0.31	1427	0.34	1316
0.31	1427	0.34	1222
0.31	1126	0.35	1738
0.31	1126	0.35	1593
0.32	1468	0.35	1447
0.32	1202	0.35	1255
0.36	1635	0.45	1572
0.36	1485	0.46	2942
0.37	1420	0.48	2532
0.37	1420	0.5	3501
0.4	1911	0.5	3501
0.4	1525	0.5	3501
0.41	1956	0.5	3293
0.43	1747	0.5	3016

(a) Graph the data. What is the relation between carat and price? Is there an outlier?
(b) What would you say to the person who purchased the diamond that was an outlier?
(c) Fit two regression models, one with all the data and the other with unusual data omitted. Estimate the slope coefficient with a 95% confidence interval in both cases. Comment on any difference.

11-90. The following table shows the population and the average count of wood storks sighted per sample period for South Carolina from 1991–2004. Fit a regression line with population as the response and the count of wood storks as the predictor. Such an analysis might be used to evaluate the relationship between storks and babies. Is regression significant at $\alpha = 0.05$? What do you conclude about the role of regression analysis to establish a cause-and-effect relationship?

Year	Population	Stork Count
1991	3559470	0.342
1992	3600576	0.291
1993	3634507	0.291
1994	3666456	0.291
1995	3699943	0.291
1996	3738974	0.509
1997	3790066	0.294
1998	3839578	0.799
1999	3885736	0.542
2000	4012012	0.495
2001	4061209	0.859
2002	4105848	0.364
2003	4148744	0.501
2004	4198068	0.656

MIND-EXPANDING EXERCISES

11-91. Suppose that we have n pairs of observations (x_i, y_i) such that the sample correlation coefficient r is unity (approximately). Now let $z_i = y_i^2$ and consider the sample correlation coefficient for the n-pairs of data (x_i, z_i). Will this sample correlation coefficient be approximately unity? Explain why or why not.

11-92. Consider the simple linear regression model $Y = \beta_0 + \beta_1 x + \epsilon$, with $E(\epsilon) = 0$, $V(\epsilon) = \sigma^2$, and the errors ϵ uncorrelated.
(a) Show that $\text{cov}(\hat{\beta}_0, \hat{\beta}_1) = -\bar{x}\sigma^2/S_{xx}$.
(b) Show that $\text{cov}(\bar{Y}, \hat{\beta}_1) = 0$.

11-93. Consider the simple linear regression model $Y = \beta_0 + \beta_1 x + \epsilon$, with $E(\epsilon) = 0$, $V(\epsilon) = \sigma^2$, and the errors ϵ uncorrelated.
(a) Show that $E(\hat{\sigma}^2) = E(MS_E) = \sigma^2$.
(b) Show that $E(MS_R) = \sigma^2 + \beta_1^2 S_{xx}$.

11-94. Suppose that we have assumed the straight-line regression model

$$Y = \beta_0 + \beta_1 x_1 + \epsilon$$

but the response is affected by a second variable x_2 such that the true regression function is

$$E(Y) = \beta_0 + \beta_1 x_1 + \beta_2 x_2$$

Is the estimator of the slope in the simple linear regression model unbiased?

11-95. Suppose that we are fitting a line and we wish to make the variance of the regression coefficient $\hat{\beta}_1$ as small as possible. Where should the observations x_i, $i = 1, 2, \ldots, n$, be taken so as to minimize $V(\hat{\beta}_1)$? Discuss the practical implications of this allocation of the x_i.

11-96. Weighted Least Squares. Suppose that we are fitting the line $Y = \beta_0 + \beta_1 x + \epsilon$, but the variance of Y depends on the level of x; that is,

$$V(Y_i|x_i) = \sigma_i^2 = \frac{\sigma^2}{w_i} \quad i = 1, 2, \ldots, n$$

where the w_i are constants, often called *weights*. Show that for an objective function in which each squared residual is multiplied by the reciprocal of the variance of the corresponding observation, the resulting **weighted least squares normal equations** are

$$\hat{\beta}_0 \sum_{i=1}^n w_i + \hat{\beta}_1 \sum_{i=1}^n w_i x_i = \sum_{i=1}^n w_i y_i$$

$$\hat{\beta}_0 \sum_{i=1}^n w_i x_i + \hat{\beta}_1 \sum_{i=1}^n w_i x_i^2 = \sum_{i=1}^n w_i x_i y_i$$

Find the solution to these normal equations. The solutions are weighted least squares estimators of β_0 and β_1.

11-97. Consider a situation where both Y and X are random variables. Let s_x and s_y be the sample standard deviations of the observed x's and y's, respectively. Show that an alternative expression for the fitted simple linear regression model $\hat{y} = \hat{\beta}_0 + \hat{\beta}_1 x$ is

$$\hat{y} = \bar{y} + r\frac{s_y}{s_x}(x - \bar{x})$$

11-98. Suppose that we are interested in fitting a simple linear regression model $Y = \beta_0 + \beta_1 x + \epsilon$, where the intercept, β_0, is known.
(a) Find the least squares estimator of β_1.
(b) What is the variance of the estimator of the slope in part (a)?
(c) Find an expression for a $100(1 - \alpha)\%$ confidence interval for the slope β_1. Is this interval longer than the corresponding interval for the case where both the intercept and slope are unknown? Justify your answer.

IMPORTANT TERMS AND CONCEPTS

Analysis of variance test in regression
Confidence interval on mean response
Correlation coefficient
Empirical model
Confidence intervals on model parameters
Intrinsically linear model
Least squares estimation of regression model parameters
Model adequacy checking
Regression analysis
Prediction interval on a future observation
Residual plots
Residuals
Scatter diagram
Significance of regression
Simple linear regression model standard errors
Statistical tests on model parameters
Transformations

13 Design and Analysis of Single-Factor Experiments: The Analysis of Variance

CHAPTER OUTLINE

13-1 DESIGNING ENGINEERING EXPERIMENTS

13-2 COMPLETELY RANDOMIZED SINGLE-FACTOR EXPERIMENT
  13-2.1 Example
  13-2.2 Analysis of Variance
  13-2.3 Multiple Comparisons Following the ANOVA
  13-2.4 Residual Analysis and Model Checking
  13-2.5 Determining Sample Size

13-3 RANDOM EFFECTS MODEL
  13-3.1 Fixed Versus Random Factors
  13-3.2 ANOVA and Variance Components

13-4 RANDOMIZED COMPLETE BLOCK DESIGN
  13-4.1 Design and Statistical Analysis
  13-4.2 Multiple Comparisons
  13-4.3 Residual Analysis and Model Checking

LEARNING OBJECTIVES

After careful study of this chapter, you should be able to do the following:

1. Design and conduct engineering experiments involving a single factor with an arbitrary number of levels
2. Understand how the analysis of variance is used to analyze the data from these experiments
3. Assess model adequacy with residual plots
4. Use multiple comparison procedures to identify specific differences between means
5. Make decisions about sample size in single-factor experiments
6. Understand the difference between fixed and random factors
7. Estimate variance components in an experiment involving random factors
8. Understand the blocking principle and how it is used to isolate the effect of nuisance factors
9. Design and conduct experiments involving the randomized complete block design

13-1 DESIGNING ENGINEERING EXPERIMENTS

Experiments are a natural part of the engineering and scientific decision-making process. Suppose, for example, that a civil engineer is investigating the effects of different curing methods on the mean compressive strength of concrete. The experiment would consist of making up several test specimens of concrete using each of the proposed curing methods and then testing the compressive strength of each specimen. The data from this experiment could be used to determine which curing method should be used to provide maximum mean compressive strength.

If there are only two curing methods of interest, this experiment could be designed and analyzed using the statistical hypothesis methods for two samples introduced in Chapter 10. That is, the experimenter has a single **factor** of interest—curing methods—and there are only two **levels** of the factor. If the experimenter is interested in determining which curing method produces the maximum compressive strength, the number of specimens to test can be determined from the operating characteristic curves in Appendix Chart VII, and the t-test can be used to decide if the two means differ.

Many single-factor experiments require that more than two levels of the factor be considered. For example, the civil engineer may want to investigate five different curing methods. In this chapter we show how the **analysis of variance** (frequently abbreviated ANOVA) can be used for comparing means when there are more than two levels of a single factor. We will also discuss **randomization** of the experimental runs and the important role this concept plays in the overall experimentation strategy. In the next chapter, we will show how to design and analyze experiments with several factors.

Statistically based experimental design techniques are particularly useful in the engineering world for improving the performance of a manufacturing process. They also have extensive applications in the development of new processes. Most processes can be described in terms of several **controllable variables**, such as temperature, pressure, and feed rate. By using designed experiments, engineers can determine which subset of the process variables has the greatest influence on process performance. The results of such an experiment can lead to

- Improved process yield
- Reduced variability in the process and closer conformance to nominal or target requirements
- Reduced design and development time
- Reduced cost of operation

Experimental design methods are also useful in **engineering design** activities, where new products are developed and existing ones are improved. Some typical applications of statistically designed experiments in engineering design include

- Evaluation and comparison of basic design configurations
- Evaluation of different materials
- Selection of design parameters so that the product will work well under a wide variety of field conditions (or so that the design will be robust)
- Determination of key product design parameters that affect product performance

The use of experimental design in the engineering design process can result in products that are easier to manufacture, products that have better field performance and reliability than their competitors, and products that can be designed, developed, and produced in less time.

Designed experiments are usually employed **sequentially**. That is, the first experiment with a complex system (perhaps a manufacturing process) that has many controllable variables

502 CHAPTER 13 DESIGN AND ANALYSIS OF SINGLE-FACTOR EXPERIMENTS: THE ANALYSIS OF VARIANCE

is often a **screening experiment** designed to determine which variables are most important. Subsequent experiments are used to refine this information and determine which adjustments to these critical variables are required to improve the process. Finally, the objective of the experimenter is optimization, that is, to determine which levels of the critical variables result in the best process performance.

Every experiment involves a sequence of activities:

1. **Conjecture**—the original hypothesis that motivates the experiment.
2. **Experiment**—the test performed to investigate the conjecture.
3. **Analysis**—the statistical analysis of the data from the experiment.
4. **Conclusion**—what has been learned about the original conjecture from the experiment. Often the experiment will lead to a revised conjecture, and a new experiment, and so forth.

The statistical methods introduced in this chapter and Chapter 14 are essential to good experimentation. **All experiments are designed experiments;** unfortunately, some of them are poorly designed, and as a result, valuable resources are used ineffectively. Statistically designed experiments permit efficiency and economy in the experimental process, and the use of statistical methods in examining the data results in **scientific objectivity** when drawing conclusions.

13-2 COMPLETELY RANDOMIZED SINGLE-FACTOR EXPERIMENT

13-2.1 Example: Tensile Strength

A manufacturer of paper used for making grocery bags is interested in improving the tensile strength of the product. Product engineering thinks that tensile strength is a function of the hardwood concentration in the pulp and that the range of hardwood concentrations of practical interest is between 5 and 20%. A team of engineers responsible for the study decides to investigate four levels of hardwood concentration: 5%, 10%, 15%, and 20%. They decide to make up six test specimens at each concentration level, using a pilot plant. All 24 specimens are tested on a laboratory tensile tester, in random order. The data from this experiment are shown in Table 13-1.

This is an example of a completely randomized single-factor experiment with four levels of the factor. The levels of the factor are sometimes called **treatments**, and each treatment has six observations or **replicates**. The role of **randomization** in this experiment is extremely

Table 13-1 Tensile Strength of Paper (psi)

Hardwood Concentration (%)	Observations						Totals	Averages
	1	2	3	4	5	6		
5	7	8	15	11	9	10	60	10.00
10	12	17	13	18	19	15	94	15.67
15	14	18	19	17	16	18	102	17.00
20	19	25	22	23	18	20	127	21.17
							383	15.96

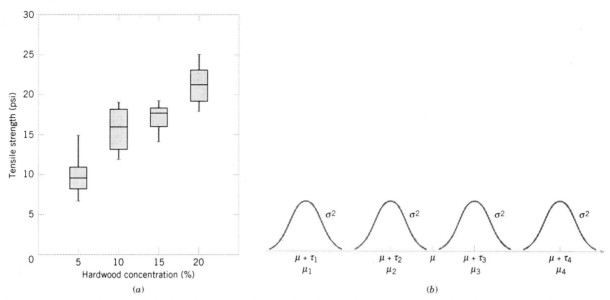

Figure 13-1 (a) Box plots of hardwood concentration data. (b) Display of the model in Equation 13-1 for the completely randomized single-factor experiment.

important. By randomizing the order of the 24 runs, the effect of any nuisance variable that may influence the observed tensile strength is approximately balanced out. For example, suppose that there is a warm-up effect on the tensile testing machine; that is, the longer the machine is on, the greater the observed tensile strength. If all 24 runs are made in order of increasing hardwood concentration (that is, all six 5% concentration specimens are tested first, followed by all six 10% concentration specimens, etc.), any observed differences in tensile strength could also be due to the warm-up effect. The role of randomization to identify causality was discussed in Section 10-1.

It is important to graphically analyze the data from a designed experiment. Figure 13-1(a) presents box plots of tensile strength at the four hardwood concentration levels. This figure indicates that changing the hardwood concentration has an effect on tensile strength; specifically, higher hardwood concentrations produce higher observed tensile strength. Furthermore, the distribution of tensile strength at a particular hardwood level is reasonably symmetric, and the variability in tensile strength does not change dramatically as the hardwood concentration changes.

Graphical interpretation of the data is always useful. Box plots show the variability of the observations *within* a treatment (factor level) and the variability *between* treatments. We now discuss how the data from a single-factor randomized experiment can be analyzed statistically.

13-2.2 Analysis of Variance

Suppose we have a different levels of a single factor that we wish to compare. Sometimes, each factor level is called a treatment, a very general term that can be traced to the early applications of experimental design methodology in the agricultural sciences. The response for each of the a treatments is a random variable. The observed data would appear as shown

Table 13-2 Typical Data for a Single-Factor Experiment

Treatment	Observations				Totals	Averages
1	y_{11}	y_{12}	$\cdots$	y_{1n}	$y_{1\cdot}$	$\bar{y}_{1\cdot}$
2	y_{21}	y_{22}	$\cdots$	y_{2n}	$y_{2\cdot}$	$\bar{y}_{2\cdot}$
$\vdots$	$\vdots$	$\vdots$	$\vdots$	$\vdots$	$\vdots$	$\vdots$
a	y_{a1}	y_{a2}	$\cdots$	y_{an}	$y_{a\cdot}$	$\bar{y}_{a\cdot}$
					$y_{\cdot\cdot}$	$\bar{y}_{\cdot\cdot}$

in Table 13-2. An entry in Table 13-2, say y_{ij}, represents the jth observation taken under treatment i. We initially consider the case in which there are an equal number of observations, n, on each treatment.

We may describe the observations in Table 13-2 by the **linear statistical model**

$$Y_{ij} = \mu + \tau_i + \epsilon_{ij} \begin{cases} i = 1, 2, \ldots, a \\ j = 1, 2, \ldots, n \end{cases} \tag{13-1}$$

where Y_{ij} is a random variable denoting the (ij)th observation, μ is a parameter common to all treatments called the **overall mean,** τ_i is a parameter associated with the ith treatment called the ith **treatment effect,** and ϵ_{ij} is a random error component. Notice that the model could have been written as

$$Y_{ij} = \mu_i + \epsilon_{ij} \begin{cases} i = 1, 2, \ldots, a \\ j = 1, 2, \ldots, n \end{cases}$$

where $\mu_i = \mu + \tau_i$ is the mean of the ith treatment. In this form of the model, we see that each treatment defines a population that has mean μ_i, consisting of the overall mean μ plus an effect τ_i that is due to that particular treatment. We will assume that the errors ϵ_{ij} are normally and independently distributed with mean zero and variance σ^2. Therefore, each treatment can be thought of as a normal population with mean μ_i and variance σ^2. See Fig. 13-1(b).

Equation 13-1 is the underlying model for a single-factor experiment. Furthermore, since we require that the observations are taken in random order and that the environment (often called the experimental units) in which the treatments are used is as uniform as possible, this experimental design is called a **completely randomized design.**

The a factor levels in the experiment could have been chosen in two different ways. First, the experimenter could have specifically chosen the a treatments. In this situation, we wish to test hypotheses about the treatment means, and conclusions cannot be extended to similar treatments that were not considered. In addition, we may wish to estimate the treatment effects. This is called the **fixed-effects model.** Alternatively, the a treatments could be a random sample from a larger population of treatments. In this situation, we would like to be able to extend the conclusions (which are based on the sample of treatments) to all treatments in the population, whether or not they were explicitly considered in the experiment. Here the treatment effects τ_i are random variables, and knowledge about the particular ones investigated is relatively unimportant. Instead, we test hypotheses about the variability of the τ_i and try to estimate this variability. This is called the **random effects, or components of variance, model.**

In this section we develop the **analysis of variance** for the fixed-effects model. The analysis of variance is not new to us; it was used previously in the presentation of regression analysis. However, in this section we show how it can be used to test for equality of treatment effects. In the fixed-effects model, the treatment effects τ_i are usually defined as deviations from the overall mean μ, so that

$$\sum_{i=1}^{a} \tau_i = 0 \qquad (13\text{-}2)$$

Let $y_{i\cdot}$ represent the total of the observations under the ith treatment and $\bar{y}_{i\cdot}$ represent the average of the observations under the ith treatment. Similarly, let $y_{\cdot\cdot}$ represent the grand total of all observations and $\bar{y}_{\cdot\cdot}$ represent the grand mean of all observations. Expressed mathematically,

$$y_{i\cdot} = \sum_{j=1}^{n} y_{ij} \qquad \bar{y}_{i\cdot} = y_{i\cdot}/n \qquad i = 1, 2, \ldots, a$$

$$y_{\cdot\cdot} = \sum_{i=1}^{a} \sum_{j=1}^{n} y_{ij} \qquad \bar{y}_{\cdot\cdot} = y_{\cdot\cdot}/N \qquad (13\text{-}3)$$

where $N = an$ is the total number of observations. Thus, the "dot" subscript notation implies summation over the subscript that it replaces.

We are interested in testing the equality of the a treatment means $\mu_1, \mu_2, \ldots, \mu_a$. Using Equation 13-2, we find that this is equivalent to testing the hypotheses

$$H_0: \tau_1 = \tau_2 = \cdots = \tau_a = 0$$
$$H_1: \tau_i \neq 0 \quad \text{for at least one } i \qquad (13\text{-}4)$$

Thus, if the null hypothesis is true, each observation consists of the overall mean μ plus a realization of the random error component ϵ_{ij}. This is equivalent to saying that all N observations are taken from a normal distribution with mean μ and variance σ^2. Therefore, if the null hypothesis is true, changing the levels of the factor has no effect on the mean response.

The ANOVA partitions the total variability in the sample data into two component parts. Then, the test of the hypothesis in Equation 13-4 is based on a comparison of two independent estimates of the population variance. The total variability in the data is described by the **total sum of squares**

$$SS_T = \sum_{i=1}^{a} \sum_{j=1}^{n} (y_{ij} - \bar{y}_{\cdot\cdot})^2$$

The partition of the total sum of squares is given in the following definition.

ANOVA Sum of Squares Identity: Single Factor

The **sum of squares identity** is

$$\sum_{i=1}^{a} \sum_{j=1}^{n} (y_{ij} - \bar{y}_{\cdot\cdot})^2 = n \sum_{i=1}^{a} (\bar{y}_{i\cdot} - \bar{y}_{\cdot\cdot})^2 + \sum_{i=1}^{a} \sum_{j=1}^{n} (y_{ij} - \bar{y}_{i\cdot})^2 \qquad (13\text{-}5)$$

or symbolically

$$SS_T = SS_{\text{Treatments}} + SS_E \qquad (13\text{-}6)$$

The identity in Equation 13-5 shows that the total variability in the data, measured by the total corrected sum of squares SS_T, can be partitioned into a sum of squares of differences between treatment means and the grand mean denoted $SS_{\text{Treatments}}$ and a sum of squares of differences of observations within a treatment from the treatment mean denoted SS_E. Differences between observed treatment means and the grand mean measure the differences between treatments, while differences of observations within a treatment from the treatment mean can be due only to random error.

We can gain considerable insight into how the analysis of variance works by examining the expected values of $SS_{\text{Treatments}}$ and SS_E. This will lead us to an appropriate statistic for testing the hypothesis of no differences among treatment means (or all $\tau_i = 0$).

Expected Values of Sums of Squares: Single Factor

The expected value of the treatment sum of squares is

$$E(SS_{\text{Treatments}}) = (a-1)\sigma^2 + n\sum_{i=1}^{a}\tau_i^2$$

and the expected value of the error sum of squares is

$$E(SS_E) = a(n-1)\sigma^2$$

There is also a partition of the number of degrees of freedom that corresponds to the sum of squares identity in Equation 13-5. That is, there are $an = N$ observations; thus, SS_T has $an - 1$ degrees of freedom. There are a levels of the factor, so $SS_{\text{Treatments}}$ has $a - 1$ degrees of freedom. Finally, within any treatment there are n replicates providing $n - 1$ degrees of freedom with which to estimate the experimental error. Since there are a treatments, we have $a(n - 1)$ degrees of freedom for error. Therefore, the degrees of freedom partition is

$$an - 1 = a - 1 + a(n - 1)$$

The ratio

$$MS_{\text{Treatments}} = SS_{\text{Treatments}}/(a-1)$$

is called the **mean square for treatments.** Now if the null hypothesis $H_0: \tau_1 = \tau_2 = \cdots = \tau_a = 0$ is true, $MS_{\text{Treatments}}$ is an unbiased estimator of σ^2 because $\sum_{i=1}^{a}\tau_i = 0$. However, if H_1 is true, $MS_{\text{Treatments}}$ estimates σ^2 plus a positive term that incorporates variation due to the systematic difference in treatment means.

Note that the **error mean square**

$$MS_E = SS_E/[a(n-1)]$$

is an unbiased estimator of σ^2 regardless of whether or not H_0 is true. We can also show that $MS_{\text{Treatments}}$ and MS_E are independent. Consequently, we can show that if the null hypothesis H_0 is true, the ratio

ANOVA F-Test

$$F_0 = \frac{SS_{\text{Treatments}}/(a-1)}{SS_E/[a(n-1)]} = \frac{MS_{\text{Treatments}}}{MS_E} \quad (13\text{-}7)$$

has an F-distribution with $a - 1$ and $a(n - 1)$ degrees of freedom. Furthermore, from the expected mean squares, we know that MS_E is an unbiased estimator of σ^2. Also, under the null hypothesis, $MS_{\text{Treatments}}$ is an unbiased estimator of σ^2. However, if the null hypothesis is false, the expected value of $MS_{\text{Treatments}}$ is greater than σ^2. Therefore, under the alternative hypothesis, the expected value of the numerator of the test statistic (Equation 13-7) is greater than the expected value of the denominator. Consequently, we should reject H_0 if the statistic is large. This implies an upper-tail, one-tail critical region. Therefore, we would reject H_0 if $f_0 > f_{\alpha, a-1, a(n-1)}$ where f_0 is the computed value of F_0 from Equation 13-7.

Efficient computational formulas for the sums of squares may be obtained by expanding and simplifying the definitions of $SS_{\text{Treatments}}$ and SS_T. This yields the following results.

Computing Formulas for ANOVA: Single Factor with Equal Sample Sizes

The sums of squares computing formulas for the ANOVA with equal sample sizes in each treatment are

$$SS_T = \sum_{i=1}^{a} \sum_{j=1}^{n} y_{ij}^2 - \frac{y_{..}^2}{N} \qquad (13\text{-}8)$$

and

$$SS_{\text{Treatments}} = \sum_{i=1}^{a} \frac{y_{i.}^2}{n} - \frac{y_{..}^2}{N} \qquad (13\text{-}9)$$

The error sum of squares is obtained by subtraction as

$$SS_E = SS_T - SS_{\text{Treatments}} \qquad (13\text{-}10)$$

The computations for this test procedure are usually summarized in tabular form as shown in Table 13-3. This is called an **analysis of variance (or ANOVA) table**.

EXAMPLE 13-1
Tensile Strength ANOVA

Consider the paper tensile strength experiment described in Section 13-2.1. We can use the analysis of variance to test the hypothesis that different hardwood concentrations do not affect the mean tensile strength of the paper.

The hypotheses are

$$H_0: \tau_1 = \tau_2 = \tau_3 = \tau_4 = 0$$

$$H_1: \tau_i \neq 0 \text{ for at least one } i.$$

Table 13-3 The Analysis of Variance for a Single-Factor Experiment, Fixed-Effects Model

Source of Variation	Sum of Squares	Degrees of Freedom	Mean Square	F_0
Treatments	$SS_{\text{Treatments}}$	$a - 1$	$MS_{\text{Treatments}}$	$\dfrac{MS_{\text{Treatments}}}{MS_E}$
Error	SS_E	$a(n - 1)$	MS_E	
Total	SS_T	$an - 1$		

Table 13-4 ANOVA for the Tensile Strength Data

Source of Variation	Sum of Squares	Degrees of Freedom	Mean Square	f_0	P-value
Hardwood concentration	382.79	3	127.60	19.60	3.59 E-6
Error	130.17	20	6.51		
Total	512.96	23			

We will use $\alpha = 0.01$. The sums of squares for the analysis of variance are computed from Equations 13-8, 13-9, and 13-10 as follows:

$$SS_T = \sum_{i=1}^{4} \sum_{j=1}^{6} y_{ij}^2 - \frac{y_{..}^2}{N}$$

$$= (7)^2 + (8)^2 + \cdots + (20)^2 - \frac{(383)^2}{24} = 512.96$$

$$SS_{\text{Treatments}} = \sum_{i=1}^{4} \frac{y_{i.}^2}{n} - \frac{y_{..}^2}{N}$$

$$= \frac{(60)^2 + (94)^2 + (102)^2 + (127)^2}{6} - \frac{(383)^2}{24} = 382.79$$

$$SS_E = SS_T - SS_{\text{Treatments}}$$
$$= 512.96 - 382.79 = 130.17$$

The ANOVA is summarized in Table 13-4. Since $f_{0.01,3,20} = 4.94$, we reject H_0 and conclude that hardwood concentration in the pulp significantly affects the mean strength of the paper. We can also find a P-value for this test statistic as follows:

$$P = P(F_{3,20} > 19.60) \simeq 3.59 \times 10^{-6}$$

Since $P \simeq 3.59 \times 10^{-6}$ is considerably smaller than $\alpha = 0.01$, we have strong evidence to conclude that H_0 is not true.

Minitab Output
Many software packages have the capability to analyze data from designed experiments using the analysis of variance. Table 13-5 presents the output from the Minitab one-way analysis of variance routine for the paper tensile strength experiment in Example 13-1. The results agree closely with the manual calculations reported previously in Table 13-4.

The Minitab output also presents 95% confidence intervals on each individual treatment mean. The mean of the ith treatment is defined as

$$\mu_i = \mu + \tau_i \quad i = 1, 2, \ldots, a$$

A point estimator of μ_i is $\hat{\mu}_i = \overline{Y}_{i.}$. Now, if we assume that the errors are normally distributed, each treatment average is normally distributed with mean μ_i and variance σ^2/n. Thus, if σ^2 were known, we could use the normal distribution to construct a CI. Using MS_E as an estimator of σ^2

Table 13-5 Minitab Analysis of Variance Output for Example 13-1

One-Way ANOVA: Strength versus CONC
Analysis of Variance for Strength

Source	DF	SS	MS	F	P
Conc	3	382.79	127.60	19.61	0.000
Error	20	130.17	6.51		
Total	23	512.96			

Individual 95% CIs For Mean
Based on Pooled StDev

Level	N	Mean	StDev
5	6	10.000	2.828
10	6	15.667	2.805
15	6	17.000	1.789
20	6	21.167	2.639

```
 ---+---------+---------+---------+--
  (---*---)
            (---*---)
               (---*---)
                      (---*---)
 ---+---------+---------+---------+--
   10.0      15.0      20.0      25.0
```

Pooled StDev = 2.551

Fisher's pairwise comparisons

Family error rate = 0.192
Individual error rate = 0.0500

Critical value = 2.086

Intervals for (column level mean) − (row level mean)

	5	10	15
10	−8.739 −2.594		
15	−10.072 −3.928	−4.406 1.739	
20	−14.239 −8.094	−8.572 −2.428	−7.239 −1.094

(The square root of MS_E is the "Pooled StDev" referred to in the Minitab output), we would base the CI on the t-distribution, since

$$T = \frac{\bar{Y}_{i\cdot} - \mu_i}{\sqrt{MS_E/n}}$$

has a t-distribution with $a(n-1)$ degrees of freedom. This leads to the following definition of the confidence interval.

Confidence Interval on a Treatment Mean

A $100(1 - \alpha)$ percent confidence interval on the mean of the ith treatment μ_i is

$$\bar{y}_{i\cdot} - t_{\alpha/2, a(n-1)} \sqrt{\frac{MS_E}{n}} \leq \mu_i \leq \bar{y}_{i\cdot} + t_{\alpha/2, a(n-1)} \sqrt{\frac{MS_E}{n}} \quad (13\text{-}11)$$

Equation 13-11 is used to calculate the 95% CIs shown graphically in the Minitab output of Table 13-5. For example, at 20% hardwood the point estimate of the mean is $\bar{y}_{4\cdot} = 21.167$, $MS_E = 6.51$, and $t_{0.025,20} = 2.086$, so the 95% CI is

$$[\bar{y}_{4\cdot} \pm t_{0.025,20}\sqrt{MS_E/n}]$$
$$[21.167 \pm (2.086)\sqrt{6.51/6}]$$

or

$$19.00 \text{ psi} \leq \mu_4 \leq 23.34 \text{ psi}$$

It can also be interesting to find confidence intervals on the difference in two treatment means, say, $\mu_i - \mu_j$. The point estimator of $\mu_i - \mu_j$ is $\bar{Y}_{i\cdot} - \bar{Y}_{j\cdot}$, and the variance of this estimator is

$$V(\bar{Y}_{i\cdot} - \bar{Y}_{j\cdot}) = \frac{\sigma^2}{n} + \frac{\sigma^2}{n} = \frac{2\sigma^2}{n}$$

Now if we use MS_E to estimate σ^2,

$$T = \frac{\bar{Y}_{i\cdot} - \bar{Y}_{j\cdot} - (\mu_i - \mu_j)}{\sqrt{2MS_E/n}}$$

has a t-distribution with $a(n - 1)$ degrees of freedom. Therefore, a CI on $\mu_i - \mu_j$ may be based on the t-distribution.

Confidence Interval on a Difference in Treatment Means

A $100(1 - \alpha)$ percent confidence interval on the difference in two treatment means $\mu_i - \mu_j$ is

$$\bar{y}_{i\cdot} - \bar{y}_{j\cdot} - t_{\alpha/2,a(n-1)}\sqrt{\frac{2MS_E}{n}} \leq \mu_i - \mu_j \leq \bar{y}_{i\cdot} - \bar{y}_{j\cdot} + t_{\alpha/2,a(n-1)}\sqrt{\frac{2MS_E}{n}}$$

(13-12)

A 95% CI on the difference in means $\mu_3 - \mu_2$ is computed from Equation 13-12 as follows:

$$[\bar{y}_{3\cdot} - \bar{y}_{2\cdot} \pm t_{0.025,20}\sqrt{2MS_E/n}]$$
$$[17.00 - 15.67 \pm (2.086)\sqrt{2(6.51)/6}]$$

or

$$-1.74 \leq \mu_3 - \mu_2 \leq 4.40$$

Since the CI includes zero, we would conclude that there is no difference in mean tensile strength at these two particular hardwood levels.

The bottom portion of the computer output in Table 13-5 provides additional information concerning which specific means are different. We will discuss this in more detail in Section 13-2.3.

An Unbalanced Experiment

In some single-factor experiments, the number of observations taken under each treatment may be different. We then say that the design is **unbalanced**. In this situation, slight modifications must be made in the sums of squares formulas. Let n_i observations be taken under treatment i ($i = 1, 2, \ldots, a$), and let the total number of observations $N = \sum_{i=1}^{a} n_i$. The computational formulas for SS_T and $SS_{\text{Treatments}}$ are as shown in the following definition.

Computing Formulas for ANOVA: Single Factor with Unequal Sample Sizes

The sums of squares computing formulas for the ANOVA with unequal sample sizes n_i in each treatment are

$$SS_T = \sum_{i=1}^{a} \sum_{j=1}^{n_i} y_{ij}^2 - \frac{y_{..}^2}{N} \qquad (13\text{-}13)$$

$$SS_{\text{Treatments}} = \sum_{i=1}^{a} \frac{y_{i.}^2}{n_i} - \frac{y_{..}^2}{N} \qquad (13\text{-}14)$$

and

$$SS_E = SS_T - SS_{\text{Treatments}} \qquad (13\text{-}15)$$

Choosing a balanced design has two important advantages. First, the ANOVA is relatively insensitive to small departures from the assumption of equality of variances if the sample sizes are equal. This is not the case for unequal sample sizes. Second, the power of the test is maximized if the samples are of equal size.

13-2.3 Multiple Comparisons Following the ANOVA

When the null hypothesis H_0: $\tau_1 = \tau_2 = \cdots = \tau_a = 0$ is rejected in the ANOVA, we know that some of the treatment or factor level means are different. However, the ANOVA doesn't identify which means are different. Methods for investigating this issue are called **multiple comparisons methods**. Many of these procedures are available. Here we describe a very simple one, Fisher's **least significant difference** (LSD) method and a graphical method. Montgomery (2005) presents these and other methods and provides a comparative discussion.

The Fisher LSD method compares all pairs of means with the null hypotheses H_0: $\mu_i = \mu_j$ (for all $i \neq j$) using the t-statistic

$$t_0 = \frac{\bar{y}_{i.} - \bar{y}_{j.}}{\sqrt{\dfrac{2MS_E}{n}}}$$

Assuming a two-sided alternative hypothesis, the pair of means μ_i and μ_j would be declared significantly different if

$$|\bar{y}_{i.} - \bar{y}_{j.}| > \text{LSD}$$

where LSD, the least significant difference, is

Least Significant Difference for Multiple Comparisons

$$\text{LSD} = t_{\alpha/2, a(n-1)} \sqrt{\frac{2MS_E}{n}} \qquad (13\text{-}16)$$

Table II Cumulative Binomial Probabilities $P(X<=x)$

n	x	0.1	0.2	0.3	0.4	0.5	0.6	0.7	0.8	0.9	0.95	0.99
1	0	0.9000	0.8000	0.7000	0.6000	0.5000	0.4000	0.3000	0.2000	0.1000	0.0500	0.0100
2	0	0.8100	0.6400	0.4900	0.3600	0.2500	0.1600	0.0900	0.0400	0.0100	0.0025	0.0001
	1	0.9900	0.9600	0.9100	0.8400	0.7500	0.6400	0.5100	0.3600	0.1900	0.0975	0.0199
3	0	0.7290	0.5120	0.3430	0.2160	0.1250	0.0640	0.0270	0.0080	0.0010	0.0001	0.0000
	1	0.9720	0.8960	0.7840	0.6480	0.5000	0.3520	0.2160	0.1040	0.0280	0.0073	0.0003
	2	0.9990	0.9920	0.9730	0.9360	0.8750	0.7840	0.6570	0.4880	0.2710	0.1426	0.0297
4	0	0.6561	0.4096	0.2401	0.1296	0.0625	0.0256	0.0081	0.0016	0.0001	0.0000	0.0000
	1	0.9477	0.8192	0.6517	0.4752	0.3125	0.1792	0.0837	0.0272	0.0037	0.0005	0.0000
	2	0.9963	0.9728	0.9163	0.8208	0.6875	0.5248	0.3483	0.1808	0.0523	0.0140	0.0006
	3	0.9999	0.9984	0.9919	0.9744	0.9375	0.8704	0.7599	0.5904	0.3439	0.1855	0.0394
5	0	0.5905	0.3277	0.1681	0.0778	0.0313	0.0102	0.0024	0.0003	0.0000	0.0000	0.0000
	1	0.9185	0.7373	0.5282	0.3370	0.1875	0.0870	0.0308	0.0067	0.0005	0.0000	0.0000
	2	0.9914	0.9421	0.8369	0.6826	0.5000	0.3174	0.1631	0.0579	0.0086	0.0012	0.0000
	3	0.9995	0.9933	0.9692	0.9130	0.8125	0.6630	0.4718	0.2627	0.0815	0.0226	0.0010
	4	1.0000	0.9997	0.9976	0.9898	0.6988	0.9222	0.8319	0.6723	0.4095	0.2262	0.0490
6	0	0.5314	0.2621	0.1176	0.0467	0.0156	0.0041	0.0007	0.0001	0.0000	0.0000	0.0000
	1	0.8857	0.6554	0.4202	0.2333	0.1094	0.0410	0.0109	0.0016	0.0001	0.0000	0.0000
	2	0.9842	0.9011	0.7443	0.5443	0.3438	0.1792	0.0705	0.0170	0.0013	0.0001	0.0000
	3	0.9987	0.9830	0.9295	0.8208	0.6563	0.4557	0.2557	0.0989	0.0159	0.0022	0.0000
	4	0.9999	0.9984	0.9891	0.9590	0.9806	0.7667	0.5798	0.3446	0.1143	0.0328	0.0015
	5	1.0000	0.9999	0.9993	0.9959	0.9844	0.9533	0.8824	0.7379	0.4686	0.2649	0.0585
7	0	0.4783	0.2097	0.0824	0.0280	0.0078	0.0016	0.0002	0.0000	0.0000	0.0000	0.0000
	1	0.8503	0.5767	0.3294	0.1586	0.0625	0.0188	0.0038	0.0004	0.0000	0.0000	0.0000
	2	0.9743	0.8520	0.6471	0.4199	0.2266	0.0963	0.0288	0.0047	0.0002	0.0000	0.0000
	3	0.9973	0.9667	0.8740	0.7102	0.5000	0.2898	0.1260	0.0333	0.0027	0.0002	0.0000
	4	0.9998	0.9953	0.9712	0.9037	0.7734	0.5801	0.3529	0.1480	0.0257	0.0038	0.0000
	5	1.0000	0.9996	0.9962	0.9812	0.9375	0.8414	0.6706	0.4233	0.1497	0.0444	0.0020
	6	1.0000	1.0000	0.9998	0.9984	0.9922	0.9720	0.9176	0.7903	0.5217	0.3017	0.0679
8	0	0.4305	0.1678	0.0576	0.0168	0.0039	0.0007	0.0001	0.0000	0.0000	0.0000	0.0000
	1	0.8131	0.5033	0.2553	0.1064	0.0352	0.0085	0.0013	0.0001	0.0000	0.0000	0.0000
	2	0.9619	0.7969	0.5518	0.3154	0.1445	0.0498	0.0113	0.0012	0.0000	0.0000	0.0000
	3	0.9950	0.9437	0.8059	0.5941	0.3633	0.1737	0.0580	0.0104	0.0004	0.0000	0.0000
	4	0.9996	0.9896	0.9420	0.8263	0.6367	0.4059	0.1941	0.0563	0.0050	0.0004	0.0000
	5	1.0000	0.9988	0.9887	0.9502	0.8555	0.6846	0.4482	0.2031	0.0381	0.0058	0.0001
	6	1.0000	0.9999	0.9987	0.9915	0.9648	0.8936	0.7447	0.4967	0.1869	0.0572	0.0027
	7	1.0000	1.0000	0.9999	0.9993	0.9961	0.9832	0.9424	0.8322	0.5695	0.3366	0.0773
9	0	0.3874	0.1342	0.0404	0.0101	0.0020	0.0003	0.0000	0.0000	0.0000	0.0000	0.0000
	1	0.7748	0.4362	0.1960	0.0705	0.0195	0.0038	0.0004	0.0000	0.0000	0.0000	0.0000
	2	0.9470	0.7382	0.4628	0.2318	0.0889	0.0250	0.0043	0.0003	0.0000	0.0000	0.0000
	3	0.9917	0.9144	0.7297	0.4826	0.2539	0.0994	0.0253	0.0031	0.0001	0.0000	0.0000
	4	0.9991	0.9804	0.9012	0.7334	0.5000	0.2666	0.0988	0.0196	0.0009	0.0000	0.0000
	5	0.9999	0.9969	0.9747	0.9006	0.7461	0.5174	0.2703	0.0856	0.0083	0.0006	0.0000
	6	1.0000	0.9997	0.9957	0.9750	0.9102	0.7682	0.5372	0.2618	0.0530	0.0084	0.0001
	7	1.0000	1.0000	0.9996	0.9962	0.9805	0.9295	0.8040	0.5638	0.2252	0.0712	0.0034
	8	1.0000	1.0000	1.0000	0.9997	0.9980	0.9899	0.9596	0.8658	0.6126	0.3698	0.0865

Table II Cumulative Binomial Probabilities $P(X <= x)$ (continued)

		\multicolumn{11}{c}{P}										
n	x	0.1	0.2	0.3	0.4	0.5	0.6	0.7	0.8	0.9	0.95	0.99
10	0	0.3487	0.1074	0.0282	0.0060	0.0010	0.0001	0.0000	0.0000	0.0000	0.0000	0.0000
	1	0.7361	0.3758	0.1493	0.0464	0.0107	0.0017	0.0001	0.0000	0.0000	0.0000	0.0000
	2	0.9298	0.6778	0.3828	0.1673	0.0547	0.0123	0.0016	0.0001	0.0000	0.0000	0.0000
	3	0.9872	0.8791	0.6496	0.3823	0.1719	0.0548	0.0106	0.0009	0.0000	0.0000	0.0000
	4	0.9984	0.9672	0.8497	0.6331	0.3770	0.1662	0.0473	0.0064	0.0001	0.0000	0.0000
	5	0.9999	0.9936	0.9527	0.8338	0.6230	0.3669	0.1503	0.0328	0.0016	0.0001	0.0000
	6	1.0000	0.9991	0.9894	0.9452	0.8281	0.6177	0.3504	0.1209	0.0128	0.0010	0.0000
	7	1.0000	0.9999	0.9984	0.9877	0.9453	0.8327	0.6172	0.3222	0.0702	0.0115	0.0001
	8	1.0000	1.0000	0.9999	0.9983	0.9893	0.9536	0.8507	0.6242	0.2639	0.0861	0.0043
	9	1.0000	1.0000	1.0000	0.9999	0.9990	0.9940	0.9718	0.8926	0.6513	0.4013	0.0956
11	0	0.3138	0.0859	0.0198	0.0036	0.0005	0.0000	0.0000	0.0000	0.0000	0.0000	0.0000
	1	0.6974	0.3221	0.1130	0.0302	0.0059	0.0007	0.0000	0.0000	0.0000	0.0000	0.0000
	2	0.9104	0.6174	0.3127	0.1189	0.0327	0.0059	0.0006	0.0000	0.0000	0.0000	0.0000
	3	0.9815	0.8389	0.5696	0.2963	0.1133	0.0293	0.0043	0.0002	0.0000	0.0000	0.0000
	4	0.9972	0.9496	0.7897	0.5328	0.2744	0.0994	0.0216	0.0020	0.0000	0.0000	0.0000
	5	0.9997	0.9883	0.9218	0.7535	0.5000	0.2465	0.0782	0.0117	0.0003	0.0000	0.0000
	6	1.0000	0.9980	0.9784	0.9006	0.7256	0.4672	0.2103	0.0504	0.0028	0.0001	0.0000
	7	1.0000	0.9998	0.9957	0.9707	0.8867	0.7037	0.4304	0.1611	0.0185	0.0016	0.0000
	8	1.0000	1.0000	0.9994	0.9941	0.9673	0.8811	0.6873	0.3826	0.0896	0.0152	0.0002
	9	1.0000	1.0000	1.0000	0.9993	0.9941	0.9698	0.8870	0.6779	0.3026	0.1019	0.0052
	10	1.0000	1.0000	1.0000	1.0000	0.9995	0.9964	0.9802	0.9141	0.6862	0.4312	0.1047
12	0	0.2824	0.0687	0.0138	0.0022	0.0002	0.0000	0.0000	0.0000	0.0000	0.0000	0.0000
	1	0.6590	0.2749	0.0850	0.0196	0.0032	0.0003	0.0000	0.0000	0.0000	0.0000	0.0000
	2	0.8891	0.5583	0.2528	0.0834	0.0193	0.0028	0.0002	0.0000	0.0000	0.0000	0.0000
	3	0.9744	0.7946	0.4925	0.2253	0.0730	0.0153	0.0017	0.0001	0.0000	0.0000	0.0000
	4	0.9957	0.9274	0.7237	0.4382	0.1938	0.0573	0.0095	0.0006	0.0000	0.0000	0.0000
	5	0.9995	0.9806	0.8822	0.6652	0.3872	0.1582	0.0386	0.0039	0.0001	0.0000	0.0000
	6	0.9999	0.9961	0.9614	0.8418	0.6128	0.3348	0.1178	0.0194	0.0005	0.0000	0.0000
	7	1.0000	0.9994	0.9905	0.9427	0.8062	0.5618	0.2763	0.0726	0.0043	0.0002	0.0000
	8	1.0000	0.9999	0.9983	0.9847	0.9270	0.7747	0.5075	0.2054	0.0256	0.0022	0.0000
	9	1.0000	1.0000	0.9998	0.9972	0.9807	0.9166	0.7472	0.4417	0.1109	0.0196	0.0002
	10	1.0000	1.0000	1.0000	0.9997	0.9968	0.9804	0.9150	0.7251	0.3410	0.1184	0.0062
	11	1.0000	1.0000	1.0000	1.0000	0.9998	0.9978	0.9862	0.9313	0.7176	0.4596	0.1136
13	0	0.2542	0.0550	0.0097	0.0013	0.0001	0.0000	0.0000	0.0000	0.0000	0.0000	0.0000
	1	0.6213	0.2336	0.0637	0.0126	0.0017	0.0001	0.0000	0.0000	0.0000	0.0000	0.0000
	2	0.8661	0.5017	0.2025	0.0579	0.0112	0.0013	0.0001	0.0000	0.0000	0.0000	0.0000
	3	0.9658	0.7473	0.4206	0.1686	0.0461	0.0078	0.0007	0.0000	0.0000	0.0000	0.0000
	4	0.9935	0.9009	0.6543	0.3530	0.1334	0.0321	0.0040	0.0002	0.0000	0.0000	0.0000
	5	0.9991	0.9700	0.8346	0.5744	0.2905	0.0977	0.0182	0.0012	0.0000	0.0000	0.0000
	6	0.9999	0.9930	0.9376	0.7712	0.5000	0.2288	0.0624	0.0070	0.0001	0.0000	0.0000
	7	1.0000	0.9988	0.9818	0.9023	0.7095	0.4256	0.1654	0.0300	0.0009	0.0000	0.0000
	8	1.0000	0.9988	0.9960	0.9679	0.8666	0.6470	0.3457	0.0991	0.0065	0.0003	0.0000
	9	1.0000	1.0000	0.9993	0.9922	0.9539	0.8314	0.5794	0.2527	0.0342	0.0031	0.0000
	10	1.0000	1.0000	0.9999	0.9987	0.9888	0.9421	0.7975	0.4983	0.1339	0.0245	0.0003
	11	1.0000	1.0000	1.0000	0.9999	0.9983	0.9874	0.9363	0.7664	0.3787	0.1354	0.0072
	12	1.0000	1.0000	1.0000	1.0000	0.9999	0.9987	0.9903	0.9450	0.7458	0.4867	0.1225

Table II Cumulative Binomial Probabilities $P(X <= x)$ (continued)

						P						
n	x	0.1	0.2	0.3	0.4	0.5	0.6	0.7	0.8	0.9	0.95	0.99
14	0	0.2288	0.0440	0.0068	0.0008	0.0001	0.0000	0.0000	0.0000	0.0000	0.0000	0.0000
	1	0.5846	0.1979	0.0475	0.0081	0.0009	0.0001	0.0000	0.0000	0.0000	0.0000	0.0000
	2	0.8416	0.4481	0.1608	0.0398	0.0065	0.0006	0.0000	0.0000	0.0000	0.0000	0.0000
	3	0.9559	0.6982	0.3552	0.1243	0.0287	0.0039	0.0002	0.0000	0.0000	0.0000	0.0000
	4	0.9908	0.8702	0.5842	0.2793	0.0898	0.0175	0.0017	0.0000	0.0000	0.0000	0.0000
	5	0.9985	0.9561	0.7805	0.4859	0.2120	0.0583	0.0083	0.0004	0.0000	0.0000	0.0000
	6	0.9998	0.9884	0.9067	0.6925	0.3953	0.1501	0.0315	0.0024	0.0000	0.0000	0.0000
	7	1.0000	0.9976	0.9685	0.8499	0.6047	0.3075	0.0933	0.0116	0.0002	0.0000	0.0000
	8	1.0000	0.9996	0.9917	0.9417	0.7880	0.5141	0.2195	0.0439	0.0015	0.0000	0.0000
	9	1.0000	1.0000	0.9983	0.9825	0.9102	0.7207	0.4158	0.1298	0.0092	0.0004	0.0000
	10	1.0000	1.0000	0.9998	0.9961	0.9713	0.8757	0.6448	0.3018	0.0441	0.0042	0.0000
	11	1.0000	1.0000	1.0000	0.9994	0.9935	0.9602	0.8392	0.5519	0.1584	0.0301	0.0003
	12	1.0000	1.0000	1.0000	0.9999	0.9991	0.9919	0.9525	0.8021	0.4154	0.1530	0.0084
	13	1.0000	1.0000	1.0000	1.0000	0.9999	0.9992	0.9932	0.9560	0.7712	0.5123	0.1313
15	0	0.2059	0.0352	0.0047	0.0005	0.0000	0.0000	0.0000	0.0000	0.0000	0.0000	0.0000
	1	0.5490	0.1671	0.0353	0.0052	0.0005	0.0000	0.0000	0.0000	0.0000	0.0000	0.0000
	2	0.8159	0.3980	0.1268	0.0271	0.0037	0.0003	0.0000	0.0000	0.0000	0.0000	0.0000
	3	0.9444	0.6482	0.2969	0.0905	0.0176	0.0019	0.0001	0.0000	0.0000	0.0000	0.0000
	4	0.9873	0.8358	0.5155	0.2173	0.0592	0.0093	0.0007	0.0000	0.0000	0.0000	0.0000
	5	0.9978	0.9389	0.7216	0.4032	0.1509	0.0338	0.0037	0.0001	0.0000	0.0000	0.0000
	6	0.9997	0.9819	0.8689	0.6098	0.3036	0.0950	0.0152	0.0008	0.0000	0.0000	0.0000
	7	1.0000	0.9958	0.9500	0.7869	0.5000	0.2131	0.0500	0.0042	0.0000	0.0000	0.0000
	8	1.0000	0.9992	0.9848	0.9050	0.6964	0.3902	0.1311	0.0181	0.0003	0.0000	0.0000
	9	1.0000	0.9999	0.9963	0.9662	0.8491	0.5968	0.2784	0.0611	0.0022	0.0001	0.0000
	10	1.0000	1.0000	0.9993	0.9907	0.9408	0.7827	0.4845	0.1642	0.0127	0.0006	0.0000
	11	1.0000	1.0000	0.9999	0.9981	0.9824	0.9095	0.7031	0.3518	0.0556	0.0055	0.0000
	12	1.0000	1.0000	1.0000	0.9997	0.9963	0.9729	0.8732	0.6020	01841	0.0362	0.0004
	13	1.0000	1.0000	1.0000	1.0000	0.9995	0.9948	0.9647	0.8329	0.4510	0.1710	0.0096
	14	1.0000	1.0000	1.0000	1.0000	1.0000	0.9995	0.9953	0.9648	0.7941	0.5367	0.1399
20	0	0.1216	0.0115	0.0008	0.0000	0.0000	0.0000	0.0000	0.0000	0.0000	0.0000	0.0000
	1	0.3917	0.0692	0.0076	0.0005	0.0000	0.0000	0.0000	0.0000	0.0000	0.0000	0.0000
	2	0.6769	0.2061	0.0355	0.0036	0.0002	0.0000	0.0000	0.0000	0.0000	0.0000	0.0000
	3	0.8670	0.4114	0.1071	0.0160	0.0013	0.0000	0.0000	0.0000	0.0000	0.0000	0.0000
	4	0.9568	0.6296	0.2375	0.0510	0.0059	0.0003	0.0000	0.0000	0.0000	0.0000	0.0000
	5	0.9887	0.8042	0.4164	0.1256	0.0207	0.0016	0.0000	0.0000	0.0000	0.0000	0.0000
	6	0.9976	0.9133	0.6080	0.2500	0.0577	0.0065	0.0003	0.0000	0.0000	0.0000	0.0000
	7	0.9996	0.9679	0.7723	0.4159	0.1316	0.0210	0.0013	0.0000	0.0000	0.0000	0.0000
	8	0.9999	0.9900	0.8867	0.5956	0.2517	0.0565	0.0051	0.0001	0.0000	0.0000	0.0000
	9	1.0000	0.9974	0.9520	0.7553	0.4119	0.1275	0.0171	0.0006	0.0000	0.0000	0.0000
	10	1.0000	0.9994	0.9829	0.8725	0.5881	0.2447	0.0480	0.0026	0.0000	0.0000	0.0000
	11	1.0000	0.9999	0.9949	0.9435	0.7483	0.4044	0.1133	0.0100	0.0001	0.0000	0.0000
	12	1.0000	1.0000	0.9987	0.9790	0.8684	0.5841	0.2277	0.0321	0.0004	0.0000	0.0000
	13	1.0000	1.0000	0.9997	0.9935	0.9423	0.7500	0.3920	0.0867	0.0024	0.0000	0.0000
	14	1.0000	1.0000	1.0000	0.9984	0.9793	0.8744	0.5836	0.1958	0.0113	0.0003	0.0000
	15	1.0000	1.0000	1.0000	0.9997	0.9941	0.9490	0.7625	0.3704	0.0432	0.0026	0.0000
	16	1.0000	1.0000	1.0000	1.0000	0.9987	0.9840	0.8929	0.5886	0.1330	0.0159	0.0000
	17	1.0000	1.0000	1.0000	1.0000	0.9998	0.9964	0.9645	0.7939	0.3231	0.0755	0.0010
	18	1.0000	1.0000	1.0000	1.0000	1.0000	0.9995	0.9924	0.9308	0.6083	0.2642	0.0169
	19	1.0000	1.0000	1.0000	1.0000	1.0000	1.0000	0.9992	0.9885	0.8784	0.6415	0.1821

$$\Phi(z) = P(Z \le z) = \int_{-\infty}^{z} \frac{1}{\sqrt{2\pi}} e^{-\frac{1}{2}u^2} du$$

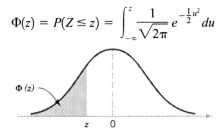

Table III Cumulative Standard Normal Distribution

z	−0.09	−0.08	−0.07	−0.06	−0.05	−0.04	−0.03	−0.02	−0.01	−0.00
−3.9	0.000033	0.000034	0.000036	0.000037	0.000039	0.000041	0.000042	0.000044	0.000046	0.000048
−3.8	0.000050	0.000052	0.000054	0.000057	0.000059	0.000062	0.000064	0.000067	0.000069	0.000072
−3.7	0.000075	0.000078	0.000082	0.000085	0.000088	0.000092	0.000096	0.000100	0.000104	0.000108
−3.6	0.000112	0.000117	0.000121	0.000126	0.000131	0.000136	0.000142	0.000147	0.000153	0.000159
−3.5	0.000165	0.000172	0.000179	0.000185	0.000193	0.000200	0.000208	0.000216	0.000224	0.000233
−3.4	0.000242	0.000251	0.000260	0.000270	0.000280	0.000291	0.000302	0.000313	0.000325	0.000337
−3.3	0.000350	0.000362	0.000376	0.000390	0.000404	0.000419	0.000434	0.000450	0.000467	0.000483
−3.2	0.000501	0.000519	0.000538	0.000557	0.000577	0.000598	0.000619	0.000641	0.000664	0.000687
−3.1	0.000711	0.000736	0.000762	0.000789	0.000816	0.000845	0.000874	0.000904	0.000935	0.000968
−3.0	0.001001	0.001035	0.001070	0.001107	0.001144	0.001183	0.001223	0.001264	0.001306	0.001350
−2.9	0.001395	0.001441	0.001489	0.001538	0.001589	0.001641	0.001695	0.001750	0.001807	0.001866
−2.8	0.001926	0.001988	0.002052	0.002118	0.002186	0.002256	0.002327	0.002401	0.002477	0.002555
−2.7	0.002635	0.002718	0.002803	0.002890	0.002980	0.003072	0.003167	0.003264	0.003364	0.003467
−2.6	0.003573	0.003681	0.003793	0.003907	0.004025	0.004145	0.004269	0.004396	0.004527	0.004661
−2.5	0.004799	0.004940	0.005085	0.005234	0.005386	0.005543	0.005703	0.005868	0.006037	0.006210
−2.4	0.006387	0.006569	0.006756	0.006947	0.007143	0.007344	0.007549	0.007760	0.007976	0.008198
−2.3	0.008424	0.008656	0.008894	0.009137	0.009387	0.009642	0.009903	0.010170	0.010444	0.010724
−2.2	0.011011	0.011304	0.011604	0.011911	0.012224	0.012545	0.012874	0.013209	0.013553	0.013903
−2.1	0.014262	0.014629	0.015003	0.015386	0.015778	0.016177	0.016586	0.017003	0.017429	0.017864
−2.0	0.018309	0.018763	0.019226	0.019699	0.020182	0.020675	0.021178	0.021692	0.022216	0.022750
−1.9	0.023295	0.023852	0.024419	0.024998	0.025588	0.026190	0.026803	0.027429	0.028067	0.028717
−1.8	0.029379	0.030054	0.030742	0.031443	0.032157	0.032884	0.033625	0.034379	0.035148	0.035930
−1.7	0.036727	0.037538	0.038364	0.039204	0.040059	0.040929	0.041815	0.042716	0.043633	0.044565
−1.6	0.045514	0.046479	0.047460	0.048457	0.049471	0.050503	0.051551	0.052616	0.053699	0.054799
−1.5	0.055917	0.057053	0.058208	0.059380	0.060571	0.061780	0.063008	0.064256	0.065522	0.066807
−1.4	0.068112	0.069437	0.070781	0.072145	0.073529	0.074934	0.076359	0.077804	0.079270	0.080757
−1.3	0.082264	0.083793	0.085343	0.086915	0.088508	0.090123	0.091759	0.093418	0.095098	0.096801
−1.2	0.098525	0.100273	0.102042	0.103835	0.105650	0.107488	0.109349	0.111233	0.113140	0.115070
−1.1	0.117023	0.119000	0.121001	0.123024	0.125072	0.127143	0.129238	0.131357	0.133500	0.135666
−1.0	0.137857	0.140071	0.142310	0.144572	0.146859	0.149170	0.151505	0.153864	0.156248	0.158655
−0.9	0.161087	0.163543	0.166023	0.168528	0.171056	0.173609	0.176185	0.178786	0.181411	0.184060
−0.8	0.186733	0.189430	0.192150	0.194894	0.197662	0.200454	0.203269	0.206108	0.208970	0.211855
−0.7	0.214764	0.217695	0.220650	0.223627	0.226627	0.229650	0.232695	0.235762	0.238852	0.241964
−0.6	0.245097	0.248252	0.251429	0.254627	0.257846	0.261086	0.264347	0.267629	0.270931	0.274253
−0.5	0.277595	0.280957	0.284339	0.287740	0.291160	0.294599	0.298056	0.301532	0.305026	0.308538
−0.4	0.312067	0.315614	0.319178	0.322758	0.326355	0.329969	0.333598	0.337243	0.340903	0.344578
−0.3	0.348268	0.351973	0.355691	0.359424	0.363169	0.366928	0.370700	0.374484	0.378281	0.382089
−0.2	0.385908	0.389739	0.393580	0.397432	0.401294	0.405165	0.409046	0.412936	0.416834	0.420740
−0.1	0.424655	0.428576	0.432505	0.436441	0.440382	0.444330	0.448283	0.452242	0.456205	0.460172
0.0	0.464144	0.468119	0.472097	0.476078	0.480061	0.484047	0.488033	0.492022	0.496011	0.500000

$$\Phi(z) = P(Z \le z) = \int_{-\infty}^{z} \frac{1}{\sqrt{2\pi}} e^{-\frac{1}{2}u^2} du$$

Table III Cumulative Standard Normal Distribution (*continued*)

z	0.00	0.01	0.02	0.03	0.04	0.05	0.06	0.07	0.08	0.09
0.0	0.500000	0.503989	0.507978	0.511967	0.515953	0.519939	0.532922	0.527903	0.531881	0.535856
0.1	0.539828	0.543795	0.547758	0.551717	0.555760	0.559618	0.563559	0.567495	0.571424	0.575345
0.2	0.579260	0.583166	0.587064	0.590954	0.594835	0.598706	0.602568	0.606420	0.610261	0.614092
0.3	0.617911	0.621719	0.625516	0.629300	0.633072	0.636831	0.640576	0.644309	0.648027	0.651732
0.4	0.655422	0.659097	0.662757	0.666402	0.670031	0.673645	0.677242	0.680822	0.684386	0.687933
0.5	0.691462	0.694974	0.698468	0.701944	0.705401	0.708840	0.712260	0.715661	0.719043	0.722405
0.6	0.725747	0.729069	0.732371	0.735653	0.738914	0.742154	0.745373	0.748571	0.751748	0.754903
0.7	0.758036	0.761148	0.764238	0.767305	0.770350	0.773373	0.776373	0.779350	0.782305	0.785236
0.8	0.788145	0.791030	0.793892	0.796731	0.799546	0.802338	0.805106	0.807850	0.810570	0.813267
0.9	0.815940	0.818589	0.821214	0.823815	0.826391	0.828944	0.831472	0.833977	0.836457	0.838913
1.0	0.841345	0.843752	0.846136	0.848495	0.850830	0.853141	0.855428	0.857690	0.859929	0.862143
1.1	0.864334	0.866500	0.868643	0.870762	0.872857	0.874928	0.876976	0.878999	0.881000	0.882977
1.2	0.884930	0.886860	0.888767	0.890651	0.892512	0.894350	0.896165	0.897958	0.899727	0.901475
1.3	0.903199	0.904902	0.906582	0.908241	0.909877	0.911492	0.913085	0.914657	0.916207	0.917736
1.4	0.919243	0.920730	0.922196	0.923641	0.925066	0.926471	0.927855	0.929219	0.930563	0.931888
1.5	0.933193	0.934478	0.935744	0.936992	0.938220	0.939429	0.940620	0.941792	0.942947	0.944083
1.6	0.945201	0.946301	0.947384	0.948449	0.949497	0.950529	0.951543	0.952540	0.953521	0.954486
1.7	0.955435	0.956367	0.957284	0.958185	0.959071	0.959941	0.960796	0.961636	0.962462	0.963273
1.8	0.964070	0.964852	0.965621	0.966375	0.967116	0.967843	0.968557	0.969258	0.969946	0.970621
1.9	0.971283	0.971933	0.972571	0.973197	0.973810	0.974412	0.975002	0.975581	0.976148	0.976705
2.0	0.977250	0.977784	0.978308	0.978822	0.979325	0.979818	0.980301	0.980774	0.981237	0.981691
2.1	0.982136	0.982571	0.982997	0.983414	0.983823	0.984222	0.984614	0.984997	0.985371	0.985738
2.2	0.986097	0.986447	0.986791	0.987126	0.987455	0.987776	0.988089	0.988396	0.988696	0.988989
2.3	0.989276	0.989556	0.989830	0.990097	0.990358	0.990613	0.990863	0.991106	0.991344	0.991576
2.4	0.991802	0.992024	0.992240	0.992451	0.992656	0.992857	0.993053	0.993244	0.993431	0.993613
2.5	0.993790	0.993963	0.994132	0.994297	0.994457	0.994614	0.994766	0.994915	0.995060	0.995201
2.6	0.995339	0.995473	0.995604	0.995731	0.995855	0.995975	0.996093	0.996207	0.996319	0.996427
2.7	0.996533	0.996636	0.996736	0.996833	0.996928	0.997020	0.997110	0.997197	0.997282	0.997365
2.8	0.997445	0.997523	0.997599	0.997673	0.997744	0.997814	0.997882	0.997948	0.998012	0.998074
2.9	0.998134	0.998193	0.998250	0.998305	0.998359	0.998411	0.998462	0.998511	0.998559	0.998605
3.0	0.998650	0.998694	0.998736	0.998777	0.998817	0.998856	0.998893	0.998930	0.998965	0.998999
3.1	0.999032	0.999065	0.999096	0.999126	0.999155	0.999184	0.999211	0.999238	0.999264	0.999289
3.2	0.999313	0.999336	0.999359	0.999381	0.999402	0.999423	0.999443	0.999462	0.999481	0.999499
3.3	0.999517	0.999533	0.999550	0.999566	0.999581	0.999596	0.999610	0.999624	0.999638	0.999650
3.4	0.999663	0.999675	0.999687	0.999698	0.999709	0.999720	0.999730	0.999740	0.999749	0.999758
3.5	0.999767	0.999776	0.999784	0.999792	0.999800	0.999807	0.999815	0.999821	0.999828	0.999835
3.6	0.999841	0.999847	0.999853	0.999858	0.999864	0.999869	0.999874	0.999879	0.999883	0.999888
3.7	0.999892	0.999896	0.999900	0.999904	0.999908	0.999912	0.999915	0.999918	0.999922	0.999925
3.8	0.999928	0.999931	0.999933	0.999936	0.999938	0.999941	0.999943	0.999946	0.999948	0.999950
3.9	0.999952	0.999954	0.999956	0.999958	0.999959	0.999961	0.999963	0.999964	0.999966	0.999967

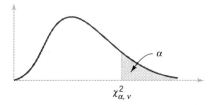

Table IV Percentage Points $\chi^2_{\alpha,\nu}$ of the Chi-Squared Distribution

ν \ α	.995	.990	.975	.950	.900	.500	.100	.050	.025	.010	.005
1	.00+	.00+	.00+	.00+	.02	.45	2.71	3.84	5.02	6.63	7.88
2	.01	.02	.05	.10	.21	1.39	4.61	5.99	7.38	9.21	10.60
3	.07	.11	.22	.35	.58	2.37	6.25	7.81	9.35	11.34	12.84
4	.21	.30	.48	.71	1.06	3.36	7.78	9.49	11.14	13.28	14.86
5	.41	.55	.83	1.15	1.61	4.35	9.24	11.07	12.83	15.09	16.75
6	.68	.87	1.24	1.64	2.20	5.35	10.65	12.59	14.45	16.81	18.55
7	.99	1.24	1.69	2.17	2.83	6.35	12.02	14.07	16.01	18.48	20.28
8	1.34	1.65	2.18	2.73	3.49	7.34	13.36	15.51	17.53	20.09	21.96
9	1.73	2.09	2.70	3.33	4.17	8.34	14.68	16.92	19.02	21.67	23.59
10	2.16	2.56	3.25	3.94	4.87	9.34	15.99	18.31	20.48	23.21	25.19
11	2.60	3.05	3.82	4.57	5.58	10.34	17.28	19.68	21.92	24.72	26.76
12	3.07	3.57	4.40	5.23	6.30	11.34	18.55	21.03	23.34	26.22	28.30
13	3.57	4.11	5.01	5.89	7.04	12.34	19.81	22.36	24.74	27.69	29.82
14	4.07	4.66	5.63	6.57	7.79	13.34	21.06	23.68	26.12	29.14	31.32
15	4.60	5.23	6.27	7.26	8.55	14.34	22.31	25.00	27.49	30.58	32.80
16	5.14	5.81	6.91	7.96	9.31	15.34	23.54	26.30	28.85	32.00	34.27
17	5.70	6.41	7.56	8.67	10.09	16.34	24.77	27.59	30.19	33.41	35.72
18	6.26	7.01	8.23	9.39	10.87	17.34	25.99	28.87	31.53	34.81	37.16
19	6.84	7.63	8.91	10.12	11.65	18.34	27.20	30.14	32.85	36.19	38.58
20	7.43	8.26	9.59	10.85	12.44	19.34	28.41	31.41	34.17	37.57	40.00
21	8.03	8.90	10.28	11.59	13.24	20.34	29.62	32.67	35.48	38.93	41.40
22	8.64	9.54	10.98	12.34	14.04	21.34	30.81	33.92	36.78	40.29	42.80
23	9.26	10.20	11.69	13.09	14.85	22.34	32.01	35.17	38.08	41.64	44.18
24	9.89	10.86	12.40	13.85	15.66	23.34	33.20	36.42	39.36	42.98	45.56
25	10.52	11.52	13.12	14.61	16.47	24.34	34.28	37.65	40.65	44.31	46.93
26	11.16	12.20	13.84	15.38	17.29	25.34	35.56	38.89	41.92	45.64	48.29
27	11.81	12.88	14.57	16.15	18.11	26.34	36.74	40.11	43.19	46.96	49.65
28	12.46	13.57	15.31	16.93	18.94	27.34	37.92	41.34	44.46	48.28	50.99
29	13.12	14.26	16.05	17.71	19.77	28.34	39.09	42.56	45.72	49.59	52.34
30	13.79	14.95	16.79	18.49	20.60	29.34	40.26	43.77	46.98	50.89	53.67
40	20.71	22.16	24.43	26.51	29.05	39.34	51.81	55.76	59.34	63.69	66.77
50	27.99	29.71	32.36	34.76	37.69	49.33	63.17	67.50	71.42	76.15	79.49
60	35.53	37.48	40.48	43.19	46.46	59.33	74.40	79.08	83.30	88.38	91.95
70	43.28	45.44	48.76	51.74	55.33	69.33	85.53	90.53	95.02	100.42	104.22
80	51.17	53.54	57.15	60.39	64.28	79.33	96.58	101.88	106.63	112.33	116.32
90	59.20	61.75	65.65	69.13	73.29	89.33	107.57	113.14	118.14	124.12	128.30
100	67.33	70.06	74.22	77.93	82.36	99.33	118.50	124.34	129.56	135.81	140.17

ν = degrees of freedom.

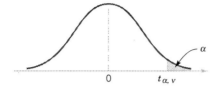

Table V Percentage Points $t_{\alpha,\nu}$ of the t-Distribution

ν \ α	.40	.25	.10	.05	.025	.01	.005	.0025	.001	.0005
1	.325	1.000	3.078	6.314	12.706	31.821	63.657	127.32	318.31	636.62
2	.289	.816	1.886	2.920	4.303	6.965	9.925	14.089	23.326	31.598
3	.277	.765	1.638	2.353	3.182	4.541	5.841	7.453	10.213	12.924
4	.271	.741	1.533	2.132	2.776	3.747	4.604	5.598	7.173	8.610
5	.267	.727	1.476	2.015	2.571	3.365	4.032	4.773	5.893	6.869
6	.265	.718	1.440	1.943	2.447	3.143	3.707	4.317	5.208	5.959
7	.263	.711	1.415	1.895	2.365	2.998	3.499	4.029	4.785	5.408
8	.262	.706	1.397	1.860	2.306	2.896	3.355	3.833	4.501	5.041
9	.261	.703	1.383	1.833	2.262	2.821	3.250	3.690	4.297	4.781
10	.260	.700	1.372	1.812	2.228	2.764	3.169	3.581	4.144	4.587
11	.260	.697	1.363	1.796	2.201	2.718	3.106	3.497	4.025	4.437
12	.259	.695	1.356	1.782	2.179	2.681	3.055	3.428	3.930	4.318
13	.259	.694	1.350	1.771	2.160	2.650	3.012	3.372	3.852	4.221
14	.258	.692	1.345	1.761	2.145	2.624	2.977	3.326	3.787	4.140
15	.258	.691	1.341	1.753	2.131	2.602	2.947	3.286	3.733	4.073
16	.258	.690	1.337	1.746	2.120	2.583	2.921	3.252	3.686	4.015
17	.257	.689	1.333	1.740	2.110	2.567	2.898	3.222	3.646	3.965
18	.257	.688	1.330	1.734	2.101	2.552	2.878	3.197	3.610	3.922
19	.257	.688	1.328	1.729	2.093	2.539	2.861	3.174	3.579	3.883
20	.257	.687	1.325	1.725	2.086	2.528	2.845	3.153	3.552	3.850
21	.257	.686	1.323	1.721	2.080	2.518	2.831	3.135	3.527	3.819
22	.256	.686	1.321	1.717	2.074	2.508	2.819	3.119	3.505	3.792
23	.256	.685	1.319	1.714	2.069	2.500	2.807	3.104	3.485	3.767
24	.256	.685	1.318	1.711	2.064	2.492	2.797	3.091	3.467	3.745
25	.256	.684	1.316	1.708	2.060	2.485	2.787	3.078	3.450	3.725
26	.256	.684	1.315	1.706	2.056	2.479	2.779	3.067	3.435	3.707
27	.256	.684	1.314	1.703	2.052	2.473	2.771	3.057	3.421	3.690
28	.256	.683	1.313	1.701	2.048	2.467	2.763	3.047	3.408	3.674
29	.256	.683	1.311	1.699	2.045	2.462	2.756	3.038	3.396	3.659
30	.256	.683	1.310	1.697	2.042	2.457	2.750	3.030	3.385	3.646
40	.255	.681	1.303	1.684	2.021	2.423	2.704	2.971	3.307	3.551
60	.254	.679	1.296	1.671	2.000	2.390	2.660	2.915	3.232	3.460
120	.254	.677	1.289	1.658	1.980	2.358	2.617	2.860	3.160	3.373
∞	.253	.674	1.282	1.645	1.960	2.326	2.576	2.807	3.090	3.291

ν = degrees of freedom.

Table VI Percentage Points f_{α,v_1,v_2} of the F-Distribution

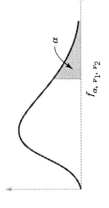

$f_{0.25,v_1,v_2}$

v_2 \ v_1	1	2	3	4	5	6	7	8	9	10	12	15	20	24	30	40	60	120	∞
1	5.83	7.50	8.20	8.58	8.82	8.98	9.10	9.19	9.26	9.32	9.41	9.49	9.58	9.63	9.67	9.71	9.76	9.80	9.85
2	2.57	3.00	3.15	3.23	3.28	3.31	3.34	3.35	3.37	3.38	3.39	3.41	3.43	3.43	3.44	3.45	3.46	3.47	3.48
3	2.02	2.28	2.36	2.39	2.41	2.42	2.43	2.44	2.44	2.44	2.45	2.46	2.46	2.46	2.47	2.47	2.47	2.47	2.47
4	1.81	2.00	2.05	2.06	2.07	2.08	2.08	2.08	2.08	2.08	2.08	2.08	2.08	2.08	2.08	2.08	2.08	2.08	2.08
5	1.69	1.85	1.88	1.89	1.89	1.89	1.89	1.89	1.89	1.89	1.89	1.89	1.88	1.88	1.88	1.88	1.87	1.87	1.87
6	1.62	1.76	1.78	1.79	1.79	1.78	1.78	1.78	1.77	1.77	1.77	1.76	1.76	1.75	1.75	1.75	1.74	1.74	1.74
7	1.57	1.70	1.72	1.72	1.71	1.71	1.70	1.70	1.70	1.69	1.68	1.68	1.67	1.67	1.66	1.66	1.65	1.65	1.65
8	1.54	1.66	1.67	1.66	1.66	1.65	1.64	1.64	1.63	1.63	1.62	1.62	1.61	1.60	1.60	1.59	1.59	1.58	1.58
9	1.51	1.62	1.63	1.63	1.62	1.61	1.60	1.60	1.59	1.59	1.58	1.57	1.56	1.56	1.55	1.54	1.54	1.53	1.53
10	1.49	1.60	1.60	1.59	1.59	1.58	1.57	1.56	1.56	1.55	1.54	1.53	1.52	1.52	1.51	1.51	1.50	1.49	1.48
11	1.47	1.58	1.58	1.57	1.56	1.55	1.54	1.53	1.53	1.52	1.51	1.50	1.49	1.49	1.48	1.47	1.47	1.46	1.45
12	1.46	1.56	1.56	1.55	1.54	1.53	1.52	1.51	1.51	1.50	1.49	1.48	1.47	1.46	1.45	1.45	1.44	1.43	1.42
13	1.45	1.55	1.55	1.53	1.52	1.51	1.50	1.49	1.49	1.48	1.47	1.46	1.45	1.44	1.43	1.42	1.42	1.41	1.40
14	1.44	1.53	1.53	1.52	1.51	1.50	1.49	1.48	1.47	1.46	1.45	1.44	1.43	1.42	1.41	1.41	1.40	1.39	1.38
15	1.43	1.52	1.52	1.51	1.49	1.48	1.47	1.46	1.46	1.45	1.44	1.43	1.41	1.41	1.40	1.39	1.38	1.37	1.36
16	1.42	1.51	1.51	1.50	1.48	1.47	1.46	1.45	1.44	1.44	1.43	1.41	1.40	1.39	1.38	1.37	1.36	1.35	1.34
17	1.42	1.51	1.50	1.49	1.47	1.46	1.45	1.44	1.43	1.43	1.41	1.40	1.39	1.38	1.37	1.36	1.35	1.34	1.33
18	1.41	1.50	1.49	1.48	1.46	1.45	1.44	1.43	1.42	1.42	1.40	1.39	1.38	1.37	1.36	1.35	1.34	1.33	1.32
19	1.41	1.49	1.49	1.47	1.46	1.44	1.43	1.42	1.41	1.41	1.40	1.38	1.37	1.36	1.35	1.34	1.33	1.32	1.30
20	1.40	1.49	1.48	1.47	1.45	1.44	1.43	1.42	1.41	1.40	1.39	1.37	1.36	1.35	1.34	1.33	1.32	1.31	1.29
21	1.40	1.48	1.48	1.46	1.44	1.43	1.42	1.41	1.40	1.39	1.38	1.37	1.35	1.34	1.33	1.32	1.31	1.30	1.28
22	1.40	1.48	1.47	1.45	1.44	1.42	1.41	1.40	1.39	1.39	1.37	1.36	1.34	1.33	1.32	1.31	1.30	1.29	1.28
23	1.39	1.47	1.47	1.45	1.43	1.42	1.41	1.40	1.39	1.38	1.37	1.35	1.34	1.33	1.32	1.31	1.30	1.28	1.27
24	1.39	1.47	1.46	1.44	1.43	1.41	1.40	1.39	1.38	1.38	1.36	1.35	1.33	1.32	1.31	1.30	1.29	1.28	1.26
25	1.39	1.47	1.46	1.44	1.42	1.41	1.40	1.39	1.38	1.37	1.36	1.34	1.33	1.32	1.31	1.29	1.28	1.27	1.25
26	1.38	1.46	1.45	1.44	1.42	1.41	1.39	1.38	1.37	1.37	1.35	1.34	1.32	1.31	1.30	1.29	1.28	1.26	1.25
27	1.38	1.46	1.45	1.43	1.42	1.40	1.39	1.38	1.37	1.36	1.35	1.33	1.32	1.31	1.30	1.28	1.27	1.26	1.24
28	1.38	1.46	1.45	1.43	1.41	1.40	1.39	1.38	1.37	1.36	1.34	1.33	1.31	1.30	1.29	1.28	1.27	1.25	1.24
29	1.38	1.45	1.45	1.43	1.41	1.40	1.38	1.37	1.36	1.35	1.34	1.32	1.31	1.30	1.29	1.27	1.26	1.25	1.23
30	1.38	1.45	1.44	1.42	1.41	1.39	1.38	1.37	1.36	1.35	1.34	1.32	1.30	1.29	1.28	1.27	1.26	1.24	1.23
40	1.36	1.44	1.42	1.40	1.39	1.37	1.36	1.35	1.34	1.33	1.31	1.30	1.28	1.26	1.25	1.24	1.22	1.21	1.19
60	1.35	1.42	1.41	1.38	1.37	1.35	1.33	1.32	1.31	1.30	1.29	1.27	1.25	1.24	1.22	1.21	1.19	1.17	1.15
120	1.34	1.40	1.39	1.37	1.35	1.33	1.31	1.30	1.29	1.28	1.26	1.24	1.22	1.21	1.19	1.18	1.16	1.13	1.10
∞	1.32	1.39	1.37	1.35	1.33	1.31	1.29	1.28	1.27	1.25	1.24	1.22	1.19	1.18	1.16	1.14	1.12	1.08	1.00

Table VI Percentage Points of the F-Distribution (continued)

$$f_{0.10, v_1, v_2}$$

Degrees of freedom for the numerator (v_1)

v_2 \ v_1	1	2	3	4	5	6	7	8	9	10	12	15	20	24	30	40	60	120	∞
1	39.86	49.50	53.59	55.83	57.24	58.20	58.91	59.44	59.86	60.19	60.71	61.22	61.74	62.00	62.26	62.53	62.79	63.06	63.33
2	8.53	9.00	9.16	9.24	9.29	9.33	9.35	9.37	9.38	9.39	9.41	9.42	9.44	9.45	9.46	9.47	9.47	9.48	9.49
3	5.54	5.46	5.39	5.34	5.31	5.28	5.27	5.25	5.24	5.23	5.22	5.20	5.18	5.18	5.17	5.16	5.15	5.14	5.13
4	4.54	4.32	4.19	4.11	4.05	4.01	3.98	3.95	3.94	3.92	3.90	3.87	3.84	3.83	3.82	3.80	3.79	3.78	3.76
5	4.06	3.78	3.62	3.52	3.45	3.40	3.37	3.34	3.32	3.30	3.27	3.24	3.21	3.19	3.17	3.16	3.14	3.12	3.10
6	3.78	3.46	3.29	3.18	3.11	3.05	3.01	2.98	2.96	2.94	2.90	2.87	2.84	2.82	2.80	2.78	2.76	2.74	2.72
7	3.59	3.26	3.07	2.96	2.88	2.83	2.78	2.75	2.72	2.70	2.67	2.63	2.59	2.58	2.56	2.54	2.51	2.49	2.47
8	3.46	3.11	2.92	2.81	2.73	2.67	2.62	2.59	2.56	2.54	2.50	2.46	2.42	2.40	2.38	2.36	2.34	2.32	2.29
9	3.36	3.01	2.81	2.69	2.61	2.55	2.51	2.47	2.44	2.42	2.38	2.34	2.30	2.28	2.25	2.23	2.21	2.18	2.16
10	3.29	2.92	2.73	2.61	2.52	2.46	2.41	2.38	2.35	2.32	2.28	2.24	2.20	2.18	2.16	2.13	2.11	2.08	2.06
11	3.23	2.86	2.66	2.54	2.45	2.39	2.34	2.30	2.27	2.25	2.21	2.17	2.12	2.10	2.08	2.05	2.03	2.00	1.97
12	3.18	2.81	2.61	2.48	2.39	2.33	2.28	2.24	2.21	2.19	2.15	2.10	2.06	2.04	2.01	1.99	1.96	1.93	1.90
13	3.14	2.76	2.56	2.43	2.35	2.28	2.23	2.20	2.16	2.14	2.10	2.05	2.01	1.98	1.96	1.93	1.90	1.88	1.85
14	3.10	2.73	2.52	2.39	2.31	2.24	2.19	2.15	2.12	2.10	2.05	2.01	1.96	1.94	1.91	1.89	1.86	1.83	1.80
15	3.07	2.70	2.49	2.36	2.27	2.21	2.16	2.12	2.09	2.06	2.02	1.97	1.92	1.90	1.87	1.85	1.82	1.79	1.76
16	3.05	2.67	2.46	2.33	2.24	2.18	2.13	2.09	2.06	2.03	1.99	1.94	1.89	1.87	1.84	1.81	1.78	1.75	1.72
17	3.03	2.64	2.44	2.31	2.22	2.15	2.10	2.06	2.03	2.00	1.96	1.91	1.86	1.84	1.81	1.78	1.75	1.72	1.69
18	3.01	2.62	2.42	2.29	2.20	2.13	2.08	2.04	2.00	1.98	1.93	1.89	1.84	1.81	1.78	1.75	1.72	1.69	1.66
19	2.99	2.61	2.40	2.27	2.18	2.11	2.06	2.02	1.98	1.96	1.91	1.86	1.81	1.79	1.76	1.73	1.70	1.67	1.63
20	2.97	2.59	2.38	2.25	2.16	2.09	2.04	2.00	1.96	1.94	1.89	1.84	1.79	1.77	1.74	1.71	1.68	1.64	1.61
21	2.96	2.57	2.36	2.23	2.14	2.08	2.02	1.98	1.95	1.92	1.87	1.83	1.78	1.75	1.72	1.69	1.66	1.62	1.59
22	2.95	2.56	2.35	2.22	2.13	2.06	2.01	1.97	1.93	1.90	1.86	1.81	1.76	1.73	1.70	1.67	1.64	1.60	1.57
23	2.94	2.55	2.34	2.21	2.11	2.05	1.99	1.95	1.92	1.89	1.84	1.80	1.74	1.72	1.69	1.66	1.62	1.59	1.55
24	2.93	2.54	2.33	2.19	2.10	2.04	1.98	1.94	1.91	1.88	1.83	1.78	1.73	1.70	1.67	1.64	1.61	1.57	1.53
25	2.92	2.53	2.32	2.18	2.09	2.02	1.97	1.93	1.89	1.87	1.82	1.77	1.72	1.69	1.66	1.63	1.59	1.56	1.52
26	2.91	2.52	2.31	2.17	2.08	2.01	1.96	1.92	1.88	1.86	1.81	1.76	1.71	1.68	1.65	1.61	1.58	1.54	1.50
27	2.90	2.51	2.30	2.17	2.07	2.00	1.95	1.91	1.87	1.85	1.80	1.75	1.70	1.67	1.64	1.60	1.57	1.53	1.49
28	2.89	2.50	2.29	2.16	2.06	2.00	1.94	1.90	1.87	1.84	1.79	1.74	1.69	1.66	1.63	1.59	1.56	1.52	1.48
29	2.89	2.50	2.28	2.15	2.06	1.99	1.93	1.89	1.86	1.83	1.78	1.73	1.68	1.65	1.62	1.58	1.55	1.51	1.47
30	2.88	2.49	2.28	2.14	2.03	1.98	1.93	1.88	1.85	1.82	1.77	1.72	1.67	1.64	1.61	1.57	1.54	1.50	1.46
40	2.84	2.44	2.23	2.09	2.00	1.93	1.87	1.83	1.79	1.76	1.71	1.66	1.61	1.57	1.54	1.51	1.47	1.42	1.38
60	2.79	2.39	2.18	2.04	1.95	1.87	1.82	1.77	1.74	1.71	1.66	1.60	1.54	1.51	1.48	1.44	1.40	1.35	1.29
120	2.75	2.35	2.13	1.99	1.90	1.82	1.77	1.72	1.68	1.65	1.60	1.55	1.48	1.45	1.41	1.37	1.32	1.26	1.19
∞	2.71	2.30	2.08	1.94	1.85	1.77	1.72	1.67	1.63	1.60	1.55	1.49	1.42	1.38	1.34	1.30	1.24	1.17	1.00

Degrees of freedom for the denominator (v_2)

Table VI Percentage Points of the F-Distribution (continued)

$$f_{0.05,\nu_1,\nu_2}$$

$\nu_2 \backslash \nu_1$	1	2	3	4	5	6	7	8	9	10	12	15	20	24	30	40	60	120	∞
1	161.4	199.5	215.7	224.6	230.2	234.0	236.8	238.9	240.5	241.9	243.9	245.9	248.0	249.1	250.1	251.1	252.2	253.3	254.3
2	18.51	19.00	19.16	19.25	19.30	19.33	19.35	19.37	19.38	19.40	19.41	19.43	19.45	19.45	19.46	19.47	19.48	19.49	19.50
3	10.13	9.55	9.28	9.12	9.01	8.94	8.89	8.85	8.81	8.79	8.74	8.70	8.66	8.64	8.62	8.59	8.57	8.55	8.53
4	7.71	6.94	6.59	6.39	6.26	6.16	6.09	6.04	6.00	5.96	5.91	5.86	5.80	5.77	5.75	5.72	5.69	5.66	5.63
5	6.61	5.79	5.41	5.19	5.05	4.95	4.88	4.82	4.77	4.74	4.68	4.62	4.56	4.53	4.50	4.46	4.43	4.40	4.36
6	5.99	5.14	4.76	4.53	4.39	4.28	4.21	4.15	4.10	4.06	4.00	3.94	3.87	3.84	3.81	3.77	3.74	3.70	3.67
7	5.59	4.74	4.35	4.12	3.97	3.87	3.79	3.73	3.68	3.64	3.57	3.51	3.44	3.41	3.38	3.34	3.30	3.27	3.23
8	5.32	4.46	4.07	3.84	3.69	3.58	3.50	3.44	3.39	3.35	3.28	3.22	3.15	3.12	3.08	3.04	3.01	2.97	2.93
9	5.12	4.26	3.86	3.63	3.48	3.37	3.29	3.23	3.18	3.14	3.07	3.01	2.94	2.90	2.86	2.83	2.79	2.75	2.71
10	4.96	4.10	3.71	3.48	3.33	3.22	3.14	3.07	3.02	2.98	2.91	2.85	2.77	2.74	2.70	2.66	2.62	2.58	2.54
11	4.84	3.98	3.59	3.36	3.20	3.09	3.01	2.95	2.90	2.85	2.79	2.72	2.65	2.61	2.57	2.53	2.49	2.45	2.40
12	4.75	3.89	3.49	3.26	3.11	3.00	2.91	2.85	2.80	2.75	2.69	2.62	2.54	2.51	2.47	2.43	2.38	2.34	2.30
13	4.67	3.81	3.41	3.18	3.03	2.92	2.83	2.77	2.71	2.67	2.60	2.53	2.46	2.42	2.38	2.34	2.30	2.25	2.21
14	4.60	3.74	3.34	3.11	2.96	2.85	2.76	2.70	2.65	2.60	2.53	2.46	2.39	2.35	2.31	2.27	2.22	2.18	2.13
15	4.54	3.68	3.29	3.06	2.90	2.79	2.71	2.64	2.59	2.54	2.48	2.40	2.33	2.29	2.25	2.20	2.16	2.11	2.07
16	4.49	3.63	3.24	3.01	2.85	2.74	2.66	2.59	2.54	2.49	2.42	2.35	2.28	2.24	2.19	2.15	2.11	2.06	2.01
17	4.45	3.59	3.20	2.96	2.81	2.70	2.61	2.55	2.49	2.45	2.38	2.31	2.23	2.19	2.15	2.10	2.06	2.01	1.96
18	4.41	3.55	3.16	2.93	2.77	2.66	2.58	2.51	2.46	2.41	2.34	2.27	2.19	2.15	2.11	2.06	2.02	1.97	1.92
19	4.38	3.52	3.13	2.90	2.74	2.63	2.54	2.48	2.42	2.38	2.31	2.23	2.16	2.11	2.07	2.03	1.98	1.93	1.88
20	4.35	3.49	3.10	2.87	2.71	2.60	2.51	2.45	2.39	2.35	2.28	2.20	2.12	2.08	2.04	1.99	1.95	1.90	1.84
21	4.32	3.47	3.07	2.84	2.68	2.57	2.49	2.42	2.37	2.32	2.25	2.18	2.10	2.05	2.01	1.96	1.92	1.87	1.81
22	4.30	3.44	3.05	2.82	2.66	2.55	2.46	2.40	2.34	2.30	2.23	2.15	2.07	2.03	1.98	1.94	1.89	1.84	1.78
23	4.28	3.42	3.03	2.80	2.64	2.53	2.44	2.37	2.32	2.27	2.20	2.13	2.05	2.01	1.96	1.91	1.86	1.81	1.76
24	4.26	3.40	3.01	2.78	2.62	2.51	2.42	2.36	2.30	2.25	2.18	2.11	2.03	1.98	1.94	1.89	1.84	1.79	1.73
25	4.24	3.39	2.99	2.76	2.60	2.49	2.40	2.34	2.28	2.24	2.16	2.09	2.01	1.96	1.92	1.87	1.82	1.77	1.71
26	4.23	3.37	2.98	2.74	2.59	2.47	2.39	2.32	2.27	2.22	2.15	2.07	1.99	1.95	1.90	1.85	1.80	1.75	1.69
27	4.21	3.35	2.96	2.73	2.57	2.46	2.37	2.31	2.25	2.20	2.13	2.06	1.97	1.93	1.88	1.84	1.79	1.73	1.67
28	4.20	3.34	2.95	2.71	2.56	2.45	2.36	2.29	2.24	2.19	2.12	2.04	1.96	1.91	1.87	1.82	1.77	1.71	1.65
29	4.18	3.33	2.93	2.70	2.55	2.43	2.35	2.28	2.22	2.18	2.10	2.03	1.94	1.90	1.85	1.81	1.75	1.70	1.64
30	4.17	3.32	2.92	2.69	2.53	2.42	2.33	2.27	2.21	2.16	2.09	2.01	1.93	1.89	1.84	1.79	1.74	1.68	1.62
40	4.08	3.23	2.84	2.61	2.45	2.34	2.25	2.18	2.12	2.08	2.00	1.92	1.84	1.79	1.74	1.69	1.64	1.58	1.51
60	4.00	3.15	2.76	2.53	2.37	2.25	2.17	2.10	2.04	1.99	1.92	1.84	1.75	1.70	1.65	1.59	1.53	1.47	1.39
120	3.92	3.07	2.68	2.45	2.29	2.17	2.09	2.02	1.96	1.91	1.83	1.75	1.66	1.61	1.55	1.55	1.43	1.35	1.25
∞	3.84	3.00	2.60	2.37	2.21	2.10	2.01	1.94	1.88	1.83	1.75	1.67	1.57	1.52	1.46	1.39	1.32	1.22	1.00

Degrees of freedom for the numerator (ν_1)

Degrees of freedom for the denominator (ν_2)

Table VI Percentage Points of the F-Distribution (continued)

$$f_{0.025, v_1, v_2}$$

v_2 \ v_1	1	2	3	4	5	6	7	8	9	10	12	15	20	24	30	40	60	120	∞
1	647.8	799.5	864.2	899.6	921.8	937.1	948.2	956.7	963.3	968.6	976.7	984.9	993.1	997.2	1001	1006	1010	1014	1018
2	38.51	39.00	39.17	39.25	39.30	39.33	39.36	39.37	39.39	39.40	39.41	39.43	39.45	39.46	39.46	39.47	39.48	39.49	39.50
3	17.44	16.04	15.44	15.10	14.88	14.73	14.62	14.54	14.47	14.42	14.34	14.25	14.17	14.12	14.08	14.04	13.99	13.95	13.90
4	12.22	10.65	9.98	9.60	9.36	9.20	9.07	8.98	8.90	8.84	8.75	8.66	8.56	8.51	8.46	8.41	8.36	8.31	8.26
5	10.01	8.43	7.76	7.39	7.15	6.98	6.85	6.76	6.68	6.62	6.52	6.43	6.33	6.28	6.23	6.18	6.12	6.07	6.02
6	8.81	7.26	6.60	6.23	5.99	5.82	5.70	5.60	5.52	5.46	5.37	5.27	5.17	5.12	5.07	5.01	4.96	4.90	4.85
7	8.07	6.54	5.89	5.52	5.29	5.12	4.99	4.90	4.82	4.76	4.67	4.57	4.47	4.42	4.36	4.31	4.25	4.20	4.14
8	7.57	6.06	5.42	5.05	4.82	4.65	4.53	4.43	4.36	4.30	4.20	4.10	4.00	3.95	3.89	3.84	3.78	3.73	3.67
9	7.21	5.71	5.08	4.72	4.48	4.32	4.20	4.10	4.03	3.96	3.87	3.77	3.67	3.61	3.56	3.51	3.45	3.39	3.33
10	6.94	5.46	4.83	4.47	4.24	4.07	3.95	3.85	3.78	3.72	3.62	3.52	3.42	3.37	3.31	3.26	3.20	3.14	3.08
11	6.72	5.26	4.63	4.28	4.04	3.88	3.76	3.66	3.59	3.53	3.43	3.33	3.23	3.17	3.12	3.06	3.00	2.94	2.88
12	6.55	5.10	4.47	4.12	3.89	3.73	3.61	3.51	3.44	3.37	3.28	3.18	3.07	3.02	2.96	2.91	2.85	2.79	2.72
13	6.41	4.97	4.35	4.00	3.77	3.60	3.48	3.39	3.31	3.25	3.15	3.05	2.95	2.89	2.84	2.78	2.72	2.66	2.60
14	6.30	4.86	4.24	3.89	3.66	3.50	3.38	3.29	3.21	3.15	3.05	2.95	2.84	2.79	2.73	2.67	2.61	2.55	2.49
15	6.20	4.77	4.15	3.80	3.58	3.41	3.29	3.20	3.12	3.06	2.96	2.86	2.76	2.70	2.64	2.59	2.52	2.46	2.40
16	6.12	4.69	4.08	3.73	3.50	3.34	3.22	3.12	3.05	2.99	2.89	2.79	2.68	2.63	2.57	2.51	2.45	2.38	2.32
17	6.04	4.62	4.01	3.66	3.44	3.28	3.16	3.06	2.98	2.92	2.82	2.72	2.62	2.56	2.50	2.44	2.38	2.32	2.25
18	5.98	4.56	3.95	3.61	3.38	3.22	3.10	3.01	2.93	2.87	2.77	2.67	2.56	2.50	2.44	2.38	2.32	2.26	2.19
19	5.92	4.51	3.90	3.56	3.33	3.17	3.05	2.96	2.88	2.82	2.72	2.62	2.51	2.45	2.39	2.33	2.27	2.20	2.13
20	5.87	4.46	3.86	3.51	3.29	3.13	3.01	2.91	2.84	2.77	2.68	2.57	2.46	2.41	2.35	2.29	2.22	2.16	2.09
21	5.83	4.42	3.82	3.48	3.25	3.09	2.97	2.87	2.80	2.73	2.64	2.53	2.42	2.37	2.31	2.25	2.18	2.11	2.04
22	5.79	4.38	3.78	3.44	3.22	3.05	2.93	2.84	2.76	2.70	2.60	2.50	2.39	2.33	2.27	2.21	2.14	2.08	2.00
23	5.75	4.35	3.75	3.41	3.18	3.02	2.90	2.81	2.73	2.67	2.57	2.47	2.36	2.30	2.24	2.18	2.11	2.04	1.97
24	5.72	4.32	3.72	3.38	3.15	2.99	2.87	2.78	2.70	2.64	2.54	2.44	2.33	2.27	2.21	2.15	2.08	2.01	1.94
25	5.69	4.29	3.69	3.35	3.13	2.97	2.85	2.75	2.68	2.61	2.51	2.41	2.30	2.24	2.18	2.12	2.05	1.98	1.91
26	5.66	4.27	3.67	3.33	3.10	2.94	2.82	2.73	2.65	2.59	2.49	2.39	2.28	2.22	2.16	2.09	2.03	1.95	1.88
27	5.63	4.24	3.65	3.31	3.08	2.92	2.80	2.71	2.63	2.57	2.47	2.36	2.25	2.19	2.13	2.07	2.00	1.93	1.85
28	5.61	4.22	3.63	3.29	3.06	2.90	2.78	2.69	2.61	2.55	2.45	2.34	2.23	2.17	2.11	2.05	1.98	1.91	1.83
29	5.59	4.20	3.61	3.27	3.04	2.88	2.76	2.67	2.59	2.53	2.43	2.32	2.21	2.15	2.09	2.03	1.96	1.89	1.81
30	5.57	4.18	3.59	3.25	3.03	2.87	2.75	2.65	2.57	2.51	2.41	2.31	2.20	2.14	2.07	2.01	1.94	1.87	1.79
40	5.42	4.05	3.46	3.13	2.90	2.74	2.62	2.53	2.45	2.39	2.29	2.18	2.07	2.01	1.94	1.88	1.80	1.72	1.64
60	5.29	3.93	3.34	3.01	2.79	2.63	2.51	2.41	2.33	2.27	2.17	2.06	1.94	1.88	1.82	1.74	1.67	1.58	1.48
120	5.15	3.80	3.23	2.89	2.67	2.52	2.39	2.30	2.22	2.16	2.05	1.94	1.82	1.76	1.69	1.61	1.53	1.43	1.31
∞	5.02	3.69	3.12	2.79	2.57	2.41	2.29	2.19	2.11	2.05	1.94	1.83	1.71	1.64	1.57	1.48	1.39	1.27	1.00

Degrees of freedom for the numerator (v_1)

Degrees of freedom for the denominator (v_2)

Table VI Percentage Points of the F-Distribution (*continued*)

$f_{0.01, v_1, v_2}$

v_2 \ v_1	1	2	3	4	5	6	7	8	9	10	12	15	20	24	30	40	60	120	∞
1	4052	4999.5	5403	5625	5764	5859	5928	5982	6022	6056	6106	6157	6209	6235	6261	6287	6313	6339	6366
2	98.50	99.00	99.17	99.25	99.30	99.33	99.36	99.37	99.39	99.40	99.42	99.43	99.45	99.46	99.47	99.47	99.48	99.49	99.50
3	34.12	30.82	29.46	28.71	28.24	27.91	27.67	27.49	27.35	27.23	27.05	26.87	26.69	26.60	26.50	26.41	26.32	26.22	26.13
4	21.20	18.00	16.69	15.98	15.52	15.21	14.98	14.80	14.66	14.55	14.37	14.20	14.02	13.93	13.84	13.75	13.65	13.56	13.46
5	16.26	13.27	12.06	11.39	10.97	10.67	10.46	10.29	10.16	10.05	9.89	9.72	9.55	9.47	9.38	9.29	9.20	9.11	9.02
6	13.75	10.92	9.78	9.15	8.75	8.47	8.26	8.10	7.98	7.87	7.72	7.56	7.40	7.31	7.23	7.14	7.06	6.97	6.88
7	12.25	9.55	8.45	7.85	7.46	7.19	6.99	6.84	6.72	6.62	6.47	6.31	6.16	6.07	5.99	5.91	5.82	5.74	5.65
8	11.26	8.65	7.59	7.01	6.63	6.37	6.18	6.03	5.91	5.81	5.67	5.52	5.36	5.28	5.20	5.12	5.03	4.95	4.86
9	10.56	8.02	6.99	6.42	6.06	5.80	5.61	5.47	5.35	5.26	5.11	4.96	4.81	4.73	4.65	4.57	4.48	4.40	4.31
10	10.04	7.56	6.55	5.99	5.64	5.39	5.20	5.06	4.94	4.85	4.71	4.56	4.41	4.33	4.25	4.17	4.08	4.00	3.91
11	9.65	7.21	6.22	5.67	5.32	5.07	4.89	4.74	4.63	4.54	4.40	4.25	4.10	4.02	3.94	3.86	3.78	3.69	3.60
12	9.33	6.93	5.95	5.41	5.06	4.82	4.64	4.50	4.39	4.30	4.16	4.01	3.86	3.78	3.70	3.62	3.54	3.45	3.36
13	9.07	6.70	5.74	5.21	4.86	4.62	4.44	4.30	4.19	4.10	3.96	3.82	3.66	3.59	3.51	3.43	3.34	3.25	3.17
14	8.86	6.51	5.56	5.04	4.69	4.46	4.28	4.14	4.03	3.94	3.80	3.66	3.51	3.43	3.35	3.27	3.18	3.09	3.00
15	8.68	6.36	5.42	4.89	4.56	4.32	4.14	4.00	3.89	3.80	3.67	3.52	3.37	3.29	3.21	3.13	3.05	2.96	2.87
16	8.53	6.23	5.29	4.77	4.44	4.20	4.03	3.89	3.78	3.69	3.55	3.41	3.26	3.18	3.10	3.02	2.93	2.84	2.75
17	8.40	6.11	5.18	4.67	4.34	4.10	3.93	3.79	3.68	3.59	3.46	3.31	3.16	3.08	3.00	2.92	2.83	2.75	2.65
18	8.29	6.01	5.09	4.58	4.25	4.01	3.84	3.71	3.60	3.51	3.37	3.23	3.08	3.00	2.92	2.84	2.75	2.66	2.57
19	8.18	5.93	5.01	4.50	4.17	3.94	3.77	3.63	3.52	3.43	3.30	3.15	3.00	2.92	2.84	2.76	2.67	2.58	2.49
20	8.10	5.85	4.94	4.43	4.10	3.87	3.70	3.56	3.46	3.37	3.23	3.09	2.94	2.86	2.78	2.69	2.61	2.52	2.42
21	8.02	5.78	4.87	4.37	4.04	3.81	3.64	3.51	3.40	3.31	3.17	3.03	2.88	2.80	2.72	2.64	2.55	2.46	2.36
22	7.95	5.72	4.82	4.31	3.99	3.76	3.59	3.45	3.35	3.26	3.12	2.98	2.83	2.75	2.67	2.58	2.50	2.40	2.31
23	7.88	5.66	4.76	4.26	3.94	3.71	3.54	3.41	3.30	3.21	3.07	2.93	2.78	2.70	2.62	2.54	2.45	2.35	2.26
24	7.82	5.61	4.72	4.22	3.90	3.67	3.50	3.36	3.26	3.17	3.03	2.89	2.74	2.66	2.58	2.49	2.40	2.31	2.21
25	7.77	5.57	4.68	4.18	3.85	3.63	3.46	3.32	3.22	3.13	2.99	2.85	2.70	2.62	2.54	2.45	2.36	2.27	2.17
26	7.72	5.53	4.64	4.14	3.82	3.59	3.42	3.29	3.18	3.09	2.96	2.81	2.66	2.58	2.50	2.42	2.33	2.23	2.13
27	7.68	5.49	4.60	4.11	3.78	3.56	3.39	3.26	3.15	3.06	2.93	2.78	2.63	2.55	2.47	2.38	2.29	2.20	2.10
28	7.64	5.45	4.57	4.07	3.75	3.53	3.36	3.23	3.12	3.03	2.90	2.75	2.60	2.52	2.44	2.35	2.26	2.17	2.06
29	7.60	5.42	4.54	4.04	3.73	3.50	3.33	3.20	3.09	3.00	2.87	2.73	2.57	2.49	2.41	2.33	2.23	2.14	2.03
30	7.56	5.39	4.51	4.02	3.70	3.47	3.30	3.17	3.07	2.98	2.84	2.70	2.55	2.47	2.39	2.30	2.21	2.11	2.01
40	7.31	5.18	4.31	3.83	3.51	3.29	3.12	2.99	2.89	2.80	2.66	2.52	2.37	2.29	2.20	2.11	2.02	1.92	1.80
60	7.08	4.98	4.13	3.65	3.34	3.12	2.95	2.82	2.72	2.63	2.50	2.35	2.20	2.12	2.03	1.94	1.84	1.73	1.60
120	6.85	4.79	3.95	3.48	3.17	2.96	2.79	2.66	2.56	2.47	2.34	2.19	2.03	1.95	1.86	1.76	1.66	1.53	1.38
∞	6.63	4.61	3.78	3.32	3.02	2.80	2.64	2.51	2.41	2.32	2.18	2.04	1.88	1.79	1.70	1.59	1.47	1.32	1.00

Degrees of freedom for the numerator (v_1)

Degrees of freedom for the denominator (v_2)

Table XII Factors for Tolerance Intervals

	Values of k for Two-Sided Intervals								
	Confidence Level								
	0.90			0.95			0.99		
Sample	Probability of Coverage								
Size	0.90	0.95	0.99	0.90	0.95	0.99	0.90	0.95	0.99
2	15.978	18.800	24.167	32.019	37.674	48.430	160.193	188.491	242.300
3	5.847	6.919	8.974	8.380	9.916	12.861	18.930	22.401	29.055
4	4.166	4.943	6.440	5.369	6.370	8.299	9.398	11.150	14.527
5	3.949	4.152	5.423	4.275	5.079	6.634	6.612	7.855	10.260
6	3.131	3.723	4.870	3.712	4.414	5.775	5.337	6.345	8.301
7	2.902	3.452	4.521	3.369	4.007	5.248	4.613	5.488	7.187
8	2.743	3.264	4.278	3.136	3.732	4.891	4.147	4.936	6.468
9	2.626	3.125	4.098	2.967	3.532	4.631	3.822	4.550	5.966
10	2.535	3.018	3.959	2.839	3.379	4.433	3.582	4.265	5.594
11	2.463	2.933	3.849	2.737	3.259	4.277	3.397	4.045	5.308
12	2.404	2.863	3.758	2.655	3.162	4.150	3.250	3.870	5.079
13	2.355	2.805	3.682	2.587	3.081	4.044	3.130	3.727	4.893
14	2.314	2.756	3.618	2.529	3.012	3.955	3.029	3.608	4.737
15	2.278	2.713	3.562	2.480	2.954	3.878	2.945	3.507	4.605
16	2.246	2.676	3.514	2.437	2.903	3.812	2.872	3.421	4.492
17	2.219	2.643	3.471	2.400	2.858	3.754	2.808	3.345	4.393
18	2.194	2.614	3.433	2.366	2.819	3.702	2.753	3.279	4.307
19	2.172	2.588	3.399	2.337	2.784	3.656	2.703	3.221	4.230
20	2.152	2.564	3.368	2.310	2.752	3.615	2.659	3.168	4.161
21	2.135	2.543	3.340	2.286	2.723	3.577	2.620	3.121	4.100
22	2.118	2.524	3.315	2.264	2.697	3.543	2.584	3.078	4.044
23	2.103	2.506	3.292	2.244	2.673	3.512	2.551	3.040	3.993
24	2.089	2.489	3.270	2.225	2.651	3.483	2.522	3.004	3.947
25	2.077	2.474	3.251	2.208	2.631	3.457	2.494	2.972	3.904
30	2.025	2.413	3.170	2.140	2.529	3.350	2.385	2.841	3.733
40	1.959	2.334	3.066	2.052	2.445	3.213	2.247	2.677	3.518
50	1.916	2.284	3.001	1.996	2.379	3.126	2.162	2.576	3.385
60	1.887	2.248	2.955	1.958	2.333	3.066	2.103	2.506	3.293
70	1.865	2.222	2.920	1.929	2.299	3.021	2.060	2.454	3.225
80	1.848	2.202	2.894	1.907	2.272	2.986	2.026	2.414	3.173
90	1.834	2.185	2.872	1.889	2.251	2.958	1.999	2.382	3.130
100	1.822	2.172	2.854	1.874	2.233	2.934	1.977	2.355	3.096

Table XII Factors for Tolerance Intervals (*continued*)

	Values of *k* for One-Sided Intervals								
	Confidence Level								
	0.90			0.95			0.99		
Sample Size	Probability of Coverage								
	0.90	0.95	0.99	0.90	0.95	0.99	0.90	0.95	0.99
2	10.253	13.090	18.500	20.581	26.260	37.094	103.029	131.426	185.617
3	4.258	5.311	7.340	6.155	7.656	10.553	13.995	17.370	23.896
4	3.188	3.957	5.438	4.162	5.144	7.042	7.380	9.083	12.387
5	2.742	3.400	4.666	3.407	4.203	5.741	5.362	6.578	8.939
6	2.494	3.092	4.243	3.006	3.708	5.062	4.411	5.406	7.335
7	2.333	2.894	3.972	2.755	3.399	4.642	3.859	4.728	6.412
8	2.219	2.754	3.783	2.582	3.187	4.354	3.497	4.285	5.812
9	2.133	2.650	3.641	2.454	3.031	4.143	3.240	3.972	5.389
10	2.066	2.568	3.532	2.355	2.911	3.981	3.048	3.738	5.074
11	2.011	2.503	3.443	2.275	2.815	3.852	2.898	3.556	4.829
12	1.966	2.448	3.371	2.210	2.736	3.747	2.777	3.410	4.633
13	1.928	2.402	3.309	2.155	2.671	3.659	2.677	3.290	4.472
14	1.895	2.363	3.257	2.109	2.614	3.585	2.593	3.189	4.337
15	1.867	2.329	3.212	2.068	2.566	3.520	2.521	3.102	4.222
16	1.842	2.299	3.172	2.033	2.524	3.464	2.459	3.028	4.123
17	1.819	2.272	3.137	2.002	2.486	3.414	2.405	2.963	4.037
18	1.800	2.249	3.105	1.974	2.453	3.370	2.357	2.905	3.960
19	1.782	2.227	3.077	1.949	2.423	3.331	2.314	2.854	3.892
20	1.765	2.028	3.052	1.926	2.396	3.295	2.276	2.808	3.832
21	1.750	2.190	3.028	1.905	2.371	3.263	2.241	2.766	3.777
22	1.737	2.174	3.007	1.886	2.349	3.233	2.209	2.729	3.727
23	1.724	2.159	2.987	1.869	2.328	3.206	2.180	2.694	3.681
24	1.712	2.145	2.969	1.853	2.309	3.181	2.154	2.662	3.640
25	1.702	2.132	2.952	1.838	2.292	3.158	2.129	2.633	3.601
30	1.657	2.080	2.884	1.777	2.220	3.064	2.030	2.515	3.447
40	1.598	2.010	2.793	1.697	2.125	2.941	1.902	2.364	3.249
50	1.559	1.965	2.735	1.646	2.065	2.862	1.821	2.269	3.125
60	1.532	1.933	2.694	1.609	2.022	2.807	1.764	2.202	3.038
70	1.511	1.909	2.662	1.581	1.990	2.765	1.722	2.153	2.974
80	1.495	1.890	2.638	1.559	1.964	2.733	1.688	2.114	2.924
90	1.481	1.874	2.618	1.542	1.944	2.706	1.661	2.082	2.883
100	1.470	1.861	2.601	1.527	1.927	2.684	1.639	2.056	2.850

Appendix B
Answers to Selected Exercises

CHAPTER 2

Section 2-1

2-1. Let a, b denote a part above, below the specification, respectively
$S = \{aaa, aab, aba, abb, baa, bab, bba, bbb\}$

2-3. Let a denote an acceptable power supply
Let f, m, c denote a supply with a functional, minor, or cosmetic error, respectively.
$S = \{a, f, m, c\}$

2-5. If only the number of tracks with errors is of interest, then $S = \{0, 1, 2, \ldots, 24\}$

2-7. S is the sample space of 100 possible two digit integers.

2-9. $S = \{0, 1, 2, \ldots, 1E09\}$ in ppb

2-11. $S = \{1.0, 1.1, 1.2, \ldots, 14.0\}$

2-13. $S = \{0, 1, 2, \ldots, \}$ in milliseconds

2-17. $c =$ connect, $b =$ busy, $S = \{c, bc, bbc, bbbc, bbbbc, \ldots\}$

2-21. (a) $S =$ nonnegative integers from 0 to the largest integer that can be displayed by the scale $S = \{0, 1, 2, 3, \ldots\}$
(b) S (c) $\{12, 13, 14, 15\}$
(d) $\{0, 1, 2, \ldots, 11\}$
(e) S
(f) $\{0, 1, 2, \ldots, 7\}$
(g) $\varnothing$
(h) $\varnothing$ (i) $\{8, 9, 10, \ldots\}$

2-23. Let d denoted a distorted bit and let o denote a bit that is not distorted.
(a) $S = \begin{Bmatrix} dddd, dodd, oddd, oodd, \\ dddo, dodo, oddo, oodo, \\ ddod, dood, odod, oood, \\ ddoo, dooo, odoo, oooo \end{Bmatrix}$
(b) No, for example $A_1 \cap A_2 = \{dddd, dddo, ddod, ddoo\}$
(c) $A_1 = \begin{Bmatrix} dddd, dodd, \\ dddo, dodo \\ ddod, dood \\ ddoo, dooo \end{Bmatrix}$
(d) $A'_1 = \begin{Bmatrix} oddd, oodd, \\ oddo, oodo, \\ odod, oood, \\ odoo, oooo \end{Bmatrix}$
(e) $A_1 \cap A_2 \cap A_3 \cap A_4 = \{dddd\}$
(f) $(A_1 \cap A_2) \cup (A_3 \cap A_4) = \{dddd, dodd, dddo, oddd, ddod, oodd, ddoo\}$

2-25. Let P denote being positive and let N denote being negative. The sample space is $\{PPP, PPN, PNP, NPP, PNN, NPN, NNP, NNN\}$.
(a) $A = \{PPP\}$
(b) $B = \{NNN\}$
(c) $A \cap B = \varnothing$
(d) $A \cup B = \{PPP, NNN\}$

2-27. (a) $A' \cap B = 10$, $B' = 10$, $A \cup B = 92$

2-29. (a) $A' = \{x \mid x \geq 72.5\}$
(b) $B' = \{x \mid x \leq 52.5\}$
(c) $A \cap B = \{x \mid 52.5 < x < 72.5\}$
(d) $A \cup B = \{x \mid x > 0\}$

2-31. Let g denote a good board, m a board with minor defects, and j a board with major defects.
(a) $S = \{gg, gm, gj, mg, mm, mj, jg, jm, jj\}$
(b) $S = \{gg, gm, gj, mg, mm, mj, jg, jm\}$

APPENDIX B ANSWERS TO SELECTED EXERCISES

2-35. 120
2-37. 144
2-39. 14,400
2-41. (a) 416,965,528
(b) 113,588,800
(c) 130,721,752
2-43. (a) 21 (b) 2520 (c) 720
2-45. (a) 1000 (b) 160 (c) 720
2-47. (a) 0.416 (b) 0.712
(c) 0.206
2-49. (a) 0.0082 (b) 0.0082

Section 2-2

2-51. (a) 0.4 (b) 0.8 (c) 0.6
(d) 1 (e) 0.2
2-53. (a) 1/10 (b) 5/10
2-55. (a) $S = \{1, 2, 3, 4, 5, 6, 7, 8\}$
(b) 2/8 (c) 6/8
2-57. (a) 0.83 (b) 0.85
2-59. $(1/10^3)*(1/26^3) = 5.7 \times 10^{-8}$
2-61. (a) $4 + 4 \times 3 + 4 \times 3 \times 3 = 52$
(b) 36/52 (c) No
2-63. (a) 0.30 (b) 0.77 (c) 0.70
(d) 0.22 (e) 0.85 (f) 0.92

Section 2-3

2-67. (a) 0.9 (b) 0 (c) 0
(d) 0 (e) 0.1
2-69. (a) 0.70 (b) 0.95 (c) No
2-71. (a) 350/370 (b) 362/370
(c) 358/370 (d) 345/370
2-73. (a) 13/130 (b) 0.90, No

Section 2-4

2-75. (a) 86/100 (b) 79/100
(c) 70/79 (d) 70/86
2-77. (a) 0.903 (b) 0.591
2-79. (a) 12/100 (b) 12/28
(c) 34/122
2-81. (a) 0.5625 (b) 0.1918
(c) 0.3333
2-83. (a) 20/100 (b) 19/99
(c) 0.038 (d) 0.2
2-85. (a) 0.02 (b) 0.000458
(c) 0.9547

Section 2-5

2-89. (a) 0.2 (b) 0.3
2-91. 0.014
2-93. 0.028
2-95. (a) 0.2376 (b) 0.0078
2-97. (a) 0.2 (b) 0.2

Section 2-6

2-103. (a) *not* independent. (b) yes

2-107. (a) 0.59 (b) 0.328 (c) 0.41
2-109. (a) 0.0048 (b) 0.0768
2-111. (a) 0.01 (b) 0.49 (c) 0.09
2-113. (a) 0.00003 (b) 0.00024
(c) 0.00107
2-115. 0.9702

Section 2-7

2-117. 0.89
2-119. (a) 0.97638 (b) 0.207552
2-121. (a) 0.615 (b) 0.618
(c) 0.052
2-123. (a) 0.9847 (b) 0.1184

Supplemental Exercises

2-125. 0.014
2-127. (a) 0.82 (b) 0.90 (c) 0.18
(d) 0.80 (e) 0.92 (f) 0.98
2-131. (a) 0.2 (b) 0.202
(c) 0.638 (d) 0.2
2-133. (a) 0.03 (b) 0.97 (c) 0.40
(d) 0.05 (e) 0.012 (f) 0.018
(g) 0.0605
2-135. (a) 0.18143 (b) 0.005976
(c) 0.86494
2-137. 0.000008
2-139. (a) 50 (b) 37 (c) 93
2-143. (a) 0.19 (b) 0.15 (c) 0.99
(d) 0.80 (e) 0.158
2-145. (a) No (b) No
(c) 40/240 (d) 200/240
(e) 234/240 (f) 1
2-147. (a) 0.282 (b) 0.718
2-149. 0.996
2-151. (a) 0.0037 (b) 0.8108
2-153. (a) 0.0778 (b) 0.00108
(c) 0.947
2-155. (a) 0.9764 (b) 0.3159
2-157. (a) 0.207 (b) 0.625
2-159. (a) 0.453 (b) 0.262
(c) 0.881 (d) 0.547
(e) 0.783 (f) 0.687
2-161. 1.58×10^{-7}
2-163. (a) 0.67336
(b) 2.646×10^{-8}
(c) 0.99973
2-165. (a) 36^7 (b) $70(26^6)$
(c) $100(26^5)$
2-167. (a) 0.994, 0.995
(b) 0.99, 0.985
(c) 0.998, 0.9975

Mind-Expanding Exercises

2-169. (a) $n = 3$ (b) $n = 3$
2-171. 0.306, 0.694

CHAPTER 3

Section 3-1

3-1. $\{0, 1, 2, \ldots, 1000\}$
3-3. $\{0, 1, 2, \ldots, 99999\}$
3-5. $\{1, 2, \ldots, 491\}$
3-7. $\{0, 1, 2, \ldots\}$
3-9. $\{0, 1, 2, \ldots, 15\}$
3-11. $\{0, 1, 2, \ldots, 10000\}$
3-13. $\{0, 1, 2, \ldots, 40000\}$

Section 3-2

3-15. (a) 1 (b) 7/8 (c) 3/4
(d) 1/2
3-17. (a) 9/25 (b) 4/25
(c) 12/25 (d) 1
3-19. $f(0) = 0.033, f(1) = 0.364,$
$f(2) = 0.603$
3-21. $P(X = 0) = 0.008,$
$P(X = 1) = 0.096,$
$P(X = 2) = 0.384,$
$P(X = 3) = 0.512$
3-23. $P(X = 50) = 0.5,$
$P(X = 25) = 0.3,$
$P(X = 10) = 0.2$
3-25. $P(X = 15) = 0.6,$
$P(X = 5) = 0.3,$
$P(X = -0.5) = 0.1$
3-27. $P(X = 0) = 0.00001,$
$P(X = 1) = 0.00167,$
$P(X = 2) = 0.07663,$
$P(X = 3) = 0.92169$

Section 3-3

3-29. (a) 7/8 (b) 1 (c) 3/4
(d) 3/8
3-31. $F(x) = \begin{cases} 0, & x < 0 \\ 0.008, & 0 \leq x < 1 \\ 0.104, & 1 \leq x < 2 \\ 0.488, & 2 \leq x < 3 \\ 1, & 3 \leq x \end{cases}$
where $f_x(0) = 0.008,$
$f_x(1) = 0.096,$
$f_x(2) = 0.384,$
$f_x(3) = 0.512$
3-33. $F(x) = \begin{cases} 0, & x < 10 \\ 0.2, & 10 \leq x < 25 \\ 0.5, & 25 \leq x < 50 \\ 1, & 50 \leq x \end{cases}$
where $P(X = 50 \text{ million}) = 0.5,$
$P(X = 25 \text{ million}) = 0.3,$
$P(X = 10 \text{ million}) = 0.2$
3-35. (a) 1 (b) 0.5 (c) 0.5
(d) 0.5

3-37. (a) 1 (b) 0.75 (c) 0.25
   (d) 0.25 (e) 0 (f) 0

Section 3-4

3-39. $\mu = 2, \sigma^2 = 2$
3-41. $\mu = 0, \sigma^2 = 1.5$
3-43. $\mu = 2.8, \sigma^2 = 1.36$
3-45. $\mu = 1.57, \sigma^2 = 0.311$
3-47. 24
3-49. $\mu = 0.0004, \sigma^2 = 0.00039996$
3-51. (a) $\mu = 18.694$,
   $\sigma^2 = 735.9644$,
   $\sigma = 27.1287$
   (b) $\mu = 37.172$,
   $\sigma^2 = 2947.996$,
   $\sigma = 54.2955$

Section 3-5

3-53. $\mu = 2, \sigma^2 = 0.667$
3-55. $\mu = 300, \sigma^2 = 6666.67$
3-57. (a) $\mu = 687.5, \sigma^2 = 56.25$
   (b) $\mu = 87.5, \sigma^2 = 56.25$
3-59. $E(X) = 4.5, E(Y) = 22.5$,
   $\sigma_Y = 14.36$

Section 3-6

3-63. (a) 0.9298 (b) 0
   (c) 0.0112 (d) 0.0016
3-65. (a) 2.40×10^{-8}
   (b) 0.9999
   (c) 9.91×10^{-18}
   (d) 1.138×10^{-4}
3-67. (a) 0 (b) 10
3-71. (a) 0.215 (b) 0.994 (c) 4
3-73. (a) 0.410 (b) 0.218
   (c) 0.37
3-75. (a) 1 (b) 0.999997
   (c) $E(X) = 12.244$,
   $V(X) = 2.179$
3-77. (a) Binomial, $p = 10^4/36^9$,
   $n = 1\text{E}09$ (b) 0
   (c) $E(X) = 4593.9$,
   $V(X) = 4593.9$
3-79. (a) 0.9961 (b) 0.9886

Section 3-7

3-81. (a) 0.5 (b) 0.0625
   (c) 0.0039 (d) 0.75
   (e) 0.25
3-83. (a) 5 (b) 5
3-85. (a) 0.0064 (b) 0.9984
   (c) 0.008
3-87. (a) 0.0167 (b) 0.9039
   (c) 50

3-89. (a) 0.13 (b) 0.098
   (c) $7.69 \approx 8$
3-91. (a) 3.91×10^{-19}
   (b) 200 (c) 2.56×10^{18}
3-93. (a) 3×10^8
   (b) 3×10^{16}

Section 3-8

3-97. (a) 0.4191 (b) 0
   (c) 0.001236
   (d) $E(X) = 0.8, V(X) = 0.6206$
3-101. (a) 0.1201 (b) 0.8523
3-103. (a) 0.087 (b) 0.9934
   (c) 0.297 (d) 0.9998
3-105. (a) 0.7069 (b) 0.0607
   (c) 0.2811.

Section 3-9

3-107. (a) 0.0183 (b) 0.2381
   (c) 0.1954 (d) 0.0298
3-109. $E(X) = V(X) = 2.996$.
3-111. (a) 0.264 (b) 48
3-113. (a) 0.4566 (b) 0.047
3-115. (a) 0.2 (b) 99.89%
3-117. (a) 0.6065 (b) 0.0067
   (c) $P(W = 0) = 0.0067$,
   $P(W = 1) = 0.0437$,
   $P(W \leq 1) = 0.0504$

Supplemental Exercises

3-119. $E(X) = 1/4, V(X) = 0.0104$
3-121. (a) $n = 50, p = 0.1$ (b) 0.112
   (c) $P(X \geq 49) = 4.51 \times 10^{-48}$
3-123. (a) 0.000224 (b) 0.2256
   (c) 0.4189
3-125. (a) 0.1024 (b) 0.1074
3-127. (a) 3000 (b) 1731.18
3-129. (a) $P(X = 0) = 0.0498$
   (b) 0.5768
   (c) $P(X \leq x) \geq 0.9, x = 5$
   (d) $\sigma^2 = \lambda = 6$. Not appropriate.
3-131. (a) 0.1877 (b) 0.4148
   (c) 15
3-133. (a) 0.0110 (b) 8/3
3-135. 40000
3-137. 0.1330
3-139. (a) 500 (b) 222.49
3-141. 0.1
3-143. $f_X(0) = 0.16, f_X(1) = 0.19$,
   $f_X(2) = 0.20, f_X(3) = 0.31$,
   $f_X(4) = 0.14$
3-145. $f_x(2) = 0.2, f_x(5.7) = 0.3$,
   $f_x(6.5) = 0.3, f_x(8.5) = 0.2$

3-147. (a) 0.0433 (b) 3.58
3-149. (a) $f_X(0) = 0.2357$,
   $f_X(1) = 0.3971$,
   $f_X(2) = 0.2647$,
   $f_X(3) = 0.0873$,
   $f_X(4) = 0.01424$,
   $f_X(5) = 0.00092$
   (b) $f_x(0) = 0.0546$,
   $f_x(1) = 0.1866$,
   $f_x(2) = 0.2837$,
   $f_x(3) = 0.2528$,
   $f_x(4) = 0.1463$,
   $f_x(5) = 0.0574$,
   $f_x(6) = 0.0155$,
   $f_x(7) = 0.0028$,
   $f_x(8) = 0.0003$,
   $f_x(9) = 0.0000$,
   $f_x(10) = 0.0000$
3-151. 37.8 seconds

CHAPTER 4

Section 4-2

4-1. (a) 0.3679 (b) 0.2858 (c) 0
   (d) 0.9817 (e) 0.0498
4-3. (a) 0.4375 (b) 0.7969
   (c) 0.5625 (d) 0.7031
   (e) 0.5
4-5. (a) 0.5 (b) 0.4375 (c) 0.125
   (d) 0 (e) 1 (f) 0.9655
4-7. (a) 0.5 (b) 49.8
4-9. (a) 0.1

Section 4-3

4-11. (a) 0.56 (b) 0.7 (c) 0 (d) 0
4-13. $F(x) = 0$ for $x < 0$; $0.25x^2$ for $0 \leq x < 2$; 1 for $2 \leq x$
4-15. $F(x) = 0$ for $x \leq 0$; $1 - e^{-x}$ for $x > 0$
4-17. $F(x) = 0$ for $x \leq 4$; $1 - e^{-(x-4)}$ for $x > 4$
4-19. 0.2
4-21. $f(x) = 0.2$ for $0 < x < 4$; $f(x) = 0.04$ for $4 \leq x < 9$

Section 4-4

4-23. $E(X) = 2, V(X) = 4/3$
4-25. $E(X) = 0, V(X) = 0.6$
4-27. $E(X) = 2$
4-29. (a) $E(X) = 109.39, V(X) = 33.19$
   (b) 54.70
4-31. (a) $E(X) = 5.1, V(X) = 0.01$
   (b) 0.3679

Section 4-5

4-33. (a) $E(X) = 0$, $V(X) = 0.577$
(b) 0.90 (c) $F(x) = 0$ for $x < -1$; $0.5x + 0.5$ for $-1 \le x < 1$; 1 for $1 \le x$

4-35. (a) $F(x) = 0$ for $x < 0.95$; $10x - 9.5$ for $0.95 \le x < 1.05$; 1 for $1.05 \le x$
(b) 0.3 (c) 0.96
(d) $E(X) = 1.00$, $V(X) = 0.00083$

4-37. (a) $F(x) = 0$ for $x < 0.2050$; $100x - 20.50$ for $0.2050 \le x < 0.2150$; 1 for $0.21.50 \le x$
(b) 0.25 (c) 0.2140
(d) $E(X) = 0.2100$, $V(X) = 8.33 \times 10^{-6}$

4-39. (a) $F(X) = x/90$ for $0 \le x \le 90$
(b) $E(X) = 45$, $V(X) = 675$
(c) 0.667 (d) 0.333

Section 4-6

4-41. (a) 0.90658 (b) 0.99865 (c) 0.07353 (d) 0.98422 (e) 0.95116

4-43. (a) 0.90 (b) 0.5 (c) 1.28 (d) -1.28 (e) 1.33

4-45. (a) 0.93319 (b) 0.69146 (c) 0.9545 (d) 0.00132 (e) 0.15866

4-47. (a) 0.93319 (b) 0.89435 (c) 0.38292 (d) 0.80128 (e) 0.54674

4-49. (a) 0.99379 (b) 0.13591 (c) 5835

4-51. (a) 0.0228 (b) 0.019 (c) 152.028 (d) small (less than 5%)

4-53. (a) 0.0082 (b) 0.72109 (c) 0.564

4-55. (a) 12.309 (b) 12.1545

4-57. (a) 0.00621 (b) 0.308538 (c) 133.33

4-59. (a) 0.1587 (b) 1.3936 (c) 0.9545

4-61. (a) 0.00043 (b) 6016 (c) 1/8

4-63. (a) 0.02275 (b) 0.324 (c) 11.455

Section 4-7

4-65. (a) 0.0853 (b) 0.8293 (c) 0.0575

4-67. (a) 0.1908 (b) 0.5 (c) 0.3829

4-69. (a) 0.2743 (b) 0.8417

4-71. 0.022

4-73. 0.498

4-75. (a) 0 (b) 0.156 (c) 10,233 (d) 8.1 days/year (e) 0.0071

Section 4-8

4-77. (a) 0.3679 (b) 0.1353 (c) 0.9502 (d) 0.95, $x = 29.96$

4-79. (a) 0.3679 (b) 0.2835 (c) 0.1170

4-81. (a) 0.1353 (b) 0.4866 (c) 0.2031 (d) 34.54

4-83. (a) 0.0498 (b) 0.8775

4-85. (a) 0.0025 (b) 0.6321 (c) 23.03 (d) same as part (c) (e) 6.93

4-87. (a) 15.625 (b) 0.1629 (c) 3×10^{-6}

4-89. (a) 0.2212 (b) 0.2865 (c) 0.2212 (d) 0.9179 (e) 0.2337

4-91. (a) 0.3528 (b) 0.04979 (c) 46.05 (d) 6.14×10^{-6} (e) e^{-12} (f) same

4-93. (a) 0.3679 (b) 0.1353 (c) 0.0498 (d) not depend on θ

Section 4-9

4-95. (a) 120 (b) 1.32934 (c) 11.6317

4-97. (a) Erlang $\lambda = 5$ calls/min, $r = 10$ (b) $E(X) = 2$, $V(X) = 0.4$ (c) 0.2 minute (d) 0.1755 (e) 0.2643

4-99. (a) 50000 (b) 0.6767

4-101. (a) 5×10^5 (b) $V(X) = 5 \times 10^{10}$, $\sigma = 223607$ (c) 0.0803

4-103. (a) 0.1429 (b) 0.1847

Section 4-10

4-107. $E(X) = 12000$, $V(X) = 3.61 \times 10^{10}$

4-109. 1000

4-111. (a) 803.68 hours (b) 85319.64 (c) 0.1576

4-113. (a) 443.11 (b) 53650.5 (c) 0.2212

Section 4-11

4-117. (a) 0.0016 (b) 0.0029 (c) $E(X) = 12.1825$, $V(X) = 1202455.87$

4-119. (a) 0.03593 (b) 1.65 (c) 2.7183 12.6965

4-121. (a) $\theta = 8.4056$, $\omega^2 = 1.6094$ (b) 0.2643 (c) 881.65

Supplemental Exercises

4-125. (a) 0.99379 (b) 0.621%

4-127. (a) 0.15866 (b) 90.0 (c) 0.9973 (d) $(0.9973)^{10}$ (e) 9.973

4-129. (a) 0.0217 (b) 0.9566 (c) 229.5

4-131. 0.8488

4-133. (a) 620.4 (b) 105154.9 (c) 0.4559

4-135. (a) 0.0625 (b) 0.75 (c) 0.5 (d) $F(x) = 0$ for $x < 2$; $x/4 - x + 1$ for $2 \le x < 4$; 1 for $4 \le x$ (e) $E(X) = 10/3$, $V(X) = 0.2222$

4-137. (a) 0.3935 (b) 0.9933

4-139. (a) $\theta = 3.43$, $\omega^2 = 0.96$ (b) 0.946301

4-141. (a) 0.6915 (b) 0.683 (c) 1.86

4-143. (a) 0.0062 (b) 0.012 (c) 5.33

4-145. 0.0008 to 0.0032

4-147. (a) 0.5633 (b) 737.5

4-149. (a) 0.9906 (b) 0.8790

Mind-Expanding Exercises

4-153. (a) $k\sigma^2$ (b) $k\sigma^2 + k(\mu - m)^2$

CHAPTER 5

Section 5-1

5-1. (a) 3/8 (b) 5/8 (c) 3/8 (d) 1/8 (e) $V(X) = 0.4961$ $V(Y) = 1.8594$ (f) $f(1) = 1/4, f(1.5) = 3/8, f(2.5) = 1/4, f(3) = 1/8$ (g) $f(2) = 1/3, f(3) = 2/3$ (h) 1 (i) 2 1/3 (j) Not independent

5-3. (a) 3/8 (b) 3/8 (c) 7/8 (d) 5/8 (e) $V(X) = 0.4219$ $V(Y) = 1.6875$

(f) $f(-1) = 1/8, f(-0.5) = 1/4,$
  $f(0.5) = 1/2, f(1) = 1/8$
(g) 1 (h) 1 (i) 0.5
(j) Not independent

5-5. (b) $f_X(0) = 0.970299,$
  $f_X(1) = 0.029403,$
  $f_X(2) = 0.000297,$
  $f_X(3) = 0.000001$
(c) 0.03 (d) $f(0) = 0.920824,$
  $f(1) = 0.077543, f(2) = 0.001632$ (e) 0.080807
(g) Not independent

5-7. (b) $f(0) = 2.40 \times 10^{-9},$
  $f(1) = 1.36 \times 10^{-6},$
  $f(2) = 2.899 \times 10^{-4},$
  $f(3) = 0.0274, f(4) = 0.972$
(c) 3.972 (d) equals $f(y)$
(e) 3.988 (f) 0.0120
(g) Independent

5-9. (b) $f_X(0) = 0.2511,$
  $f_X(1) = 0.0405,$
  $f_X(2) = 0.0063,$
  $f_X(3) = 0.0009,$
  $f_X(4) = 0.0001$
(c) 0.0562
(d) $f_{Y|3}(0) = 2/3, f_{Y|3}(1) = 1/3,$
  $f_{Y|3}(2) = f_{Y|3}(3) = f_{Y|3}(4) = 0$
(e) 0.0003 (f) 0.0741
(g) Not independent

5-11. (a) $p_1 = 0.05, p_2 = 0.85,$
  $p_3 = 0.10$
(d) $E(X) = 1, V(X) = 0.95$
(f) 0.07195
(g) 0.7358
(h) $E(Y) = 17$
(i) 0
(j) $P(X = 2, Y = 17) = 0.0540, P(X = 2 | Y = 17) = 0.2224$
(k) $E(X | Y = 17) = 1$

5-13. (b) 0.1944 (c) 0.0001
(e) $E(X) = 2.4$
(f) $E(Y) = 1.2$ (g) 0.7347
(h) 0
(i) $P(X = 0 | Y = 2) = 0.0204,$
  $P(X = 1 | Y = 2) = 0.2449,$
  $P(X = 2 | Y = 2) = 0.7347$
(j) 1.7143

5-15. (c) 0.308 (d) 5.7

Section 5-2

5-17. (a) 0.4444 (b) 0.6944
(c) 0.5833 (d) 0.3733
(e) 2 (f) 0

(g) $f_X(x) = 2x/9; 0 < x < 3$
(h) $f_{Y|1.5}(y) = 2y/9; 0 < y < 3$
(i) 2 (j) 4/9
(k) $f_{X|2}(x) = 2x/9; 0 < x < 3$

5-19. (a) 1/81 (b) 5/27
(c) 0.01235 (d) 16/81
(e) 12/5 (f) 8/5
(g) $f(x) = 4x^3/81; 0 < x < 3$
(h) $f_{Y|X=1}(y) = 2y; 0 < y < 1$
(i) 1 (j) 0
(k) $f_{X|Y=2}(x) = 2x/9; 0 < x < 3$

5-21. (a) 0.9879 (b) 0.0067
(c) 0.000308 (d) 0.9939
(e) 0.04 (f) 8/15
(g) $f(x) = 5e^{-5x}; x > 0$
(h) $f_{Y|X=1}(y) = 3e^{3-3y}; 1 < y$
(i) 4/3
(j) $0.9502, f_Y(2) = 15e^{-6}/2;$
  $0 < y$
(k) $f_{X|Y=2}(y) = 2e^{-2x}; 0 < x < 2$

5-23. (a) 1/30 (b) 1/12 (c) 19/9
(d) 97/45 (g) 1 (h) 0.25

5-25. (a) $P(X > 5, Y > 5) = 0.0439,$
  $P(X > 10, Y > 10) = 0.0019$
(b) 0.0655

5-27. (a) 0.25 (b) 0.0625 (c) 1
(d) 1 (e) 2/3 (f) 0.25
(g) 0.0625
(h) $f_{X|YZ}(y, z) = 2x; 0 < x < 1$
(i) 0.25

5-29. (a) 0.75 (b) 3/4 (c) 0.875
(d) 0.25 (g) 1 for $x > 0$

5-31. (a) 0.032 (b) 0.0267

Section 5-3

5-33. 0.8851
5-35. $c = 1/36, \rho = -0.0435$
5-37. $\rho = -0.5$, negative
5-39. $c = 8/81, \rho = 0.4924$
5-41. $\sigma_{XY} = \rho_{XY} = 0$
5-43. $\sigma_{XY} = \rho_{XY} = 0$

Section 5-4

5-47. (a) 0.7887 (b) 0.7887
(c) 0.6220
5-49. 0.8270
5-55. (a) 18 (b) 77 (c) 0.5
(d) 0.873

Section 5-5

5-57. (a) $E(T) = 4, \sigma_T = 0.1414$
5-59. (a) 0 (b) 1
5-61. $E(X) = 1290, V(X) = 19600$

5-63. (a) 0.002 (b) $n = 6$
(c) 0.9612
5-65. (a) 0.0027 (b) No (c) 0

Section 5-6

5-67. $f_Y(y) = \frac{1}{4}; y = 3, 5, 7, 9$
5-69. (b) 18

Supplemental Exercises

5-75. (a) 3/8 (b) 3/4 (c) 3/4
(d) 3/8
(e) $E(X) = 7/8, V(X) = 39/64,$
  $E(Y) = 7/8, V(Y) = 39/64$
(h) 2/3 (i) Not independent
(j) 0.7949

5-77. (a) 0.0560 (b) $Z \sim \text{Bin}(20, 0.1)$
(c) 2 (d) 0.863
(g) Not independent (f) 1

5-79. (a) 1/108 (b) 0.5787
(c) 3/4 (d) 0.2199
(e) 9/4 (f) 4/3

5-81. 3/4

5-83. (a) 0.085 (b) $Z \sim \text{Bin}(10, 0.3)$
(c) 3

5-85. (a) 0.499 (b) 0.5

5-87. (a) 0.057 (b) 0.057

5-91. (a) $E(T) = 1.5, V(T) = 0.078$
(b) 0.0367
(c) $E(P) = 4, V(P) = 0.568$

CHAPTER 6

Section 6-1

6-1. $\bar{x} = 74.0044, s = 0.00473$
6-3. $\bar{x} = 7068.1, s = 226.5$
6-5. $\bar{x} = 43.975, s = 12.294$
6-7. $\mu = 5.44$
6-9. $\bar{x}_1 = 287.89, s_1 = 160.154$
  $\bar{x}_2 = 325.01, s_2 = 121.20$
6-11. $\bar{x} = 7.184, s = 0.02066$
6-13. (a) $\bar{x} = 65.86, s = 12.16$
(b) $\bar{x} = 66.86, s = 10.74$

Section 6-2

6-15. $\tilde{x} = 1436.5$, lower quartile:
  $Q_1 = 1097.8$, and upper
  quartile: $Q_3 = 1735.0$
6-17. $\tilde{x} = 89.250$, lower quartile:
  $Q_1 = 86.100$, and upper
  quartile: $Q_3 = 93.125$
6-19. median: $\tilde{x} = 1436.5$, mode:
  1102, 1315, and 1750, mean:
  $\bar{x} = 1403.7$
6-21. $\bar{x} = 366.57, s = 940.02,$
  and $\tilde{x} = 41.455$

6-23. 95th percentile = 5479
6-25. $\bar{x} = 260.3$, $s = 13.41$, $\tilde{x} = 260.85$, and 90th percentile = 277.2
6-27. $\bar{x} = 89.45$, $s = 2.8$, $\tilde{x} = 90$, and proportion = 22/40 = 55%

Section 6-4

6-45. (a) $\bar{x} = 2.415$, $s^2 = 0.285$, and $s = 0.543$
6-47. (a) $\bar{x} = 952.44$, $s^2 = 9.55$, and $s = 3.09$
(b) $\tilde{x} = 953$
6-49. (a) $\tilde{x} = 67.50$, lower quartile: $Q_1 = 58.50$, and upper quartile: $Q_3 = 75.00$
(b) $\tilde{x} = 68.00$, lower quartile: $Q_1 = 60.00$, and upper quartile: $Q_3 = 75.00$

Supplemental Exercises

6-79. (a) Sample 1 Range = 4
Sample 2 Range = 4
(b) Sample 1: $s = 1.604$
Sample 2: $s = 1.852$
6-85. (b) $\bar{x} = 9.325$, $s = 4.4858$

Mind-Expanding Exercises

6-95. s^2 (old) = 50.61, s^2 (new) = 5061.1
6-99. $\bar{y} = 431.89$, $s_y^2 = 34.028$
6-101. $\tilde{x} = 69$
6-103. (a) $\bar{x} = 89.29$ (b) $\bar{x} = 89.19$

CHAPTER 7

Section 7-2

7-1. 0.8186
7-3. 0.4306
7-5. 0.191
7-7. $n = 12$
7-9. 0.2312
7-11. (a) 0.5885 (b) 0.1759
7-13. 0.983
7-15. $V(\hat{\theta}_1) = \sigma^2/7$ is smallest
7-21. Bias = σ^2/n
7-23. (a) 423.33 (b) 9.08 (c) 1.85 (d) 424 (e) 0.2917
7-27. (d) 0.01 (e) 0.0413

Section 7-3

7-43. (a) 5.046 (b) 5.05

Section 7-4

7-29. $\bar{x}$
7-33. unbiased
7-35. (a) $\hat{\theta} = \Sigma x_i^2/(2n)$
(b) same as part (a)
7-41. (b) $\lambda_0(m + x + 1)/(m + \lambda_0 + 1)$

Supplemental Exercises

7-49. 0.8664
7-51. 5.6569
7-53. $n = 100$
7-55. $\hat{\theta} = \bar{x}/3$

CHAPTER 8

Section 8-2

8-1. (a) 96.76% (b) 98.72% (c) 93.56%
8-3. (a) 1.29 (b) 1.65 (c) 2.33
8-5. (a) 4 (b) 7
8-7. (a) Longer (b) No (c) Yes
8-9. [87.85, 93.11]
8-11. (a) [74.0353, 74.0367]
(b) [1005, ∞)
8-13. (a) [3232.11.3267.89]
(b) [3226.4, 3273.6]
8-15. 267
8-17. 2^2
8-19. (a) [13.383, 14.157]
(b) [13.521, ∞)
(c) 1 (d) 2

Section 8-3

8-21. (a) 2.179 (b) 2.064 (c) 3.012 (d) 4.073
8-23. [58197.33, 62082.07]
8-25. [1.094, 1.106]
8-27. (−∞, 125.312)
8-29. [443.520, 528.080]
8-31. [301.06, 333.34]
8-33. (b) [2237.3, 2282.5]
(c) [2241.4, ∞)
8-35. (b) [4.051, 4.575]
(c) [4.099, ∞)
8-37. (b) [2.813, 2.991]

Section 8-4

8-39. [0.00003075, ∞)
8-41. [0.31, 0.46]
8-43. [0.626, 1.926]
8-45. [0.0061. 0.0131]

Section 8-5

8-47. (a) [0.02029, 0.06637]
(b) [−∞, 0.0627]

8-49. (a) [0.501, 0.571]
(b) [0.506, ∞)
8-51. (a) [0.225, 0.575] (b) 2305
(c) 2401
8-53. 666

Section 8-7

8-55. [52131.1, 68148.3]
8-57. [1.068, 1.13]
8-59. [292.049, 679.551]
8-61. [263.7, 370.7]
8-63. [2193.5, 2326.5]
8-65. 90% PI = [2.71, 3.09]
90% CI = [2.85, 2.95]
90% CI = [2.81, 2.99]
8-67. [49555.54, 70723.86]
8-69. [1.06, 1.14]
8-71. TI = [237.18, 734.42]
CI = [443.42, 528.08]
8-73. TI = [247.60, 386.60]
CI = [301.06, 333.34]
8-75. TI = [2.49, 3.31]
CI = [2.84, 2.96]

Supplemental Exercises

8-79. (a) 0.0997 and 0.064
(b) 0.044 and 0.014
(c) 0.0051 and < 0.001
8-81. (a) Normality
(b) [16.99, ∞)
(c) [16.99, 33.25]
(d) (−∞, 343.74]
(e) [28.23, 343.74]
(f) $16.91 < \mu < 29.09$
$15.85 < \sigma^2 < 192.97$
(g) mean: [16.88, 33.12], variance: [28.16, 342.94]
8-83. (a) [13.74, 16.92]
(b) [13.24, 17.42]
8-85. (a) Yes
(b) [197.84, 208.56]
(c) [185.41, 220.99]
(d) [171.21, 235.19]
8-87. [0.0956, ∞)
8-89. (a) Yes
(b) [1.501, 1.557]
(c) [1370, 1.688]
(d) [1.339, 1.719]
8-91. (a) [0.0004505, 0.009549]
(b) 518
(c) 26002
8-93. (a) Normality
(c) [18.478, 26.982]
(e) [19.565, 123.289]

Mind-Expanding Exercises
8-95. (b) [28.62, 101.98]
8-97. (a) 46 (b) [10.19, 10.41], $p = 0.6004$
8-99. 950 of CIs and 0.9963

CHAPTER 9

Section 9-1
9-1. (a) Yes (b) No (c) No (d) No (e) No
9-3. (a) $H_0: \mu = 20$nm $H_1: \mu < 20$nm (b) No
9-5. (a) $\alpha = 0.02275$.
  (b) $\beta = 0.15866$
  (c) $\beta = 0.5$
9-7. (a) $11.4175 \leq \overline{X}_c \leq 11.42$
  (b) $11.5875 \leq \overline{X}_c \leq 11.59$
  (c) $11.7087 \leq \overline{X}_c \leq 11.71$
  (d) $11.7937 \leq \overline{X}_c \leq 11.84$
9-9. (a) P-value = 0.0135
  (b) P-value $\leq$ 0.000034
  (c) P-value = 0.158655
9-11. (a) 0.09296
  (b) $\beta = 0.04648$
  (c) $\beta = 0.00005$
9-13. (a) $\beta \cong 0.005543$
  (b) $\beta \cong 0.082264$
  (c) As n increases, β decreases
9-15. (a) $\alpha = 0.05705$ (b) $\beta = 0.5$
  (c) $\beta = 0.05705$.
9-17. (a) $\mu \leq 191.40$
  (b) $\mu \leq 185.37$
  (c) $\mu \leq 186.6$
  (d) $\mu \leq 183.2$
9-19. (a) P-value = 0.2148
  (b) P-value = 0.008894
  (c) P-value = 0.785236
9-21. (a) $\alpha = 0.0164$
  (b) $1 - \beta = 0.21186$
  (c) α will increase and the power will increase with increased sample size.
9-23. (a) P-value = 0.238
  (b) P-value = 0.0007
  (c) P-value = 0.2585
9-25. (a) $\alpha = 0.29372$
  (b) $\beta = 0.25721$
9-27. (a) $\alpha \approx 0$
  (b) $\beta = 0.99506$

Section 9-2
9-29. (a) $H_0: \mu = 10, H_1: \mu > 10$
  (b) $H_0: \mu = 7, H_1: \mu \neq 7$
  (c) $H_0: \mu = 5, H_1: \mu < 5$

9-31. (a) $a = z_\alpha \cong -2.33$
  (b) $a = z_\alpha \cong -1.64$
  (c) $a = z_\alpha \cong -1.29$
9-33. (a) P-value $\cong$ 0.04
  (b) P-value $\cong$ 0.066
  (c) P-value $\cong$ 0.69
9-35. (a) P-value $\cong$ 0.98
  (b) P-value $\cong$ 0.03
  (c) P-value $\cong$ 0.65
9-37. (a) $z_0 = -0.95 > -1.96$, do not reject H_0
  (b) $\beta = 0.80939$
  (c) $n \cong 16$
9-39. (a) $z_0 = 1.26 < 1.65$ do not reject H_0
  (b) P-value = 0.1038
  (c) $\beta \cong 0.000325$
  (d) $n \cong 1$
  (e) $39.85 \leq \mu$
9-41. (a) $z_0 = 1.56 > -1.65$, do not reject H_0
  (b) P-value $\cong$ 0.94
  (c) Power = 0.97062
  (d) $n \cong 5$
  (e) $\mu \leq 104.53$
9-43. (a) $z_0 = 1.77 > 1.65$, reject H_0
  (b) P-value $\cong$ 0.04
  (c) Power = 1
  (d) $n \cong 2$
  (e) $4.003 \leq \mu$
9-45. (a) critical value > 2.539
  (b) critical value > 1.796
  (c) critical value > 1.345
9-47. (a) $0.05 \leq p \leq 0.1$
  (b) $0.05 \leq p \leq 0.1$
  (c) $0.5 \leq p \leq 0.8$
9-49. (a) $0.95 \leq p \leq 0.975$
  (b) $0.025 \leq p \leq 0.05$
  (c) $0.6 \leq p \leq 0.75$

Section 9-3
9-51. (a) $|t_0| = 3.48 > 2.064$ reject; H_0 P-value = 0.002
  (b) Yes
  (c) power $\cong$ 1
  (d) $n = 20$
  (e) $98.065 \leq \mu \leq 98.463$
9-53. (a) $|t_0| = 1.456 < 2.064$, do not reject H_0; $0.1 < P$-value < 0.2.
  (b) Yes
  (c) power = 0.80
  (d) $n = 100$
  (e) $129.406 \leq \mu \leq 130.100$

9-55. (a) $|t_0| = 1.55 < 2.861$, do not reject H_0; $0.10 < P$-value < 0.20
  (b) Yes, see normal probability plot
  (c) power = 0.30
  (d) $n = 40$
  (e) $1.9 \leq \mu \leq 4.62$
9-57. (a) $t_0 = 0.15 < 1.753$., do not reject H_0; P-value > 0.40.
  (b) $n = 4$. Yes
9-59. $t_0 = 3.46 > 1.833$, reject H_0 $0.0025 < P$-value < 0.005
9-61. (a) $t_0 = -14.69 < 1.6604$, do not reject H_0; P-value > 0.995
  (b) Yes
  (c) power = 1
  (d) $n = 15$

Section 9-4
9-63. (a) critical values 6.84 and 38.58
  (b) critical values 3.82 and 21.92
  (c) critical values 6.57 and 23.68
9-65. (a) $\chi^2_{1-\alpha,n-1} = 7.63$
  (b) $\chi^2_{1-\alpha,n-1} = 4.57$
  (c) $\chi^2_{1-\alpha,n-1} = 7.79$
9-67. (a) $0.5 < P$-value < 0.9
  (b) $0.5 < P$-value < 0.9
  (c) $0.005 < P$-value < 0.01
9-69. (a) $\chi^2_0 = 0.23 < 26.30$ do not reject H_0; P-value > 0.995
  (b) $0.07 \leq \sigma$
9-71. (a) $\chi^2_0 = 109.52 > 79.49$, reject H_0; P-value < 0.01
  (b) $0.31 < \sigma < 0.46$
9-73. (a) $\chi^2_0 = 12.46 > 7.26$, do not reject H_0; $0.1 < P$-value < 0.4
  (b) $\sigma \leq 5240$
9-75. (a) $\chi^2_0 = 11.52 < 19.02$ do not reject H_0; $0.2 < P$-value
  (b) $n = 30$
  (c) $10.90 \leq \sigma^2 \leq 76.80$

Section 9-5
9-77. (a) $z_0 = -1.31 > -1.65$, do not reject H_0; P-value = 0.095
  (b) $p \leq 0.0303$
9-79. (a) $z_0 = 2.06 > 1.65$, reject H_0; P-value = 0.0196
  (b) $0.7926 \leq p$

9-81. (a) $z_0 = -0.94 < 2.33$ do not reject H_0; P-value $= 0.826$
(b) $0.035 \le p$

9-83. $z_0 = 1.58 > -2.33$ do not reject H_0

9-85. (a) $z_0 = 0.54 < 1.65$, do not reject H_0; P-value $= 0.295$
(b) $\beta = 0.639, n \cong 118$

Section 9-7

9-87. (a) $\chi_0^2 = 6.955 < 15.09$ do not reject H_0
(b) P-value $= 0.2237$ (from Minitab)

9-89. (a) $\chi_0^2 = 10.39 > 7.81$ reject H_0
(b) P-value $= 0.0155$

9-91. (a) $\chi_0^2 = 769.57 >>> 36.42$, reject H_0
(b) P-value $= 0$

Section 9-8

9-93. $\chi_0^2 \not> \chi_{0.05,6}^2$, do not reject H_0; P-value $= 0.070$ (from Minitab)

9-95. $\chi_0^2 > \chi_{0.01,9}^2$ reject H_0; P-value $= 0.002$

9-97. $\chi_0^2 \not> \chi_{0.01,3}^2$, do not reject H_0; P-value $= 0.013$

9-99. (a) $\chi_0^2 > \chi_{0.05,1}^2$
(b) P-value < 0.005

Supplemental Exercises

9-101. (a) 50, Normal, p, $\dfrac{p(1-p)}{50}$
(b) 80, Normal, p, $\dfrac{p(1-p)}{80}$
(c) 100, Normal, p, $\dfrac{p(1-p)}{100}$
(d) As the sample size increases, the variance of the sampling distribution decreases.

9-103. (a) $\beta = 0.564$
(b) $\beta = 0.161$
(c) $\beta = 0.0116$
(d) β, decreases as the sample size increases; larger n

9-105. (a) $\beta = 0.38974$
power $= 1 - \beta = 0.61026$
(b) $\beta = 0.00494$
power $= 1 - \beta = 0.995$
(c) $\beta = 0.00118$
power $= 1 - \beta = 0.9988$
(d) β decreases, power increases

9-107. (a) Place what we are trying to demonstrate in the alternative hypothesis.
(b) $0.05 < P$-value < 0.10

9-109. (a) $\chi_0^2 = 1.75 > 1.24$, do not reject H_0
(b) 0.00001 could have occurred as a result of sampling variation.

9-111. (a) $\chi_0^2 = 17.929 > 14.07$ reject H_0
(b) P-value $= 0.0123$

9-113. $\chi_o^2 > \chi_{0.05,5}^2$, reject normality

9-115. (a) $z_0 = -7.23 < -1.65$, reject H_0
(b) P-value $\cong 0$
(c) $\chi_o^2 = 12 > \chi_{0.05,5}^2 = 11.07$ reject normality

9-117. (b) $t_0 = 1.608 < 2.093$, do not reject H_0
(c) $0.1 < P$-value < 0.2
(d) $2.270 \le \mu \le 4.260$

9-119. (a) $t_0 = 0.667 < 2.059$, do not reject H_0
(b) $0.5 < P$-value < 0.8
(c) $246.84 \le \mu \le 404.15$

9-121. (a) $2.6 < \chi_0^2 = 10.53 < 26.76$ do not reject H_0
(b) $0.2 < P$-value
(c) $0.236 \le \sigma^2 \le 2.43$

9-123. (a) $t_0 = 0.47 < 2.228$, do not reject the H_0
(b) $0.5 < P$-value < 0.8
(c) n should be at least 50.
(d) $\beta \approx 0.1$

9-125. (a) $0.1 < P$-value < 0.5, do not reject H_0
(b) $0.01 < P$-value < 0.025, reject H_0

Mind-Expanding Exercises

9-129. (a) $z_0 = 4.82 > -1.65$, do not reject H_0
(c) $\hat{\theta} = 282.36$, $z_0 = -4.57 > 1.65$, do not reject H_0

CHAPTER 10

Section 10-2

10-1. (a) $-1.96 < z_0 = -0.9 < 1.96$, do not reject H_0; P-value $= 0.368$
(b) $-9.79 \le \mu_1 - \mu_2 \le 3.59$

(c) Power $= 0.14$
(d) $n_1 = n_2 = 180$

10-3. (a) $z_0 = 0.937 < 2.325$, do not reject H_0; P-value $= 0.174$
(b) $\mu_1 - \mu_2 \ge -4.74$
(c) Power $= 0.04$
(d) Use $n_1 = n_2 = 339$

10-5. (a) $z_0 = -5.84 < 1.645$, do not reject H_0; P-value $= 1$
(b) $\mu_1 - \mu_2 \ge 6.8$
(c) Power $= 0.9988$
(d) The sample size is adequate

10-7. (a) $z_0 = -7.25 < -1.645$ reject H_0; P-value $\cong 0$
(b) $-3.684 \le \mu_1 - \mu_2 \le -2.116$
(c) $n_1 = n_2 = 11$

10-9. (a) $-5.83 \le \mu_1 - \mu_2 \le -0.57$; P-value $= 0.0173$
(b) Yes
(c) Power $= 0.9616; n \cong 10$
(d) Normal
(e) Normal

10-11. (a) $t_0 = -1.94 < -1.701$
$0.025 < P$-value < 0.05
(b) $\mu_1 - \mu_2 \le -0.196$
(c) Power $= 0.95$
(d) $n = n_1 = n_2 = 21$

Section 10-3

10-13. (a) $-2.042 < t_0 = 0.230 < 2.042$, do not reject H_0; P-value > 0.80
(b) $-0.394 \le \mu_1 - \mu_2 \le 0.494$

10-15. (a) $t_0 = -3.11 < -2.485$, reject H_0
(b) $-5.688 \le \mu_1 - \mu_2 \le -0.3122$

10-17. (a) Assumptions verified
(b) $t_0 = -2.83 < -2.101$ reject H_0; $0.010 < P$-value < 0.020
(c) $-0.7495 \le \mu_1 - \mu_2 \le -0.1105$

10-19. (a) $t_0 = -5.498 < -2.021$ reject H_0; P-value < 0.0010
(b) $n_1 = n_2 = 38$

10-21. (a) $t_0 = 3.03 > 2.056$ reject H_0; $0.005 < P$-value < 0.010
(b) $t_0 = 3.03 > 1.706$, reject H_0

10-23. (a) $t_0 = 7.0 > 2.048$, reject H_0; P-value $\cong 0$

(b) $14.93 \leq \mu_1 - \mu_2 \leq 27.28$
(c) $n > 8$
10-25. (a) $t_0 = 2.82 > 2.326$ reject H_0; P-value $\cong 0.025$
(b) $\mu_1 - \mu_2 \geq 0.178$
10-27. (a) Normal
(b) $t_0 = 2.558 > 2.101$ reject H_0; P-value ≈ 0.02
(c) $1.86 \leq \mu_1 - \mu_2 \leq 18.94$
(d) Power = 0.05
(e) $n = 51$

Section 10-4
10-29. (a) $0.1699 \leq \mu_d \leq 0.3776$
(b) t-test is appropriate.
10-31. $-727.46 \leq \mu_d \leq 2464.21$
10-33. (a) $t_0 = 5.465 > 1.761$ reject H_0; P-value $\cong 0$
(b) $18.20 \leq \mu_d$
10-35. (a) $t_0 = 8.387 > 1.833$ reject H_0
(b) $t_0 = 3.45 > 1.833$ reject H_0
(c) Yes
10-37. (a) Normal
(b) $-0.379 \leq \mu_d \leq 0.349$
(c) $6 \leq n$

Section 10-5
10-39. (a) $f_{0.25,5,10} = 1.59$
(b) $f_{0.10,24,9} = 2.28$
(c) $f_{0.05,8,15} = 2.64$
(d) $f_{0.75,5,10} = 0.529$
(e) $f_{0.90,24,9} = 0.525$
(f) $f_{0.95,8,15} = 0.311$
10-41. $f_0 = 0.805 > 0.166$, do not reject H_0;
$0.22 \leq \dfrac{\sigma_1^2}{\sigma_2^2}$
10-43. (a) $f_0 = 1.21 > 0.333$, do not reject H_0;
$0.403 \leq \dfrac{\sigma_1^2}{\sigma_2^2} \leq 3.63$
(b) Power = 0.65
(c) $n \approx 31$
10-45. (a) $f_0 = 0.923 > 0.365$, do not reject H_0
(b) $0.3369 \leq \dfrac{\sigma_1^2}{\sigma_2^2} \leq 2.640$
10-47. (a) $0.6004 \leq \dfrac{\sigma_1}{\sigma_2} \leq 1.428$
(b) $0.5468 \leq \dfrac{\sigma_1}{\sigma_2} \leq 1.5710$

(c) $0.661 \leq \dfrac{\sigma_1}{\sigma_2}$
10-49. $0.4058 < f_0 = 1.78 < 2.46$, do not reject H_0
10-51. $0.248 < f_0 = 0.640 < 4.04$, do not reject H_0;
$0.159 \leq \dfrac{\sigma_1^2}{\sigma_2^2} \leq 2.579$
10-53. $0.333 < f_0 = 1.35 < 3$, do not reject H_0;
$0.45 \leq \dfrac{\sigma_1^2}{\sigma_2^2} \leq 4.05$
10-55. (a) $0.248 < f_0 = 3.337 < 4.03$, do not reject H_0
(b) No

Section 10-6
10-57. (a) $z_0 = 4.45 > 1.96$ reject H_0; P-value ≈ 0
(b) $0.039 \leq p_1 - p_2 \leq 0.1$
10-59. (a) $z_0 = 5.36 > 2.58$ reject H_0; P-value $= 0$

Supplemental Exercises
10-61. (a) normality, equality of variance, and independence of the observations.
(b) $1.40 \leq \mu_1 - \mu_2 \leq 8.36$
(c) Yes
(d) $0.1582 \leq \dfrac{\sigma_1^2}{\sigma_2^2} \leq 5.157$
(e) No
10-63. (a) $t_0 = 2.554 > 1.895$, reject H_0
(b) $t_0 = 2.554 < 2.998$, do not reject H_0
(c) $t_0 = -1.986 < 1.895$, do not reject H_0
(d) $t_0 = -1.986 < 2.998$, do not reject H_0
10-65. (a) $z_0 = 6.55 > 1.96$, reject H_0
(b) $z_0 = 6.55 > 2.58$, reject H_0
(c) z_0 is so large
10-67. (a) $-0.0335 \leq p_1 - p_2 \leq 0.0329$
(b) $-0.0282 \leq p_1 - p_2 \leq 0.0276$
(c) 95% CI: $-0.0238 \leq p_1 - p_2 \leq 0.0232$
90% CI: $-0.0201 \leq p_1 - p_2 \leq 0.0195$
10-69. (a) Yes
(b) Yes if similar populations

10-71. (a) $0.0987 \leq \mu_1 - \mu_2 \leq 0.2813$
(b) $0.0812 \leq \mu_1 - \mu_2 \leq 0.299$
(c) $\mu_1 - \mu_2 \leq 0.2813$
(d) $z_0 = 3.42 > 1.96$, reject H_0; P-value = 0.00062
(e) $n_1 = n_2 = 9$
10-73. (a) $z_0 = -5.36 < -2.58$, reject H_0
(b) conclusions are the same
(c) $n = 60$
10-75. (a) No
(b) data appear normal with equal variances
(c) It is more apparent the data follow normal distributions.
(d) $18.114 < \dfrac{\sigma_V^2}{\sigma_M^2} < 294.35$
(e) $f_0 = 72.78 > 4.03$, reject H_0
10-77. (a) Normality appears valid.
(b) $0.50 < P$-value < 0.80, do not reject H_0
(c) $n = 30$
10-79. (a) It may not be assumed that $\sigma_1^2 = \sigma_2^2$
(b) $t_0 = -2.74 < -2.131$, reject H_0
(c) Power = 0.95
(d) $n = 26$

Mind-Expanding Exercises
10-85. (c) $-1.317 \leq \theta \leq 4.157$

CHAPTER 11

Section 11-2
11-1. (a) $\hat{\beta}_0 = 48.013$, $\hat{\beta}_1 = -2.330$
(b) 37.99 (c) 39.39 (d) 6.71
11-3. (a) $\hat{\beta}_0 = -5.56$, $\hat{\beta}_1 = 12.652$, $\sigma^2 = 32.6$
(b) 89.33 (c) 12.652 (d) 0.79
(e) -7.56
11-5. (a) $\hat{\beta}_0 = -6.3355$, $\hat{\beta}_1 = 9.20836$, $\sigma^2 = 3.7746$
(b) 500.124 (c) 9.20836
(d) -1.618
11-7. (a) $\hat{\beta}_0 = -16.5093$, $\hat{\beta}_1 = 0.0693554$, $\sigma^2 = 7.3212$
(b) 1.39592 (c) 49.38
11-9. (b) $\hat{\beta}_0 = 234.071$, $\hat{\beta}_1 = -3.50856$, $\sigma^2 = 398.25$

(c) 128.814
(d) 156.833 and 15.1175
11-11. (b) $\hat{\beta}_0 = 2625.39$,
$\hat{\beta}_1 = 36.962, \sigma^2 = 9811.2$
(c) 1886.15
11-13. (a) $\hat{\beta}_0 = 0.658, \hat{\beta}_1 = 0.178$,
$\sigma^2 = 0.083$
(b) 3.328 (c) 0.534
(d) 1.726 and 0.174
11-15. (b) $\hat{\beta}_0 = 2.02, \hat{\beta}_1 = 0.0287$,
$\sigma^2 = 0.0253$
11-17. (a) $\hat{y} = 39.2 - 0.0025x$
(b) $\hat{\beta}_1 = -0.0025$
11-19. (b) $\hat{\beta}_0^* = 2132.41$,
$\hat{\beta}_1^* = -36.9618$

Section 11-4
11-21. (a) $f_0 = 74.63$, P-value = 0.000002, reject H_0
(b) $\hat{\sigma}^2 = 1.8436$,
$se(\hat{\beta}_1) = 0.2696$
(c) $se(\hat{\beta}_0) = 0.9043$
11-23. (a) $f_0 = 103.53$, P-value = 0.0000, reject H_0
(b) $\hat{\sigma}^2 = 32.63$,
$se(\hat{\beta}_1) = 1.243$
$se(\hat{\beta}_0) = 9.159$
(c) $t_0 = 2.134$, do not reject H_0
11-25. (a) $f_0 = 74334.4$, P-value ≈ 0, reject H_0
(b) $se(\hat{\beta}_0) = 1.66765$,
$se(\hat{\beta}_1) = 0.0337744$
(c) $t_0 = -23.37$, P-value = 0.000, reject H_0
(d) $t_0 = -3.8$, P-value < 0.005, reject H_0
11-27. (a) $f_0 = 44.0279$, P-value = 0.00004, reject H_0
(b) $se(\hat{\beta}_0) = 9.84346$,
$se(\hat{\beta}_1) = 0.0104524$
(c) $t_0 = -1.67718$, P-value = 0.12166, fail to reject H_0
11-29. (a) $f_0 = 155.2$, P-value < 0.00001, reject H_0
(b) $se(\hat{\beta}_0) = 2.96681$,
$se(\hat{\beta}_1) = 45.3468$
(c) $t_0 = -2.3466$, P-value = 0.0306, fail to reject H_0
(d) $t_0 = 57.8957$, P-value < 0.00001, reject H_0
(e) $t_0 = 2.7651$, P-value = 0.0064, reject H_0
11-31. (a) P-Value = 0.0000, reject H_0

(b) $\hat{\sigma}^2 = 0.083$
$se(\hat{\beta}_0) = 0.1657$,
$se(\hat{\beta}_1) = 0.014$
(c) Reject H_0
11-33. (a) P-value = 0.310, No
(b) $\hat{\sigma}^2 = 30.69$
(c) $se(\hat{\beta}_0) = 9.141$
11-35. 0.70

Sections 11-5 and 11-6
11-37. (a) $(-2.9713, -1.7423)$
(b) (46.7145, 49.3115)
(c) (41.3293, 43.0477)
(d) (39.1275, 45.2513)
11-39. (a) (10.106, 15.198)
(b) $(-24.316, 13.200)$
(c) $\hat{\mu} = 95.656, (92.903, 98.409)$
(d) (83.642, 107.67)
11-41. (a) (9.10130, 9.31543)
(b) $(-11.6219, -1.04911)$
(c) (498.72024, 501.52776)
(d) (495.57344, 504.67456)
11-43. (a) (0.03689, 0.10183)
(b) $(-47.0877, 14.0691)$
(c) (44.0897, 49.1185)
(d) (37.8298, 55.3784)
11-45. (a) (201.552, 266.590)
(b) $(-4.67015, -2.34696)$
(c) (111.8339, 145.7941)
11-47. (a) $(-43.1964, -30.7272)$
(b) (2530.09, 2720.68)
(c) (1823.7833, 1948.5247)
(d) (1668.9013, 2103.4067)
11-49. (a) (0.1325, 0.2235)
(b) (0.119, 1.196)
(c) (1.87, 2.29)

Section 11-7
11-51. (a) $R^2 = 78.7\%$
11-53. (a) $R^2 = 99.986\%$
11-55. (a) $R^2 = 87.94\%$
11-57. (a) $R^2 = 85.22\%$
11-59. (a) $R^2 = 89.6081\%$
(c) $R^2 = 95.73\%$
(d) $\hat{\sigma}^2$ old = 9811.21,
$\hat{\sigma}^2$ new = 4022.93
11-63. (a) $f_0 = 207$, reject H_0

Section 11-8
11-65. (a) $t_0 = 4.81$, P-value < 0.0005, reject H_0
(b) $z_0 = 1.747$, P-value = 0.04, reject H_0

(c) $\rho \geq 2.26$, reject H_0
11-67. (a) $t_0 = 5.475$, P-value = 0.000, reject H_0
(b) (0.3358, 0.8007)
(c) Yes
11-69. (a) $\hat{y} = -0.0280411 + 0.990987x$
(b) $f_0 = 79.838$, reject H_0
(c) 0.903
(d) $t_0 = 8.9345$, reject H_0
(e) $z_0 = 3.879$, reject H_0
(f) (0.7677, 0.9615)
11-71. (a) $\hat{y} = 5.50 + 6.73x$
(b) 0.948
(c) $t_0 = 8.425$, reject H_0
(d) (0.7898, 0.9879)
11-73. (a) $r = 0.887$
(b) $t_0 = 10.164$, reject H_0,
P-value < 0.005
(c) (0.774, 0.945)
(d) $z_0 = 2.808$, reject H_0,
P-value = 0.005

Section 11-9
11-75. (a) Yes (b) No (c) Yes
(d) Yes
11-77. (b) $\hat{y} = -0.8819 + 0.00385x$
(c) $f_0 = 122.03$, reject H_0
11-79. (b) $\hat{y} = -0.966824 + 1.54376x$
(c) $f_0 = 2555263.9$, P-value ≈ 0, reject H_0
(d) $[-0.97800, -0.95565]$
(e) $t_0 = -1.9934$, reject H_0

Supplemental Exercises
11-81. $y^* = 1.2232 + 0.5075x$ where $y^* = 1/y$
11-83. $\hat{y} = 0.7916x$
11-85. (b) $\hat{y} = 0.6714 + -2964x$
(c) $R^2 = 21.5\%$
11-87. (b) $\hat{y} = -0.699 + 1.66x$
(c) $f_0 = 22.75$, reject H_0,
P-value = 0.001
(d) (3.399, 5.114)
11-89. (c) all data: (7741.74, 10956.26),
outlier removed: (8345.22, 11272.79)

CHAPTER 12

Section 12-1
12-1. (b) $\hat{y} = 171.055 + 3.714x_1 - 1.126x_2$
(c) 189.49

12-3. (b) 2

12-5. (a) $\hat{y} = 49.90 - 0.01045x_1 - 0.0012x_2 - 0.00324x_3 + 0.292x_4 - 3.855x_5 + 0.1897x_6$
(b) $\hat{\sigma}^2 = 4.965$
$se(\hat{\beta}_0) = 19.67$,
$se(\hat{\beta}_1) = 0.02338$,
$se(\hat{\beta}_2) = 0.01631$,
$se(\hat{\beta}_3) = 0.0009459$,
$se(\hat{\beta}_4) = 1.765$,
$se(\hat{\beta}_5) = 1.329$,
$se(\hat{\beta}_6) = 0.273$
(c) 29.867

12-7. (a) $\hat{y} = 47.82 - 9.604x_1 + 0.44152x_2 + 18.294x_3$
(b) $\hat{\sigma}^2 = 12.3$
(c) $se(\hat{\beta}_0) = 49.94$,
$se(\hat{\beta}_1) = 3.723$,
$se(\hat{\beta}_2) = 0.2261$, and
$se(\hat{\beta}_3) = 1.323$
(d) 91.38

12-9. (a) $\hat{y} = -440.39 + 19.147x_1 + 68.080x_2$
(b) $\hat{\sigma}^2 = 55563$ $se(\hat{\beta}_0) = 94.20$,
$se(\hat{\beta}_1) = 3.460$, and
$se(\hat{\beta}_2) = 5.241$
(c) 186.675

12-11. (a) $\hat{y} = -0.1105 + 0.4072x_1 + 2.108x_2$
(b) $\hat{\sigma}^2 = 0.00008$
$se(\hat{\beta}_0) = 0.2501$,
$se(\hat{\beta}_1) = 0.1682$, and
$se(\hat{\beta}_2) = 5.834$
(c) 0.97074

12-13. (a) $\hat{y} = 238.56 + 0.3339x_1 - 2.7167x_2$
(b) $\hat{\sigma}^2 = 1321$
(c) $se(\hat{\beta}_0) = 45.23$,
$se(\hat{\beta}_1) = 0.6763$, and
$se(\hat{\beta}_2) = 0.6887$
(d) 61.5195

12-15. (a) $\hat{y} = 3.22 + 1.224x_3 + 4.423x_7 - 4.091x_{10}$
(b) $\hat{\sigma}^2 = 3.7$
(c) $se(\hat{\beta}_0) = 6.519$,
$se(\hat{\beta}_3) = 0.1080$,
$se(\hat{\beta}_7) = 0.2799$, and
$se(\hat{\beta}_{10}) = 0.4953$
(d) 82.079

12-17. (a) $\hat{y} = 383.80 - 3.6381x_1 - 0.1119x_2$
(b) $\hat{\sigma}^2 = 153.0$,
$se(\hat{\beta}_0) = 36.22$,
$se(\hat{\beta}_1) = 0.5665$,
$se(\hat{\beta}_2) = 0.04338$
(c) 180.95
(d) $\hat{y} = 484.0 - 7.656x_1 - 0.222x_2 - 0.0041x_{12}$
(e) $\hat{\sigma}^2 = 147.0$,
$se(\hat{\beta}_0) = 101.3$,
$se(\hat{\beta}_1) = 3.846$,
$se(\hat{\beta}_2) = 0.113$,
$se(\hat{\beta}_{12}) = 0.0039$
(f) 173.1

Section 12-2

12-19. (a) $f_0 = 184.25$,
P-value = 0.000, Reject H_0
(b) $t_0(\hat{\beta}_1) = 16.21$,
P-value < 0.001, reject H_0
$t_0(\hat{\beta}_2) = -11.04$,
P-value < 0.001, reject H_0

12-21. (a) P-value = 8.16 E-5, 3.82 E-8, and 0.3378
(b) $t_0 = 0.98$, do not reject H_0

12-23. (a) $f_0 = 19.53$, reject H_0

12-25. (a) $f_0 = 828.31$, reject H_0
(b) $t_0 = -2.58$, reject H_0
$t_0 = 1.84$, do not reject H_0
$t_0 = 13.82$, reject H_0

12-27. (a) $f_0 = 99.67$, reject H_0,
P-value ≈ 0
(b) $t_0 = 5.539$, reject H_0
$t_0 = 12.99$, reject H_0

12-29. (a) $f_0 = 9.28$, reject H_0,
P-value = 0.015
(b) $t_0 = 0.49$, do not reject H_0
$t_0 = -3.94$, reject H_0

12-31. (a) $f_0 = 379.07$, reject H_0,
P-value ≈ 0
(b) $t_0 = 11.33$, reject H_0
$t_0 = 15.80$, reject H_0
$t_0 = -8.26$, reject H_0
(c) $f_0 = 249.1$, reject H_0

12-33. (a) $f_0 = 97.59$, reject H_0,
P-value = 0.002
(b) $t_0 = -6.42$, reject H_0
$t_0 = -2.57$, do not reject H_0
(c) $f_0 = 6.629$, do not reject H_0
(d) $f_0 = 7.714$, do not reject H_0
(e) $f_0 = 1.11$, do not reject H_0
(f) 147.0

Sections 12-3 and 12-4

12-35. (a) $(49.927 < \beta_0 < 292.183)$
$(0.033 < \beta_1 < 7.393)$
$(-2.765 < \beta_2 < 0.513)$
(b) $(159.58, 219.362)$
(c) $(127.646, 251.296)$

12-37. (a) $(-20.477 < \beta_1 < 1.269)$
$(-0.245 < \beta_2 < 1.076)$
$(14.428 < \beta_3 < 22.159)$
(b) $(77.582, 105.162)$
(c) $(82.133, 100.611)$

12-39. (a) $(-6.9467 < \beta_1 < -0.3295)$
$(-0.3651 < \beta_2 < 0.1417)$
(b) $(-45.8276 < \beta_1 < 30.5156)$
$(-1.3426 < \beta_2 < 0.8984)$
$(-0.03433 < \beta_3 < 0.04251)$

12-41. (a) $(12.1363 < \beta_1 < 26.1577)$
$(57.4607 < \beta_2 < 78.6993)$
(b) $(-233.4, 63.2)$
(c) $(-742.09, 571.89)$

12-43. (a) $(0.0943 < \beta_1 < 0.7201)$
$(-8.743 < \beta_2 < 12.959)$
(b) $(0.861, 0.896)$
(c) $(0.855, 0.903)$

12-45. (a) $(-2.173 < \beta_1 < 2.841)$
$(-5.270 < \beta_2 < -0.164)$
(b) $(-36.7, 125.8)$
(c) $(-112.8, 202.0)$
(d) CI: $(107.4, 267.2)$
PI: $(30.7, 344.0)$

12-47. (a) $-10.183 < \beta_0 < 16.623$
$(1.001 < \beta_3 < 1.447)$
$(3.845 < \beta_7 < 5.001)$
$(-5.113 < \beta_{10} < -3.069)$
(b) 0.3877
(c) $(81.2788, 82.8792)$

12-49. (a) $-8.658 \leq \beta_0 \leq 108.458$
$-0.08 \leq \beta_2 \leq 0.059$
$-0.05 \leq \beta_3 \leq 0.047$
$-0.006 \leq \beta_7 \leq 0$
$-4.962 \leq \beta_8 \leq 5.546$
$-7.811 \leq \beta_9 \leq 0.101$
$-1.102 \leq \beta_{10} \leq -0.523$
(c) $\hat{y} = 61.001 - 0.0208x_2 - 0.0035x_7 - 3.457x_9$
$50.855 \leq \beta_0 \leq 71.147$
$-0.036 \leq \beta_2 \leq -0.006$
$-0.006 \leq \beta_7 \leq -0.001$
$-6.485 \leq \beta_9 \leq -0.429$

Section 12-5

12-51. (a) 0.893
12-53. (a) 0.978
12-55. (a) 0.843
12-57. (a) 0.997
12-59. (a) 0.756
12-61. (a) 0.985   (b) 0.99

12-63. (b) 0.9937, 0.9925
12-65. (a) 0.12 (b) Yes

Section 12-6

12-67. (a) $\hat{y} = -1.633 + 1.232x - 1.495x^2$
(b) $f_0 = 1.858613$, reject H_0
(c) $t_0 = -601.64$, reject H_0

12-69. (a) $\hat{y} = -1.769 + 0.421x_1 + 0.222x_2 - 0.128x_3 - 0.02x_{12} + 0.009x_{13} + 0.003x_{23} - 0.019x_1^2 - 0.007x_2^2 + 0.001x_3^2$
(b) $f_0 = 19.628$, reject H_0
(d) $f_0 = 1.612$, do not reject H_0

12-71. (b) $\hat{y} = 56.677 - 0.1457x_1 - 0.00525x_2 - 0.138x_3 - 4.179x_4$

12-73. (a) Min C_p: x_1, x_2
$C_p = 3.0$, $MS_E = 55563.92$
$\hat{y} = -440.39 + 19.147x_1 + 68.080x_2$
Min MS_E is same as Min C_p
(b) Same as part (a)
(c) Same as part (a)
(d) Same as part (a)
(e) All models are the same

12-75. (a) Min C_p: x_1
$C_p = 1.1$, $MS_E = 0.0000705$
$\hat{y} = -0.20052 + 0.467864x_1$
Min MS_E is same as Min C_p
(b) Same as part (a)
(c) Same as part (a)
(d) Same as part (a)
(e) All models are the same

12-77. (a) Min C_p: x_2
$C_p = 1.2$, $MS_E = 1178.55$
$\hat{y} = 253.06 - 2.5453x_2$
Min MS_E is same as Min C_p
(b) Same as part (a)
(c) Same as part (a)
(d) Same as part (a)
(e) All models are the same

12-79. (a) Min C_p: x_1, x_2
$C_p = 2.9$, $MS_E = 10.49$
$\hat{y} = -50.4 + 0.671x_1 + 1.30x_2$
Min MS_E is same as Min C_p
(b) Same as part (a)
(c) Same as part (a)
(d) Same as part (a)
(e) All models are the same

12-81. (a) Min. $C_p = 5.4$
$\hat{y} = 3.2992 + 0.82871x_{pctcomp} + 0.00056x_{yds} + 3.828x_{yds/att} - 0.08458x_{td} + 3.715x_{pcttd} - 4.26352x_{pct\ int}$
Min. $MS_E = 0.0227$
$\hat{y} = 0.9196 + 0.005207x_{alt} + 0.83061x_{pctcomp} + 4.008689x_{yds/att} - 0.05211x_{td} + 3.5494x_{pcttd} - 0.0793x_{int} - 3.9205x_{pct\ int}$
(b) $\hat{y} = 3.158 + 4.048x_{yds/att} - 4.209x_{pct\ int} + 3.358x_{pcttd} + 0.83x_{pctcomp}$
(c) Same as part (b)
(d) Same as min MSE model in part (a)

12-83. (a) Min $C_p = 1.3$:
$\hat{y} = 61.001 - 0.02076x_{cid} - 0.00354x_{etw} - 3.457x_{axle}$
Min $MS_E = 4.0228$
$\hat{y} = 49.5 - 0.017547x_{cid} - 0.0034252x_{etw} + 1.29x_{cmp} - 3.184x_{axle} - 0.0096x_{c02}$
(b) $\hat{y} = 63.31 - 0.0178x_{cid} - 0.00375x_{etw} - 3.3x_{axle} - 0.0084x_{c02}$
(c) Same as Min MSE model in part (a)
(d) $\hat{y} = 45.18 - 0.00321x_{etw} - 4.4x_{axle} + 0.385x_{n/v}$
(e) Min C_p model is preferred
(f) Min $C_p = 4.0$,
Min $MS_E = 2.267$
$\hat{y} = 10 - 0.0038023x_{etw} + 3.936x_{cmp} + 15.216x_{co} - 0.011118x_{c02} - 7.401x_{trans} + 3.6131x_{drv1} + 2.342x_{drv2}$
Stepwise:
$\hat{y} = 39.12 - 0.0044x_{etw} + 0.271x_{n/v} - 4.5x_{trns} + 3.2x_{drv1} + 1.7x_{drv2}$
Forward selection:
$\hat{y} = 41.12 - 0.00377x_{etw} + 0.336x_{n/v} - 2.1x_{axle} - 3.4x_{trans} + 2.1x_{drv1} + 2x_{drv2}$
Backward elimination is same as Min C_p and Min MS_E

12-85. (a) $\hat{y} = -0.304 + 0.083x_1 - 0.031x_3 + 0.004x_2^2$
$C_p = 4.04$ $MS_E = 0.004$
(b) $\hat{y} = -0.256 + 0.078x_1 + 0.022x_2 - 0.042x_3 + 0.0008x_3^2$
$C_p = 4.66$ $MS_E = 0.004$
(c) Prefer the model in part (a)

Supplemental Exercises

12-89. (a) $f_0 = 1321.39$, reject H_0, P-value < 0.00001
(b) $t_0 = -1.45$, do not reject H_0
$t_0 = 19.95$, reject H_0
$t_0 = 2.53$, fail to reject H_0

12-91. (a) $\hat{y} = 4.87 + 6.12x_1^* - 6.53x_2^* - 3.56x_3^* - 1.44x_4^*$
(b) $f_0 = 21.79$, reject H_0
$t_0 = 5.76$, reject H_0
$t_0 = -5.96$, reject H_0
$t_0 = -2.90$, reject H_0
$t_0 = -4.99$, reject H_0

12-93. (a) $\hat{y}^* = 21.068 - 1.404x_3^* + 0.0055x_4 + 0.000418x_5$
$MS_E = 0.013156$ $C_p = 4.0$
(b) Same as part (a)
(c) x_4, x_5 with $C_p = 4.1$ and $MS_E = 0.0134$
(d) The part (c) model is preferable
(e) Yes

12-95. (a) $\hat{y} = 300.0 + 0.85x_1 + 10.4x_2$, $\hat{y} = 405.8$
(b) $f_0 = 55.37$, reject H_0
(c) 0.9022
(d) $MS_E = 10.65$
(e) $f_0 = 0.291$, do not reject H_0

12-97. (a) $f_0 = 18.28$, reject H_0
(b) $f_0 = 2$, do not reject H_0
(c) $MS_E(\text{reduced}) = 0.005$
$MS_E(\text{Full}) = 0.004$

Mind-Expanding Exercises

12-99. $R^2 \geq 0.449$

CHAPTER 13

Section 13-2

13-1. (a) $f_0 = 14.76$, reject H_0
13-3. (a) $f_0 = 12.73$, reject H_0
(b) P-value $\cong 0$
13-5. (a) $f_0 = 16.35$, reject H_0
(c) 95%: (140.71, 149.29)
99%: (7.36, 24.14)
13-7. (a) $f_0 = 1.86$, do not reject H_0
(b) P-value = 0.214